EISENHOWER PUBLIC LIBRARY
3 1134 00301 6938
Y0-AER-029

Advances in Food Diagnostics

Advances in Food Diagnostics

Editors
Leo M. L. Nollet
Fidel Toldrá

Administrative Editor
Y. H. Hui

Leo M.L. Nollet, PhD, is Professor of Biochemistry, Aquatic Ecology, and Ecotoxicology at the Department of Applied Engineering Sciences of University College Ghent (Hogeschool Gent), Ghent, Belgium. His research interests are in the domain of food analysis, chromatography, and analysis of environmental parameters.

Fidel Toldrá, PhD, is a Research Professor at the Department of Food Science, Instituto de Agroquímica y Tecnología de Alimentos (CSIC), Spain, and serves as European editor of *Trends in Food Science and Technology* and editor-in-chief of *Current Nutrition & Food Science*. His research is focused on food (bio)chemistry and analysis.

Administrative Editor **Y.H. Hui**, PhD, West Sacramento, California, is a consultant to the food industry and has served as the author, editor, or editor-in-chief of numerous books in food science, technology, engineering, medicine, and law, including *Handbook of Food Science, Technology and Engineering* (4 volumes); *Food Processing: Principles and Applications*; and *Handbook of Fruits and Fruit Processing*.

Blackwell Publishing Professional
2121 State Avenue, Ames, Iowa 50014, USA

Orders: 1-800-862-6657
Office: 1-515-292-0140
Fax: 1-515-292-3348
Web site: www.blackwellprofessional.com

Blackwell Publishing Ltd
9600 Garsington Road, Oxford OX4 2DQ, UK
Tel.: +44 (0)1865 776868

Blackwell Publishing Asia
550 Swanston Street, Carlton, Victoria 3053, Australia
Tel.: +61 (0)3 8359 1011

First edition, 2007

Library of Congress Cataloging-in-Publication Data

Nollet, Leo M. L., 1948–
Advances in food diagnostics/Leo M. L. Nollet and Fidel Toldrá; [editor, Y. H. Hui].—1st ed.
p. cm.
Includes bibliographical references and index.
ISBN-13: 978-0-8138-2221-1 (alk. paper)
ISBN-10: 0-8138-2221-1
1. Food—Analysis. 2. Food adulteration and inspection. 3. Food—Quality. 4. Food—Safety measures. I. Toldrá, Fidel. II. Hui, Y. H. (Yiu H.) III. Title.

TX541.N65 2007
664′.07—dc22

2006036121

The last digit is the print number: 9 8 7 6 5 4 3 2 1

Contents

Contributors, vii
Preface, xi

1. Assuring Safety and Quality along the Food Chain, 1
Gerhard Schiefer

2. Methodologies for Improved Quality Control Assessment of Food Products, 11
Manuel A. Coimbra, Sílvia M. Rocha, and António S. Barros

3. Application of Microwaves for On-line Quality Assessment, 49
Ruth De los Reyes, Marta Castro-Giráldez, Pedro Fito, and Elías De los Reyes

4. Ultrasounds for Quality Assurance, 81
Bosen Zhao, Otman A. Basir, and Gauri S. Mittal

5. NMR for Food Quality and Traceability, 101
Raffaele Sacchi and Livio Paolillo

6. Electronic Nose for Quality and Safety Control, 119
Naresh Magan and Natasha Sahgal

7. Rapid Microbiological Methods in Food Diagnostics, 131
Daniel Y. C. Fung

8. Molecular Technologies for Detecting and Characterizing Pathogens, 155
Geraldine Duffy and Terese Catarame

9. DNA-Based Detection of GM Ingredients, 175
Alexandra Ehlert, Francisco Moreano, Ulrich Busch, and Karl-Heinz Engel

10. Protein-Based Detection of GM Ingredients, 199
A. Rotthier, M. Eeckhout, N. Gryson, K. Dewettinck, and K. Messens

11. Immunodiagnostic Technology and Its Applications, 211
Didier Levieux

12. Rapid Liquid Chromatographic Techniques for Detection of Key (Bio)Chemical Markers, 229
M-Concepción Aristoy, Milagro Reig, and Fidel Toldrá

13. Sampling Procedures with Special Focus on Automatization 253
K. K. Kleeberg, D. Dobberstein, N. Hinrichsen, A. Müller, P. Weber, and H. Steinhart

14. Data Processing, 295
Riccardo Leardi

15. Data Handling, 323
Philippe Girard, Sofiane Lariani, and Sébastien Populaire

16. The Market for Diagnostic Devices in the Food Industry, 347
Hans Hoogland and Huub Lelieveld

Index, 359

Contributors

M.-Concepción Aristoy
Instituto de Agroquímica y Tecnología de Alimentos (C.S.I.C.)
46100 Burjassot (Valencia)
Spain
Chapter 12

António S. Barros
Department of Chemistry
University of Aveiro
3810-193 Aveiro
Portugal
Chapter 2

Otman A. Basir
Department of Electrical and Computer Engineering
University of Waterloo
Waterloo, Ontario, N2L 3G1
Canada
Chapter 4

Ulrich Busch
Bavarian Health and Food Safety Authority LGL
Veterinärstr. 2
85764 Oberschleissheim
Germany
Chapter 9

Marta Castro-Giráldez
Food Science and Technology Department
Food Engineering Institute for Development
Polytechnical University of Valencia
46022 Valencia
Spain
Chapter 3

Terese Catarame
Teagasc, National Food Centre
Ashtown
Dublin 15
Ireland
Chapter 8

Manuel A. Coimbra
Department of Chemistry
University of Aveiro
3810-193 Aveiro
Portugal
Chapter 2

Elías De los Reyes
Food Science and Technology Department
Food Engineering Institute for Development
Polytechnical University of Valencia
46022 Valencia
Spain
Chapter 3

Ruth De los Reyes
Food Science and Technology Department
Food Engineering Institute for Development
Polytechnical University of Valencia
46022 Valencia
Spain
Chapter 3

K. Dewettinck
Laboratory of Food Technology and Engineering
Department of Food Safety and Food Quality
Faculty of Bioscience Engineering
University of Ghent
Ghent
Belgium
Chapter 10

D. Dobberstein
Institute of Biochemistry and Food Chemistry
University of Hamburg
20146 Hamburg
Germany
Chapter 13

Geraldine Duffy
Teagasc, National Food Centre
Ashtown
Dublin 15
Ireland
Chapter 8

M. Eeckhout
Department of Food Science and Technology
Faculty of Biotechnological Sciences
University College Ghent
Ghent University Association
Voskenslaan 270
B9000 Ghent
Belgium
Chapter 10

Alexandra Ehlert
Technical University of Munich
Center of Food and Life Sciences
85350 Freising-Weihenstephan
Germany
Chapter 9

Karl-Heinz Engel
Technical University of Munich
Center of Food and Life Sciences
85350 Freising-Weihenstephan
Germany
Chapter 9

Pedro Fito
Food Science and Technology Department
Food Engineering Institute for Development
Polytechnical University of Valencia
46022 Valencia
Spain
Chapter 3

Daniel Y. C. Fung
Department of Animal Sciences and Industry
Kansas State University
Manhattan, Kansas 66506
USA
Chapter 7

Philippe Girard
Quality Management, Nestec SA
Avenue Nestlé 55
CH-1800 Vevey
Switzerland
Chapter 15

N. Gryson
Department of Food Science and Technology
Faculty of Biotechnological Sciences
University College Ghent
Ghent University Association
Voskenslaan 270
B9000 Ghent
Belgium
Chapter 10

N. Hinrichsen
Institute of Biochemistry and Food Chemistry
University of Hamburg
20146 Hamburg
Germany
Chapter 13

Hans Hoogland
Unilever Research Lab
3130 AC, Vlaardingen
Netherlands
Chapter 16

K. K. Kleeberg
Institute of Biochemistry and Food Chemistry
University of Hamburg
20146 Hamburg
Germany
Chapter 13

Sofiane Lariani
BioAnalytical Science Department
Nestlé Research Center
Vers-Chez-les-Blanc
CH-1000 Lausanne 26
Switzerland
Chapter 15

Riccardo Leardi
Department of Chemistry and Food and Pharmaceutical Technologies
University of Genova
Via Brigata Salerno (ponte)
I-16147 Genova
Italy
Chapter 14

Huub Lelieveld
Ensahlaan 11
3723 HT Bilthoven
Netherlands
Chapter 16

Didier Levieux
SRV Immunochimie INRA Theix
63122 St.-Genès-Champanelle
France
Chapter 11

Naresh Magan
Applied Mycology Group
Biotechnology Center
Cranfield University
Silsoe, Bedfordshire MK45 4DT
United Kingdom
Chapter 6

K. Messens
Department of Food Science and Technology
Faculty of Biotechnological Sciences
University College Ghent
Ghent University Association
Voskenslaan 270
B9000 Ghent
Belgium
Chapter 10

Gauri S. Mittal
School of Engineering
University of Guelph
Guelph, Ontario, N1G 2W1
Canada
Chapter 4

Francisco Moreano
Bavarian Health and Food Safety Authority LGL
Veterinärstr. 2
85764 Oberschleissheim
Germany
Chapter 9

A. Müller
Institute of Biochemistry and Food Chemistry
University of Hamburg
20146 Hamburg
Germany
Chapter 13

Livio Paolillo
Department of Chemistry
University of Naples Federico II
Via Cintia
80126 Naples
Italy
Chapter 5

Sébastien Populaire
Quality Assurance Department
Nestlé Research Center
Vers-Chez-les-Blanc
CH-1000 Lausanne 26
Switzerland
Chapter 15

Milagro Reig
Instituto de Agroquímica y Tecnología de Alimentos (C.S.I.C.)
46100 Burjassot
(Valencia)
Spain
Chapter 12

Sílvia M. Rocha
Department of Chemistry
University of Aveiro
3810-193 Aveiro
Portugal
Chapter 2

A. Rotthier
Department of Food Science and Technology
Faculty of Biotechnological Sciences
University College Ghent
Ghent University Association
Voskenslaan 270
B9000 Ghent
Belgium
Chapter 10

Raffaele Sacchi
Department of Food Science
University of Naples Federico II
Faculty of Agriculture
80055 Portici
Italy
Chapter 5

Natasha Sahgal
Applied Mycology Group
Biotechnology Center
Cranfield University
Silsoe, Bedfordshire MK45 4DT
United Kingdom
Chapter 6

Gerhard Schiefer
University of Bonn
Meckenheimer Allee 174
D-53115 Bonn
Germany
Chapter 1

H. Steinhart
Institute of Biochemistry and Food Chemistry
University of Hamburg
20146 Hamburg
Germany
Chapter 13

Fidel Toldrá
Department of Food Science
Instituto de Agroquímica y Tecnología de Alimentos (CSIC)
46100 Burjassot
(Valencia)
Spain
Chapter 12

P. Weber
Institute of Biochemistry and Food Chemistry
University of Hamburg
20146 Hamburg
Germany
Chapter 13

Bosen Zhao
Department of Electrical and Computer Engineering
University of Waterloo
Waterloo, Ontario, N2L 3G1
Canada
Chapter 4

Preface

The main goal of the book *Advances in Food Diagnostics* is to provide the reader with a comprehensive resource, covering the field of diagnostics in the food industry.

While covering conventional (typically lab-based) methods of analysis, the book focuses on leading-edge technologies that are being (or about to be) introduced. The book looks at areas such as food quality assurance, safety, and traceability. Issues such as improved quality control, monitoring pesticide and herbicide residues in food, determining the nutritional content of food, and distinguishing between GM and "conventional" foodstuffs are covered.

In the first chapter issues about safety and quality along the food chain are discussed. Increased globalization, industrialization, and sophistication of food production and trade increase the need for improved process control, process management, and communication inside enterprises but especially between enterprises along the vertical food production chain.

Chapter 2 shows how Fourier transform infrared (FT-IR) spectroscopy combined with principal component analysis (PCA) can be used as a rapid tool for the analysis of polysaccharide food additives and to trace food adulterations. PCA and partial least squares (PLS1) regression are chemometric methodologies that allow significant improvements in data analysis when compared with univariate analysis. Chapter 2 also discusses how screening and distinction of coffee brews can be done based on the combined headspace (HS)–solid phase microextraction (SPME)–gas chromatography (GC)–PCA (HS-SPME-GC-PCA) methodology.

Finally, the last part of chapter 2 shows a new methodological approach based on cyclic voltammetry for assessment of quality control for cork stoppers, recently proposed for the rapid screening of cork-wine model interactions in order to determine if the cork stoppers are able to contaminate a wine.

Wide experience has now been accumulated in the field of microwave applications. Sensor systems based upon microwaves can be a viable nondestructive method of on-line control.

The aim of chapter 3 is to provide an overview of the current applications of low-power microwaves for food quality control and to offer a detailed examination of the current research in this field.

Chapter 4 details applications of ultrasounds of foods for quality assurance, including evaluations of texture, viscosity, density, and particle size.

Chapter 5 is not a full review of all nuclear magnetic resonance (NMR) applications in food science but shows some selected examples of recent applications of this versatile and powerful technique closely focused on "food quality and traceability." In particular, examples of application of NMR in the fields of food quality (nutritional, sensory, freshness) and traceability (geographical origin, botanical origin, animal species, process technology applied to foods) have been selected.

Rapid early detection of mold activity in food throughout the food chain is required as part of a quality assurance program and to enable critical control points to be effectively monitored. The rapid development of electronic nose (e.nose) technology has resulted in examination of the potential of using this qualitative approach to decide the status of raw materials and processed food, including bakery products. The recent developments of electronic nose and also of electronic tongue technology with applications for different foods are summarized in chapter 6.

"Rapid microbiological methods," as discussed in chapter 7, is a dynamic area in applied microbiology dealing with the study of improved methods in the isolation, early detection, characterization, and enumeration of microorganisms and their products in clinical, food, industrial, and environmental samples. In the past 15 years this field has emerged into an important subdivision of the general field of applied microbiology and is gaining momentum nationally and internationally as an area of research and application to monitor the numbers, kinds, and metabolites of microorganisms related to food spoilage, food preservation, food fermentation, food safety, and food-borne pathogens.

Molecular diagnostic techniques, the contents of chapter 8, are based on the detection of a fragment of genetic material (nucleic acids, i.e., DNA or RNA) that is unique to the target organism, and as such they are highly specific. One of the most practical and useful applications of molecular tools is their specificity as they target genetic regions unique to the organism, and depending on the gene target, they can also yield valuable information about virulence properties of the organism. They are also invaluable in detecting and identifying infectious agents.

Chapter 9 reviews applications of DNA-based methods to the analysis of foods. In the last decade the need for methods to detect and to quantify DNA from genetically modified organisms (GMOs) has been a major driver for the development and optimization of PCR-based techniques. Therefore, this field of application is used to outline principles, challenges, and current developments of DNA analysis in foods. In addition, DNA-based approaches in the areas of food authentication, detection of food-borne pathogenic microorganisms, and screening for food allergens are reviewed.

In chapter 10, the advantages and limitations of protein detection for GMOs are discussed. Specific attention is also focused on the use of these methods in the traceability of GMOs along the production chain.

In comparison to other analytical methods such as HPLC, electrophoresis, gas chromatography, or mass spectrometry, immunoassays can provide highly sensitive and specific analyses that are rapid, economical, and relatively simple to perform. Chapter 11 describes the general principles of a wide variety of immunoassays formats. Then, a critical appraisal of their applications in food analysis is given.

High-performance liquid chromatography (HPLC) constitutes a technique that has become widely use for the analysis of foods. It is being used for a large number of applications, which include the analysis of nutrients, chemical and biochemical contaminants, markers for processing control, detection of adulterations, and control of raw materials and products. HPLC has been traditionally used for almost all these tasks, but the actual challenge is to improve the throughput to better compete with other techniques that have been appearing lately. The fundamentals of this technique and a summarized description of these applications are described in chapter 12. As sample preparation is the most tedious and time-consuming step in food analysis, this chapter is mainly focused on those methodologies with less sample manipulation before HPLC analysis.

Chapter 13 gives an overview of modern sample preparation techniques with regard to subsequent analytical methods. In the majority of cases, sampling procedures cannot be discussed separately but have to be considered as a combination of analyte extraction and measurement. The most important extraction methods are presented based on aroma and flavor analytics, since related compounds cover a wide range of different chemical classes and properties. Further sections focus on specific extraction and analytical methods for the main food ingredients: lipids, proteins, and carbohydrates. These methods are often characterized by conventional extraction procedures followed by specific analytical techniques.

The goal of chapter 14 is simply that it be read and understood by the majority of the readers of this book. This goal will be completely achieved if some of them, after having read it, could say: "Chemometrics is easy and powerful indeed, and from now on I will always think in a multivariate way."

Data handling, as discussed in chapter 15, is all activities dealing with data, from their collection (How should experiments be set up? How many samples/measurements are required?) to their transformation into useful information (Which method should be used?) so as to answer a specific question.

Finally chapter 16 discusses the market for diagnostic devices in the food industry. The industry is in general driven by consumer demands and by governmental legislation. Over the last decade an immense change with consumers has taken place. Food has changed from an energy and nutrient source to a product that influences the well-being of individuals. Consumers realize that food may have a significant impact on their lives, and consequently they demand higher-quality, fresher products. Moreover, it is now known that a large part of the population is allergic to some foods. In rare cases allergens may have severe—even life-threatening—consequences. Consumers therefore need to know what is in the food they buy, and regulators try to provide this information because consumers assume that governments ascertain that food is always safe. This puts severe pressures on the industry that now needs to comply with accurate labeling and stringent tracking and tracing systems to be able to respond instantly to any (real, potential, or perceived) incident. On top of these requirements, the industry is faced with the complexity of today's supply chains. Ingredients are sourced from all over the world (globalization), spreading food-related hazards as fast as the ingredients and products move. Consequently, surveillance must be stepped up to be able to keep hazards with microorganisms, toxins, allergens (labeling), and chemical contamination under control.

The editors wish to thank all the contributors for their hard work and excellent results, now the chapters of this book. We would also like to thank Susan Engelken of Blackwell Publishing for her endless patience.

Leo M. L. Nollet and Fidel Toldrá

Advances in Food Diagnostics

1 Assuring Safety and Quality along the Food Chain

Gerhard Schiefer

Quality and Safety: Issues

The term *quality* has become a focus point in all discussions regarding the production and provision of food products to markets and consumers—quality in the broad sense of serving the consumers' needs (see also Oakland 1998) by providing them with the right product at the right time and with the right service. In today's competitive food markets the quality approach is a precondition for sustainable market acceptance. It is a core pillar in enterprises' and the sector's sustainability, which builds on economic viability, quality orientation, and an appropriate embedment in its environment.

In an enterprise, a **sustainable delivery of quality** is a result of a comprehensive effort. It involves the implementation of a quality approach at all levels of activities, reaching from enterprise management to process organization, process management, and product control. Enterprise quality systems build on routine quality assurance and improvement activities that might encompass one or several of these levels. However, most food quality systems focus on system activities at several levels, involving process organization, process management, and product control.

Food safety is an inherent element of quality. It receives special attention not only by enterprises but also by policy and legislation because of its key importance for consumers' health and the responsibility for food safety by enterprises and policy alike. Globalization and industrialization in the production and provision of food have increased the potential risk in food safety and initiated increased efforts and controls in food safety assurance.

The efficient **transportation** of quality from the farm and any of the subsequent stages of processing and trade to the consumer as the final customer requires cooperation along the chain. The dependence of food quality and safety on activities of all stages in the chain makes chain cooperation a prerequisite of any advanced quality assurance scheme. The cooperation might build on individual arrangements, sector agreements, or any other way that avoids the loss and supports the gain of quality along the chain.

Chain cooperation has become a crucial element in quality assurance and especially food safety initiatives in the food sector. However, in the food sector, chains usually develop dynamically in a network of interconnected enterprises with constantly changing lines of supplier-customer relationships. In this scenario, chain cooperation is based on network cooperation or, in other words, on sector agreements.

The **quality guarantee** that one can derive from the implementation of a quality system depends on the evaluation of the system as a whole. Quality and food safety deficiencies at any stage might remain with the product throughout the remaining stages until it reaches the consumer. The most crucial need for guarantees involves guarantees for food safety.

They constitute the baseline guarantee level and the prerequisite for consumers' trust and market acceptance (Henson and Hooker 2001; Verbeke 2005).

The delivery of quality guarantees is based on **controls**, both in the organization of processes (process controls) and in process management (management controls). However, for the delivery of guarantees, these controls need to be integrated into a comprehensive scheme (**quality program**) that could serve as a cooperation platform for enterprises within supply chains and networks, as well as provide a basis for communication with consumers.

Key issues involve agreements on **chain-encompassing quality assurance** schemes and the ability to clearly identify the product flow through the production chain by linking the different product entities that are being produced and traded at the different stages of the chain from the farm to the consumer as the final customer and their quality status (**tracking and tracing capability**).

The following chapters cover the development path from tracking and tracing toward quality assurance in food chains, the organizational concepts and quality programs for implementation, and the role of information and communication systems for operational efficiency.

Tracking and Tracing through Chains and Networks

The tracking and tracing of food products throughout the food chain has become a dominant issue in discussions on food quality and especially on the assurance of food safety (Luning and Marcelis 2005). In the European Union, enterprises in the food sector are legally required to implement appropriate tracking and tracing capabilities (EU 2002).

They allow, for any product and from any stage within the chain, identification of the initial source (backward tracing) and, eventually, its final destination (forward tracing). This supports the (backward) identification of sources of product deficiencies and the (forward) isolation of any other product that might have been affected by these sources. Tracking and tracing capabilities support consumer protection in case of food contamination. Furthermore, they support the communication of the quality status of products on their way through the food chain and provide the basis for the delivery of quality guarantees at each stage of the chain and toward the consumers at the final stage.

However, it should be noted that beyond this discussion line, the organization of tracking and tracing schemes (TT schemes) also has a **managerial dimension** in supporting the efficiency in the logistics chain (supply chain) from the source (farms) to the final destination (the consumer). In fact, the managerial dimension was at the center point of initial discussions on tracking and tracing schemes not just in the food sector but in other sectors as well (Golan et al. 2004).

This emphasizes the global relevance of tracking and tracing schemes and their role as a baseline feature for not only the delivery of guarantees for food safety and quality but also for logistics efficiency, which is at the core of enterprises' economics interests.

From a historical point of view, the TT schemes evolved from internal efforts and were subsequently extended to supply chains and networks. This historic development also characterizes a path of increasing complexity. The identification of product units and the monitoring of their movements inside an enterprise require fewer coordination efforts than are necessary in supply chains, and especially in a sector as a whole with its larger number of enterprises and the different and ever-changing trade relationships.

The identification of product units and the monitoring of their movements are an easy way to solve a problem, if product modification during the various stages of a supply chain process does not affect the composition of the product. The most complex TT scenario concerns commodity products where an individual "product unit" cannot be based on a physical product element (e.g., a piece of grain) but needs to be based on **logistics elements (batches)** that might involve production plots, transportation trucks, or storage units of any kind (Golan et al. 2004; Schiefer 2006). The linkage of these different batches in a **batch sequence** generates the production flow with its modifications and provides the basis for tracking and tracing activities.

Food Safety—the Baseline

The general assurance of food safety is a prime concern and responsibility of society. Traditionally, food safety rests on the formulation and implementation of standards regarding the measurable quality of products, such as the quantity of substances in the product with potentially negative effects on human health.

This approach is presently being supplemented (not replaced) by a proactive approach that intends to prevent food safety deficiencies from the beginning through regulations on the appropriate organization and management of processes in production, trade, and distribution.

During recent years, policy discussions and legislative actions concerning proactive food safety improvement initiatives concentrated on

1. the assurance of tracking and tracing of products and
2. the implementation of the Hazardous Analysis and Critical Control Point (HACCP) principles (USDA 1997; Luning et al. 2002).

However, as both of these initiatives require enterprise activities for implementation, any regulations regarding their utilization in the food sector require cooperation by enterprises. This is a crucial point in food safety assurance. Society (represented by policy) has responsibilities in the provision of food safety guarantees to its members but has to rely on activities by enterprises to substantiate the guarantees (fig. 1.1).

In this scenario, the "value" of society's guarantees depends on its ability to assure enterprises' cooperation (i.e., on the effectiveness of the sector control systems).

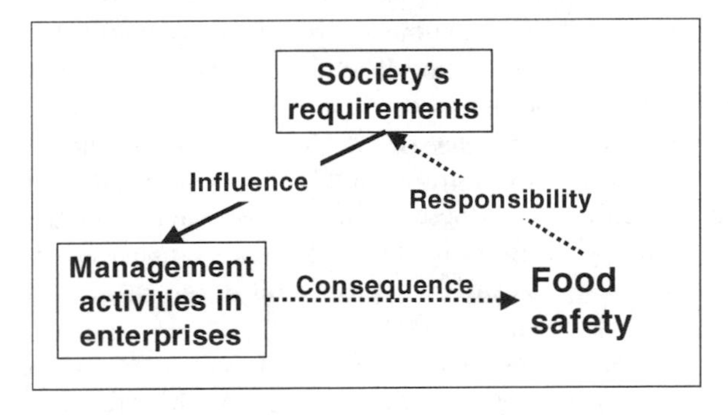

Fig. 1.1. Chain of influence in food safety assurance.

However, the enforcement of enterprises' cooperation through appropriate control systems has consequences for trade and constitutes, in principle, nontariff trade barriers that have to adhere to the European and international trade agreements. At the international level, the World Trade Organization (WTO) provides the umbrella for trade regulations. It allows introducing trade-related regulations that avoid food safety hazards if backed by sufficient scientific evidence. An important reference in this context is the Codex Alimentarius Commission (FAO/WHO 2001; Luning et al. 2002), a joint initiative by the Food and Agriculture Organization (FAO) and the World Health Organization (WHO). In its "Codes of Practice" and "Guidelines" it addresses aspects of process management including, as its most prominent recommendation, the utilization of the HACCP principles.

Food Quality—Delivery Concepts

In enterprises and food chains, the delivery of quality and quality guarantees that reach beyond food safety traditionally builds on four principal areas of quality activities integrated into a systematic process of continuous improvement. They include

1. the quality of enterprise management, as exemplified by the concept of "total quality management" (TQM) (Oakland and Porter 1998; Gottlieb-Petersen and Knudsen 2002),
2. the quality of process organization, frequently captured in the phrase "good practice,"
3. the quality of process management, usually phrased as "quality management," and
4. the quality of products that could be captured through sensor technology and so forth.

Present discussions on the assurance of food quality in the food sector concentrate primarily on the quality of **process organization** and **process management** and combine it with specific requirements on product quality characteristics. This integrated view is based on the understanding that not all food product characteristics with relevance for quality could be identified and competitively evaluated through inspection of the final product. It refocuses attention from traditional product inspection to the prevention of deficiencies in food quality.

However, it should be noted that successful quality initiatives of enterprises usually build on leadership initiatives related to the TQM approach and with a strong focus on continuous improvement activities. In this scenario, quality-oriented process management is an integral part of the more comprehensive management approach and not a "standalone" solution for the elimination of quality problems.

A quality-oriented process management is characterized by management routines, such as audit activities, that support the organization and control of processes to assure desired process outputs with little or no deviation from output specifications (**process quality**). The integration and specification of these routines constitute a "management system" or, with a view on the quality-focused objectives, a "quality management system." Well-known examples include the standard series ISO9000 or the HACCP principles (USDA 1997; ISO 2001; Krieger and Schiefer 2003).

The traditional view of quality assurance in supply chains of any kind builds on the isolated implementation of quality management systems in individual enterprises and assumes a sufficient consideration of quality objectives through the chain of supplier-customer relationships in which each supplier focuses on the best possible fulfillment of quality expectations of its immediate customers (Spiegel 2004).

However, this traditional view does not match with the specifics of food production and the requirements of quality assurance in the food sector. These specifics suggest that substantial improvements can only be reached through an increased cooperation between stages regarding the specification of quality levels, agreements on process controls, and the utilization of quality management schemes. This requires agreements on information exchange and the establishment of appropriate communication schemes.

Initiatives toward integrated food supply chains have been a focus of developments during the 1990s, especially in export-oriented countries like the Netherlands and Denmark (Spiegel 2004).

These developments were primarily initiated for gaining competitive advantage in a quality-oriented competitive market environment; improvements in the sector's food quality situation were initially of secondary concern.

Quality Programs—Steps toward Sector Quality Agreements

Overview

During the last decade, a variety of initiatives in different countries have focused on the formulation of comprehensive quality programs that ask for the simultaneous implementation of a set of activities in process organization and process management that assures a certain level of food quality and safety in enterprises and food chains. These programs, also referred to as *quality systems* or (if restricted to process management) *quality management systems*, were of a universal, national, or regional scope.

Principal examples with focus on food chains include (Schiefer 2003).

1. initiatives on the basis of rather closed supply chains such as the Dutch "IKB chains" (IKB for Integrated Chain Management) (Wierenga et al. 1997) and
2. sector-encompassing approaches that have little requirements on focused organizational linkages between enterprises such as the German Q&S system (Nienhoff 2003).

Specific alternatives are programs that evolved from retail trade, such as the "International Food Standard" (FCD 2005; Krieger and Schiefer, forthcoming). They do not involve the supply chain as a whole but function as a quality filter for deliveries from supplier enterprises and the food chains these are connected to.

A Closed System Concept—the Case of IKB

The IKB concept is a chain management concept for food supply chains that was designed in the Netherlands in the 1980s for improvements in the efficiency and quality of food production. Its initial focus was on closed production chains with a central coordinating body linked to the processing industry (Wierenga et al. 1997). Product deliveries into the IKB chains are restricted to enterprises that conform to certain quality requirements. A key example involves conformity to the Dutch standard series good manufacturing practice (GMP)(Luning et al. 2002). Today's developments open the closed chain approach and move it closer toward a network system.

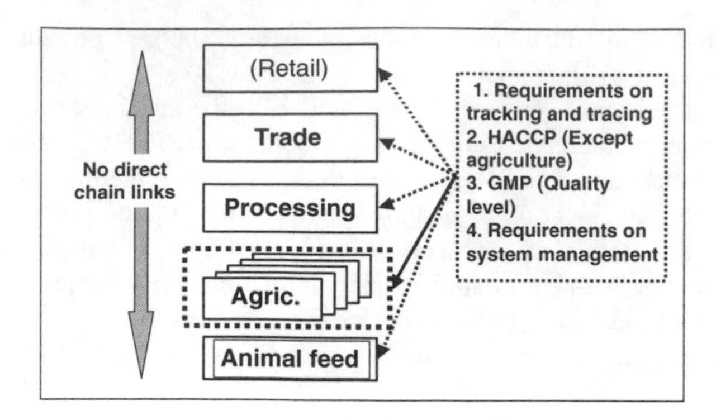

Fig. 1.2. Q&S system organization.

An Open Sector System Concept—the Case of Q&S

The system of Q&S addresses all stages of the vertical supply chain. However, it can be implemented by each individual enterprise at each stage, with the exception of agricultural enterprises that can only act as a group (fig. 1.2), and without any further coordination with its suppliers and/or customers.

The Q&S system is an open system, and its coordination is determined, in principle, by common agreements on the different stages' quality responsibility. The approach tries to best adapt the food quality control activities to the actual market infrastructure that builds on open supply networks with continuously changing trade relationships. It does not place new organizational requirements on enterprise cooperation or restrictions on the development of individual market relationships within the supply chain.

The system preserves flexibility in market relationships between enterprises but, as an open flexible system, does require substantial efforts to move the whole system to higher quality levels. Furthermore, the approach does not support the implementation of more advanced quality assurance systems of individual groups within the general system environment. Such efforts would reduce the guarantee value of the general system for the remaining participants and contradict the interest of the system as a whole.

Trade Initiatives

The retail sector has designed its own "standards" for requirements on quality activities in its supplier enterprises, including those from agriculture that deliver directly to the retail stage (for an overview see Hofwegen et al. 2005). Examples include the international active standard **EurepGAP**, which focuses on agricultural enterprises (GAP: Good Agricultural Practice), initially in the production of fruits and vegetables, today in most agricultural production lines; **IFS**, the International Food Standard, with a stronghold in Germany and France; and **BRC**, the standard of the British Retail Consortium, which has influenced many quality initiatives in food supply chains in the UK and elsewhere.

Furthermore, a global retail initiative, the Global Food Safety Initiative (GFSI) has formulated requirements on food safety assurance activities for retailer-based standards, which, if requirements are met, receive formal acceptance status by the GFSI (fig. 1.3).

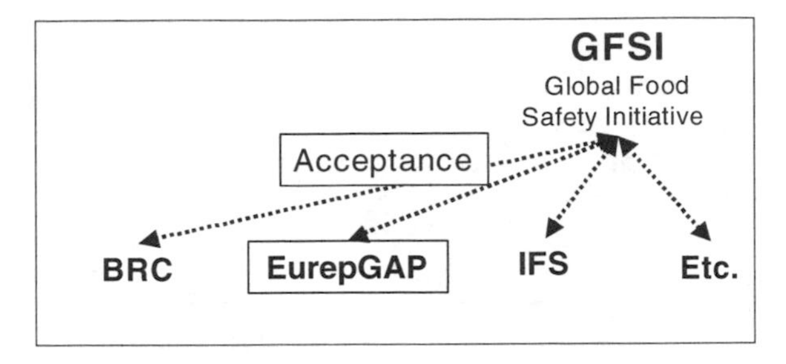

Fig. 1.3. Relationships between retail quality initiatives.

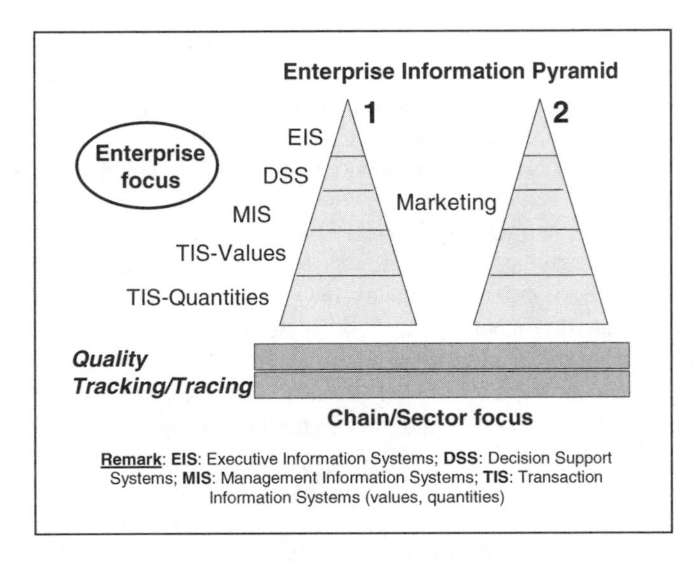

Fig. 1.4. Information layers with enterprise (1,2) and chain/sector focus.

The Information Challenge

Information Clusters

Both tracking and tracing capabilities as well as the fulfillment of quality expectations at the consumers' end depend on activities in enterprises throughout the supply chain and, as a consequence, on the collection of information from chain participants and its communication throughout the chain, with the consumers as the final recipients. This requires the availability of a feasible sector-encompassing communication infrastructure.

Traditionally, the organization of information in enterprises builds on a number of information layers that correspond with the different levels of business management and decision support. They reach from transaction information at the lowest level to executive information at the highest level (Turban et al. 2001). These layers are presently being complemented by two additional layers at the transaction level that incorporate information for tracking and tracing as well as for quality assurance and improvement activities (fig. 1.4).

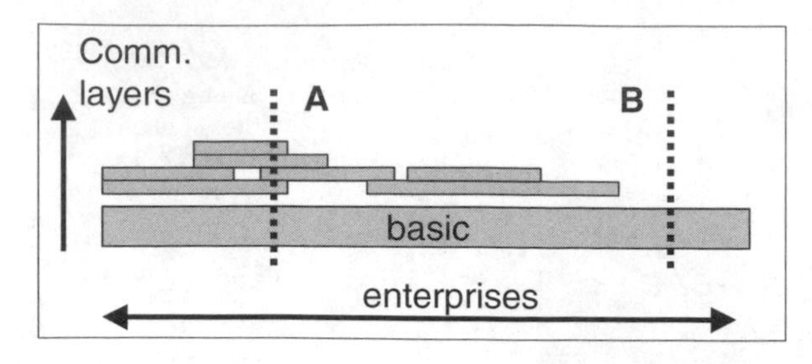

Fig. 1.5. Agreed communication clusters with participation of enterprise A in five and enterprise B in one of the clusters.

These new layers differ from traditional enterprise information layers by their focus. Their focus is not the individual enterprise but the vertical chain of production and trade. They are linked to the flow of goods and connect, in principle, the different stages of production and trade with each other and the consumer. Their realization depends on agreements between trading partners on responsibilities, content, organization, and technologies.

The layers were initiated by requirements from legislation (EU 2002; Luning et al. 2002) and markets for tracking and tracing capabilities and by increasing expectations of consumers regarding the quality of products and production processes.

A sector encompassing general agreement is restricted to the lowest level of legal requirements. Any communication agreements beyond this level are subject to specific business interests, which might be limited to clusters of enterprises with common trading interests. In a network environment, individual enterprises might be members of different clusters, resulting in a future patchwork of interrelated and overlapping communication clusters (fig. 1.5).

The content of quality communication layers depends on the quality requirements of enterprises and consumers. However, the diversity of interests in a sector could generate an almost unlimited number of possible requirement sets or, in other words, needs for communication clusters. This is not a feasible approach.

In this situation, the quality requirements of quality programs could serve as a basic reference for the separation of communication clusters. First initiatives toward this end are under way. These developments will separate the sector's food production into different segments with different quality guarantees.

Organizational Alternatives

The principal alternatives for sectorwide information infrastructures focus on two different dimensions. The information may be communicated between enterprises directly, or it may be communicated between enterprises through a common data network that is linked with enterprises' internal information systems (fig. 1.6) (see, e.g., the example in Hannus et al. 2003). These approaches mirror classical network approaches, such as bus or ring network topologies (Turban et al. 2001).

However, there is an additional communication alternative that avoids the communication of data but communicates assurances that certain information is true. If enterprises are assured that their suppliers fulfill the requirements of a certain quality system,

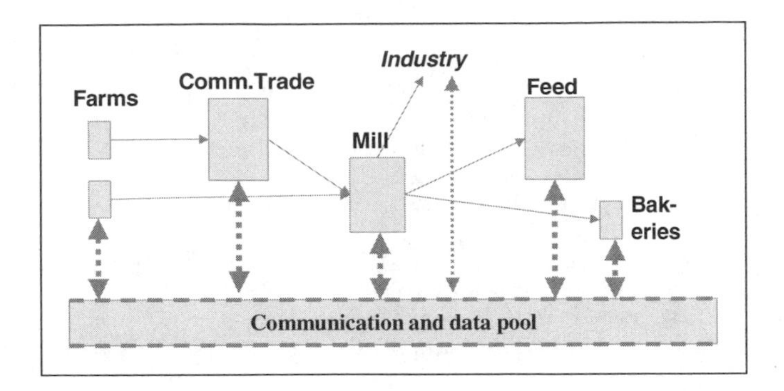

Fig. 1.6. Sectorwide communication and data pool (example: grain chain).

information linked to the requirements does not have to be communicated—the assurance (e.g., in terms of a certificate) is sufficient. As information infrastructures for quality assurance are not yet established sufficiently, this last approach is attractive for the time being and utilized with a number of quality programs (Reardon et al. 2001).

Added Value of Emerging Information Infrastructures

The interest in quality of customers and consumers, the chain efficiency aspect, and the legal requirements on the tracking and tracing capability of the food chain together provide the argument for the establishment of a sectorwide information infrastructure. However, newly emerging aspects of quality communication schemes involve the potential for possible added values that these infrastructures could provide. As an example, chain-focused extension services might utilize information from various stages to arrive at recommendations for improvements in chain quality performance or chain efficiency.

Conclusion

Initiatives to improve tracking and tracing capabilities as well as the delivery of trustworthy and stable quality products are means to control risks and to assure and develop markets. From this point of view, they are prerequisites for enterprises in the food market to be sustainable economically. Considerations of public health and legal requirements support development and are not contradictory.

Increased globalization, industrialization, and sophistication of food production and trade increase the need for improved process control, process management, and communication inside enterprises but especially between enterprises along the vertical food production chain. This requires substantial investments in, among others, the design of new quality assurance concepts, cooperation agreements throughout the sector, the identification of accepted quality levels, the allocation of quality assurance responsibilities, the design and implementation of communication systems, and the distribution of investment and operations costs.

This makes the move from the traditional view on quality production to today's requirements difficult and a challenge for the sector—but a challenge that needs to be met.

References

EU. 2002. Regulation (EC) No. 178/2002 of the European Parliament and the Council. Official Journal of the European Communities.

FAO/WHO. 2001. Codex Alimentarius Commission, Report, Joint FAO/WHO Food Standards Programme, Rome.

FCD. 2005. IFS compendium of doctrine. Report, Federation des Entreprises du Commerce et de la distribution (http://www.food-care.org).

Golan, E., Krissoff, B., Kuchler, F., Calvin, L., Nelson, K., Price, G. 2004. Traceability in the US food supply: economic theory and industry studies. Report AER-830, USDA/ERS.

Gottlieb-Petersen, C., Knudsen, S. 2002. From product to total quality management in the food supply chain. Proceedings, 13th International Conference, International Food and Agribusiness Management Association (IFMA), Wageningen.

Hannus, T., Poignée, O., Schiefer, G. 2003. The implementation of a web based supply chain information system. In: Harnos, Z., Herdon, M., Wiwczaroski, T.B., eds., Information technology for a better agri-food sector, environment and rural living, Proceedings, EFITA2003, Debrecen-Budapest.

Henson, S., Hooker, N.H. 2001. Private sector management of food safety: public regulation and the role of private controls. International Food and Agribusiness Management Review 4 (1), pp. 7–18.

Hofwegen, van G., Becx, G., Broek, van den J. 2005. Drivers for competitiveness in agro-food chains: a comparative analysis of 10 EU food product chains. Report, Wageningen University, (http://www.eumercopol.org).

ISO. 2001. ISO standards compendium: ISO 9000—quality management. ISO-Publisher, Geneva.

Krieger, St., Schiefer, G. 2003. Quality management schemes in Europe and beyond. In: Schiefer, G., Rickert, U., eds., Quality assurance, risk management and environmental control in agriculture and food supply networks, pp. 35–50, University of Bonn/ILB, Bonn.

———. Forthcoming. Costs and benefits of food safety and quality improvements in food chains: methodology and applications. Technical report, FAO, Rome.

Luning, P.A., Marcelis, W.J. 2005. Food quality management and innovation. In: Jongen, W.M.F, Meulenberg, M.T.G., eds., Innovation in agri-food systems, Wangeningen Academic Publishers, Wangeninen, the Netherlands.

Luning, P.A., Marcelis, W.J., Jongen, W.M.F. 2002. Food quality management—a techno-managerial approach. Wageningen.

Nienhoff, H.J. 2003. QS quality and safety: a netchain quality management approach. In: Schiefer, G., Rickert, U., eds., Quality assurance, risk management and environmental control in agriculture and food supply networks, pp. 627–30, University of Bonn/ILB, Bonn.

Oakland, J.S., Porter, J. 1998. Total quality management. Butterworth's, Heineman, London.

Reardon, T., Codron, J.M., Busch, L., Bingen, J., Harris, C. 2001. Change in agrifood grades and standards: agribusiness strategic responses in developing countries. International Food and Agribusiness Management Review 2 (3/4), pp. 421–35.

Schiefer, G. 2003. From enterprise activity "quality management" to sector initiative "quality assurance": development, situation and perspectives. In: Schiefer, G., Rickert, U., eds., Quality assurance, risk management and environmental control in agriculture and food supply networks, pp. 3–22. University of Bonn/ILB, Bonn.

———. 2006. Computer support for tracking, tracing and quality assurance schemes in commodities. Journal for Consumer Protection and Food Safety 1 (2).

Schiefer, G., Rickert, U., eds. 2003. Quality assurance, risk management and environmental control in agriculture and food supply networks. University of Bonn/ILB, Bonn.

Spiegel, van der M. 2004. Measuring effectiveness of food quality management. PhD dissertation, Wageningen University.

Turban, E., McLean, E., Wetherbe, J. 2001. Information technology for strategic management. 2nd ed. John Wiley and Sons, New York.

USDA. 1997. Hazard analysis and critical control point principles and application guidelines. Washington, D.C. (http://www.cfsam.fda.gov).

Verbeke, W. 2005. Agriculture and the food industry in the information age. European Review of Agricultural Economics 32 (3), pp. 347–68.

Wierenga, B., Tilburg, A. van, Grunert, K.G., Steenkamp, J.-B.E.M., Wedel, M., eds. 1997. Agricultural marketing and consumer behavior in a changing world. Kluwer, Boston.

2 Methodologies for Improved Quality Control Assessment of Food Products

Manuel A. Coimbra, Sílvia M. Rocha, and António S. Barros

Introduction

The authentication of food is a major concern for the consumer and for the food industry at all levels of the food-processing chain, from raw materials to final products. The search and development of fast and reliable methods are nowadays at a premium. Midinfrared spectroscopy, chromatographic techniques associated with solid phase microextraction, and cyclic voltammetry, tandem with chemometry, are examples of methodologies that can be applied for the improvement of quality control assessment of food products.

Among the complex food constituents, the identification of the added polysaccharides could be a key factor if a rapid and reliable method is attainable. Classical chemical methods of polysaccharide determination are time-consuming and not always straightforward for a widespread routine application in the food industry. The first part of this chapter shows how Fourier transform infrared (FT-IR) spectroscopy combined with principal component analysis (PCA) can be used as a rapid tool for the analysis of polysaccharide food additives and to trace food adulterations. PCA and partial least squares (PLS1) regression are chemometric methodologies that allow significant improvements in data analysis when compared with univariate analysis. However, for the analysis of complex matrices, such as those of food products, other approaches are still required. The second part of the chapter presents one example of the use of combined regions of the FT-IR spectra for quantification purposes by applying PLS1 regression to an outer product (OP) matrix, and an example of the application of orthogonal signal correction (OSC)-PLS1 regression to minimize the matrix effect in the FT-IR spectra.

Screening and distinction of coffee brews can be done based on the combined headspace (HS)–solid phase microextraction (SPME)–gas chromatography (GC)–PCA (HS-SPME-GC-PCA) methodology. Using this methodological approach, presented in the third part of the chapter, diagnostic global volatile profiles of coffees brews can be obtained allowing their distinction to preclude the use of mass spectrometry for the identification of the volatile compounds.

Finally, the fourth part of the chapter shows a new methodological approach based on cyclic voltammetry for assessment of quality control for cork stoppers, recently proposed for the rapid screening of cork-wine model interactions in order to determine if the cork stoppers are able to contaminate a wine.

Use of Fourier Transform Infrared (FT-IR) Spectroscopy as a Tool for the Analysis of Polysaccharide Food Additives

Polysaccharides and their derivatives are widely used in food-processing technologies as gelling agents and thickeners. Starch, carrageenan, and pectin are examples of the most used polysaccharides in the food industry.

Starch, an important thickening and binding agent, is a mixture of two main glucan constituents, amylose, a linear polysaccharide composed of (1 → 4)-α-D-linked glucopiranose residues (fig. 2.1a) and amylopectin, a branched polysaccharide composed of (1 → 4)- and (1 → 4,6)-α-D-glucopiranose residues (fig. 2.1b).

Carrageenan utilization in food processing is based on its ability to gel, to increase the solution viscosity, and to stabilize emulsions and various dispersions. The carrageenans are characterized by an alternating repeating (1 → 4)-linked disaccharide structure consisting of 3,6-anhydro-α-D-galactopyranosyl-(1 → 3)-β-D-galactopyranosyl residues. A sulphate group at positions C2, C4, or C6 can substitute each residue. The carrageenans, depending on the sulphate substitutions, can be defined as kappa (κ), β-D-Gal*p*-4-sulphate and 3,6-anhydro-α-D-Gal*p* (fig. 2.2a); iota (ι), β-D-Gal*p*-4-sulphate and 3,6-anhydro-α-D-Gal*p*-2-sulphate (fig. 2.2b); and lambda (λ), nongelling agent consisting of β-D-Gal*p*-2-sulphate and α-D-Gal*p*-2,6-disulphate (fig. 2.2c).

Pectins are polysaccharides composed of a linear backbone of (1 → 4)-α-D-Gal*p*A interspersed by α-(1 → 2)-Rha*p* residues and with side chains constituted mainly of β-D-Gal*p* and α-L-Ara*f* residues (fig. 2.3a). Pectin with high ester content (fig. 2.3b) forms gels in the presence of sucrose, as in marmalades, and low ester pectin can set into a gel in the presence of Ca^{2+} (Belitz et al. 2004).

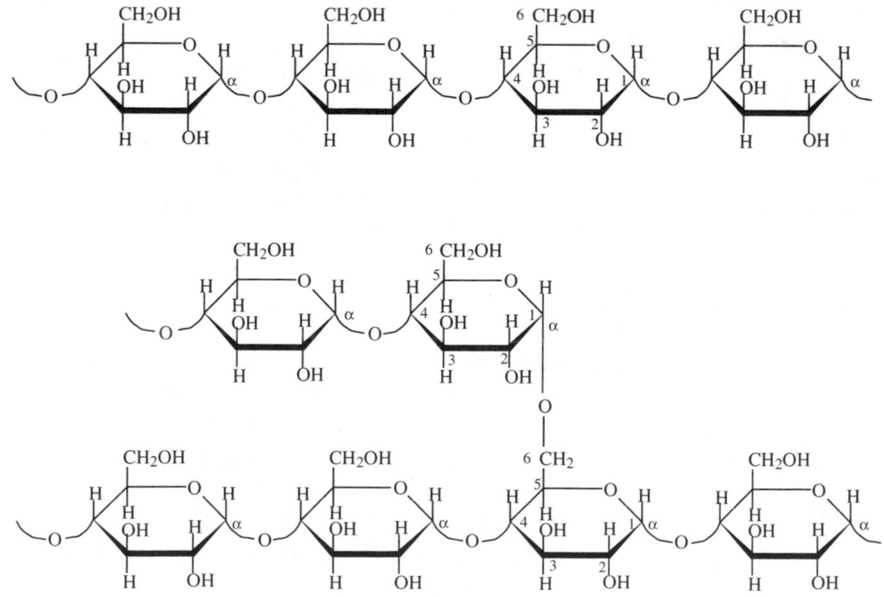

Fig. 2.1. Schematic representation of the glycosidic structure of the polysaccharides constituents of starch: **(a)** amylose and **(b)** amylopectin.

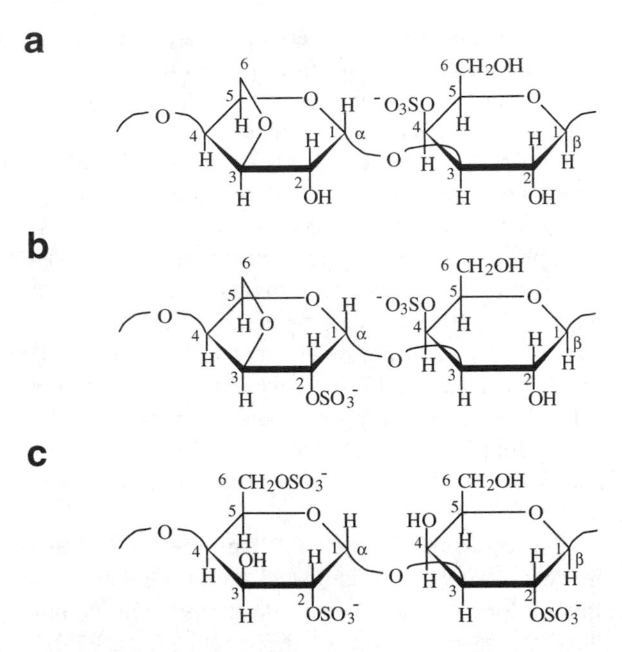

Fig. 2.2. Schematic representation of the glycosidic structure of the disaccharide repetitive unit of carrageenans: **(a)** kappa (κ), **(b)** iota (ι), and **(c)** lambda (λ).

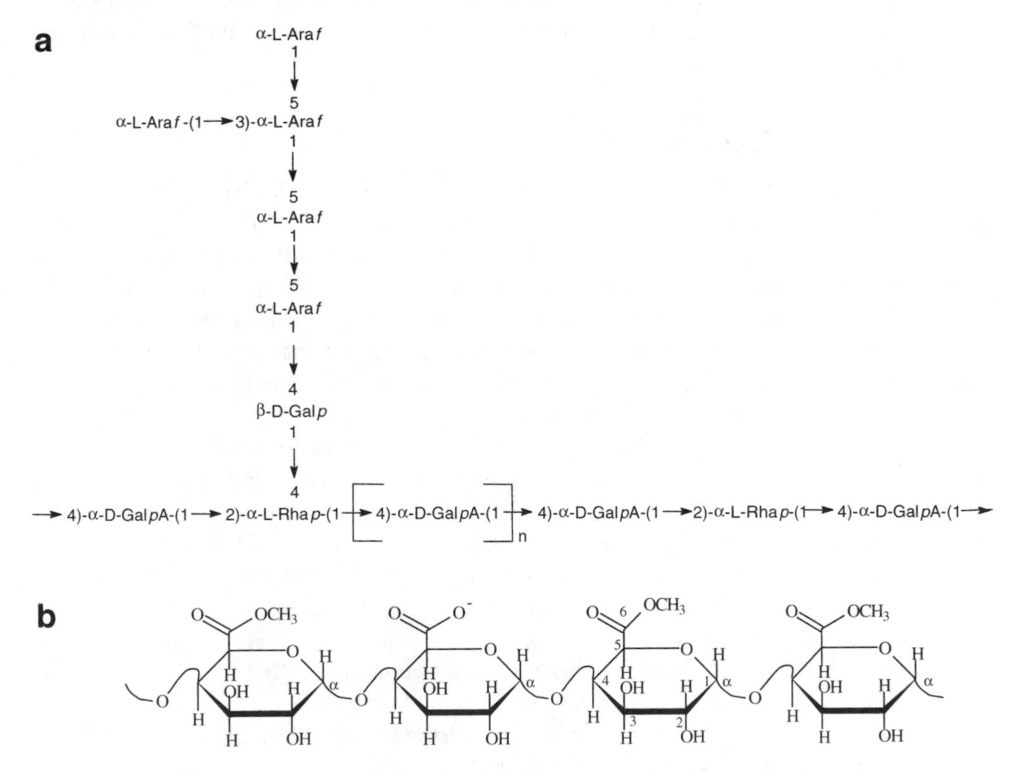

Fig. 2.3. Schematic representation of the glycosidic structure of pectic polysaccharides: **(a)** branched pectic polysaccharide and **(b)** pectin with high ester content.

The authentication of food is a major concern for the consumer and for the food industry at all levels of the food-processing chain, from raw materials to final products. Among the complex food constituents, the identification of the added polysaccharides could be a key factor if a rapid and reliable method is attainable.

Vibrational spectroscopy has been found to have important applications in the analysis and identification of sugars in food industries (Mathlouthi and Koenig 1986). Particularly, midinfrared spectroscopy has been shown to be a rapid, versatile, and sensitive tool for elucidating the structure, physical properties, and interactions of carbohydrates (Kačuráková and Wilson 2001); to study pectic polysaccharides and hemicelluloses extracted from plants (Kačuráková et al. 2000); and to detect structural and compositional changes occurring in the cell walls of grapes during processing (Femenia et al. 1998). The carbohydrates show high absorbances in the 1200–950 cm^{-1} region, which is within the so-called fingerprint region (fig. 2.4), where the position and intensity of the bands are specific for every polysaccharide (Filippov 1992). Due to severe band overlapping in this region, it has been very difficult to assign the absorbances at specific wavenumbers to specific bonds or functional groups.

The application of chemometrics to the FT-IR spectra has been shown to be a reliable and fast method for characterization of amidated pectins (Engelsen and Norgaard 1996), and for classification of corn starches (Dupuy et al. 1997) and commercial carrageenans (Jurasek and Phillips 1998). Among many other applications, FT-IR and chemometrics have also been used for a quick evaluation of cell wall monosaccharide composition of polysaccharides of pectic (Coimbra et al. 1998) and hemicellulosic origin (Coimbra et al. 1999), screening of *Arabidopsis* cell wall mutants (Chen et al. 1998), and evaluation of structural and compositional changes in the cell walls of pears with sun-dried processing (Ferreira et al. 2001).

Identification of Polysaccharide Food Additives by FT-IR Spectroscopy

Figure 2.5a shows the PC1 versus PC2 scores scatter plot of the FT-IR spectra in the 1200–800 cm^{-1} region (fig. 2.4b) of 27 carbohydrate standards: six monosaccharides (arabinose—Ara, fructose—Fru, galactose—Gal, galacturonic acid—GalA, glucose—Glc, and mannose—Man), three disaccharides (lactose, maltose, and sucrose), four glucans (amylose, amylopectin, barley β-glucan, and starch), five carrageenans (ι-, λ-, κ-carrageenan, commercial carrageenan, and commercial carrageenan–pectin mixture), three galactans (arabic gum, arabinogalactan, and galactan), and six pectins having different degrees of methylesterification.

The distribution of the samples in the PC1 axis is as a function of the composition in glucose (PC1 negative) and galactose (PC1 positive). Glucose-rich compounds (starch, amylose, amylopectin, β-glucan, maltose, sucrose, and glucose) were all located in PC1 negative, independent of their monomeric or polymeric nature. On the other hand, polysaccharides such as the carrageenans (except λ-carrageenan) and galactans, and the monosaccharides galactose, fructose, and galacturonic acid, were placed in PC1 positive. Based on the scores scatter plot, the positive absorption band that can be observed in the loadings plot (fig. 2.5b) in the 1100–1030 cm^{-1} range, with maximum at 1068 cm^{-1}, can be attributed to Gal, and the band in the range 1030–944 cm^{-1}, with minimum at 998 cm^{-1}, can be ascribed to Glc, in accordance also with other published data (Kačuráková and Mathlouthi 1996, Kačuráková et al. 2000).

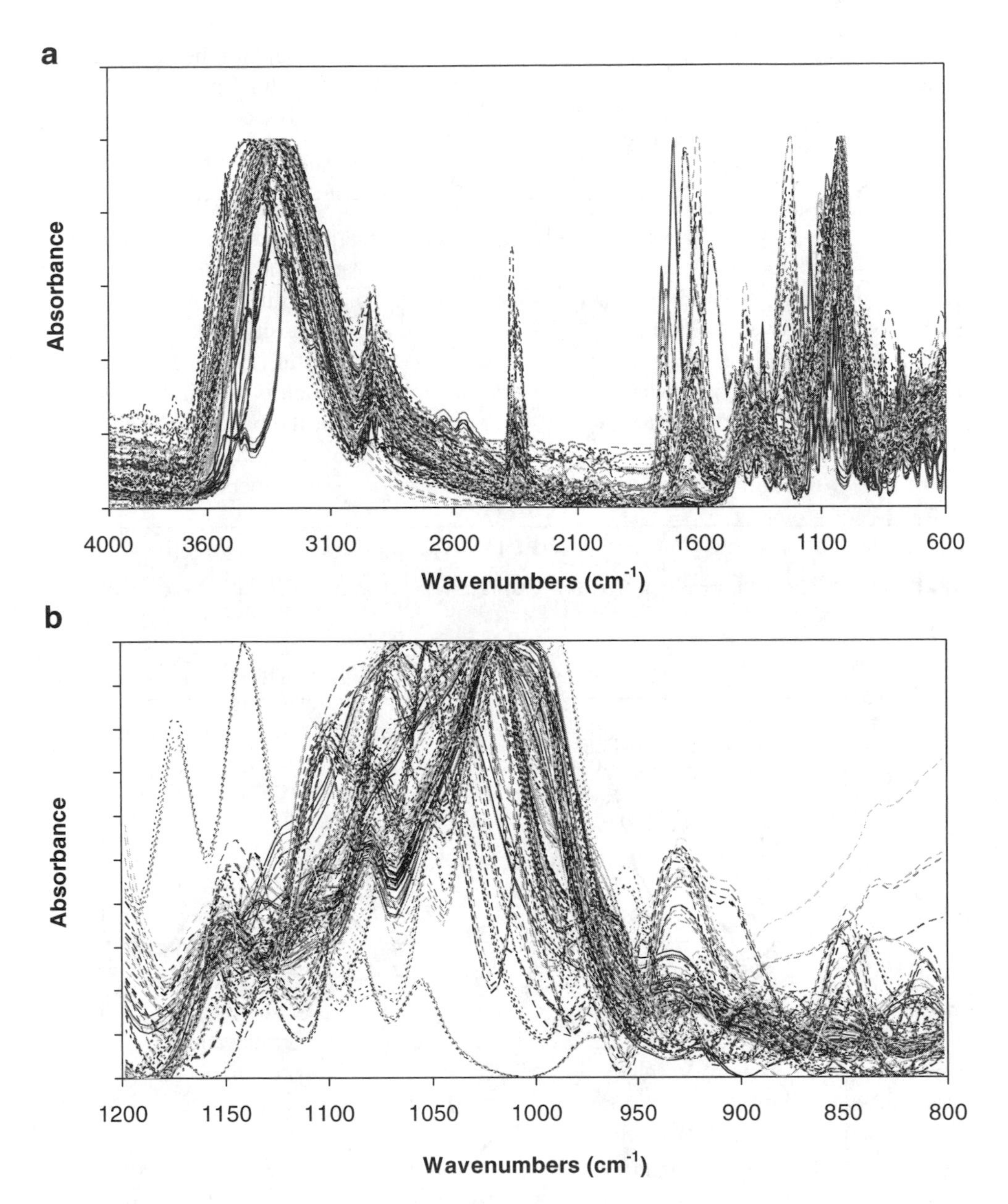

Fig. 2.4. FT-IR normalized spectra of replicates of 27 carbohydrate standards: **(a)** 4000–600 cm^{-1} region; **(b)** 1200–800 cm^{-1} region.

PC2 distinguishes the spectra of the pectin samples and GalA (PC2 negative) from all the others, especially the carrageenans and sucrose (PC2 positive). Pectins with different degrees of methylesterification were observed together in the same region, not clearly separated using this spectral region. Carrageenans (PC1 and PC2 positive) were placed differently from Gal (PC1 positive and PC2 negative) in the scores plan, except λ-carrageenan, which was located in PC1 negative and PC2 positive. This may be due to

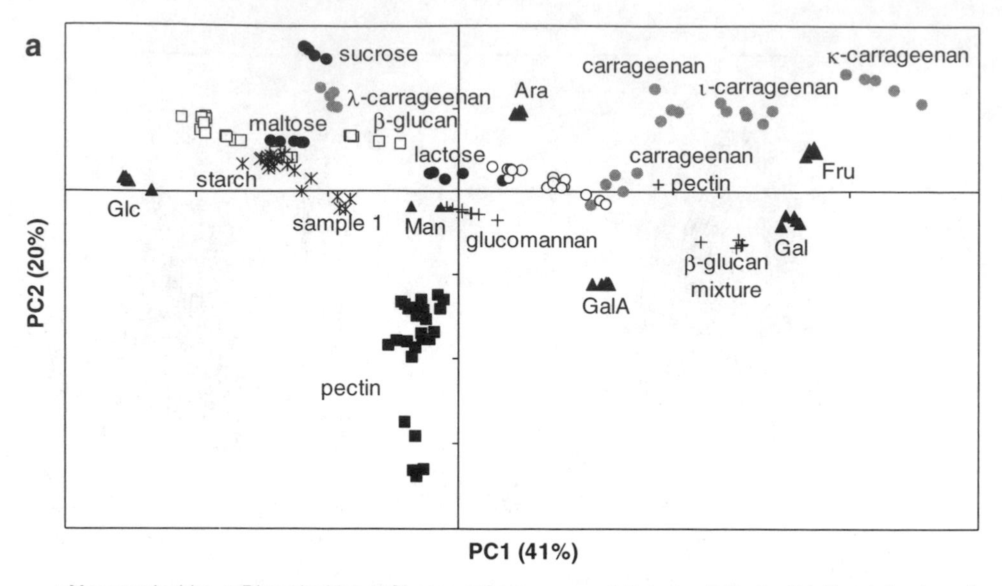

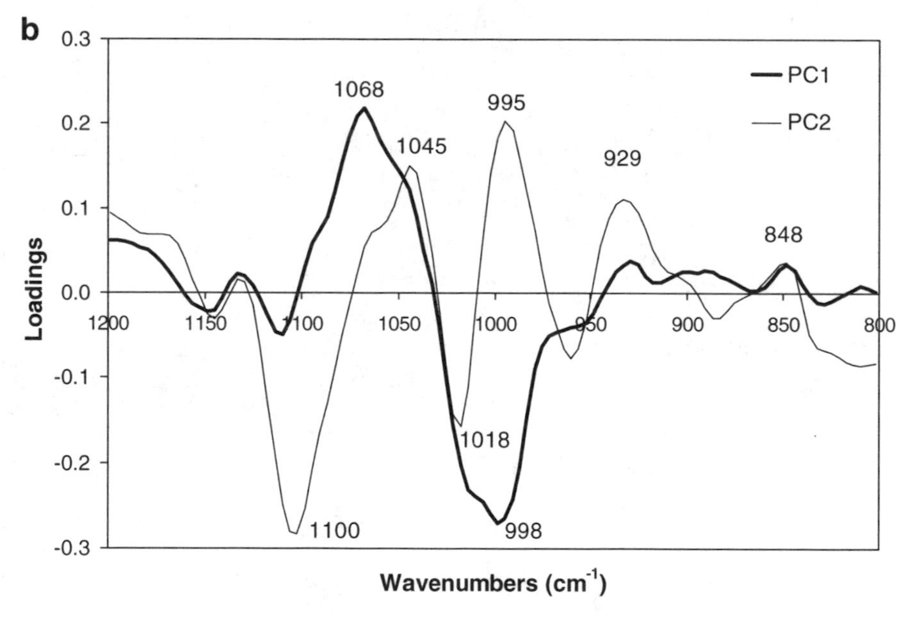

Fig. 2.5. PC1 × PC2 scores scatter plot **(a)** and loadings plot **(b)** of the FT-IR spectra of mono-, di-, and polysaccharide standards, confectionery jelly polysaccharides, and food supplements (glucomannan and β-glucan mixture) in the 1200–800 cm^{-1} region (Černá et al. 2003).

the higher content in sulphate and absence of 3,6-anhydro-Gal in this carrageenan when compared to the others. The spectra analysis suggests that the commercial carrageenan is a κ-carrageenan. The distinction of the pectic samples can also be seen in the loadings plot of PC2 by the absorbances at the negative side at 1145, 1100, 1018, and 960 cm^{-1} and

Table 2.1. Characteristics of studied samples, according to the manufacturer labels.

Sample *N*	Sample Specification	Sample Components
1	Pectin jelly with fruit flavor	Sugar, glucose syrup, water, pectin, citric acid, acidity regulator (sodium citrate), flavors, colors
2	Peach-flavored jellies	Sugar, glucose syrup, water, pectin, citric acid, sodium citrate, natural peach flavor, color
3	Fruit-flavored gums with real fruit juices	Sugar, glucose syrup, water, pectin, acids (citric acid, lactic acid), acidity regulators (potassium sodium tartrate, trisodium citrate, calcium lactate, sodium polyphosphate), fruit juices
4	Fruit jelly	Sugar, glucose syrup, water, pectin, citric acid, sodium citrate, natural flavors, natural colors
5	Pectin jelly with fruit flavor	Sugar, glucose syrup, water, pectin, citric acid, acidity regulator (sodium citrate), moisturizer (sorbitol), flavor, colors

Source: Černá et al. 2003.

by the absorbances at the positive side at 1064 and 1045 cm^{-1}, as have been described for pectic polysaccharides (Coimbra et al. 1998, 1999).

Isolated polysaccharides from confectionery jellies were placed in the PC1 negative side (fig. 2.5a), near starch, albeit not compatible with the product label of the manufacturers, who claimed that confectionery jellies contained pectin in addition to other sugars (table 2.1). The occurrence of starch, later confirmed by other methods (Černá et al. 2003), is an elucidative example of an application of FT-IR spectroscopy in the 1200–800 cm^{-1} region as a very reliable and quick tool for food authentication of carbohydrate-based food additives.

Influence of Hydration on FT-IR Spectra of Food Additive Polysaccharides

FT-IR spectroscopy is very sensitive to the carbohydrate changes in conformation and to the constraints imposed by the hydrogen bonding with water (Kačuráková and Mathlouthi 1996). However, at least within a certain hydration range, distinction between samples is still possible, as can be seen for amylose in figure 2.6.

Figure 2.7 shows the PCA scores scatter plot and loadings plot of the FT-IR spectra in the 1200–800 cm^{-1} region of mono-, di-, and polysaccharide samples in a dry form and containing 6–40 percent water (hydrated). It is possible to see that the distribution of the samples in the PC1 × PC2 plan was invariant to the hydrated status of the sample. The exception was mannose (containing 48 percent water) that was shifted from PC1 negative to PC1 positive. Although the hydration of the samples gave broader bands, ascribed to molecular rearrangements and disappearance of the crystalline structures (Kačuráková et al. 1998), the presence of water in the given amount in these samples did not change the overall spectral characteristics that allowed their distinction, as observed by the loadings plot similar to the one shown in figure 2.5b.

When the samples were measured dissolved in water, the distribution of the samples in the PC1 axis (fig. 2.8a) was done as a function of the water content in the samples, as all samples measured in dry or hydrated forms (less than 48 percent of water) were placed in PC1 negative and the majority of the samples measured in solution were placed

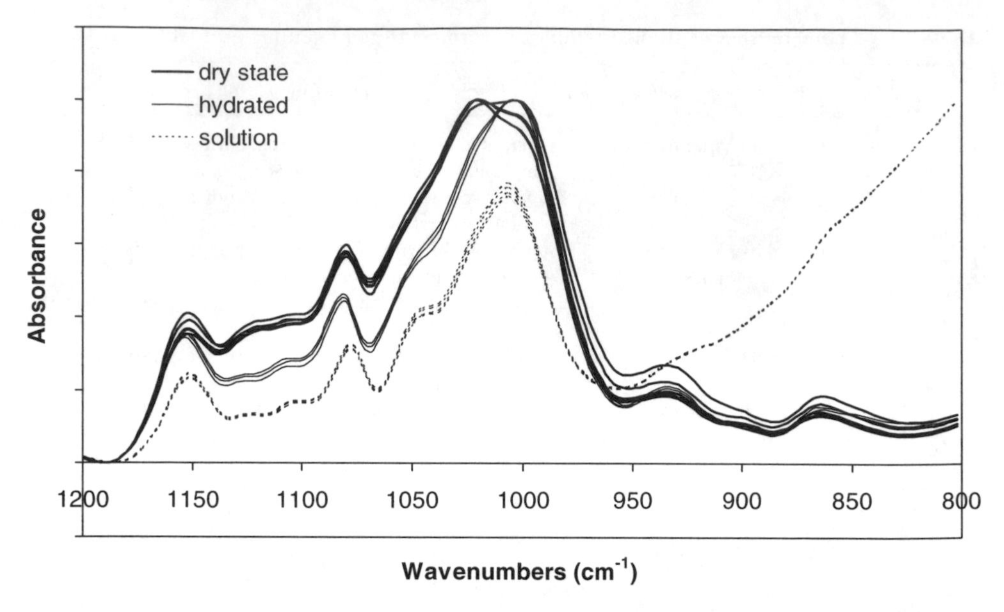

Fig. 2.6. FT-IR spectra in the 1200–800 cm^{-1} region of amylose containing different amounts of water.

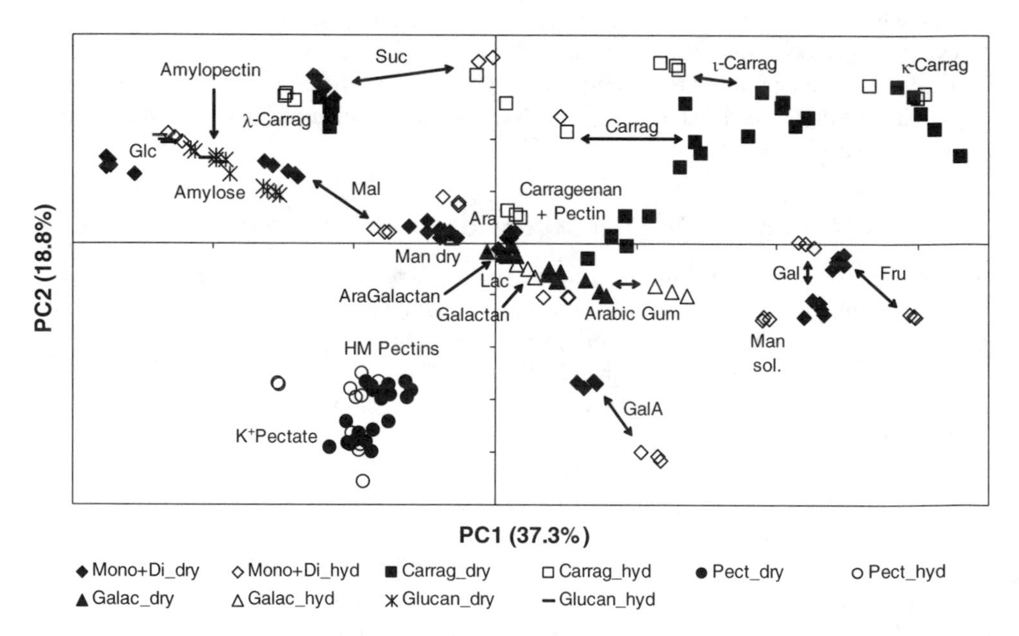

Fig. 2.7. Scores scatter plot (PC1 × PC2—axes cross each other at the origin) of FT-IR spectra in the 1200–800 cm^{-1} region of mono-, di-, and polysaccharide standards in dry and hydrated forms (Čopíková et al. 2006).

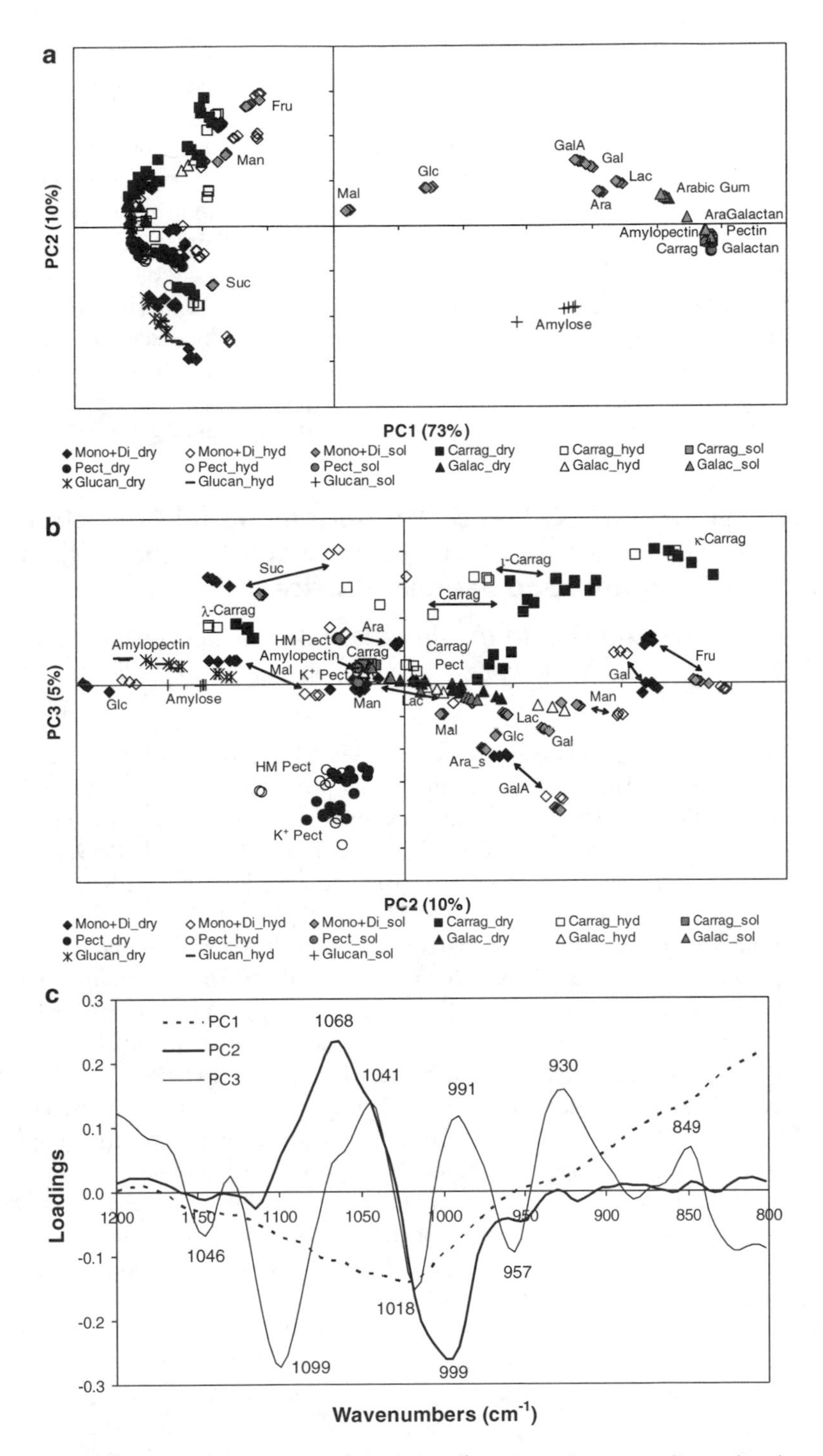

Fig. 2.8. PCA of FT-IR spectra in the 1200–800 cm^{-1} region of mono-, di-, and polysaccharide standards in dry, hydrated, and solution forms. **(a)** PC1 × PC2 scores scatter plot; **(b)** PC2 × PC3 scores scatter plot; **(c)** loadings plot (Čopíková et al. 2006).

in PC1 positive (Čopíková et al. 2006). The smooth curve of the loadings profile of PC1 (fig. 2.8c) is related to the spectra of samples that have been distinguished only by their different amount of sugars (Coimbra et al. 2002).

Figure 2.8b shows that the distinction between the samples can be obtained by the PC2 × PC3 scores scatter plot for all dry and hydrated samples in a scores scatter plot similar to that obtained for PC1 × PC2 (fig. 2.8a). Also, the loadings plot of PC2 and PC3 was similar to the loadings plot of PC1 and PC2, respectively, of figure 2.5b, which shows that the variability of the samples not related to the water effects could be recovered in PC2 and PC3 when samples in solution are included. For the majority of the samples analyzed in solution, significant shifts can be observed. With the exception of amylose, whose spectra did not change significantly with the water content, the spectra of all other polysaccharides (amylopectin, carrageenans, pectins and pectates, and galactans) were placed near the PC2 and PC3 origin, precluding their distinction.

Use of Outer Product (OP) and Orthogonal Signal Correction (OSC) PLS1 Regressions in FT-IR Spectroscopy for Quantification Purposes of Complex Food Sample Matrices

OP-PLS1 Regression Applied to the Prediction of the Degree of Methylesterification of Pectic Polysaccharides in Extracts of Olive and Pear Pulps

Pectic polysaccharides are involved in the complex fibrillar network of plant cell wall structure that defines the mechanical and functional properties of the cell wall (Cosgrove 2001; Roberts 2001). As structural components, pectic polysaccharides influence the texture of fruits on ripening (Martin-Cabrejas et al. 1994; Paull et al. 1999; Vierhuis et al. 2000; Jiménez et al. 2001; Mafra et al. 2001), storage (Bartley and Knee 1982), and processing (Femenia et al. 1998). As already discussed, pectic polysaccharides are also of great importance in the food industry due to their gelling ability in jams and jellies, as well as for being used in fruit preparations for dairies, as stabilizers in fruit and milk beverages (Claus et al. 1998), and as dietary fibers (Sun et al. 1998; Sun and Hughes 1999).

Pectic polysaccharides, which have a backbone constituted mainly of galacturonic acid (GalA) residues, can be partially esterified with methanol (fig. 2.3). The degree of methylesterification (DM) is defined as the percentage of carboxyl groups esterified with methanol (Voragen et al. 1995). The presence of methylester groups affects the cross-linking of pectic polysaccharides by Ca^{2+}, which plays an important role in the organization of polysaccharides in plant cell walls (Brett and Waldron 1996; Wellner et al. 1998) and, consequently, may influence the texture properties of fruits during ripening and processing. The gelation mechanisms of pectins are also dependent on the DM (Grant et al. 1973; Walkinshaw and Arnott 1981; Lopes da Silva et al. 1995).

Several analytical methods have been proposed for the determination of the DM of pectic polysaccharides. They include alkali hydrolysis of the methylester groups and subsequent determination of the DM by titration (Mizote et al. 1975) in galacturonic-acid-rich samples. The independent quantifications of (1) the total amount of uronic acids colorimetrically and (2) the methanol released after alkali hydrolysis by gas chromatography (Knee 1978; McFeeters and Armstrong 1984; Waldron and Selvendran 1990), high-performance liquid chromatography (HPLC) (Voragen et al. 1986), and enzymatic oxidation (Klavons and Bennett 1986) are used when the pectic polysaccharides also contain neutral sugars. Another approach is the reduction of pectin methylester groups of GalA to galactose (Gal) and the determination of the DM either by the increase in the amount of Gal or by the change in the amount of GalA, quantified by gas chromatography and colorimetric analysis, respectively (Maness et al. 1990). Instrumental techniques such as ^{1}H-NMR (Grasdalen et al. 1988; Renard and Jarvis 1999), ^{13}C-NMR (Pfeffer et al. 1981), and FT-IR (Chatjigakis et al. 1998) spectroscopies have also been proposed.

The use of infrared spectroscopy on pectic substances was previously applied to distinguish between high- and low-methoxyl contents (Reintjes et al. 1962) and was proven to be a useful tool to distinguish and evaluate the methoxyl content of different commercial pectins with high and low levels of esterification (Haas and Jager 1986). FT-IR spectroscopy, as proposed by Chatjigakis et al. (1998), is a simple, quick, and nondestructive method of DM evaluation in cell wall material extracts. The estimation of DM is based on a calibration curve using samples of standard pectins with known degrees of esterification and the spectral bands at 1749 cm^{-1} and 1630 cm^{-1}, assigned, respectively, to the absorption of the esterified and nonesterified carboxyl groups of pectin molecules. However, this methodology has shown not to be suitable for the determination of the DM of the pectic polysaccharides when other carboxylates and carbonylester groups, such as those from cell wall phenolics, are present.

As FT-IR spectroscopy in the wavenumber region between 1200 and 850 cm^{-1} has been used as a reliable and fast method for the evaluation of polysaccharide composition (Coimbra et al. 1998, 1999; Ferreira et al. 2001), the application of a methodology for the determination of the DM of pectic polysaccharides present in raw cell wall extracts using the combination of the 1800–1500 cm^{-1} and 1200–850 cm^{-1} regions of the FT-IR spectra, by means of an outer product analysis, was proposed (Barros et al. 2002).

To acquire two sets of signals for the same samples and analyze how they vary simultaneously as a function of some property, one possibility is to apply statistical techniques to the n outer product matrices calculated, for each of the n samples. The procedure starts by calculating the products of the intensities in the two signal domains for each sample. All the intensities of one domain are multiplied by all intensities in the other domain, resulting in a data matrix containing all possible combinations of the intensities in the two domains (fig. 2.9a). The outer product of two signal-vectors of lengths r and c for the n samples gave n (r rows by c columns) matrices, which are then unfolded to give n ($r \times c$)-long row-vectors (fig. 2.9b). This procedure corresponds to a mutual weighting of each signal by the other (Barros et al. 1997; Barros 1999; Rutledge et al. 2001):

1. if the intensities are high simultaneously in the two domains, the product is higher;
2. if the intensities are low simultaneously in the two domains, the product is lower;
3. if one is high and the other low, the resulting product tends to an intermediate value.

After analysis of the set of n ($r \times c$)-long row-vectors, each vector of calculated statistical parameters was folded back to give a (r rows by c columns) matrix, which may be easily examined to detect the relations between the two domains.

In the present example, the two considered domains belonged to the midinfrared region (homospectral analysis): the first one to the region 1800–1500 cm^{-1} (79 wavenumbers) and the second one to the region 1200–850 cm^{-1} (91 wavenumbers). The outer product of these two regions gave a vector with (79 × 91) (7189 elements) for each sample. All the samples vectors were then concatenated into an **X** matrix, which was then used in partial

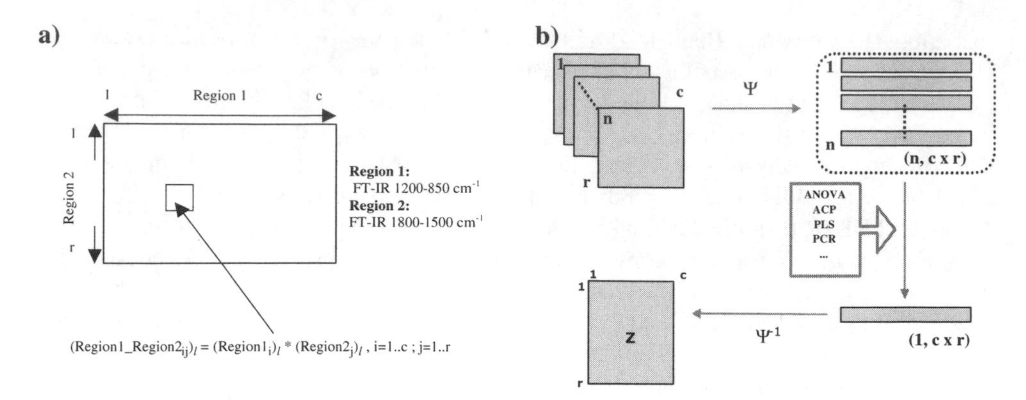

Fig. 2.9. **(a)** Calculation of the outer product between the two FT-IR regions; **(b)** unfolding of the outer product matrices, concatenation of the vectors, statistical analysis of the resulting matrix, and refolding of the vectors of calculated values (Barros et al. 2002).

least squares (PLS1) regression to model the DM. The obtained **b** vector, which established the relationship between the **X** variables and the **y** vector, was therefore composed of 7189 elements. In order to facilitate the interpretation of this vector, it was folded back to give a matrix **B** (79 × 91), which highlighted the links between the variables' (wavenumbers') interactions in the two regions.

In this example, pectic-polysaccharide-rich samples with a galacturonic acid content greater than 52 mol%, obtained from olive pulp and pear matrices after extraction using different aqueous solutions, were used (Barros et al. 2002). The relative amount of polymeric sugars of the samples was 48–85 percent, and the degree of methylesterification, estimated by gas chromatography from the amount of methanol released after saponification, ranged from 5 to 91 percent.

The classical multivariate approach for the determination of the degree of methylesterification in the region 1800–1500 cm^{-1}, using the bands located at 1750 and 1630 cm^{-1}, does not allow a regression model for olive pulp and pear pectic polysaccharide extracts with acceptable predictive power (a model with nine latent variables (LV) with a root mean square error of prediction (RMSEP) of 14.7 percent and a coefficient of determination (R^2) of 0.79). This could be due to the presence of esters and carboxylate groups from phenolics in the samples (presence of UV-absorbing materials and total sugar content less than 85 percent). To relate more precisely the esters and carboxylate groups with the amount of GalA present in the samples, the combination of the absorbance in the regions 1800–1500 and 1200–850 cm^{-1} can be done by means of an OP matrix. Figure 2.10a shows the linear relationship between the estimated DM values, obtained by OP-PLS1, and those observed by gas chromatography direct injection of the saponified solution of the polysaccharide extract. The folded **b** vector is shown in figure 2.10b as a 2D grey-level map. This 2D map allowed establishing relationships between the wavenumbers of the two FT-IR regions. The most important links between variables in the two domains, represented by darker spots, are shown in table 2.2. The positive relationships are related to those variables links that are correlated to the DM; conversely, the negative ones are anticorrelated to the DM. According to figure 2.10b and table 2.2, the most important wavenumbers in the 1800–1500 cm^{-1} region are 1746 cm^{-1} (assigned to carbonylesters) and 1626 and

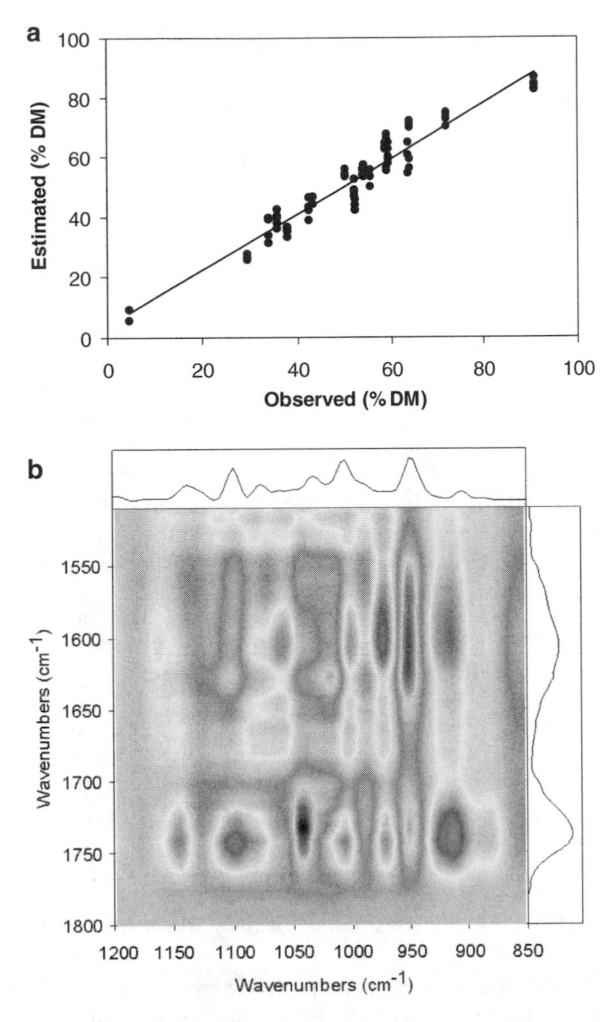

Fig. 2.10. **(a)** OP-PLS1 calibration curve plot for determination of the degree of methylesterification of pectic polysaccharides; **(b)** OP-PLS1 2D **b** vector map for determination of the DM of pectic polysaccharides; the variance profiles of each region are shown in the left and top sides (Barros et al. 2002).

$1603\,cm^{-1}$ (assigned to carboxylates). The **b** vector profiles of these wavenumbers in the region $1200–850\,cm^{-1}$ occurred at 1100 and $1018\,cm^{-1}$, which correlated positively to the ester vibration and negatively to the carboxylate vibrations. These wavenumbers have been assigned to GalA (Coimbra et al. 1998; Kacuráková et al. 2000; Ferreira et al. 2001). The ester band was also positively related to the band at $1145\,cm^{-1}$ and negatively related to the band at $1041\,cm^{-1}$. The carboxylate bands were negatively correlated to 1100 and $1018\,cm^{-1}$ and positively correlated to $1060\,cm^{-1}$, in accordance with the relative absorbance of GalA and neutral sugars (Coimbra et al. 1998; Ferreira et al. 2001). These results show that the absorbance values of the pectic polysaccharides in the region $1200–850\,cm^{-1}$ are positively correlated with the absorbance of the ester

Table 2.2. b vector variables—main relationships as a function of the DM.

$1800–1500\ cm^{-1}$ region	$1200–850\ cm^{-1}$ region
Positive relationship	
1746 (carbonylester)	913 (s), 971 (m), 1006 (m), 1100 (s), 1145 (m)
1603 (antisymmetric carboxylate stretching)	913 (m), 975 (s), 1002 (m), 1056 (m)
Negative relationship	
1746 (carbonylester)	952 (m), 1044 (s)
1603 (antisymmetrical carboxylate stretching)	952 (s)
1626 (carboxylates)	1018 (m), 1100 (m)

Source: Barros et al. 2002.
Note: (s) = strong; (m) = medium.

Table 2.3. Prediction of DM values for commercial pectins using OP-PLS1 (nine LV).

		DM (mol%)		
Sample Number	GalA (mol%)	Observed[a]	Predicted	CV (%)
41	81	93	94	0.3
42	74	67	72	1.7
43	66	26	23	2.3

Source: Barros et al. 2002.
[a]Degree of methylesterification of pectic polysaccharides determined by gas chromatography.

groups and anticorrelated with the absorbance of the carboxylate groups in the estimation of their DM.

This OP-PLS1 calibration model, when applied to predict the DM of three commercial citrus pectins of defined DM (93, 67, and 26 percent), gave very similar values to those present in the samples (table 2.3), with an RMSEP of 5.6 percent and R^2 of 0.99 and also very low coefficients of variation (CVs) for the spectra repetitions. This shows that this model could be used to predict the DM of pectic samples due to the low prediction error and to the high linearity observed over a wide range of DM in pure pectic samples.

For complex samples, the prediction error RMSEP was 21.8 percent with an R^2 of 0.82 (fig. 2.11a). Despite the observed variability, the OP-PLS1 could predict with some reliability the DM of pectic polysaccharides of samples with low purity and from different sources. Using the classical $1800–1500\,cm^{-1}$ region for the prediction of the DM, the obtained prediction error is much higher (RMSEP = 72.5 percent) with a very low R^2 (0.63). The prediction curve plot for this region (fig. 2.11b.) shows that using solely this region, it is not possible to quantify the DM in complex pectic samples.

In conclusion, the use of the combination of FT-IR spectra in the region $1200–850\,cm^{-1}$ with the ester and carboxylate regions ($1800–1500\,cm^{-1}$) allowed the prediction of the DM of pectic polysaccharides in extracts of olive and pear pulps even in the presence of phenolic compounds.

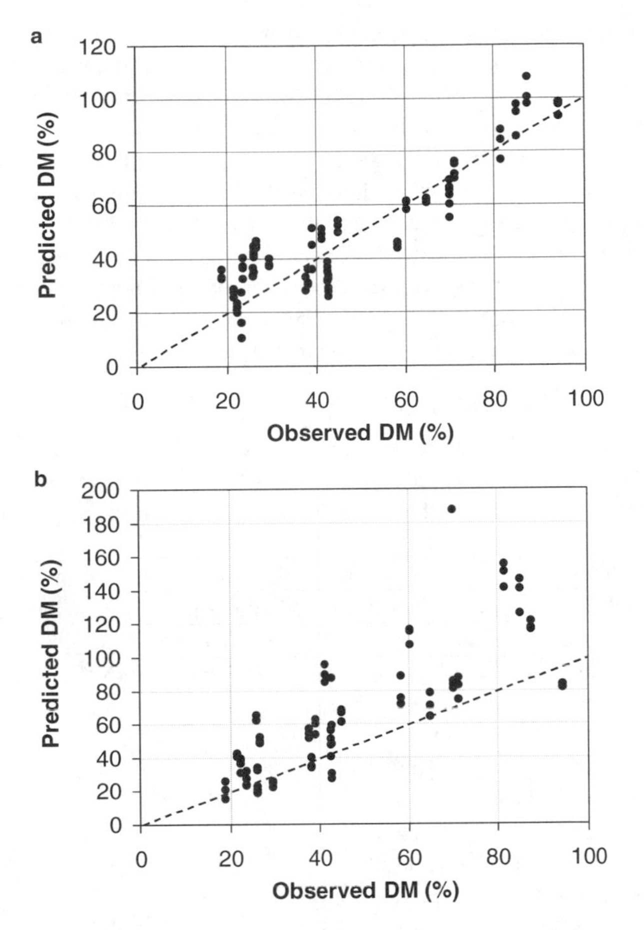

Fig. 2.11. Prediction of the DM of pectic polysaccharides for an external data set of olive samples (dotted line represents the optimal prediction). **(a)** Using the OP-PLS1 regression model; **(b)** using PLS1 in the classical 1800–1500 cm^{-1} region (Barros et al. 2002).

OSC-PLS1 Regression Applied to White and Red Wine Polymeric Material Extracts

Polysaccharides make up one group of wine macromolecules that, depending on their composition, structure, and concentration, is relevant for explaining and controlling wine stability (Segarra et al. 1995) and retention of aroma compounds (Goubet et al. 1998). They originated from both grapes and microorganisms. Arabinans (Belleville et al. 1993), type II arabinogalactans (Pellerin et al. 1995), and rhamnogalacturonans and galacturonans (Pellerin et al. 1996) arise from native cell wall pectic polysaccharides of grape berry after degradation by pectic enzymes during grape maturation and during the first steps of wine making. Yeasts produce mannans and mannoproteins during and after fermentation (Waters et al. 1994), whereas glucans are produced by *Botrytis cinerea*, which may infect grape berries (Dubourdieu and Rebereau-Gayon 1981).

In wine, FT-IR has been proposed and implemented for routine analysis of a large number of parameters, such as ethanol, volatile acidity, pH, tartaric acid, lactic acid, SO_2, glucose and fructose, acetic acid, citric acid, and polyphenols (Dubernet et al. 2000). Also, a rapid method for discrimination of red wine cultivars based on FT-IR spectra of the phenolic extracts in the 1640–950 cm^{-1} region was proposed (Edelmann et al. 2001), as well as for identification of polysaccharides (Coimbra et al. 2002). By a PLS1 regression model based on the FT-IR spectral region between 1200 and 800 cm^{-1}, it was possible to quantify mannose from mannoproteins in purified white wine extracts (Coimbra et al. 2002). However, this methodology does not allow a predictive ability for a wider range of samples, including samples from different and more complex matrices, such as the whole polymeric material, and from red wines, as the inherent complicated variations can hinder relevant information (Coimbra et al. 2005). Therefore, preprocessing techniques should be used to enhance or sort out the relevant information, making the models simpler and easier to interpret.

Orthogonal signal correction (OSC) is one of many preprocessing filters that aim to remove strong systematic variations in a given independent set of variables that are not correlated with the dependent variables. That is, the filter removes from the dependent variables (**y**) structures that are orthogonal to the independent set of variables (**X**) (Wold et al. 1998).

Figure 2.12a shows typical FT-IR spectra of the wine polymeric material extracts. The spectra show high absorbance at wavenumbers characteristic of wine polysaccharides, namely, the OH absorbance at 3440 cm^{-1} and the carbohydrate absorbance at 1200–800 cm^{-1} (Coimbra et al. 2002). Slight differences are observed in the carbohydrate region between the white and red wine (fig. 2.12b).

Using the 1200–800 cm^{-1} region of the FT-IR spectra, a PLS1 applied to 156 FT-IR spectra of white wine extracts allowed a calibration model with four latent variables (fig. 2.13a). The observed Man content, when compared with the amount estimated by gas chromatography of the acetylated mannitol derivatives obtained by acid hydrolysis, reduction, and acetylation of mannose residues, showed a linear relationship between 3 and 88 mol% of Man, with a root mean square error of cross-validation (RMSECV) of 17 percent

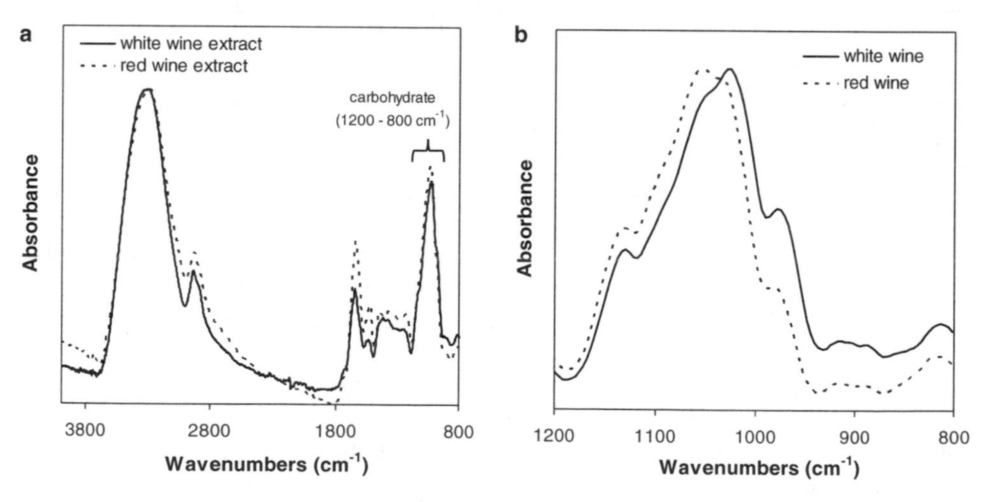

Fig. 2.12. FT-IR spectra of the wine polymeric material. **(a)** 4000–800 cm^{-1} region; **(b)** 1200–800 cm^{-1} region (Coimbra et al. 2005).

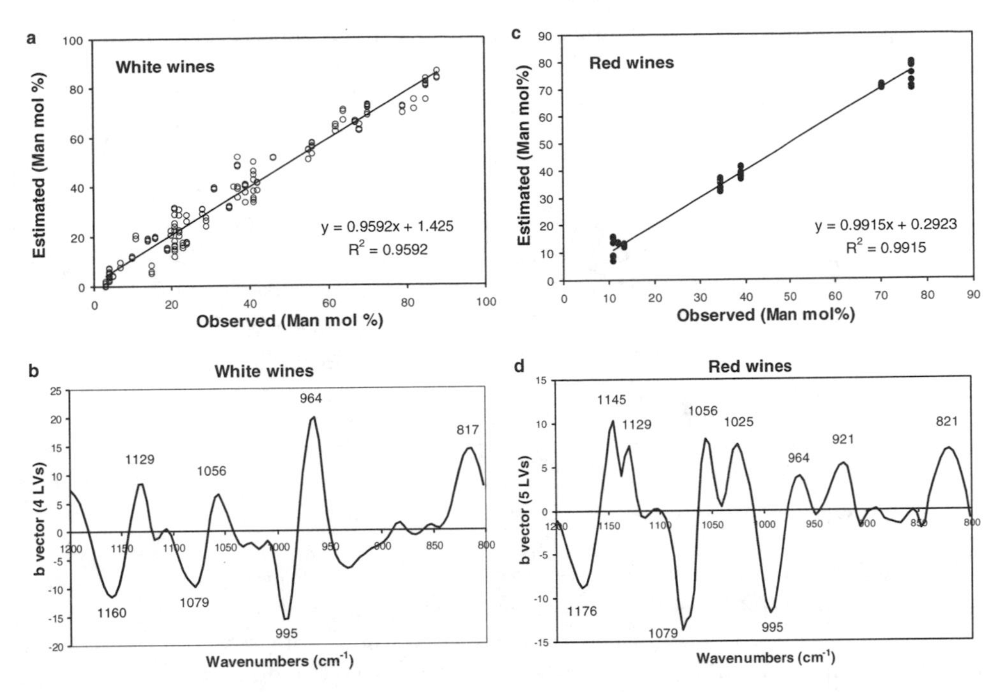

Fig. 2.13. PLS1 regression model for the estimation of polymeric mannose in white (four LVs) and in red (five LVs) wine extracts. **(a)** and **(c)** relationship plot between observed versus estimated amount of mannose, **(b)** and **(d)** **b** vector profiles (Coimbra et al. 2005).

with an R^2 of 0.96. The **b** vector profile of the calibration model (fig. 2.13b) shows that the most important wavenumbers related to the variability of Man were the bands located at 1129, 1056, 964, and 817 cm^{-1}, which increase as the Man content increases, and the bands located at 1160, 1079, and 995 cm^{-1}, which increase as the Man content decreases. For the 24 FT-IR spectra of red wines it was necessary to have a calibration model with five LVs to have a predictive power. The observed versus estimated Man relationships plot is represented in figure 2.13c, showing a good linear relationship between a wide range (11 to 77 mol%) of Man. The RMSECV obtained was 14 percent, with an R^2 of 0.99. The **b** vector profile of the calibration model (figure 2.13d) shows that the most important wavenumbers related to the variability of Man were the bands located at 1145, 1129, 1056, 1025, 964, 921, and 821 cm^{-1}, which increase as the Man content increases, and the bands located at 1176, 1079, and 995 cm^{-1}, which increase as the Man content decreases.

When a PLS1 calibration model for Man using both types of wines (red and white) was built, the optimal dimensionality was found to be a model with two LVs with a RMSECV of 36.2 percent and with a plot of the actual versus estimated values for the polymeric Man as shown in figure 2.14a, showing that the regression model does not give satisfactory prediction ability. This lower-predictive calibration model was possibly related to the heterogeneous nature of the noncarbohydrate material present in the red and white wine extracts. The compounds of this material can originate FT-IR signal complexities that can introduce major sources of variation that had very small or no predictive ability to the criterion of interest (Man quantification). In these cases, one common approach is to use different data pretreatments to reduce the effect of variabilities not related to the factor of

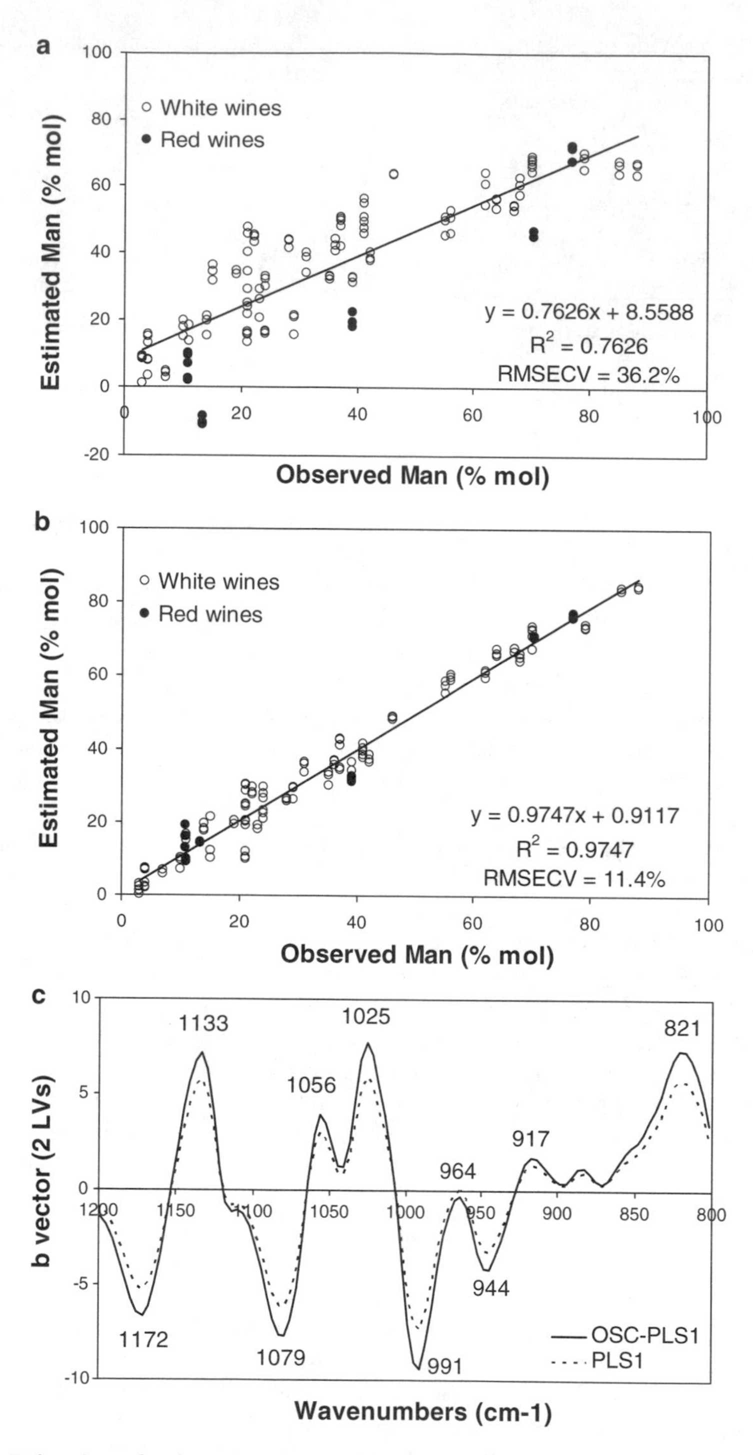

Fig. 2.14. Estimation of polymeric mannose in white and red wine extracts (two LVs). **(a)** PLS1 regression model; **(b)** OSC-PLS1 regression model; **(c) b** vector profiles (Coimbra et al. 2005).

Table 2.4. Percentage of explained variability for PLS1 and OSC-PLS1 optimal models.

	PLS1		OSC-PLS1	
	X	y	x	y
LV1	32.9	58.7	40.5	69.4
LV2	55.4	17.6	59.4	28.0
Total	88.3	76.3	99.9	97.4

Source: Coimbra et al. 2005.

interest. Among several methodologies that can be used, OSC is one of the most promising, as it can be used to remove from the spectra a certain number of factors that are orthogonal (not related) to the criterion of interest (Man) (Wold et al. 1998; Fearn 2000).

The methodology for selecting the optimal model dimensionality using the OSC was as follows: (1) The OSC procedure was applied to the calibration data set by removing 1, 2, to *n* factors. (2) After each factor was removed, a PLS1 regression was applied to the corrected data set, and the data's predictive ability was assessed by internal cross-validation (leave-3-out). By removing from the calibration set 11 factors an optimal PLS1 model was created with two LVs given a RMSECV of 11.4 percent (fig. 2.14b), showing a significant increase in the model predictive power when compared to the PLS1 calibration model, with a better linear trend in the plot of the observed versus estimated values for the polymeric Man using the OSC-PLS1 model. The amount of explained variability for each LV for both PLS1 and OSC-PLS1 proposed calibration models is shown in table 2.4. The OSC-PLS1 explains 97.4 percent of the total variability contained in the **y** vector (polymeric Man values), whereas PLS1 only explains around 76.3 percent. Figure 2.14c represents the **b** vector plot for the PLS1 and OSC-PLS1 with two LVs, where it can be seen that the only difference of the **b** vector plot between PLS1 and OSC-PLS1 is the intensity of the bands, not their positions. The characteristic bands for the quantification of polymeric Man can be identified as the bands located at 1133, 1056, 1025, 917, and 821 cm^{-1}, which are positively related to Man content, and the bands located at 1172, 1079, and 991 cm^{-1}, which are negatively related to Man content.

These results show that OSC-PLS1 is a procedure that can be used to improve the predictive ability of the model quantification of mannose from mannoproteins based on the FT-IR spectral region between 1200 and 800 cm^{-1}, allowing it to be used for a wider range of samples, including white and red wine extracts from complex matrices.

Screening and Distinction of Coffee Brews Based on Headspace–Solid Phase Microextraction–Gas Chromatography–Principal Component Analysis (HS-SPME-GC-PCA)

The two major species of coffee, *Coffea arabica* and *C. canephora*, var. Robusta, differ considerably in price, quality, and consumer acceptance. The washed Arabica coffees are characterized by some acidity and intense aroma, while natural, dry-processed Arabica coffees are less acidic and have a less-marked aroma but a richer body. Robusta coffees are characterized by their bitterness and a typical earthy and woody flavor. Blending, which can be done before or after roasting, has the purpose of obtaining coffee brews with a higher quality when compared to their individual counterparts. In espresso, washed coffees bring a fine, intense aroma, and natural coffees add body (Illy and Viani 1995). Torrefacto coffee is widely used in Argentina, Costa Rica, and Spain, where its consumption represents 83 percent of the commercially available coffees in hotel trade. To obtain torrefacto coffee, at the end of the roasting process, sucrose (no more than 15 percent in weight) is added. At high temperatures, sucrose is converted into caramel, forming a burnt film around the coffee bean, making it more bitter and less odorous (Sanz et al. 2002).

The volatile fraction of roasted coffee has been analyzed by many authors (Dart and Nursten 1985; Flament 1989, 1991; Maarse and Visscher 1989; Nijssen et al. 1996), who have identified about 850 compounds (Flament and Bessiére-Thomas 2002). Various methods of extraction have been used to study the aroma fraction of coffee brews. As an alternative to injection of an organic solvent extract, the vapor phase surrounding the brew (headspace) can be directly analyzed. This alternative approach gives the most accurate composition of flavors. However, when a large volume of headspace gas is injected, the carrier gas dilutes the sample. This problem can be solved by injecting the headspace gas directly inside a capillary column—on-column injection (Shimoda and Shibamoto 1990)—using a purge and trap system (Semmelroch and Grosch 1995), using an adsorbent (Pollien et al. 1997), or using a static headspace sampler (Sanz et al. 2001, 2002). In 1990, the solid phase microextraction (SPME) technique was developed for headspace sampling (Arthur and Pawliszyn 1990). It is a simple, rapid, solvent-free, and not very expensive method (fig. 2.15) that has been shown to be suitable for use in coffees (Yang and Peppard 1994; Bicchi et al. 1997, 2002; Ramos et al. 1998; Roberts et al. 2000).

The sensory characteristics of coffee brew depend on the method of extraction used. Petracco (2001) classified the extraction methods from a qualitative perspective: the grouping criterion chosen takes into account both the mode and the time of coffee-water contact. Among the pressure methods, plunger (cafetière) coffee, where the suspension of hot water and coffee powder is pressed through a plunger, and espresso coffee are examples. Some studies have been done on the taste and mouth feel (Maeztu et al. 2001a) and flavor and aroma (Liardon and Ott 1984; Maeztu et al. 2001b) of espresso coffee. At

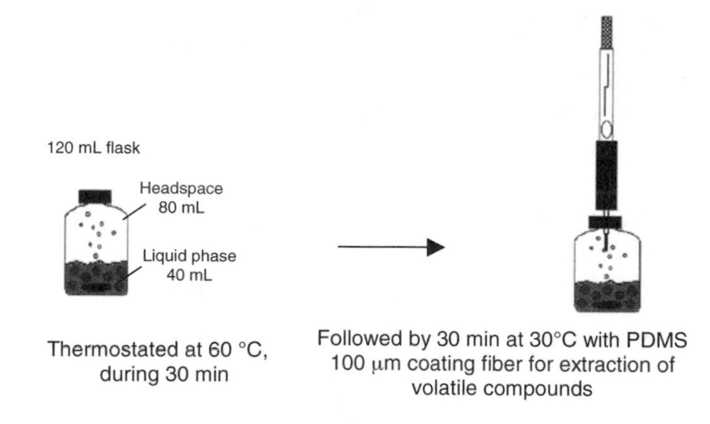

Fig. 2.15. Experimental procedure for sample preparation and extraction of volatile compounds from coffee brew by headspace–solid phase microextraction (HS-SPME).

least 28 compounds were reported as characteristic odorants of ground and brewed coffee (Shimoda and Shibamoto 1990; Blank et al. 1991, 1992; Grosch 1995; Semmelroch and Grosch 1995, 1996; Mayer et al. 2000). The change in the flavor profile from the ground coffee to the brewed is caused by a change in the concentrations of these compounds (Blank et al. 1991; Semmelroch and Grosch 1995) and not by the formation of new odorants. The sensory analysis carried out by a trained panel is too cumbersome and costly to be introduced as a routine procedure. Assessors in sensory panels cannot always, consistently and objectively, identify the sample, specially a blend. Their perception of the aroma of the coffee will depend on physiological and psychological factors. The coffee industry needs a simple, quick, and objective method to classify, especially, the botanical varieties of coffees (Arabica or Robusta) and/or the type of blends used in brewed coffee preparations.

Aiming for screening and distinction of coffee brews based on the combined technique of headspace–solid phase microextraction–gas chromatography–principal component analysis (HS-SPME-GC-PCA), a methodology has been proposed, based on the definition of the global volatile profiles of coffee brews, that is complementary to that obtained by sensory evaluation and that precludes the identification of the volatile compounds by mass spectrometry (Rocha et al. 2004a). Coffee brews were (1) a blend of natural roasted 80 percent Robusta and 20 percent Arabica (R80 : A20), (2) a 50 percent torrefacto of 80 percent Robusta and 20 percent Arabica (R80 : A20 torrefacto), and (3) a natural roasted 100 percent Arabica (A100).

Volatile profiles, expressed as a relative percentage of the GC peak area for the different chemical classes of volatile compounds ensuing from the headspace SPME analysis by gas chromatography–mass spectrometry (GC-MS), are shown in figure 2.16 (Rocha et al. 2004a). The relative percentage of GC peak area for the different chemical classes of the R80 : A20 natural and torrefacto coffees were similar for both espresso and plunger coffees, although the natural R80 : A20 coffee had a more intense aroma than the torrefacto. The decrease of volatile compounds in thc torrcfacto brew may be due to the fact that, in this blend, a fraction (6 percent) of the coffee had been replaced with sugar. Also, torrefacto coffee is usually submitted to a lower degree of roast to avoid flavors produced by the burnt sugar. The presence of sucrose during the roast may also contribute to this decrease by hindering the volatile compounds in a caramel interface. The relative percentage of the GC peak area for furans, in both espresso and plunger coffees, was higher in A100 than in R80 : A20 coffees, and conversely, the relative percentage of GC peak areas for pyrazines was lower. Pyrazines are products obtained from the Maillard reactions, and they are more abundant in Robusta coffee (Sanz et al. 2002; Ho et al. 1993), which is consistent with the lower content of free amino acids in Arabica coffees (Illy and Viani 1995).

Figure 2.17 shows the PCA scores scatter plots of PC1 × PC2 (fig. 2.17a) and PC1 × PC3 (fig. 2.17b), which contain 82 percent of the total variability of SPME volatile profiles of espresso and plunger coffees.

Figure 2.18 represents the corresponding loadings plots that establish the relative importance of each volatile component according to its retention time. The PC1 × PC2 scores plot (fig. 2.17a) shows the distinction between the Arabica coffees (PC1 negative and PC2 positive) and Robusta coffees, both R80 : A20 and R80 : A20 torrefacto (PC1 positive and PC2 negative).

According to the corresponding loadings plots (fig. 2.18a, b), the Arabica brews are characterized by the compound with the retention time of 44.9 min (furfurylacetate), and

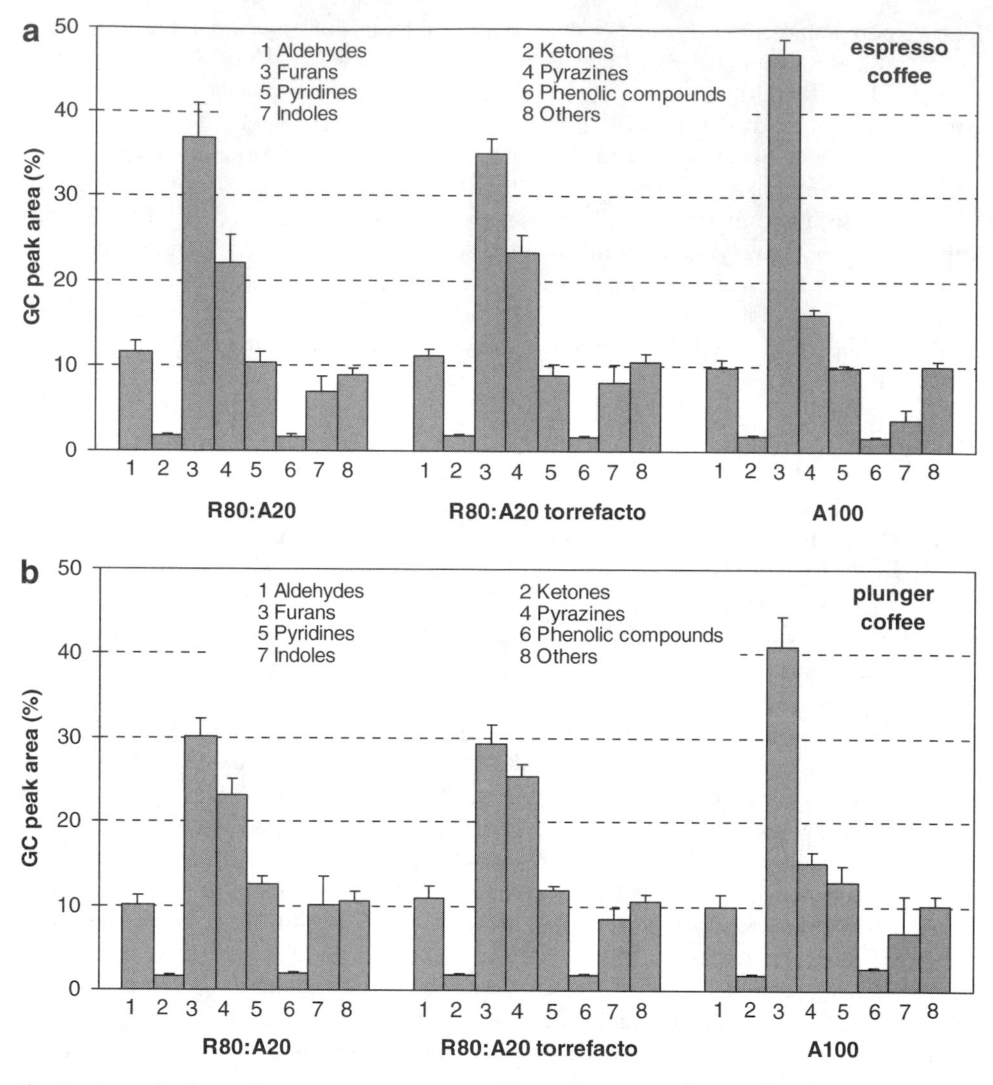

Fig. 2.16. Volatile profile, expressed as a relative percentage of the GC peak area for the different chemical classes of volatile compounds, of **(a)** espresso and **(b)** plunger coffees using a polydimethylsiloxane (PDMS) SPME coating. R80:A20—80 percent Robusta and 20 percent Arabica blend; R80:A20 torrefacto—80 percent Robusta torrefacto and 20 percent Arabica blend; A100—100 percent Arabica coffee (Rocha et al. 2004a).

Robusta brews are characterized by the compounds with the retention times of 12 min (2-methylbutanal), 38.2 min (2-ethyl-5-methylpyrazine), 39.3 min (trimethylpyrazine), and 41.1 min (3-ethyl-2,5-dimethylpyrazine). These results are in accordance with the distinction between Robusta and Arabica espresso coffee based on the amount of aldehydes and pyrazines (Petracco 2001), as furfurylacetate (PC1 negative and PC2 positive) was twice higher in Arabica than in Robusta brews and 2-methylbutanal, 2-ethyl-5-methylpyrazine, trimethylpyrazine, and 3-ethyl-2,5-dimethylpyrazine (PC1 positive and PC2 negative) were up to three times higher in Robusta than in Arabica brews.

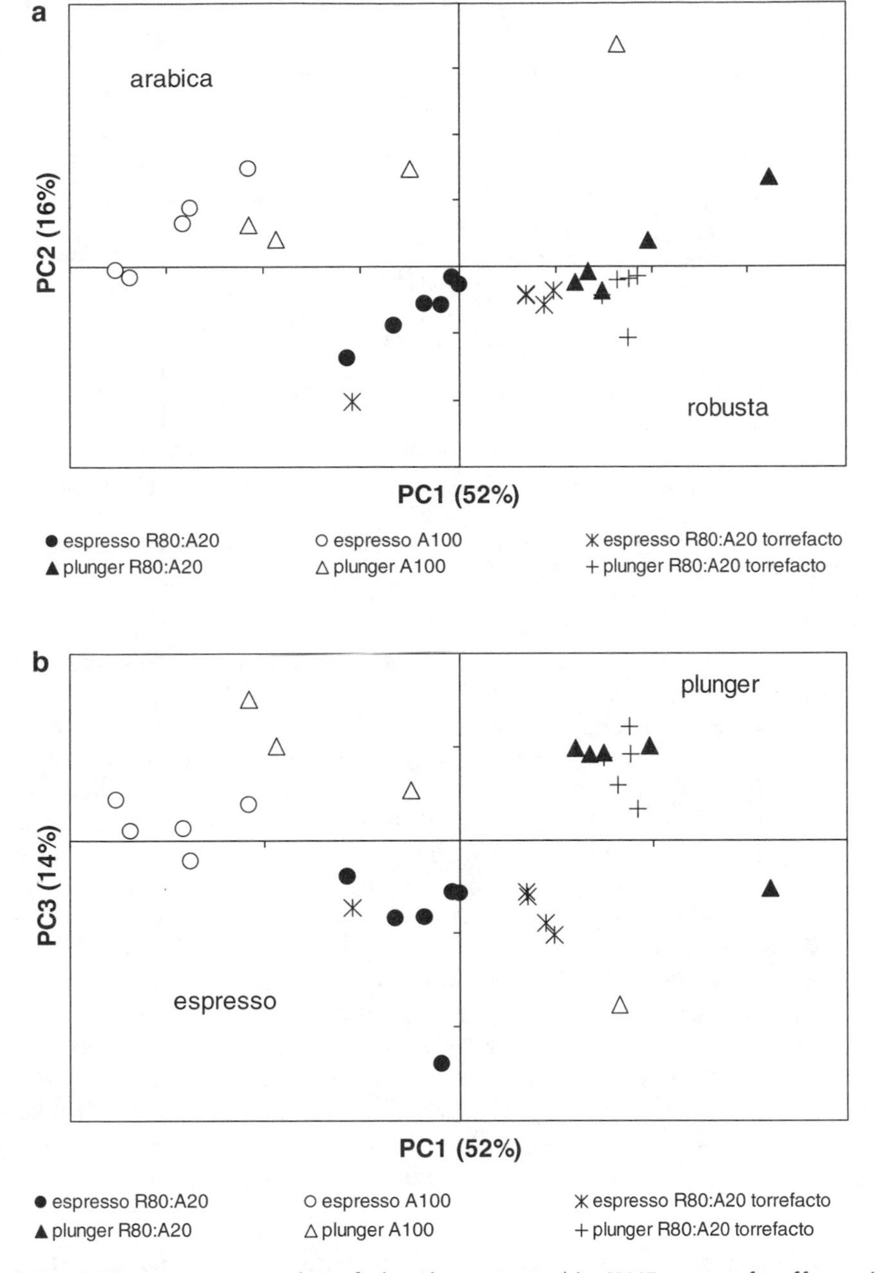

Fig. 2.17. PCA scores scatter plot of the chromatographic SPME areas of coffee volatile compounds. **(a)** PC1 × PC2 and **(b)** PC1 × PC3 (axes cross each other at the origin) (Rocha et al. 2004a).

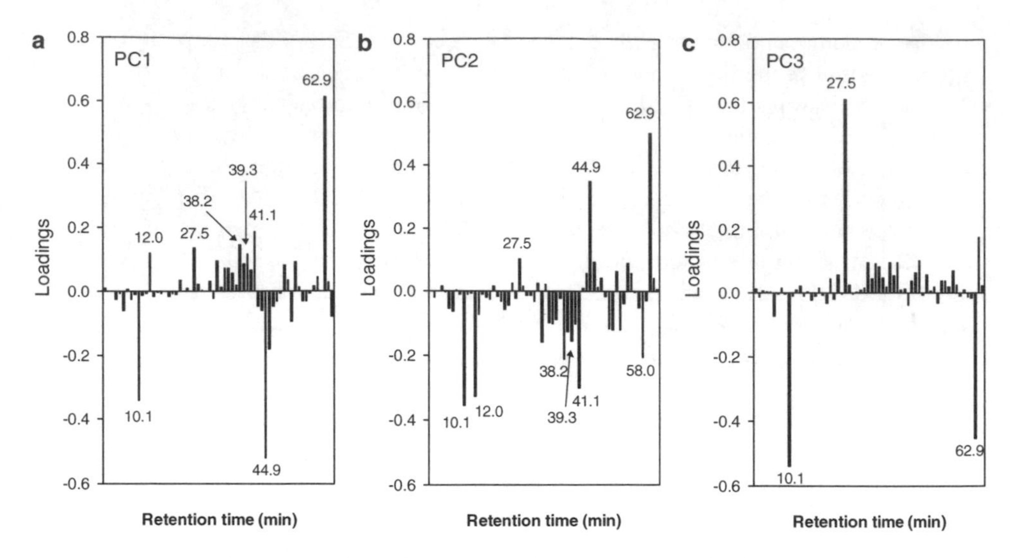

Fig. 2.18. PCA loadings plot of the chromatographic SPME areas of coffee volatile compounds. **(a)** PC1, **(b)** PC2, and **(c)** PC3 (Rocha et al. 2004a).

The PC1 × PC3 scores plot (fig. 2.17b) shows the distinction between plunger (PC1 positive and PC3 positive) and espresso (PC1 negative and PC3 negative) coffee brews. According to the corresponding loadings plots (fig. 2.18a, c), the plunger coffee brews are characterized mainly by the compound with the retention time of 27.5 min (pyridine) and espresso coffee brews are characterized by the compound with the retention time of 10.1 min (2-methylfuran). For the three coffees studied, pyridine (PC1 positive and PC3 positive) was, in fact, 23–43 percent higher in plunger than in espresso coffee brews, and 2-methylfuran (PC1 negative and PC3 negative) was 10–62 percent higher in espresso than in plunger coffee brews.

The volatile profile of espresso and plunger coffee brews obtained by SPME-GC-MS seems to be established mostly by the botanical variety (Arabica or Robusta) than by the process of preparation of the brews (espresso or plunger). Furthermore, the use of the variability given just by the GC areas and respective retention times, combined with the PCA, allowed for the observed distinction. The combined technique of HS-SPME-GC-PCA, when compared with the conventional techniques based on GC-MS identification of volatile compounds, can be proposed as a lower-cost, fast, and reliable technique for screening and distinction of coffee brews (Rocha et al. 2004a).

Study of Cork (from *Quercus suber* L.)–Wine Model Interactions Based on Voltammetric Multivariate Analysis

The cork from *Quercus suber* L. is the premium raw material used to produce wine-bottling stoppers. The cork plays an important role in determining wine quality due to its peculiar features: impermeability to air and liquids (preventing wine oxidation), ability to adhere to a glass surface, compressibility, resilience, and chemical inertness (Simpson et al. 1986). If the cork stoppers are in direct contact with the wine, volatile and nonvolatile compounds soluble in ethanol/water can migrate, thus contributing to the wine's sensorial properties. However, being a natural product, cork can be attacked and contaminated in ways that could promote differences in its properties.

Several studies on the off-flavors associated with cork stoppers have been carried out (Simpson et al. 1986; Rocha et al. 1996; Pollnitz et al. 1996), with the 2,4,6-trichloroanisole reported as the main agent responsible for cork-related off-flavors (Capone et al. 2002). New and rapid methodological approaches have been developed in order to study the volatile fraction associated with the wine cork taint (Boudaoud and Eveleigh 2003; Juanola et al. 2004; Zalacain et al. 2004). Conversely, few studies have been carried out concerning the soluble cork fraction that can migrate to wine (Varea et al. 2001), which can have sensorial effects in wines and may form complexes with wine anthocyanins, thus influencing the astringency (Singleton and Trousdale 1992).

The material unbonded or loosely bonded to the cork cell wall (i.e., the low-molecular-weight material) may be extracted with ethanol/water. This fraction is composed, mainly, of phenolic compounds and exhibits only about 2 percent carbohydrates (Rocha 1997; Rocha et al. 2004b). The major sugar component is xylose (53 mol%), glucose accounts for 17 percent, and uronic acids and arabinose represent 13 and 10 percent, respectively. Only trace amounts of deoxyhexoses can be detected (Rocha et al. 2004b). Low-molecular-weight polyphenols, such as ellagic acid, gallic acid, protocatechuic acid, cafeic acid, vanillic acid and vanillin, and ellagitannins, have been reported as the cork phenols susceptible to migrating into wine (Varea et al. 2001).

A specific cork contamination is the defect known in the industry as *mancha amarela*, or y*ellow spot* (MA). This cork contamination is represented by modifications in the cork's mechanical, structural, and optical properties and is potentially able to cause off-flavors in wine. Studies by scanning electron microscopy carried out on healthy cork (S) and MA cork (fig. 2.19) showed that the cellular structures of the infected and healthy tissues are different and that the attacked tissues were composed of deformed and wrinkly cells with cell wall separation at the middle lamella level (Rocha et al. 2000). These changes were related to the degradation of lignin and of pectic polysaccharides, as could be inferred by the deposition of calcium in the intercellular space of the attacked cells (Rocha et al. 2000).

In order to evaluate if the cork stoppers were able to contaminate a wine, it would be useful to use a screening technique for monitoring cork prior to being in contact with wine. As an initial approach, the resulting solutions of the matrix ethanol/water (10 percent v/v) set in contact both with a standard cork (S) and with a contaminated cork (MA) were studied by voltammetric techniques (Rocha et al. 2005). The need for a fast and reliable methodology for monitoring modifications in a wine model solution promoted by contact with contaminated cork prompted the application of voltammetric techniques, such as cyclic voltammetry (CV) and/or square wave voltammetry (SWV).

Voltammetric methods are relatively simple, rapid, sensitive, and low cost, requiring minimal preparation of samples and, thus, can be appropriated as screening techniques. Furthermore, voltammetry is suitable for the determination of several redox active organic compounds, namely phenolics, including methoxyphenols, flavonoids, and other antioxidant molecules of interest in diverse areas (Wheeler et al. 1990; Evtuguin et al. 2000; Filipiak 2001; Papanikos et al. 2002). All these compounds can be detected by electrooxidation at glassy carbon electrodes. Cyclic voltammetry has also been used to identify phenolic compounds in beer (Filipiak 2001), tea and coffee (Kilmartin and Hsu

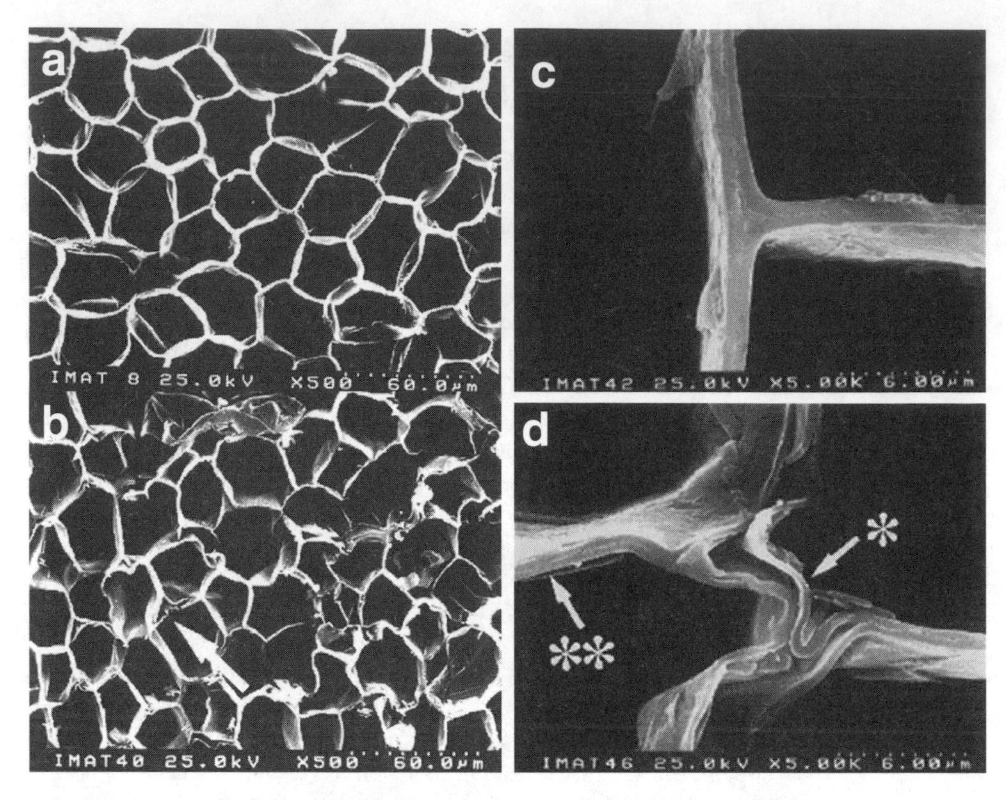

Fig. 2.19. SEM photographs of tangential section of reproduction cork showing "honeycomb"-type arrangement of cells of standard cork **(a)** and of cork with mancha amarela showing cellular separation **(b)**. Cell wall of standard cork **(c)** and of cork with mancha amarela, showing cellular separation (*) and thinning of the middle lamella (**) **(d)** (Rocha et al. 2000).

2003; Roginsky et al. 2003), and wines (Zou et al. 2002; Kilmartin et al. 2002). On the other hand, sugars cannot be detected by voltammetry at conventional carbon electrodes due to low-redox-reaction kinetics, more positive potential of oxidation (more positive than the limit of the working potential window), or self-poisoning of the electrode surface (Wittstock et al. 1998). A new methodological approach will be described for the rapid screening of cork-wine model interactions in order to determine if the cork stoppers were able to contaminate a wine (Rocha et al. 2005).

Evaluation of the Voltammetric Analysis in What Concerns the Cyclic and Square Wave Techniques

Figure 2.20 displays the voltammetric signature of cyclic voltammograms (fig. 2.20a) and square wave voltammograms (fig. 2.20b, c) of the S and MA cork extracts with 15 days of contact with the wine model matrix, diluted with the NaCl electrolyte (no pH adjustment).

The voltammograms of both samples showed the presence of oxidizable compounds within the potential window +200 to +800 mV; the overall responses are the sum of the various species present. For the S cork extracts the cyclic voltammograms exhibited a major oxidation peak at about +400 mV, whereas for the MA cork samples a second, more

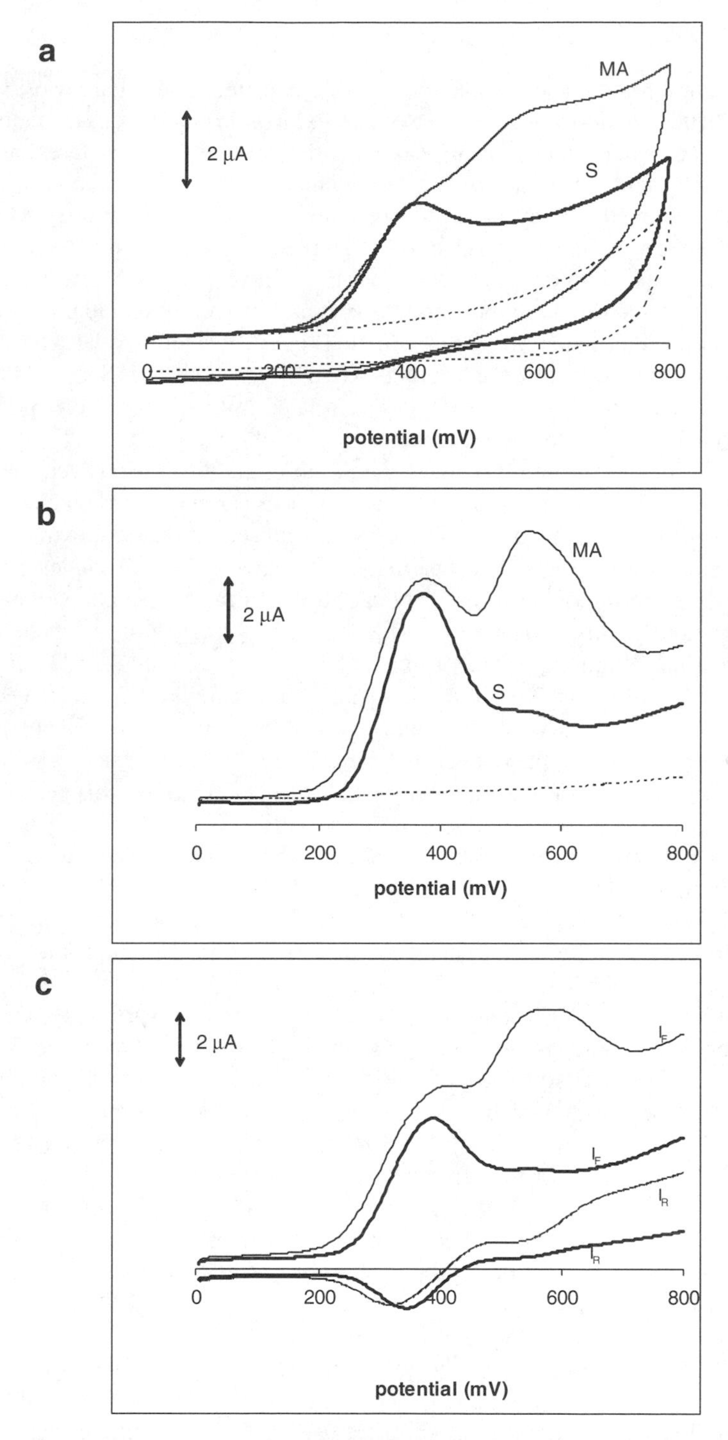

Fig. 2.20. Cyclic voltammogram **(a)** and square wave voltammograms **(b** and **c)** of cork extracts with 15 days of extraction time, diluted in NaCl without pH adjustment, for corks S and MA (pH extracts S = 4.4, and pH extracts MA = 5.1). Dotted curves are the background voltammograms. Voltammograms in **c** represent the forward (F) and reverse (R) current components of voltammograms in **b**. Cyclic voltammetry with scan rate of 100 mV/s and square wave voltammetry with frequency 15 Hz (Rocha et al. 2005).

intense, oxidation peak at about +580 mV was also detected. The square wave voltammograms (fig. 2.20b, c) confirm the presence of two oxidizable populations at about +400 mV and 560 mV. This small deviation of peak potentials is due to the differential nature of the square wave signal. The first oxidizable population appears in both cork extracts, S and MA, but the population at more positive potential seems to be mostly related to the cork MA. The cyclic voltammograms of both sample extracts showed lack of reversibility because the cathodic counterparts were absent in the reverse scans. However, the analysis of the individual current components of the square wave voltammograms (fig. 2.20c) allowed the detection of small reverse (reduction) peaks for both oxidation processes.

The first anodic peak observed at 400–410 mV for both the S and MA samples ($pH_{extracts\ S} = 4.4$, and $pH_{extracts\ MA} = 5.1$) may be related to the presence of a population of phenolics containing *ortho*-diphenol groups or triphenol (galloyl) groups (Zou et al. 2002) and eventually some flavonol glycosides (Kilmartin et al. 2002). However, the oxidation of *ortho*-diphenols is generally a fully reversible process at glassy carbon electrodes (Zou et al. 2002; Kilmartin et al. 2002). Consequently, if present in the cork extracts, this class of compounds is a smaller fraction. The major difference between samples S and MA is the occurrence of an important peak at about 580 mV in the MA cyclic voltammograms. This peak at more positive potential may be due to the presence of other phenolics with a lower antioxidant strength, such as vanillic (Kilmartin et al. 2002) and coumaric acids (Jorgensen and Skibsted 1998; Kilmartin et al. 2002), cafeic acid, protocatechuic acid (Filipiak 2001), vanillyl alcohol (Evtuguin et al. 2000) (all isolated phenols), or *meta*-diphenols on the A-ring of flavonoids (Kilmartin et al. 2002). Considering that some of these compounds are lignin-related compounds (Rocha et al. 1996), this peak can be proposed as a possible marker to follow lignin degradation. The fact that this peak at about 580 mV is characteristic of the MA also confirms, as expected, that lignin degradation may occur in MA cork.

Principal component analysis (PCA) was applied to the voltammetric data, in order to assess the differences in the analytical signals and to recover the main signal features that characterized the S and MA corks. The scores scatter plot of PC1 × PC2 of the cyclic voltammetry data (fig. 2.21a), containing 94 percent of the total variability, shows a clear separation between both types of samples along PC1 (which account for 77 percent of the total variability). The signal bands (fig. 2.21b) related to this separation are located at 391 mV (PC1 negative), which characterizes the S samples. At the PC1 positive side, the band located at 584 mV is related to MA samples. These results show a very different signal region contribution for the separation of samples.

The scores scatter plot of PC1 × PC2 of square wave data (fig. 2.21c), which represents 96 percent of the total variability, shows that the separation is, once more, possible along PC1 axis (contains 74 percent of the total variability). The S samples are mainly located at the PC1 negative side, except for one misplaced sample, while MA samples are found at the PC1 positive side. The PC1 loadings profile (fig. 2.21d) shows two intense bands: the one located at 336 mV seems to be characteristic of the S samples, whereas the MA samples are related to the variations linked to the 560 mV band. The slightly different location of the bands is certainly related to the use of different voltammetric techniques, namely due to the current acquisition regime of the square wave technique, which provides a differential current. Cyclic voltammetry was elected as the analytical technique because it provided slightly better distinction of the characteristic bands by PCA (Rocha et al. 2005).

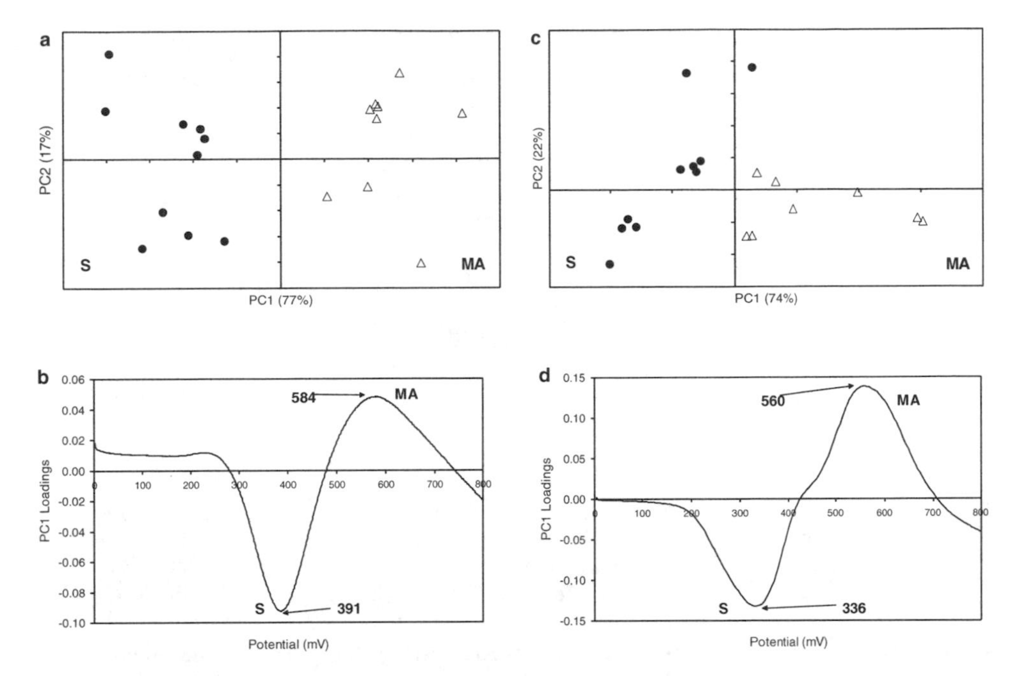

Fig. 2.21. Voltammetric PCA for cork extracts without pH adjustment, 15 days of extraction time. **(a)** Scores scatter plot from the cyclic voltammetric data (PC1 × PC2); **(b)** the corresponding loadings profiles (PC1 and PC2); **(c)** scores scatter plot from the square wave voltammetric data (PC1 × PC2); **(d)** the corresponding loadings profiles (PC1 and PC2).

Cyclic Voltammetric Analysis for Cork Classification

The extraction time of 15 days is time-consuming, and not adequate for prediction purposes. Experiments with just 1 day of extraction were devised.

Figure 2.22 displays the voltammetric signature of the 1-day extracts of S and MA cork. Two main conclusions can be drawn from these results. First, the overall current is lower than the observed for the data with 15 days of extraction (cf. fig. 2.20a).

Taking the peak at the more negative potential as a reference (at about +380 mV), the amount of extractable phenolics of high antioxidant strength (e.g., galloyl phenolics or flavonol glycosides) for the 1-day extracts may be estimated as approximately one-third of the value for the 15 days of extraction for the cork S, and two-thirds of that value for the cork MA. Therefore, this class of compounds is more rapidly extracted from the cork MA, which may be related to the high degradation of the cell tissues in this cork. Second, the voltammetric signature of the cork samples changed, especially for the MA cork (cf. fig. 2.20a), indicating qualitative changes in the population of extractables. It must be noted that there was a small shift of the peak potentials toward less positive potentials, compared to the data with 15 days of extraction, which is only a consequence of the higher pH values (increase of 0.5 pH units) of the extracts with 1 day of extraction time. The major qualitative differences were seen for the MA cork, namely in the potential region of 560–580 mV, where the characteristic peak that was detected in the 15 days extraction samples is apparently absent. Thus, it appears that the extraction kinetics of that population at more positive potential, which characterizes MA cork, is low.

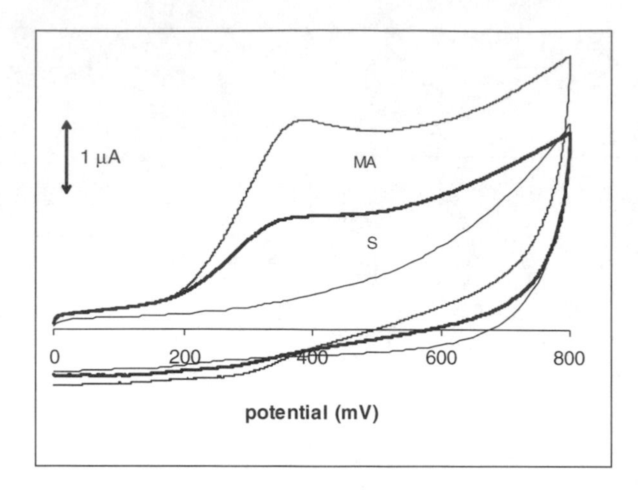

Fig. 2.22. Cyclic voltammograms of cork extracts with 1 day of extraction time, diluted in NaCl, without pH adjustment (pH extracts S = 4.9, and pH extracts MA = 5.6), for corks S and MA. Thin line curves are the background voltammogram. Scan rate of 100 mV/s (Rocha et al. 2005).

The application of a cluster analysis procedure (PLS_Cluster) to voltammetric data allows building of a discrimination model that was used to classify new samples. PLS_Cluster (Barros and Rutledge 2004) is a data-clustering method based on the PLS algorithm (Wold et al. 1982; Geladi and Kowalski 1986). This procedure provides a way to group samples based on the inner variability and/or relationships among samples and/or variables (features) and at the same time give information on the reasons for the groupings. The method is based on a self-organizing mechanism that uses the PLS1 procedure to achieve a hierarchical segregation of the samples based on the variability (or relationships) present in the **X** matrix to progressively build up a feature vector (**y**) that characterizes the relationships between the objects of the **X** matrix. PLS properties/entities such as the regression vectors, loadings X (**p**), W (**w**), and B coefficients (**b**) can be used to characterize the segregation (e.g., the chemical relevance). This method can be used in two different approaches: (1) dichotomic PLS_Cluster (DiPLS_Cluster) and (2) generalized PLS_Cluster (GenPLS_Cluster). The present work uses the DiPLS_Cluster approach for a binary segmentation of the samples.

The application of the DiPLS_Cluster method gives the segregation of the samples shown as a dendrogram in figure 2.23. One can see from this dendrogram that node 1, apart from one MA and two misplaced S samples, clearly discriminates between the MA and S known groups. The S group is mostly characterized by the variation located around 201 mV, whereas the MA group is mainly related to variations located around 377 and 530 mV. The discrimination promoted by the band at 530 mV, which was unclear in the original voltammograms, is especially important as it indicates that the population at more positive potential, clearly seen in the former voltammograms for the 15 days contact samples, is present in the 1-day contact. Again that population is characteristic of the MA cork. The variation of the potential from about 580 mV (15 days of extraction time) to 530 mV is certainly related to the higher pH of the 1-day extract solutions as well as to relative changes in the individual compounds contributing the overall oxidation band.

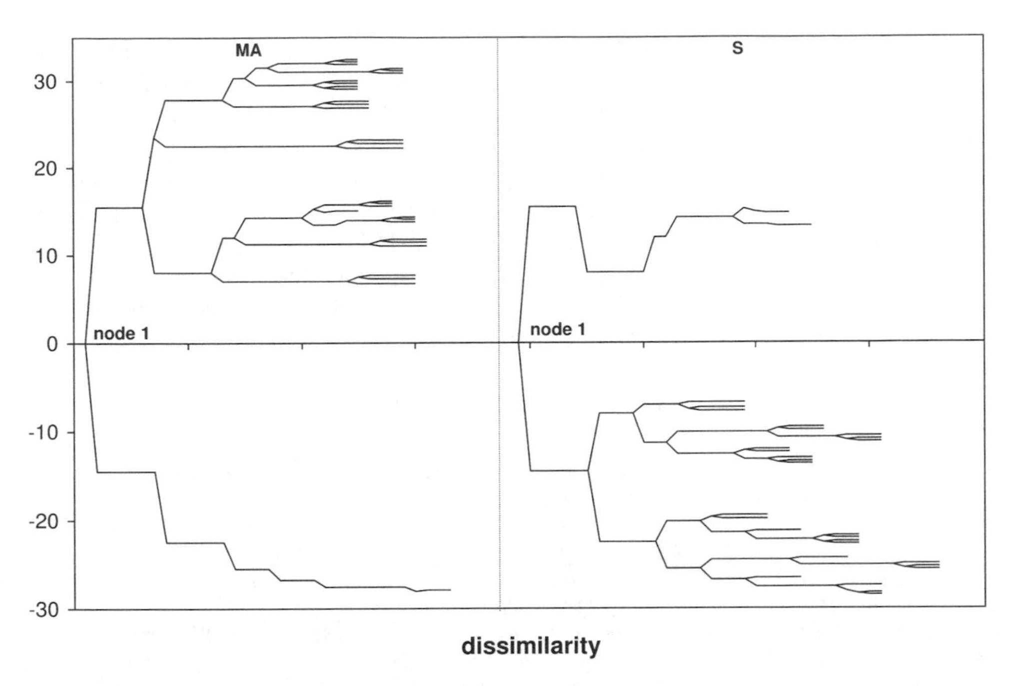

Fig. 2.23. DiPLS_Cluster dendrogram plot as a function of MA and S samples (known sample classification shifted along the *x* axis for clarity) (Rocha et al. 2005).

Validation of this model showed that 80 percent of the data set was correctly classified as S samples, whereas 90 percent of the data set was correctly classified as MA cork samples. Being a natural product and due to the various degrees of contamination, the observed classification rates can be accepted as high.

In conclusion, voltammetric methods based on the redox properties of compounds could be successfully used to establish the voltammetric signature of S and MA corks. The major difference between samples S and MA is the occurrence of an important peak at about 580 mV in the MA cyclic voltammograms. This peak at more positive potential may be assigned to lignin-related phenolics; thus it can be proposed as a possible marker to follow lignin degradation. Furthermore, the comparative analysis of the cyclic voltammograms for extraction times of 1 day and 15 days points to the existence of a population of easily oxidizable phenolics (oxidation peak at about 380 mV), which seems to be common to both cork samples S and MA. On the other hand, the class of phenolic compounds characteristic of the MA cork presented a relatively low kinetics of extraction. The application of a hierarchical clustering analysis (DiPLS_Cluster) allowed the classification of each type of cork, which allows predicting if the cork stoppers would be able to contaminate a wine. Therefore, the cyclic voltammetry associated with multivariate analysis allowed the development of a fast methodology for screening corks.

Concluding Remarks

The present chapter has shown some of the state-of-the-art applications of instrumental techniques tandem with chemometric tools for food quality assessment. Due to the complexity of food systems and owing to the fact that the instruments used nowadays allow a deeper signal screening of those systems, it is essential to use mathematical/statistical approaches to extract meaningful information for such biocomplex systems. The results shown have elucidated the advantage and widespread applicability of the instrumental/chemometric approach for studying a variety of systems in a qualitative and quantitative way. At the same time, it has been shown that this methodology is fast, which points to its wider application in the agrofood industry, providing low-cost approaches and allowing the study of ever increasingly complex systems. It is possible to foresee that for the agrofood industry these methodologies have a great advantage in terms of information recovery, speed, and low-cost implementations.

References

Arthur, C. L. and Pawliszyn, J. 1990. Solid phase microextraction with thermal desorption using fused silica optical fibers. *Analytical Chemistry* 62:2145–48.

Barros, A. S. 1999. Contribution à la sélection et la omparaison de variables caractéristiques. PhD thesis, Institut National Agronomique Paris-Grignon, France, chap. 1.

Barros, A. S. and Rutledge, D. N. 2004. PLS(-)Cluster: a novel technique for cluster analysis. *Chemometrics and Intelligent Laboratory Systems* 70:99–112.

Barros, A. S., Safar, M., Devaux, M. F., Robert, P., Bertrand, D., and Rutledge, D. N. 1997. Relations between mid-infrared and near infrared spectra detected by analysis of variance of an intervariable data matrix. *Applied Spectroscopy* 51:1384–93.

Barros, A. S., Mafra, I., Ferreira, D., Cardoso, S., Reis, A., Lopes da Silva, J. A., Delgadillo, I., Rutledge, D. N., and Coimbra, M. A. 2002. Determination of the degree of methylesterification of pectic polysaccharides by FT-IR using an outer product PLS1 regression. *Carbohydrate Polymers* 50:85–94.

Bartley, I. M. and Knee, M. 1982. The chemistry of textural changes in fruit during storage. *Food Chemistry* 9:47–58.

Belitz, H.-D., Grosch, W., and Schieberle, P. 2004. *Food Chemistry*, 3rd ed. Springer-Verlag: Berlin Heidelberg.

Belleville, M. P., Williams, P., and Brillouet, J. M. 1993. A linear arabinan from red wine. *Phytochemistry* 33:227–29.

Bicchi, C. P., Panero, O. M., Pellegrino, G. M., and Vanni, A. C. 1997. Characterization of roasted coffee and coffee beverages by solid phase microextraction-gas chromatography and principal component analysis. *Journal of Agricultural and Food Chemistry* 45:4680–86.

Bicchi, C., Iori, C., Rubiolo, P., and Sandra, P. 2002. Headspace Sorptive Extraction (HSSE), Stir Bar Sorptive Extraction (SBSE), and Solid Phase Microextraction (SPME) applied to the analysis of roasted Arabica coffee and coffee brew. *Journal of Agricultural and Food Chemistry* 50:449–59.

Blank, I., Sen, A., and Grosch, W. 1991. Aroma impact compounds of Arabica and Robusta coffee: qualitative and quantitative investigations. ASIC, 14th Colloquium, San Francisco, pp. 117–29.

———. 1992. Potent odorants of the roasted powder and brew of Arabica coffee. *Zeitschrift fur Lebensmittel-Untersuchung und-Forschung* 195:239–45.

Boudaoud, N. and Eveleigh, L. 2003. A new approach to the characterization of volatile signatures of cork wine stoppers. *Journal of Agricultural and Food Chemistry* 51:1530–33.

Brett, C. T. and Waldron, K. W. 1996. *Physiology and Biochemistry of Plant Cell Walls.* 2nd ed. Chapman and Hall: London.

Capone, D. L., Skouroumounis, G. K., and Sefton, M. A. 2002. Permeation of 2,4,6-trichloroanisole through cork closures in wine bottles. *Australian Journal of Grape and Wine Research* 8:196–99.

Černá, M., Barros, A. S., Nunes, A., Rocha, S. M., Delgadillo, I., Čopíková, J., and Coimbra, M. A. 2003. Use of FT-IR spectroscopy as a tool for the analysis of polysaccharide food additives. *Carbohydrate Polymers* 51:383–89.

Chatjigakis, A. K., Pappas, C., Proxenia, N., Kalantzi, O., Rodis, P., and Polissiou, M. 1998. FT-IR spectroscopic determination of the degree of esterification of cell wall pectins from stored peaches and correlation to textural changes. *Carbohydrate Polymers* 37:395–408.

Chen, L. M., Carpita, N. C., Reiter, W. D., Wilson, R. H., Jeffries, C., and McCann, M. C. 1998. A rapid method to screen for cell wall-mutants using discriminant analysis of Fourier transform infrared spectra. *Plant Journal* 16:385–92.

Claus, R., Nielsen, B. U., and Glahn, P.-E. 1998. Pectin. In *Polysaccharides: Structural Diversity and Functional Versatility,* Severian Dumitriu, ed., pp. 37–432. Marcel Dekker: New York.

Coimbra, M. A., Barros, A., Barros, M., Rutledge, D. N., and Delgadillo, I. 1998. Multivariate analysis of uronic acid and neutral sugars in whole pectic samples by FT-IR spectroscopy. *Carbohydrate Polymers* 37:241–48.

Coimbra, M. A., Barros, A., Rutledge, D. N., and Delgadillo, I. 1999. FTIR spectroscopy as a tool for the analysis of olive pulp cell-wall polysaccharide extracts. *Carbohydrate Research* 317:145–54.

Coimbra, M. A., Gonçalves, F., Barros, A., and Delgadillo, I. 2002. FT-IR spectroscopy and chemometric analysis of white wine polysaccharide extracts. *Journal of Agricultural and Food Chemistry* 50:3405–11.

Coimbra, M. A., Barros, A., Coelho, E., Gonçalves, F., Rocha, S. M., and Delgadillo, I. 2005. Quantification of polymeric mannose in wine extracts by FT-IR spectroscopy and OSC-PLS1 regression. *Carbohydrate Polymers* 61:434–40.

Čopíková, J., Barros, A. S., Šmídová, I., Černá, M., Teixeira, D. H., Delgadillo, I., Synytsya, A., and Coimbra, M. A. 2006. Influence of hydration of food additive polysaccharides on FT-IR spectra distinction, *Carbohydrate Polymers* 63:355–59.

Cosgrove, D. J. 2001. Wall structure and wall loosening: a look backwards and forwards. *Plant Physiology* 125:131–34.

Dart, S. K. and Nursten, H. E. 1985. Volatile components. In *Coffee Chemistry*, vol. 1, Clarke, R. J., and Macrae, R., eds., Elsevier Applied Science: London.

Dubernet, M., Dubernet, M., Dubernet, V., Coulomb, S., Lerch, M., and Traineau, I. 2000. Analyse objective de la qualité des vendages par spectrométrie infra-rouge à transformé de Fourier (IRTF) et réseaux de neurones. XXVème Congrès Mondial de la Vigne et du Vin, Paris, Section II—Enologie, pp. 215–21.

Dubourdieu, D. and Ribereau-Gayon, P. 1981. Structure of the extracellular β-D-glucan from *Botritis cinerea. Carbohydrate Research* 93:294–99.

Dupuy, N., Wojciechowski, C., Huvenne, J. P., and Legrand, P. 1997. Mid-infrared spectroscopy and chemometrics in corn starch classification. *Journal of Molecular Structure* 410–11:551–54.

Edelmann, A., Diewok, J., and Lendl, B. 2001. A rapid method for discrimination of red wine cultivars based on mid-infrared spectroscopy of phenolic wine extracts. *Journal of Agricultural and Food Chemistry* 49:1139–45.

Engelsen, S. B. and Norgaard, L. 1996. Comparative vibrational spectroscopy for determination of quality parameters in amidated pectins as evaluated by chemometrics. *Carbohydrate Polymers* 30:9–24.

Evtuguin, D. V., Pascoal Neto, C., Carapuça, H., and Soares, J. 2000. Lignin degradation in oxygen delignification catalysed by [PMo7V5O40](8-) polyanion. Part II. Study on lignin monomeric model compounds. *Holzforschung* 54:511–18.

Fearn, T. 2000. On orthogonal signal correction. *Chemometrics and Intelligent Laboratory Systems* 50:47–52.

Femenia, A., Sánchez, E. S., Simal, S., and Rosselló, C. 1998. Effects of drying pretreatments on the cell wall composition of grape tissues. *Journal of Agricultural and Food Chemistry* 46:271–76.

Ferreira, D., Barros, A., Coimbra, M. A., and Delgadillo, I. 2001. Use of FT-IR spectroscopy to follow the effect of processing in cell wall polysaccharide extracts of a sun-dried pear. *Carbohydrate Polymers* 45: 175–82.

Filipiak, M. 2001. Electrochemical analysis of polyphenolic compounds. *Analytical Sciences* 17:i1667–70.

Filippov, M. P. 1992. Practical infrared spectroscopy of pectic substances. *Food Hydrocolloids* 6:115–42.

Flament, I. 1989. Coffee, cocoa and tea. *Food Reviews International* 5:317–414.

———. 1991. Coffee, cocoa and tea. In *Volatile Compounds in Foods and Beverages*, Maarse, H., ed. pp. 617–69. Marcel Dekker: New York.

Flament, I. and Bessière-Thomas, Y. 2002. *Coffee Flavor Chemistry*. Wiley: Chichester.

Geladi, P. and Kowalski, B. R. 1986. Partial least-squares regression—a tutorial. *Analytica Chimica Acta* 185:1–17.

Goubet, I., Le Quere, J. L., and Voilley, A. J. 1998. Retention of aroma compounds by carbohydrates: influence of their physicochemical characteristics and of their physical state; a review. *Journal of Agricultural and Food Chemistry* 46:1981–90.

Grant, G. T., Morris, E. R., Rees, D. A., Smith, P. J. C., and Thom, D. 1973. Biological interactions between polysaccharides and divalent cations: the egg-box model. *FEBS Letters* 32:195–98.

Grasdalen, H., Bakoy, O. E., and Larsen, B. 1988. Determination of the degree of esterification and the distribution of methylated and free carboxyl groups in pectins by ^{1}H-N.M.R. spectroscopy. *Carbohydrate Research* 184:183–91.

Grosch, W. 1995. Instrumental and sensory analysis of coffee volatiles. ASIC, 16th Colloquium, Kyoto, pp. 147–46.

Haas, U. and Jager, M. 1986. Degree of esterification of pectins determined by photoacoustic near infrared spectroscopy. *Journal of Food Science* 51:1087–88.

Ho, C. T., Hwang, H. I., Yu, T. H., and Zhang, J. 1993. An overview of the maillard reactions related to aroma generation in coffee. 15th Colloquium ASIC, Montpellier, pp. 519–27.

Illy, A. and Viani, R. 1995. *Espresso Coffee—the Chemistry of Quality*. Academic Press: London.

Jiménez, A., Rodríguez, R., Fernández-Caro, I., Guillén, R., Fernández-Bolaños, J., and Heredia, A. 2001. Olive fruit cell wall: degradation of pectic polysaccharides during ripening. *Journal of Agricultural and Food Chemistry* 49:409–15.

Jorgensen, L. V. and Skibsted, L. H. 1998. Flavonoid deactivation of ferrylmyoglobin in relation to ease of oxidation as determined by cyclic voltammetry. *Free Radical Research* 28:335–51.

Juanola, R., Guerrero, L., Subirá, D., Salvadó, V., Insa, S., Garcia Regueiro, J. A., and Anticó, E. 2004. Relationship between sensory and instrumental analysis of 2,4,6-trichloroanisole in wine and cork stoppers. *Analytica Chimica Acta* 513:291–97.

Jurasek, P. and Phillips, G. O. 1998. The classification of natural gums. Part IX. A method to distinguish between two types of commercial carrageenan. *Food Hydrocolloids* 12:389–92.

Kačuráková, M. and Mathlouthi, M. 1996. FTIR and laser-Raman spectra of oligosaccharides in water: characterization of the glycosidic bond. *Carbohydrate Research* 284:145–57.

Kačuráková, M. and Wilson, R. H. 2001. Developments in mid-infrared FT-IR spectroscopy of selected carbohydrates. *Carbohydrate Polymers* 44:291–303.

Kačuráková, M., Belton, P.S., Wilson, R.H., Hirsch, J., and Ebringerová, A. 1998. Hydration properties of xylan-type structures: an FTIR study of xylooligosaccharides. *Journal of the Science of Food and Agriculture* 77:38–44.

Kačuráková, M., Capek, P., Sasinková, V., Wellner, N., and Ebringerová, A. 2000. FT-IR study of plant cell wall model compounds: pectic polysaccharides and hemicelluloses. *Carbohydrate Polymers* 43:195–203.

Kilmartin, P. A. and Hsu, C. F. 2003. Characterization of polyphenols in green, oolong, and black teas and in coffee, using cyclic voltammetry. *Food Chemistry* 82:501–12.

Kilmartin, P. A., Zou, H., and Waterhouse, A. L. 2002. Correlation of wine phenolic composition versus cyclic voltammetry response. *American Journal of Enology and Viticulture* 53:294–302.

Klavons, J. A. and Bennet, R. D. 1986. Determination of methanol using alcohol oxidase and its application to methyl ester content of pectins. *Journal of Agricultural and Food Chemistry* 34:597–99.

Knee, M. 1978. Properties of polygalacturonate and cell cohesion in apple fruit cortical tissue. *Phytochemistry* 17:1257–60.

Liardon, R. and Ott, U. 1984. Application of multivariate statistics for the classification of coffee headspace profiles. *Lebensmittel-Wissenschaft und-Technologie (Food Science and Technology)* 17:32–38.

Lopes da Silva, J. A., Gonçalves, M. P., and Rao, M. A. 1995. Kinetics and thermal behaviour of the structure formation process in high-methoxyl pectin/sucrose gelation. *International Journal of Biological Macromolecules* 17:25–32.

Maarse, H. and Visscher, C. A. 1989. *Volatile Compounds in Food: Qualitative and Quantitative Data*. TNO-CIVO Food Analysis Institute: Zeist, the Netherlands, pp. 661–79.

Maeztu, L., Andueza, S., Ibáñez, C., De Peña, M. P., Bello, J., and Cid, C. 2001a. Multivariate methods for differentiation of espresso coffees from different botanical varieties and types of roast by foam, taste and mouthfeel. *Journal of Agricultural and Food Chemistry* 49:4743–47.

Maeztu, L., Sanz, C., Andueza, S., De Peña, M. P., Bello, J., and Cid, C. 2001b. Characterization of espresso coffee aroma by HS-GC-MS and sensory flavor profile. *Journal of Agricultural and Food Chemistry* 49:5437–44.

Mafra, I., Lanza, B., Reis, A., Marsilio, V., Campestre, C., De Angelis, M., and Coimbra, M. A. 2001. Effect of ripening on texture, microstructure and cell wall polysaccharide composition of olive fruit (*Olea europaea*). *Physiologia Plantarum* 111:439–47.

Maness, O. N., Ryan, J. D., and Mort, A. J. 1990. Determination of the degree of methyl esterification of pectins in small samples by selective reduction of esterified galacturonic acid to galactose. *Analytical Biochemistry* 185:346–52.

Martin-Cabrejas, M. A., Waldron, K. W., Selvendran, R. R., Parker, M. L., and Moates, G. K. 1994. Ripening-related changes in the cell walls of Spanish pear (*Pyrus communis*). *Physiologia Plantarum* 91:671–79.

Mathlouthi, M. and Koenig, J. L. 1986. Vibrational spectra of carbohydrates. *Advances in Carbohydrate Chemistry and Biochemistry* 44:7–89.

Mayer, F., Czerny, M., and Grosch, W. 2000. Sensory study of the character impact aroma compounds of a coffee beverage. *European Food Research and Technology* 211:272–76.

McFeeters, R. F. and Armstrong, S. A. 1984. Measurement of pectin methylation in plant cell walls. *Analytical Biochemistry* 139:212–17.

Mizote, A., Odagiri, H., Toei, K., and Tanaka, K. 1975. Determination of residues of carboxilic acids (mainly galacturonic acid) and their degree of esterification in industrial pectins by colloid titration with cat-floc. *Analyst* 100:822–26.

Nijssen, L. M., Visscher, C. A., Maarse, H., Willemsens, L. C., and Boelens, M. H. 1996. *Volatile Compounds in Food: Qualitative and Quantitative Data*. 7th ed. TNO Nutrition and Food Research Institute: Zeist, the Netherlands.

Papanikos, A., Eklund, J., Jackson, W. R., Kenche, V. B., Campi, E. M., Robertson, A. D., Jarrott, B., Beart, P. M., Munro, F. E., and Callaway, J. K. 2002. Cyclic voltammetry as an indicator of antioxidant activity. *Australian Journal of Chemistry* 55:205–12.

Paull, R. E., Gross, K., and Qiu, Y. 1999. Changes in papaya cell walls during fruit ripening. *Postharvested Biology and Technology* 16:79–89.

Pellerin, P., Vidal, S., Williams, P., and Brillouet, J-P. 1995. Characterisation of five type II arabinogalactan-protein fractions from red wine of increasing uronic acid content. *Carbohydrate Research* 277:135–43.

Pellerin, P., Doco, T., Vidal, S., Williams, P., Brillouet, J-P., and O'Neill, M. A. 1996. Structural characterisation of red wine rhamnogalacturonan II. *Carbohydrate Research* 290:183–97.

Petracco, M. 2001. Technology IV: beverage preparation; brewing trends for the new millenium. In *Coffee: Recent Developments*, Clarke, R. J. and Vitzthum, O., eds. Blackwell Science: Oxford.

Pfeffer, P. E., Doner, L. W., Hoagland, P. D., and McDonald, G. G. 1981. Molecular interactions with dietary fiber components: investigation of the possible association of pectin and bile acids. *Journal of Agricultural and Food Chemistry* 29:455–61.

Pollien, P., Krebs, Y., and Chaintreau, A. 1997. Comparison of a brew and an instant coffee using a new GC-olfatometric method. ASIC, 17th Colloquium, Nairobi, pp. 191–96.

Pollnitz, A. P., Pardon, K. H., Liacopoulos, D., Skouroumounis, G. K., and Sefton, M. A. 1996. The analysis of 2,4,6-trichloroanisole and other chloroanisoles in tainted wines and corks. *Australian Journal of Grape and Wine Research* 2:184–90.

Ramos, E., Valero, E., Ibañez, E., Reglero, G., and Tabera, J. 1998. Obtention of a brewed coffee aroma extract by an optimized supercritical CO_2-based process. *Journal of Agricultural and Food Chemistry* 46:4011–16.

Reintjes, M., Musco, D. D., and Joseph, G. H. 1962. Infrared spectra of some pectic substances. *Journal of Food Science* 27:441–45.

Renard, C. M. G. C. and Jarvis, M. C. 1999. Acetylation and methylation of homogalacturonans 1: optimisation of the reaction and characterisation of the products. *Carbohydrate Polymers* 39:201–7.

Roberts, D. D., Polien, P., and Milo, C. 2000. Solid-phase microextraction method development for headspace analysis of volatile flavor compounds. *Journal of Agricultural and Food Chemistry* 48:2430–37.

Roberts, K. 2001. How the cell wall acquired a cellular context. *Plant Physiology* 125:127–30.

Rocha, S. 1997. Study of the chemical composition, cellular structure and volatile components of cork from *Quercus suber* L. I—Concerning the microbiological attacks. II—Concerning the autoclave procedure. PhD thesis, Department of Chemistry, University of Aveiro, Portugal.

Rocha, S., Delgadillo, I., and Ferrer Correia, A. J. 1996. GC-MS study of volatiles of normal and microbiologically attacked cork from *Quercus suber* L. *Journal of Agricultural and Food Chemistry* 44:865–71.

Rocha, S., Coimbra, M. A., and Delgadillo, I. 2000. Demonstration of pectic polysaccharides in cork cell wall from *Quercus suber* L. *Journal of Agricultural and Food Chemistry* 48:2003–7.

Rocha, S., Maeztu, L., Barros, A., Cid, C., and Coimbra, M. A. 2004a. Screening and distinction of coffee brews based on the headspace-SPME-GC-PCA technique. *Journal of the Science of Food and Agriculture* 84:43–51.

Rocha, S., Coimbra, M. A., and Delgadillo, I. 2004b. Occurrence of furfuraldeydes during the processing of *Quercus suber* L. cork: simultaneous determination of furfural, 5-hydroxymethylfurfural and 5-methylfurfural and their relation with cork carbohydrates. *Carbohydrate Polymers* 56:287–93.

Rocha, S., Ganito, S., Barros, A., Carapuça H. M., and Delgadillo, I. 2005. Study of cork (from *Quercus suber* L.)—wine model interactions based on voltammetric multivariate analysis. *Analytica Chimica Acta* 528:147–56.

Roginsky, V., Barsukova, T., Hsu, C. F., and Kilmartin, P. A. 2003. Chain-breaking antioxidant activity and cyclic voltammetry characterization of polyphenols in a range of green, oolong, and black teas. *Journal of Agricultural and Food Chemistry* 51:5798–5802.

Rutledge, D. N., Barros, A. S., and Giangiacomo, R. 2001. Interpreting near infrared spectra of solutions by outer product analysis with time domain-NMR. In *Advances in Magnetic Resonance in Food Science*, Belton, P., Delgadillo, I., Gil, A. M., and Webb, G. A., eds., pp. 179–92. Royal Society of Chemistry.

Sanz, C., Ansorena, D., Bello, J., and Cid, C. 2001. Optimizing headspace temperature and time sampling for identification of volatile compounds in ground roasted Arabica coffee. *Journal of Agricultural and Food Chemistry* 49:1364–69.

Sanz, C., Maeztu, L., Zapelena, M. J., Bello, J., and Cid, C. 2002. Profiles of volatile compounds and sensory analysis of three blends of coffee: influence of different percentages of Arabica and Robusta and influence of roasting coffee with sugar. *Journal of the Science of Food and Agriculture* 82:840–47.

Segarra, I., Lao, C., López-Tamames, E., and Torre-Boronat, M. C. 1995. Spectrophotometric methods for the analysis of polysaccharide levels in winemaking products. *American Journal of Enology and Vitiviniculture* 46:564–70.

Semmelroch, P. and Grosch, W. 1995. Analysis of roasted coffee powders and brews by gas chromatography-olfactometry of headspace samples. *Lebensmittel-Wissenschaft und-Technologie (Food Science and Technology)* 28:310–13.

———. 1996. Studies on character impact odorants of coffee brews. *Journal of Agricultural and Food Chemistry* 44:537–43.

Shimoda, M. and Shibamoto, T. 1990. Isolation and identification of headspace volatiles from brewed coffee with an on-column GC/MS method. *Journal of Agricultural and Food Chemistry* 38:802–4.

Simpson, R. F., Amon, J. M., and Daw, A. J. 1986. Off-flavor in wine caused by guaiacol. *Food Technology in Australia* 38:31–33.

Singleton, V. L. and Trousdale, E. K. 1992. Anthocyanin-tannin interactions explaining differences in polymeric phenols between white and red wines. *American Journal of Enology and Viticulture* 43:63–70.

Sun, R. C. and Hughes, S. 1999. Fractional isolation and physico-chemical characterization of alkali-soluble polysaccharides from sugar beet pulp. *Carbohydrate Polymers* 38:273–81.

Sun, R. C., Fang, J. M., Goodwin, A., Lawther, J. M., and Bolton, A. J. 1998. Isolation and characterization of polysaccharides from abaca fibre. *Journal of Agricultural and Food Chemistry* 46:2817–22.

Varea, S., García-Vallejo, M. C., and Cadahía, E. 2001. Polyphenols susceptible to migrate from cork stoppers to wine. *European Food Research and Technology* 231:56–61.

Vierhuis, E., Schols, H. A., Beldman, G., and Voragen, A. G. J. 2000. Isolation and characterization of cell wall material from olive fruit (*Olea europaea* cv koronieki) at different ripening stages. *Carbohydrate Polymers* 43:11–21.

Voragen, A. G. J., Schols, H. A., and Pilnik, W. 1986. Determination of the degree of methylation and acetylation of pectins by h.p.l.c. *Food Hydrocolloids* 1:65–70.

Voragen, A. G. J., Pilnik, W., Thibault, J.-F., Axelos, M. A. V., and Renard, C. M. G. C. 1995. Pectins. In *Food Polysaccharides and Their Applications*, Sephen, A. M., ed., pp. 287–339. Marcel Dekker: New York.

Waldron, K. W. and Selvendran, R. R. 1990. Composition of the cell walls of different asparagus (*Asparagus officinalis*) tissues. *Physiologia Plantarum* 80:568–75.

Walkinshaw, M. D. and Arnott, S. 1981. Conformations and interactions of pectins. II. Models for junction zones in pectinic acid and calcium pectate gels. *Journal of Molecular Biology* 153:1075–85.

Waters, E. J., Pellerin, P., and Brillouet, J. M. 1994. A *Saccharomyces* mannoprotein that protects wine from protein haze. *Carbohydrate Polymers* 23:185–91.

Wellner, N., Kacuráková, M., Malovíková, A., Wilson, R. H., and Belton, P. S. 1998. FT-IR study of pectate and pectinate gels formed by divalent cations. *Carbohydrate Research* 308:123–31.

Wheeler, S. K., Coury Jr., L. A., and Heineman, W. R. 1990. Fouling-resistant, polymer-modified graphite-electrodes. *Analytica Chimica Acta* 237:141–48.

Wittstock, G., Strubing, A., Szargan, R., and Werner, G. 1998. Glucose oxidation at bismuth-modified platinum electrodes. *Journal of Electroanalytical Chemistry* 444:61–73.

Wold, S., Martens, H., and Wold, H. A. 1982. *Lecture Notes in Mathematics.* Proceedings of the Conference on Matrix Pencils, March, Ruhe, A. and Kägström, B., eds., Heidelberg: Springer-Verlag, pp. 286–93.

Wold, S., Antii, H., Lindgren, F., and Öhman, J. 1998. Orthogonal signal correction of near-infrared spectra. *Chemometrics and Intelligent Laboratory Systems* 44:175–85.

Yang, X. and Peppard, T. 1994. Solid-phase microextraction for flavor analysis. *Journal of Agricultural and Food Chemistry* 42:1925–30.

Zalacain, A., Alonso, G. L., Lorenzo, C., Iñiguez, M., and Salinas, M. R. 2004. Stir bar sorptive extraction for the analysis of wine cork taint. *Journal of Chromatography A* 1033:173–78.

Zou, H., Kilmartin, P. A., Inglis, M. J., and Frost, A. 2002. Extraction of phenolic compounds during vinification of Pinot Noir wine examined by HPLC and cyclic voltammetry. *Australian Journal of Grape and Wine Research* 8:163–74.

3 Application of Microwaves for On-line Quality Assessment

Ruth De los Reyes, Marta Castro-Giráldez, Pedro Fito, and Elías De los Reyes

Introduction

The food industry has a growing interest in improving sensing techniques for quality control based on the behavior of food when subjected to an electromagnetic field. Such electromagnetic parameters as complex permittivity, amplitude and phase, and scattering parameters can be related to certain quality process parameters (Nelson et al. 1998; Shiinoki et al. 1998; Nelson 1999; Clerjon and Damez 2005). For instance, electrical properties and moisture content, especially free water, are closely related to each other; dielectric measurements can therefore be used to analyze water activity in foodstuffs (Clerjon et al. 2003). Also, instruments for measuring electromagnetic parameters are now available and, when precisely calibrated, can be used for rapid and nondestructive control measurements.

Microwaves are a common designation for electromagnetic waves at frequencies between 300 MHz and 300 GHz (fig. 3.1). The interest in food industry application of microwaves for dielectric heating appeared in the years following the end of World War II, but the development of microwaves stopped due to technological reasons and the high cost of investment (Decareau 1985). At the beginning of the 1980s, the possibilities of microwave applications and their considerable advantages were recognized (Decareau and Peterson 1986), and microwave ovens became more popular. This increase in the use of domestic microwave ovens gave rise to a reduction in the cost of the relatively high-power magnetron. However, the cost of this element increases exponentially when the power is on an industrial scale (Fito et al. 2004). At present, domestic microwave ovens are universally accepted by consumers, and other microwave heating applications are widely used in industry; baking, drying, blanching, thawing, tempering, and packaging are the most important. Therefore wide experience has now been accumulated in this field, and sensor systems based upon microwaves can be a viable nondestructive method of on-line control. The availability of components, affordable prices, speed of acquisition, and processing of data make them even more attractive.

The aim of this chapter is to provide an overview of the current applications of low-power microwaves for food quality control and to offer a detailed examination of the current research in this field.

Microwave-Material Interaction Aspects

Complex permittivity (ε_r) (equation 3.1) is the dielectric property that describes food behavior under an electromagnetic field (Metaxas and Meredith 1993; Datta and

Anantheswaran 2001). The real part of complex permittivity is called the dielectric constant (ε'), and the imaginary part is called effective loss factor (ε''_{eff}). The subscript r indicates that values are related to air, and the variable is therefore dimensionless.

$$\varepsilon_r = \varepsilon' - j \cdot \varepsilon''_{eff} \tag{3.1}$$

Under a microwave field, the charges of certain food components (water, salts, etc.) try to displace from their equilibrium positions to orientate themselves following the field, storing microwave energy that is released when the applied field stops. This behavior is called *polarization;* ε' reports the material's ability to store this electromagnetic energy (or the ability to be polarized). Only a perfect dielectric can store and release wave energy without absorbing it. The parameter ε''_{eff} is related to absorption and dissipation of electromagnetic energy from the field (Decareau 1985). Such energy absorptions are caused by different factors that depend on structure, composition, and measurement frequency, thus ε_{eff}'' can be expressed by equation 3.2 (Metaxas and Meredith 1993).

$$\varepsilon_{eff}'' = \varepsilon_d'' + \varepsilon_{MW}'' + \varepsilon_e'' + \varepsilon_a'' + \sigma/\varepsilon_o\omega \tag{3.2}$$

In this equation the last term is called ionic losses. σ, ε_o, and ω are material conductivity, air dielectric constant, and angular frequency, respectively. Subscripts d, MW, e, and a indicate dipolar, Maxwell-Wagner, electronic, and atomic losses, respectively. The different contributing mechanisms to the loss factor of a moist material are schematically represented in figure 3.1.

Under a microwave field, molecules with an asymmetric charge distribution (permanent dipoles) rotate to align with the electric field storing energy (dipolar polarization, also called orientation polarization). Dipolar contribution to total losses occurs when a phase lag between the dipole alignment and microwave field appears. When this phenomenon occurs, the material starts to lose stored energy and the dielectric constant decreases; before this moment, the dielectric constant value was ε_s (static dielectric constant). When the frequency of the field is too high for dipole rotation, the dipolar losses finish and the dielectric constant reaches a constant value (ε_∞). In figure 3.2, dipolar losses and dielectric constant versus logarithm of frequency are represented following the ideal Debye model (Debye 1929). In the same figure, the Debye model is also represented in the complex plane (ε'' versus ε') as a semicircle with locus of points ranging from ($\varepsilon' = \varepsilon_s$, $\varepsilon'' = 0$) at the low-frequency limit to ($\varepsilon' = \varepsilon_\infty$, $\varepsilon'' = 0$) at the high frequency limit. Such a representation is called Cole and Cole diagram.

The highest value of dipolar losses ($\varepsilon_d''_{max}$) is produced at the relaxation frequency; this frequency is the inverse of relaxation time (Hewlett-Packard 1992a; Metaxas and Meredith 1993). Deviations from the pure Debye theory have been studied by different authors (Metaxas and Meredith 1993; Datta and Anantheswaran 2001). The dipolar contribution to total losses is one of the most important at microwave frequencies due to the fact that water is an abundant and common component of foods. Otherwise, as frequency is increased (the highest microwave frequencies and above them), the electromagnetic field can affect smaller particles, inducing dipoles even in neutral molecules (atomic polarization) and neutral atoms (electronic polarization). Atomic and electronic losses have behavior similar to permanent dipolar losses.

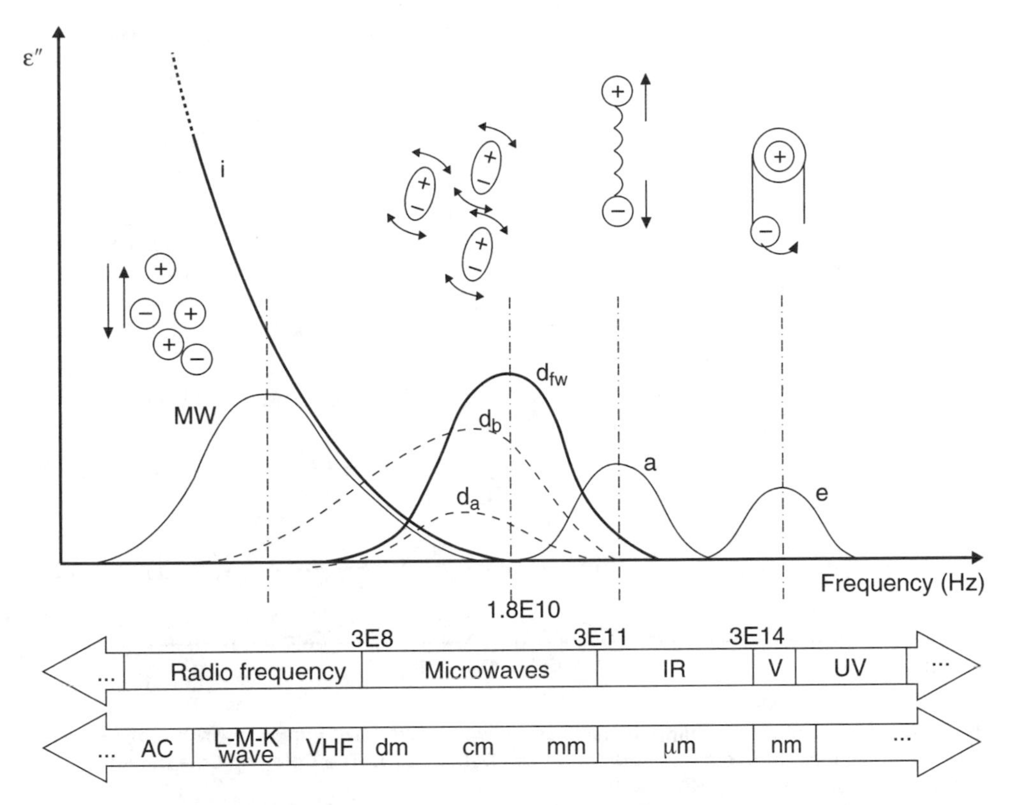

Fig. 3.1. Schematic representation over the electromagnetic spectrum (in logarithm scale) of the different effects that contribute to effective loss factor. i: ionic losses; MW: Maxwell-Wagner effect; d_{fw}: dipolar losses of free water; d_b: dipolar losses of bound water; d_a: dipolar losses of isopropyl alcohol; a: atomic losses; e: electronic losses.

At radio frequencies and the lowest microwave frequencies, charged atoms and molecules (ions) are affected by the field. Such ions move trying to follow the changes in the electric field. In the case that ions do not find any impediment (aqueous solutions, conducting materials), ionic conductivity gives rise to an increment in effective losses. At these frequencies the ionic losses are the main contributors to the loss factor (supposing ions to be present in the material).

Foods are complex systems and usually present conducting regions surrounded by nonconducting regions; for example, foods with a cellular structure have cytoplasm (conducting region) surrounded by the membrane (nonconducting region). In these cases, ions are trapped by the interfaces (nonconducting regions), and as the ion movement is limited, the charges are accumulated, increasing the overall capacitance of the food (Hewlett-Packard 1992a) and the dielectric constant (Maxwell-Wagner polarization). This phenomenon is produced at low frequencies at which the charges have enough time to accumulate at the borders of the conducting regions.

The Maxwell-Wagner losses curve versus frequency has the same shape as the dipolar losses curve (fig. 3.1). At higher frequencies, the charges do not have enough time to accumulate, and the polarization of the conducting region does not occur. At frequencies

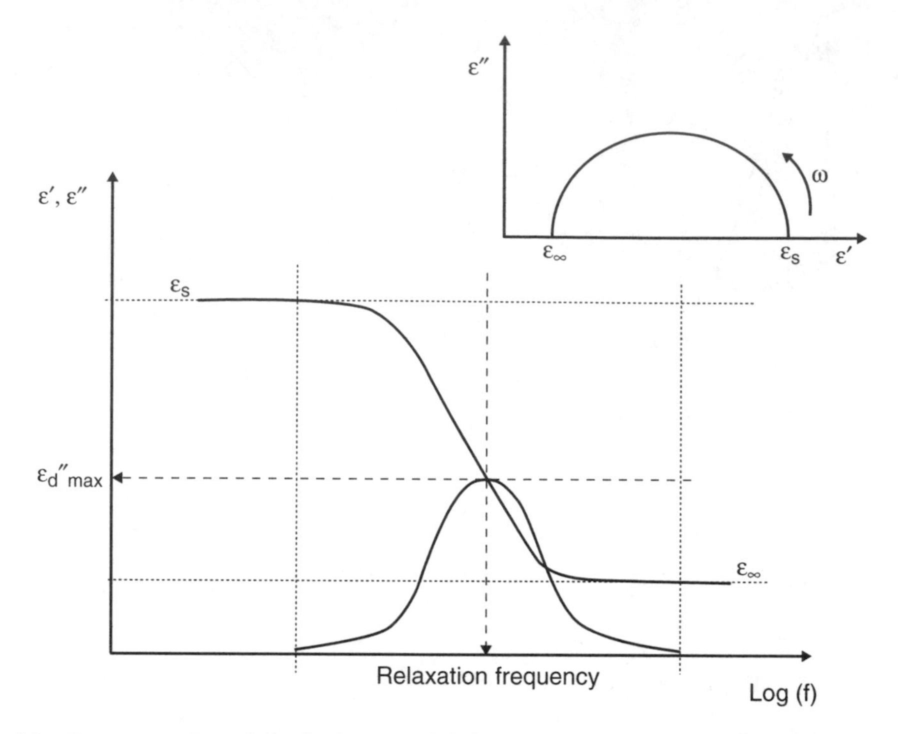

Fig. 3.2. Representation of dipolar losses and dielectric constant versus logarithm of frequency following the ideal Debye model (adapted from Debye 1929). The Cole and Cole diagram is also represented at the top right of the figure.

above the Maxwell-Wagner relaxation frequency, both ionic losses and the Maxwell-Wagner effect are difficult to distinguish due to the fact that both effects exhibit the same slope (1/f).

Foods are multicomponent and multiphase systems; therefore, more than one single mechanism contributes to the combined effects. Figure 3.3 shows different shape variations in effective loss factor curves versus frequency for the case of combined dipolar and ionic losses. Type 0 represents a typical pure dipolar loss factor curve, σ increases between type 0 and 4 curves, $\varepsilon_d''_{max}$ is the highest value of dipolar losses, and relaxation frequency is the inverse of relaxation time (Metaxas and Meredith 1993; De los Reyes et al. 2005b).

In general, foods are materials with high losses, and under a microwave field, they can absorb part of the wave energy. The power that can be dissipated in a given material volume (Pv) is related to ε_{eff}'' by equation 3.3, in which E is the electric field strength (Metaxas and Meredith 1993; Datta et al. 2005).

$$Pv = 2\pi f \cdot \varepsilon_o \cdot \varepsilon_{eff}'' \cdot E^2 \ (W/cm^3) \tag{3.3}$$

High-power dissipation in foods is one of the reasons for the numerous high-power heating applications that have been developed since the 1950s. The interest in improving heating applications has provided a great deal of knowledge on the dielectric properties of materials and wave parameters measurements. This detailed knowledge has been very

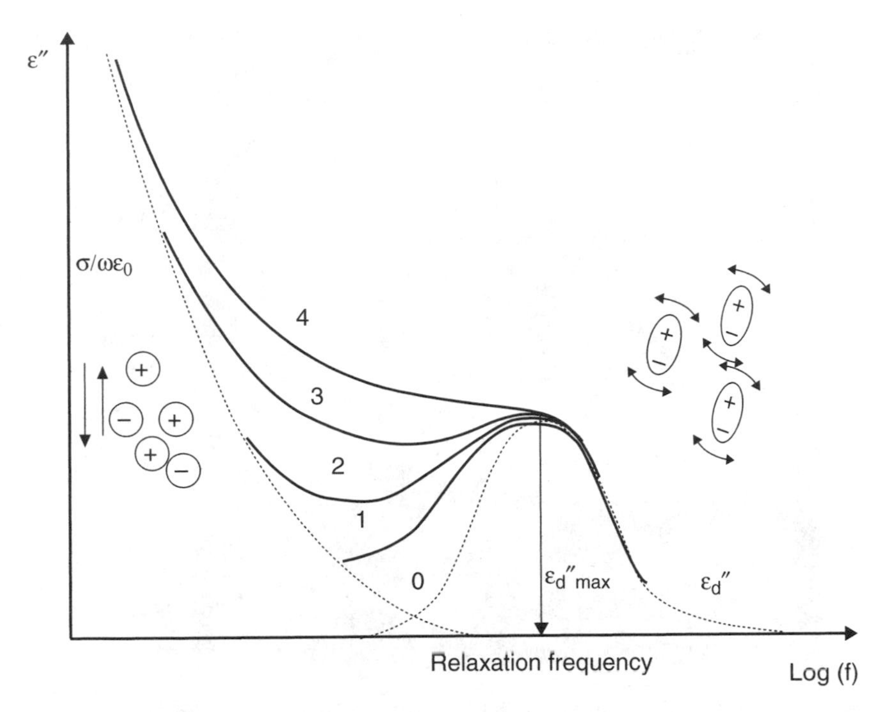

Fig. 3.3. Influence of salt content in systems with different proportions of dipoles (water) and ions (salts) in the shape of effective loss factor curve. Salt content increases in curves from 0 (without salt, only water) and 4 (saturated solution) (De los Reyes et al. 2005b).

useful in further research into new low-power on-line sensors that relate these properties or parameters to process variables of the food industry. The main factors that can affect dielectric properties are specifically explained in this chapter. The following section gives an overview of measurement techniques and sensors.

Measurement Techniques

The dielectric properties of foods can be determined by several techniques using different microwave-measuring sensors, depending on the frequency range of interest, the type of target food, and the degree of accuracy required. A vector network analyzer (VNA) is very useful and versatile for detailed studies. At microwave frequencies, generally about 1 GHz and higher, transmission line, resonant cavity, and free-space techniques have been very useful (Nelson 1998).

Microwave dielectric properties measurement methods can be classified as reflection or transmission measurements using resonant or nonresonant systems, with open or closed structures for sensing the properties of material samples (Kraszewski 1980). Transmission measurements are made by at least two sensors: emitter sensors send the signal through the material, and receiver sensors capture the resulting signal. Reflected measurements are made when the same sensor emits and receives the signal after it is reflected by the material.

The different measuring techniques are explained next.

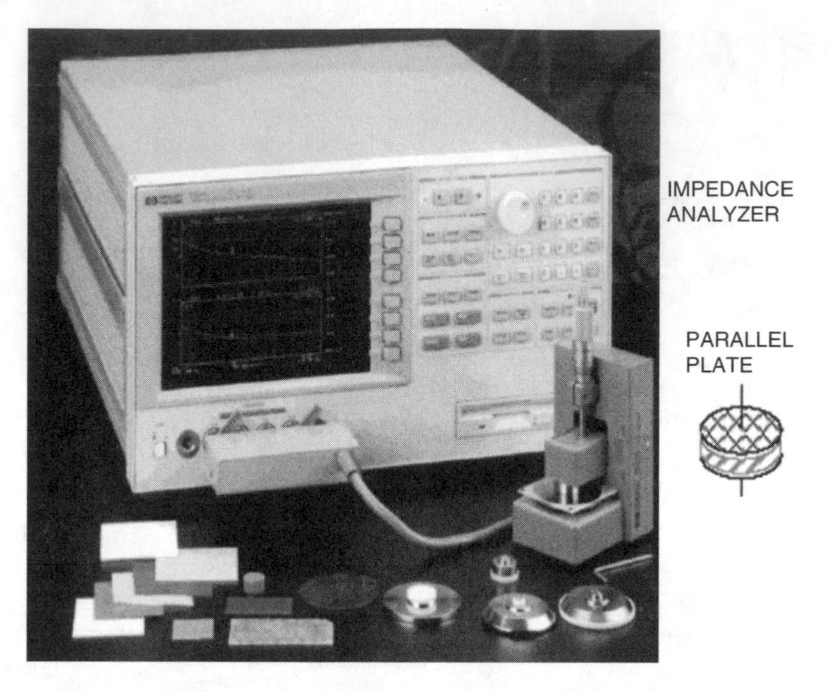

Fig. 3.4. Impedance analyzer and parallel plate (from Agilent Technologies).

Parallel Plate Technique

The parallel plate method requires placing a thin sheet of material between two electrodes to form a capacitor (fig. 3.4). This method is also called the capacitance technique and uses an LCR meter or impedance analyzer to measure capacitance and dissipation. This method is typically used at low frequencies (<100 MHz).

Open-Ended Probe Technique

A typical coaxial probe system consists of a vector network analyzer, a coaxial probe (fig. 3.5), an external computer, and software to calculate permittivity from calibrated S-parameter measurements. This method is widely used for liquids or semisolids, although solids with a flat surface can also be measured, which makes it ideal for many foodstuffs. The material is measured by immersing the probe into the liquid or semisolid samples, or by touching the flat surface of the solid. This method is nondestructive and easy to use and does not need sample preparation. It uses a frequency range from 200 MHz to 20 GHz and requires a sample thickness of >1 cm (Hewlett-Packard 1992b). Actually the sample has to be thick enough to assume that it is endless. Air gaps or bubbles between a solid and the probe must be avoided since they produce errors. This is a reflection measurement method in which the same coaxial probe emits and receives the signal.

Resonant Cavity Technique

Cavities are characterized by the central frequency (fc) and quality factor (Q). Permittivity is calculated from the changes in these properties due to the presence of the sample. The

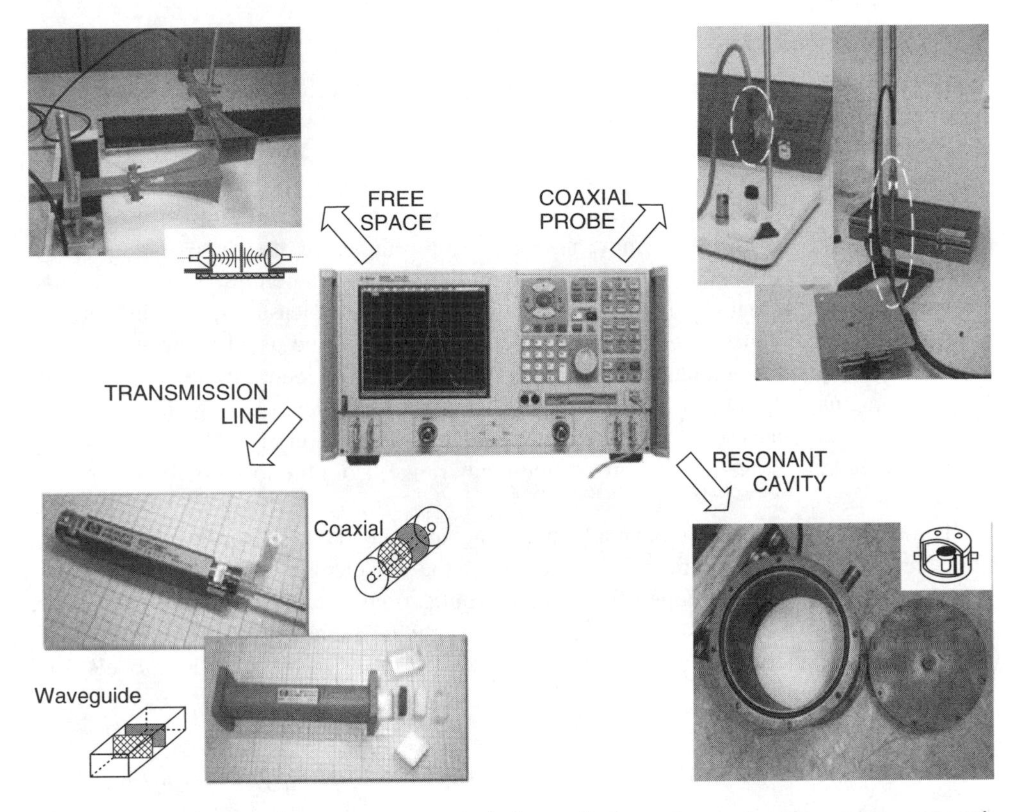

Fig. 3.5. Overview of the measurement techniques that can be used with a vector network analyzer.

sample is placed in the center of a waveguide that has been made into a cavity. The sample volume must be precisely known. It has good resolution for low-loss materials and small samples. This technique uses a single frequency and is commonly used for measuring the dielectric properties of homogeneous food materials, since it is simple, accurate, and capable of operating at high temperatures (Sucher and Fox 1963; de Loor and Meijboom 1966; Bengtsson and Risman 1971; Metaxas and Meredith 1993).

Transmission Line Technique

This method uses a waveguide or coaxial transmission line with a vector network analyzer (VNA). Free space is sometimes considered a transmission line technique. In this case at least two antennas are needed.

In waveguide and coaxial transmission line techniques, the most important factor is the sample preparation. Rectangular samples and annular samples have to be prepared for waveguide and coaxial lines, respectively. ε' and ε'' can be determined by measuring the phase and amplitude of microwave signals reflected from or transmitted through the target material. This method is useful for hard machineable solids and requires a precise sample shape, and is therefore usually a destructive method.

Free-Space Technique

Free-space technique requires a large, flat, thin, parallel-faced sample and special calibration considerations (Hewlett-Packard 1992b). It does not need special sample preparation and presents certain advantages due to the fact that it is a nondestructive and noncontact measuring method and can be implemented in industrial applications for on-line process control. This technique can also be used at high temperatures. The sample is placed in front of one or between two or more antennas. In the first case, the same antenna transmits and receives the signal. In the second case, there are two antennas, a transmitter and a receiver. The attenuation and phase shift of the signal are measured, and the data are processed in a computer to obtain the dielectric properties. The usual assumption made for this technique is that a uniform plane wave is normally incident on the flat surface of a homogeneous material and that the sample size must be sufficiently large to neglect the diffraction effects caused by the edges of the sample (Trabelsi et al. 1998).

This technique is useful for a broad frequency range, from the low-microwave region to mm-wave.

Tomography images can be made by using an antenna system and a reconstruction algorithm. Pixel definition is a function of the wavelength inside the food. Other techniques have been developed for medical applications, such as confocal microwave imaging (CMI), which focuses backscattered signals to create images that indicate regions of significant scattering (Fear et al. 2002). Figure 3.6 shows simple breast models used in feasibility testing.

Time Domain Reflectometry (Also Known as Spectroscopy) Technique

Time domain reflectometry methods were developed in the 1980s. This technique measures the complex permittivity of dielectric materials over a wide frequency range, from 10 MHz to 10 GHz. It is a rapid, accurate, and nondestructive method. It utilizes the reflection characteristic of the sample to obtain the dielectric properties. It must be emphasized that the sample size must be small and the material must be homogeneous.

Microstrip Transmission Line Technique

The effective permittivity, represented by a combination of the substrate permittivity and the permittivity of the material above the line, of a microstrip transmission line is strongly dependent on the permittivity of the region above the line. This effect has been utilized in implementing microwave circuits and, to a lesser extent, in the investigation of dielectric permittivity (fig. 3.7).

Factors Affecting the Electrical Properties of Food

The composition, phases, and structure of food materials determine their electrical properties at a given temperature and measurement frequency. Molecules present in the system, like water, fatty acids, and soluble solids, as well as their mobility and polarity, induce different interactions with the field. Water's and solutes' activity may be used as thermodynamic variables useful for being correlated with the experimental values of dielectrics properties to obtain adequate predictive equations. In the case of the water (usually the

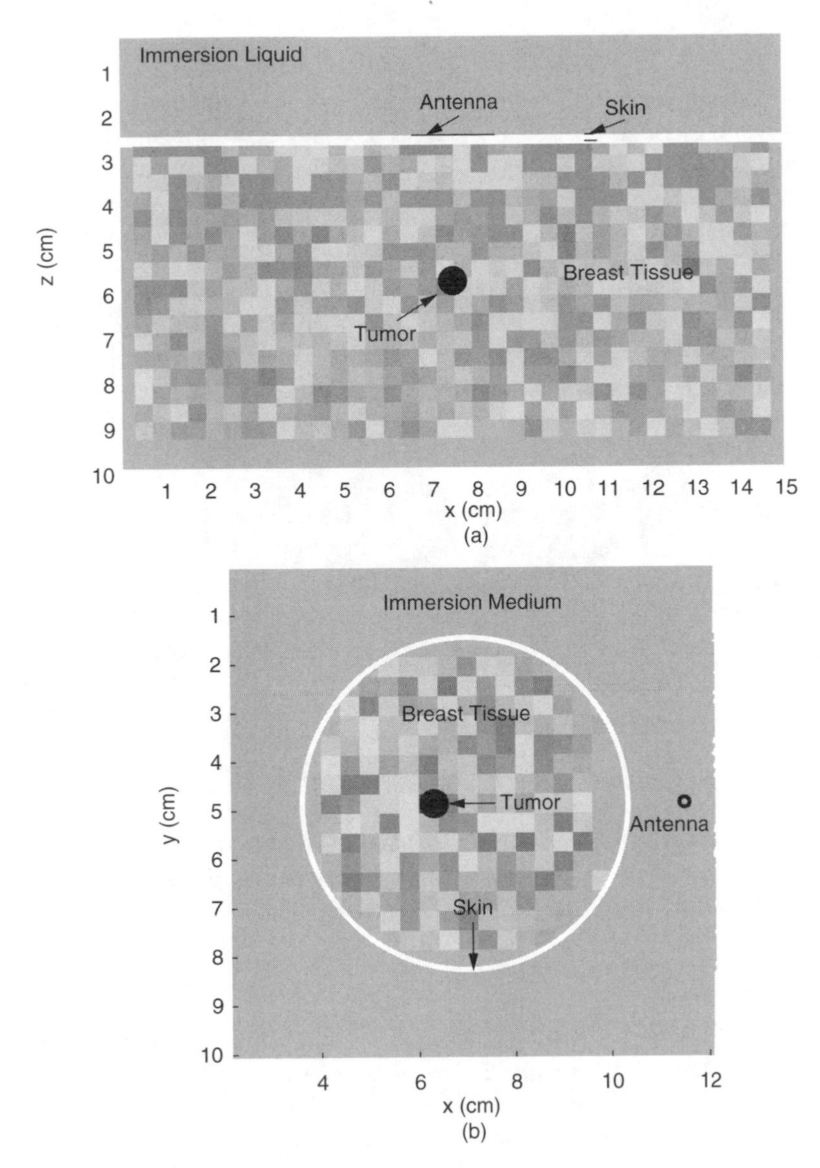

Fig. 3.6. Simple models used for simulations of confocal microwave imaging: **(a)** planar model, **(b)** cylindrical model.

major factor responsible for dielectric food properties) the use of sorption isotherms may help to explain the food-microwave interaction.

Food density can indirectly be correlated to the dielectric behavior of food due to the fact that the food-microwave interaction is related to a given volume. The lower the density, the lower the number of molecules able to interact with the wave-per-volume unit. The dielectric properties of granular and powdered foods, such as grains and flours, are highly affected by bulk density, and porosity has a strong influence on the dielectric properties of porous foods.

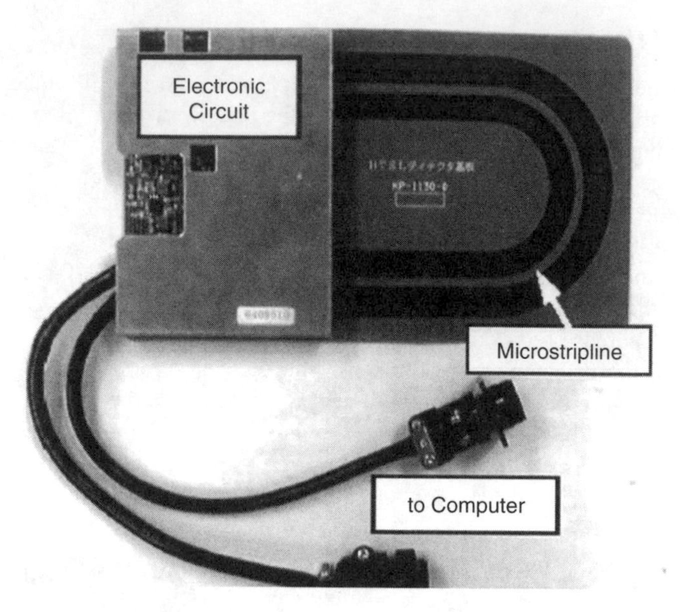

Fig. 3.7. Microstrip transmission line.

In general, foods are very complex and heterogeneous systems, so structural factors also affect the dielectric behavior. The specific cases of fruits, fish and meat products, and dairy products are analyzed and related to industrial applications in the next section, "Applications of Microwaves' Measurements in Food Quality Control."

Frequency Effects

The different mechanisms of polarization and losses in relation to the frequency have been explained previously (figs. 3.1 and 3.2). It is important to note that the correct selection of the measuring frequencies is critical in order to be able to perceive the phenomena that are occurring during food manufacture/processing—for instance, in the denaturalization of proteins (Bircan and Barringer 2002) or the postmortem changes in meat (Clerjon and Damez 2005).

Water and salts are the components that have the most influence in the dielectric response of food under a microwave field (figs. 3.1 and 3.2); moreover, when they appear together, the combined effect changes the dielectric spectra, as has been shown already (fig. 3.3). The dielectric response of these components is also affected by the structure of foods, due to the fact that the solid matrix interacts with the water molecules and ions, diminishing their thermodynamic activity.

For this and other reasons, such as economy or equipment availability, a previous spectroscopic study is important to select the optimal frequency in order to obtain good correlations between the dielectric properties and the food variables under study.

Examples of spectra variation can be seen in figures 3.8, 3.10–3.13, 3.17, and 3.19.

Temperature Effects

Temperature generally increases particles' mobility, which in turn produces an increase in polarization ability (ε′) and relaxation frequency. The loss factor is also affected by temperature changes, but the trend of this modification depends more on factors such as salt content or measurement frequency. As has been explained previously, the presence of ions increases the loss factor at low frequencies (fig. 3.1). At high temperatures this effect is greater due to the increment in particles' mobility.

Normally, a single frequency is selected for dielectric measurements; therefore, depending on whether this frequency is above or below the relaxation frequency, the loss factor will increase or decrease respectively with temperature increase (Kent 2001).

In figure 3.8, the effect of temperature on the dielectric properties of water can be observed. In this figure the experimental dielectric spectra have been completed with the Debye model (broken lines). The curves of the figure are accorded with the data reported already by other authors (Datta and Anantheswaran 2001; Kaatze 1989). The relaxation frequency increment with temperature can be appreciated in the figure.

Taking into account that the relaxation frequency is also modified by temperature, either nonlinear or linear trends can be expected for the loss factor, depending on the measuring frequency. The presence of ions can also change the loss factor trend at a single frequency below the relaxation frequency. In this case, when temperature increases, the effective loss factor decreases until the temperature at which the ionic losses' contribution increases the effective loss factor, causing a change in the tendency and a minimum at the transition

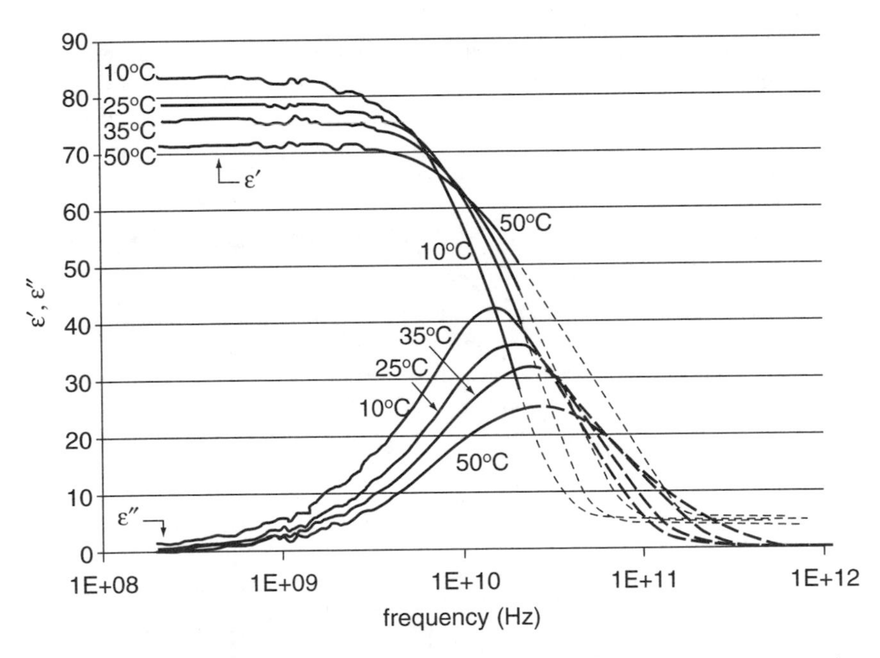

Fig. 3.8. Representation of temperature effects in dielectric properties of distilled water between 0.2 and 20 GHz. The dielectric spectra above 20 GHz are represented in broken lines, following the Debye model. The frequency is given in logarithm scale.

temperature. Many authors have found this transition temperature in foods at 2.45 GHz (Mudgett 1995). Moreover, it was obtained in a linear relation between ash content and the transition temperature of 2.45 GHz of certain fruits and vegetables (Sipahioglu and Barringer 2003).

The biggest changes in dielectric properties happen during freezing or thawing (Datta et al. 2005). Frozen water molecules are not free to align with the field, so the dielectric constant and loss factor have very low values in frozen foods (Bengtsson et al. 1963). In the transition stage, where the food is only partially frozen, dielectric properties are a strong function of salt content (ions) and unfrozen fraction. Tylose dielectric properties have been reported in this range of temperatures at 2.45 GHz (Chamchong and Datta 1999).

Water

Water in solid food matrices can appear as a water solution of the food solutes (namely free water) inside capillaries or cells (liquid phase) or in a bound state, physically adsorbed on the solid matrix. In figure 3.1 bound and free water losses are schematized. The liquid phase of foods usually contains soluble solids such as sugars or salts that hinder the rotation of water dipoles, influencing the real and imaginary parts of permittivity, as will be explained in the following. On the other hand, free water dielectric behavior can be described by the Debye ideal model (fig. 3.2) (Debye 1929). The presence of soluble solids causes Debye model deviations and dispersions (Metaxas and Meredith 1993; Kent 2001).

When a food sample is subjected to a dehydration treatment, free water can leave more easily than bound water (Bilbao Sáinz 2002; Martín Esparza 2002). These two states of water can be distinguished by using the dielectric properties (fig. 3.9). This phenomenon is mainly observed in the dielectric constant, which takes higher values in free water than in bound water. As we have said before, bound water has less freedom of movement (polarization).

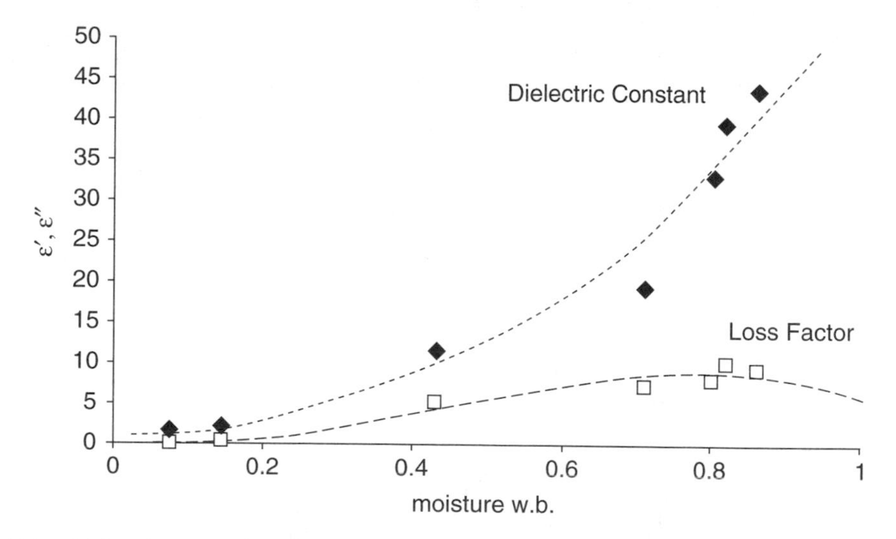

Fig. 3.9. Dielectric properties of apple in function of moisture content in wet basis (g_{water}/g_{total}) during drying process.

Soluble Solids Content

The most common water-soluble solids are salts and sugars. Both can hinder water mobility, reducing polarization ability. Sugars can also stabilize water by hydrogen bonds with their hydroxyl groups; therefore sugars reduce the relaxation frequency. In figure 3.10 it can be observed that an increase in sucrose content (expressed in Brix) reduces the relaxation frequency, but on the other hand, the increase of salts in the system does not change the relaxation frequency (as can be seen in figure 3.11).

Salt's content produces an increase in ionic losses at low frequencies and produces a combined effect in aqueous solutions (fig. 3.3). The frequency spectra of loss factor of sodium chloride solutions between 0 and 16 percent are shown in figure 3.11 (De los Reyes et al. 2005c). As sodium chloride content increases, loss factor curves change progressively from type 0 to almost type 4 curves, which were described in figure 3.3.

The size of the ions present in foods is also an important factor that influences dielectric properties, since it is directly related to their mobility. Figure 3.12 shows frequency spectra from water solutions of 1 and 2 percent of sodium chloride and calcium lactate at 25°C. Calcium lactate ions are significantly bigger than those of sodium chloride, so their effect on ionic losses is minor.

Other authors have described the dielectric properties of different salt solutions. Vicedo et al. (2002) studied the effects of ferric gluconate and calcium lactate salts in ε' and ε'' of water-salt-glucose systems. Wei et al. (1992) studied aqueous alkali-chloride systems and found a relation between the ionic radius of the cations and their effect on dielectric properties.

The liquid phase of foods can also be composed of salt and sugars together. In this case, the combined effect curves are not exactly as described in figure 3.3, since sugar content influences relaxation frequency so that the combined effect is also displaced in the same way as the relaxation frequency. Figure 3.13 shows loss factor spectra from sodium

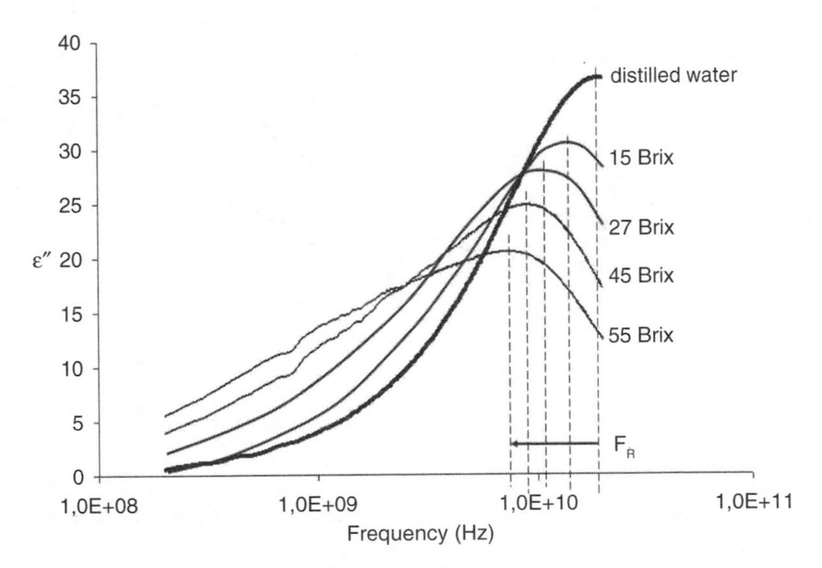

Fig. 3.10. Loss factor frequency spectra (in logarithm scale) of sucrose solutions between 0 (distilled water) and 55 Brix. Relaxation frequencies (F_R) are marked by broken lines.

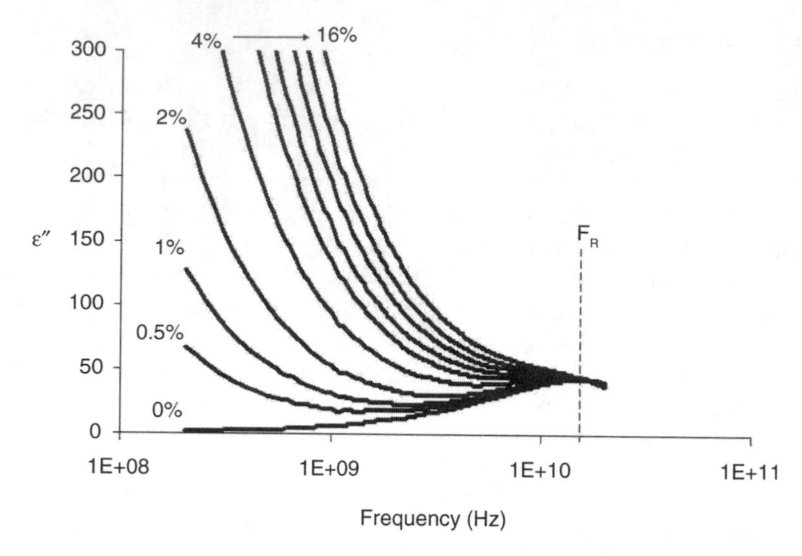

Fig. 3.11. Loss factor frequency spectra (in logarithm scale) of sodium chloride solutions between 0 and 16 percent at 12°C. The solution of 0 percent represents distilled water, and that of 16 percent refers to a water-NaCl system with 16 g of NaCl per 100 g of solution. The other percentages are 0.5, 1, 2, 4, 6, 8, 10, and 12. Relaxation frequency (F_R) is marked by broken lines.

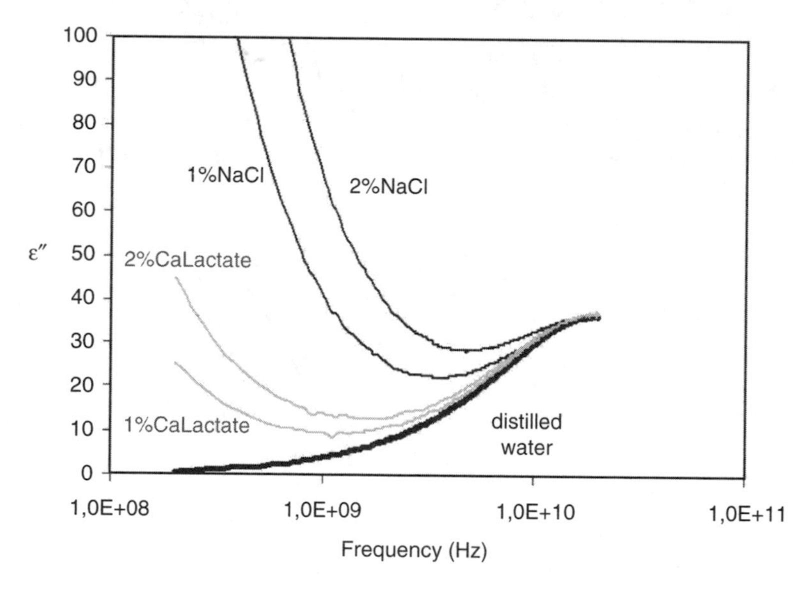

Fig. 3.12. Frequency spectra (in logarithm scale) from water solutions of 1 and 2 percent sodium chloride and calcium lactate at 25°C (adapted from De los Reyes 2005c, d).

chloride and sucrose aqueous solutions at 25°C. The composition of these aqueous solutions is detailed in table 3.1.

The dotted lines traced in figure 3.13 represent the shape of loss factor curves, taking into account only the dipolar effect displaced by the presence of sucrose. These dotted curves coincide with those of sucrose solutions of 27.5, 36.7, and 42.5 Brix (fig. 3.10). In

Table 3.1. Aqueous solutions composition.

Solution	Sucrose Content (%)	NaCl Content (%)
20% NaCl	0.0	20.0
(1 : 1)	27.5	10.0
(1 : 2)	36.7	6.7
(1 : 3)	42.5	5.0
55 Brix	55.0	0.0

Source: Adapted from De los Reyes et al. 2005d.

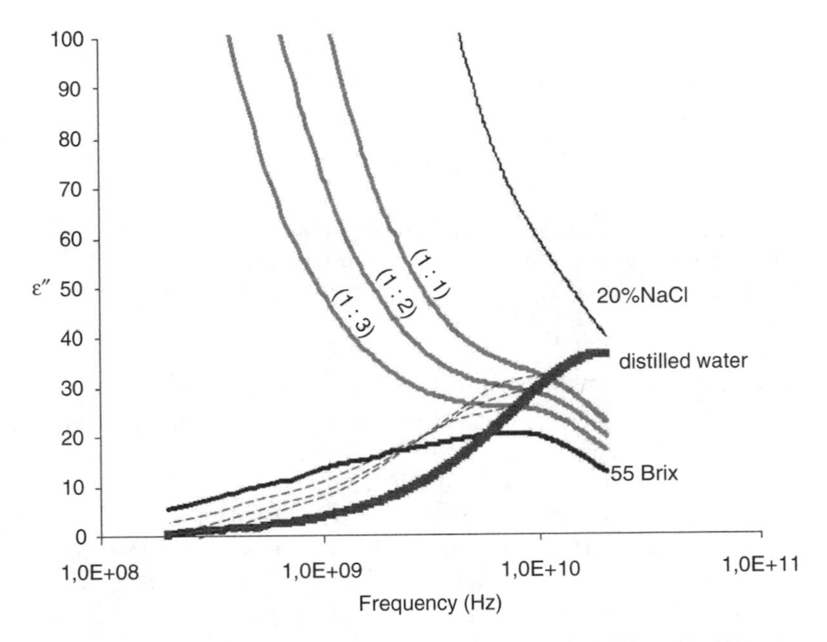

Fig. 3.13. Loss factor spectra (in logarithm scale) from sodium chloride and sucrose aqueous solutions at 25°C. Twenty percent NaCl indicates a solution of 20 g of NaCl per 100 g of solution; 55 Brix refers to a solution of 55 g of sucrose per 100 g of solution; and (1 : 1), (1 : 2), and (1 : 3) are combinations of the two previous solutions in these proportions (adapted from De los Reyes et al. 2005d).

figure 3.13 the part of the curves (1 : 1), (1 : 2), (1 : 3) related to ionic losses are displaced in the same way as the relaxation frequency.

Fruits and vegetables dehydrated in osmotic solutions, with or without pressure treatments (vacuum impregnation or osmotic dehydration, respectively), are an example of high-moisture foods with different amounts of soluble solids in their liquid phases. Some authors have found that the dielectric properties of these products are very similar to the dielectric properties of solutions with the same amount of solutes as the products have in their liquid phase (Vicedo et al. 2002; Martinez et al. 2002; Alandes et al. 2004a, 2004b; De los Reyes et al. 2005a, 2005d; Betoret et al. 2004). In figure 3.14, the loss factor values (at 2.45 GHz and 21°C) of different fruits osmotically dehydrated in sucrose solutions, sodium chloride solutions, and a combination of both are represented as a function of the soluble solids content of their liquid phases (z_{ss}).

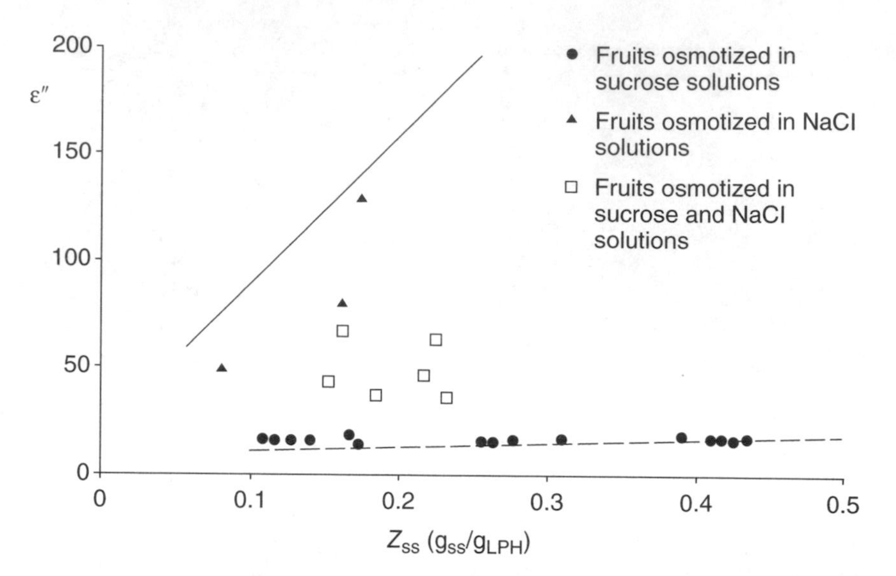

Fig. 3.14. Loss factor values (at 2.45 GHz and 21°C) of some fruits osmotically dehydrated with sucrose, sodium chloride, and combined solutions represented as a function of the soluble solids content of their liquid phases (z_{ss}) (adapted from Vicedo et al. 2002; De los Reyes et al. 2005a, 2005d).

It can be observed that the loss factor (ε″) of fruits osmotically dehydrated with sucrose solutions is almost identical to that of sucrose solutions with the same Brix (marked with a dotted line). On the other hand, due to the fact that fruits have sugars in their composition, the loss factor (ε″) of fruits osmotically dehydrated with salt solutions is different from that of salt solutions, but it is always higher than that of sucrose solutions.

The amount of salt content included in the food by osmotic dehydration depends on the process conditions, and only a drastic treatment with a high concentration of salt solution can make the dielectric properties of the osmotized fruit approach those of salt solutions (marked with a continuous line).

Long-Chain Carbohydrates

The main long-chain carbohydrates in foods are starches and gums. There are few hydroxyl groups in starches able to stabilize water, so their presence in water solutions produces a smaller reduction in dielectric properties than sugars do. Starch gelatinization also produces changes in its dielectric properties, since gelatinized starches can bind less water to their structure than ungelatinized starches can, leading to higher dielectric properties.

Gums have the ability to bind large amounts of free water in the system, reducing the number of free water molecules available to align with the field.

Fats and Proteins

Fats and proteins are the major components in foods of animal origin (meat, fish, eggs, dairy products). Fats have very low dielectric properties, and their presence in foods produces a reduction in free water per volume unit (table 3.2).

Table 3.2. Dielectric data of some commercial fats and oils.

		300 MHz			1000 MHz			3000 MHz		
Sample		25°C	48°C	82°C	25°C	48°C	82°C	25°C	48°C	82°C
Soybean salad oil	ε′	2.853	2.879	2.862	2.612	2.705	2.715	2.506	2.590	2.594
	ε″	0.159	0.138	0.092	0.168	0.174	0.140	0.138	0.168	0.160
Corn oil	ε′	2.829	2.868	2.861	2.638	2.703	2.713	2.526	2.567	2.587
	ε″	0.174	0.134	0.103	0.175	0.174	0.146	0.143	0.166	0.163
Lard	ε′	2.718	2.779	2.770	2.584	2.651	2.656	2.486	2.527	2.541
	ε″	0.153	0.137	0.109	0.158	0.159	0.137	0.127	0.154	0.148
Tallow	ε′	2.603	2.772	2.765	2.531	2.568	2.610	2.430	2.454	2.492
	ε″	0.126	0.141	0.105	0.147	0.146	0.134	0.118	0.143	0.144
Hydrogenated vegetable shortening	ε′	2.683	2.777	2.772	2.530	2.654	2.665	2.420	2.534	2.550
	ε″	0.141	0.140	0.103	0.147	0.153	0.137	0.117	0.146	0.146
Bacon fat	ε′	2.748	2.798	2.770	2.608	2.655	2.649	2.493	2.538	2.536
	ε″	0.165	0.139	0.099	0.163	0.161	0.144	0.130	0.152	0.149

Source: Adapted from Pace et al. 1968.

Proteins are complex molecules with soluble and insoluble zones. Their ionizable surface regions can bind water or salts. Large proteins are not very reactive to microwave frequencies, but the solvated or hydrated form of proteins, hydrolyzed proteins and polypeptides, are much more reactive (Datta et al. 2005). The structural changes produced during their denaturalization can greatly modify the dielectric properties, due to the release of water or salts. This change in dielectric properties can be used to detect protein denaturalization by microwave sensors (Bircan and Barringer 2002).

Applications of Microwaves' Measurements in Food Quality Control

As explained in the last section, dielectric measurements can provide important information during industrial processes due to the relationships between food properties and electromagnetic parameters. Complex permittivity can be correlated with structural as well as physical and chemical properties such as humidity, soluble solids content, porosity, characteristics of solid matrix, and density. The changes in these properties are usually related to the food treatments applied throughout the industrial process; for instance, losses of water in drying processes or losses of salts in desalting processes (De los Reyes et al. 2005c). Also, structural changes in macromolecules, such as protein denaturalization, can occur during processing, leading to a modification of the dielectric properties (Bircan and Barringer 2002). For all these reasons, the measurement of dielectric properties can be used as a tool for on-line food process control.

Low-power microwaves change their parameters (amplitude, phase) according to the food properties, and this change can be measured. This is the basic principle on which food-quality microwave sensors are based. As sensors use low-power microwaves, there are no permanent effects on the food. Also, some microwave sensors, such as coaxial probes, waveguides, or striplines, can be used in nondestructive measurement methods. There are also noncontact techniques, such as free space, which have advantages as on-line sensors.

The aim of this section is to provide an overview of the most important microwave applications as techniques in food control.

Determination of Moisture Content

Water is the main component in foods influenced by microwave energy, and therefore nowadays most methods of determining moisture content are based on electrical properties.

The determination of moisture content based on electromagnetic parameters has been used in agriculture for at least 90 years and has been in common use for about 50 years (Nelson 1977, 1991, 1999).

A lot of studies have been carried out relating the dielectric constant and loss factor with moisture content in foods (Bengtsson and Risman 1971; Roebuck and Goldblith 1972; Nelson 1978; Nelson et al. 1991; Ndife et al. 1998).

Further research in this field has occurred during recent years. Trabelsi and Nelson (2005c) studied a method of moisture sensing in grains and seeds by measurement of their dielectric properties. The reliability of the method was tested for soybeans, corn, wheat, sorghum, and barley. The frequency used was 7 GHz with the free-space technique. In the same year, the authors used the same technique at 2–18 GHz to determine the dielectric properties of cereal grains and oilseeds in order to predict the moisture content by microwave measurements (Trabelsi and Nelson 2005a). Funk et al. (2005) also studied moisture in grain. This paper presents a unified grain moisture algorithm, based on measurements of the real part of the complex permittivity of grain at 149 MHz using the transmission line method. Trabelsi and Nelson (2005b) reported the moisture content in unshelled and shelled peanuts using the free-space method at a frequency of 8 GHz.

Since most efforts have been directed to the moisture determination of different materials, commercial meters for on-line moisture measurements have already been developed. These moisture meters are based on automatic on-line calculations of the reflected wave and dielectric permittivity, yielding physico-chemical properties, such as moisture content, composition, and density, without affecting the product. For instance, Keam Holdem® Industry (Auckland, New Zealand) provides on-line moisture testing and analyzing systems. This manufacturer provides devices for measuring moisture in processed cheese, moisture and salt in butter, moisture and density in dried lumber and whole kernel grain, and fat-to-lean ratio in pork middles.

Another interesting application for on-line moisture measurement is a sensor for green tea developed by Okamura and Tsukamoto (2005), which can measure moisture content as high as 160 to 300 percent on a dry basis by use of microwaves at 3 GHz with a microstripline (fig. 3.15).

A microwave moisture meter has also been developed for continuous control of moisture in grains, sugar, and dry milk in technological processes (Lisovsky 2005). A consortium of companies from different countries, Microradar®, produces a commercial microwave moisture meter for measuring moisture in fluids, solids, and bulk materials based on this method.

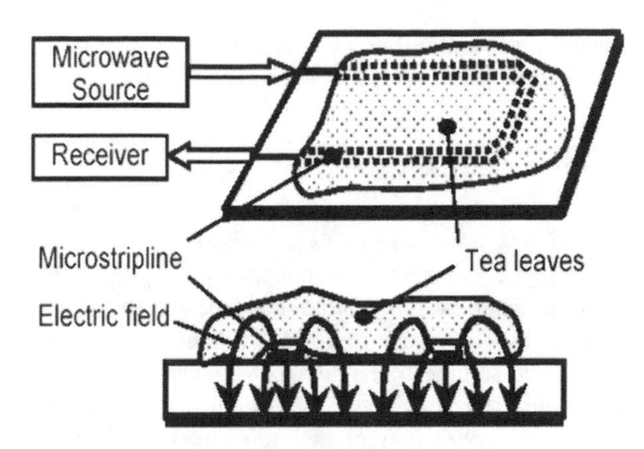

Fig. 3.15. Schema of a microstripline used for moisture measurement of tea leaves (Okamura and Tsukamoto 2005).

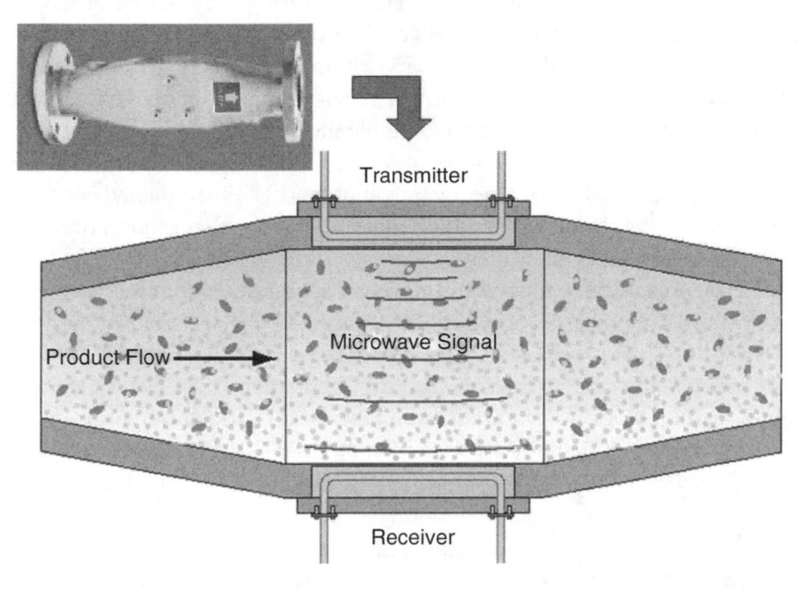

Fig. 3.16. Guided Microwave Spectometer® (Thermo Electron Corporation, United States) and its operation schema.

In 2005, K. Joshi reported a technique for on-line, time domain, nondestructive microwave aquametry (United States Patent nos. 6,204,670 and 6,407,555). The paper reports a novel technique for determining moisture levels in substances such as seeds, soil, soap, tissue paper, and milk powder.

A Guided Microwave Spectometer® (Thermo Electron Corporation, United States) has been developed for on-line measurements of multiphase products (fig. 3.16). This guide is used to measure moisture in raw materials such as corn, rice, and soybeans and in processed materials such as tomato paste and ground meat. It can also measure Brix, pH, viscosity, and acid in orange juice, soft drinks, mayonnaise, and tomato products; fat in ground meats, peanut butter, and milk and other dairy products; salt in mashed potatoes and most vegetable products; and, lastly, alcohol in beverages.

Quality Control of Fish, Seafood, and Meat

The determination of quality parameters in biological tissues based on electromagnetic parameters is still a complex topic due to its heterogeneous composition and complex matrix. The permittivity and conductivity parameters are the properties that determine the propagation of an electromagnetic wave in biological tissue. Gabriel et al. (1996) described these parameters in detail. It is important to point out that the limitation of most dielectric probes is the volume of the sample that interacts with the field. The volume has to be representative of the whole piece of meat or fish, due to the fact that the electromagnetic parameters in this kind of tissue vary in a heterogeneous way.

The dielectric properties of various meat products under different conditions and using different methods at microwave frequencies have been measured (Bengtsson and Risman 1971; Ohlsson et al. 1974; To et al. 1974). The dielectric properties of turkey meat were measured at 915 and 2450 MHz (Sipahioglu et al. 2003a). The authors developed a number of equations to correlate the real and imaginary part of permittivity with temperature, moisture, water activity, and ash. Other equations were developed to model the dielectric properties of ham as a function of temperature and composition (Sipahioglu et al. 2003b). The interaction between microwave radiation and meat products was also studied by Zhang et al. (2004). A complete study of the dielectric properties of meats and ingredients used in meat products at microwave and radio frequencies was also reported recently (Lyng et al. 2005). Some of the results given in this paper are shown in table 3.3.

The dielectric properties of fish products were also measured at microwave frequencies by different authors (Bengtsson et al. 1963; Kent 1972, 1977; Wu et al. 1988; Zheng et al. 1998).

Further investigations gave different control applications of dielectric measurements. As explained earlier, working in a frequency range (dielectric spectra) gives more information than measuring at a single frequency due to the fact that different effects can be

Table 3.3. Composition of meat.

Species (anatomical location)	Type	Moisture (%)	Protein (%)	Fat (%)	Ash (%)	Salt (%)	ε'	ε''
Beef (forequarter trimmings)	Lean	71.5	21.3	6.1	0.83	0.11	70.5	418.7
Lamb (leg)	Lean	73.0	21.9	3.6	1.48	0.14	77.9	387.2
Pork (shoulder)	Lean	73.9	20.1	4.4	1.13	0.08	69.6	392.0
Chicken (breast)	Lean	73.6	24.3	1.2	0.86	0.13	75.0	480.8
Turkey (breast)	Lean	74.5	24.1	0.4	0.98	0.08	73.5	458.4
Pork (back)	Fat	19.0	3.9	76.1	0.20	0.07	12.5	13.1

Source: Adapted from Lyng et al. 2005.

observed when dielectric spectra are being analyzed (see the temperature and frequency effects sections).

Studies by Miura et al. concluded that spectra analysis is a very useful tool for quality control of foodstuffs. Specifically, the authors studied the differences between raw, frozen, and boiled chicken at 25°C. They also studied the dielectric spectra of fish, vegetables, eggs, dairy products, and beverages (Miura et al. 2003).

It has been reported that it is possible to predict the fat composition in fish or minced meat using electromagnetic measurements (Kent 1990; Kent et al. 1993; Borgaard et al. 2003). The fat content in these foods is clearly related to the water content of the product, so that if one is known the other can be determined. A microwave instrument that mainly consists of a microstripline is currently being marketed (Distell Company, West Lothian, Scotland). This compact and nondestructive meter can measure the lipid content of certain kinds of fish, meat, and poultry products.

Another important application of microwaves in foods is to analyze fish and meat freshness. After death, muscle is not able to utilize energy by the respiratory system. Glycolysis is a method of creating energy by converting glycogen to lactate. Postmortem changes lead to a temporary rigidity of muscles. Otherwise, glycolysis lowers the pH, bringing it closer to its isoelectric point and decreasing the water-holding capacity (Hullberg 2004). The level of glycogen stored in the animal at the time of slaughter affects the texture of the future marketed meat. For all these reasons, during rigor mortis the dielectric properties are expected to change (Datta et al. 2005). A microwave polarimetric method was used to follow the changes in muscle structure during bovine meat aging (Clerjon and Damez 2005). Promising studies have been carried out to evaluate the freshness of fish products. Haddock muscle showed significant changes of its dielectric properties during rigor mortis at radio frequencies between 1 Hz and 100 KHz (Martisen et al. 2000). Kent et al. (2004b) studied the effect of storage time and temperature on the dielectric properties of thawed frozen cod (*Gadus morhua*) in order to estimate the quality of this product. The same year, Kent et al. (2004a) developed a combination of dielectric spectroscopy and multivariate analysis to determine the quality of chilled Baltic cod (*G. morhua*). These studies yielded a prototype developed by SEQUID (Seafood Quality Identification) (Knöchel et al. 2004; Kent el at 2005b) for measuring and analyzing the quality of different seafoods. The SEQUID project concentrated on the measurement of the dielectric properties of fish tissue as a function of time both in frozen and chilled storage. This project has shown that it is possible, using a combination of time domain reflectometry and multivariate analysis, to predict certain quality-related variables, both sensory and biochemical, with an accuracy comparable to existing methods. Kent et al. (2005a) have also reported a way to determine the quality of frozen hake (*Merluccius capensis*) by analyzing its changes in microwave dielectric properties. A sensor for measuring freshness in fish is already on the market (Distell Company®, West Lothian, Scotland).

The determination of added water in fish, fish products, and meat using microwave dielectric spectra was widely studied by Kent et al. (2000, 2001, 2002). The added glaze on frozen foods such as cooked and peeled prawns was determined by measuring changes in the dielectric properties (Kent and Stroud 1999).

De los Reyes et al. (2005c) verified the viability of an on-line measurement system using low-power microwaves to determine the desalting point of salted cod. Dielectric spectroscopy was performed on cod samples at different desalting stages and on its

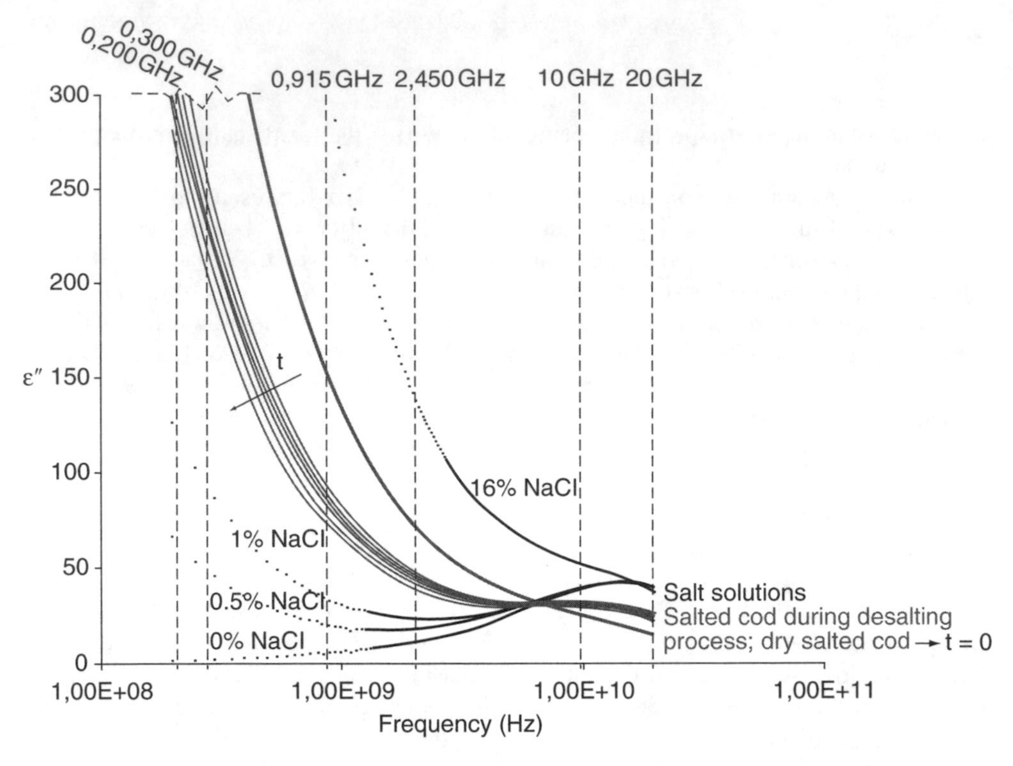

Fig. 3.17. Dielectric spectroscopy on cod samples at different desalting stages (gray continuous lines) and on salt solutions (black dotted lines). Discontinuous vertical lines mark some single frequencies, and t indicates desalting time, which increases in the arrow direction. The frequency is given in logarithm scale.

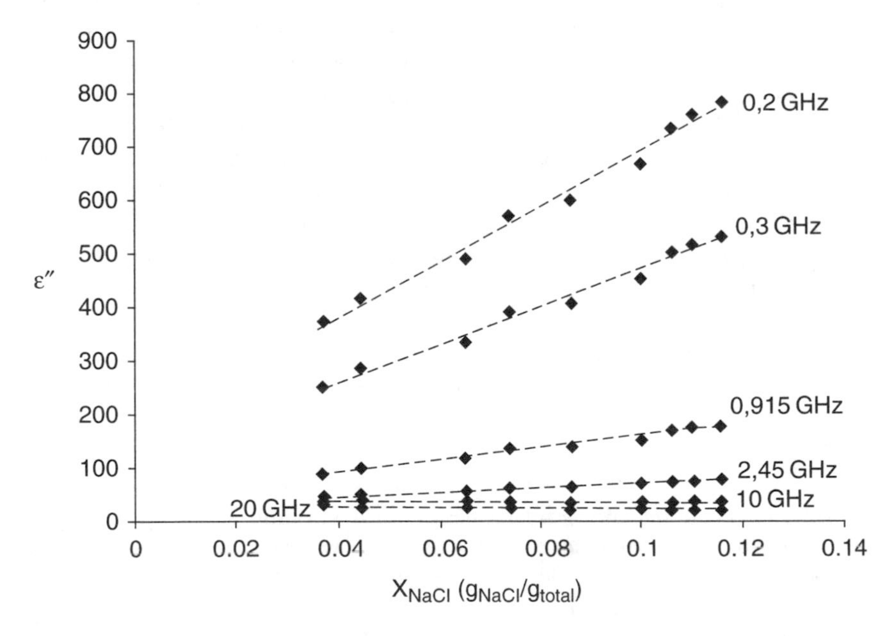

Fig. 3.18. Loss factor values versus sodium chloride content for each single frequency marked in figure 3.17.

desalting solutions in order to find an appropriate measurement frequency (or frequency range) (fig. 3.17).

Optimum frequencies were selected and dielectric properties data were related to other physico-chemical properties of cod samples measured at the same desalting stages, such as moisture and salt content. Good correlations (approximately $R^2 = 0,99$) were found between salt content in cod samples and its loss factor values at 200 and 300 MHz. These results indicated the viability of developing an on-line control system for the cod desalting process (fig. 3.18).

Applications of Dielectric Properties in Fruits and Vegetables

The dielectric properties of various fruits and vegetables have been reported many times (Tran et al. 1984; Nelson 1982; Seaman and Seals 1991; Nelson et al. 1993, 1994; Kuang and Nelson 1997; Sipahioglu and Barringer 2003).

A recent study on the frequency and temperature dependence of the permittivity of fresh fruits and vegetables was reported in ISEMA 2005 (Nelson 2005). Dielectric properties were also measured in fruits such as apples or citric fruits and related to process variables for a posterior implementation of an on-line quality control system (Romero et al. 2004; De los Reyes et al. 2005a, 2005b). A summary of dielectric properties of different fruits is shown in table 3.4.

Other studies were recently carried out to correlate dielectric properties with some interesting variables of vacuum-impregnated squash. Complex permittivity of fresh squash samples, sucrose and sodium chloride solutions, and squash samples vacuum impregnated with these solutions were measured with a coaxial probe (HP85070E) connected to a networks analyzer (E8362B) by De los Reyes et al. (2005a). ε' and ε'' values of sucrose and sodium chloride solutions were correlated with water activity. The authors found that the same equations can also be used to correlate permittivity and water activity of vacuum-impregnated squash samples. Structural and physico–chemical properties of these samples were also determined and qualitatively related to dielectric properties.

De los Reyes et al. (2005b) studied the suitability of using low-power microwaves for on-line nondestructive measurement of dielectric properties of citric fruits. The authors tried to relate the dielectric properties to process variables. The dielectric properties of citric fruits were measured using a coaxial probe and a network analyzer in the range of 800 MHz to 10 GHz frequencies. The process variables measured were water content, density, Brix, ash content, and water activity. Spectral analysis revealed good correlation among the different measuring frequencies with certain process properties, especially free salts and free water.

Dairy Products

The dielectric properties of dairy products have hardly been studied. Rzepecka and Pereira (1974) studied the dielectric properties of whey and skimmed-milk powders. Mudgett et al. (1974, 1980) modeled the dielectric properties of aqueous solutions of nonfat dried milk. Representative dielectric properties of milk and its constituents are given in table 3.5.

Other studies on dielectric properties of butter have been reported (Sone et al. 1970; Rzepecka and Pereira 1974). The permittivity value decreased rapidly at temperatures

Table 3.4. Permittivities of fresh fruits and vegetables at 23°C.

Fruit/Vegetable	Moisture Content %(w.b.)[a]	Tissue Density (g cm^{-3})	Dielectric Constant (ε′) (frequency, MHz)		Dielectric Loss Factor (ε″) (frequency, MHz)	
			915	2450	915	2450
Apple	88	0.76	57	54	8	10
Avocado	71	0.99	47	45	16	12
Banana	78	0.94	64	60	19	18
Cantaloupe	92	0.93	68	66	14	13
Carrot	87	0.99	59	56	18	15
Cucumber	97	0.85	71	69	11	12
Grape	82	1.10	69	65	15	17
Grapefruit	91	0.83	75	73	14	15
Honeydew	89	0.95	72	69	18	17
Kiwifruit	87	0.99	70	66	18	17
Lemon	91	0.88	73	71	15	14
Lime	90	0.97	72	70	18	15
Mandarin juice	88	0,96	68	65	12	14
Mango	86	0.96	64	61	13	14
Onion	92	0.97	61	64	12	14
Orange	87	0.92	73	69	14	16
Papaya	88	0.96	69	67	10	14
Peach	90	0.92	70	67	12	14
Pear	84	0.94	67	64	11	13
Potato	79	1.03	62	57	22	17
Radish	96	0.76	68	67	20	15
Squash	95	0.70	63	62	15	13
Strawberry	92	0.76	73	71	14	14
Sweet potato	80	0.95	55	52	16	14
Tomato	91	1.02	75	71	18	16
Turnip	92	0.89	63	61	13	12

Source: Adapted from Nelson et al. 1994; Betoret et al. 2004; De los Reyes et al. 2005a, 2005b, 2005d.
[a]w.b. = wet basis.

below freezing point due to the fact that free water crystallizes. On the other hand, above 30°C the permittivity increase might be due to the disintegration of the emulsion (Venkatesh and Raghavan 2004). A linear increase in dielectric constant with moisture content was shown by Prakash and Armstrong (1970) in a small range of moisture content. A method for evaluating the moisture and salt contents of salted butter was presented by Doi et al. (1991). The dielectric properties of salted water-in-oil (W/O) emulsion as a model of salted butter were studied by Shiinoki et al. (1998). The authors reported the relationship between the dielectric properties of a W/O emulsion and such microwave properties, and discussed the viability of the microwave transmission method for monitoring on-line the moisture and salt contents in a continuous salted-butter-making process. Dielectric properties of cheese were studied by some authors (Datta et al. 2005; Green 1997; Herve et al. 1998). It was shown that the dielectric properties of cheese are dependent on the composition. Dielectric properties showed an increase when moisture content

Table 3.5. Dielectric properties of milk and its constituents at 2.45 GHz and 20°C.

Description	Fat (%)	Protein (%)	Lactose (%)	Moisture (%)	ε′	ε″
1% Milk	0.94	3.31	4.93	90.11	70.60	17.60
3.25% Milk	3.17	3.25	4.79	88.13	68.00	17.60
Water + lactose I	0.00	0.00	4.00	96.00	78.20	13.80
Water + lactose II	0.00	0.00	7.00	93.00	77.30	14.40
Water + lactose III	0.00	0.00	10.00	90.00	76.30	14.90
Water + sodium caseinate I	0.00	3.33	0.00	96.67	74.60	15.50
Water + sodium caseinate II	0.00	6.48	0.00	93.62	73.00	15.70
Water + sodium caseinate III	0.00	8.71	0.00	91.29	71.40	15.90
Lactose (solid)	0.00	0.00	100.00	0.00	1.90	0.00
Sodium caseinate (solid)	0.00	100.00	0.00	0.00	1.60	0.00
Milk fat (solid)	100.00	0.00	0.00	0.00	2.60	0.20
Water (distilled)	0.00	0.00	0.00	100.00	78.00	13.40

Source: Adapted from Kudra et al. 1992.

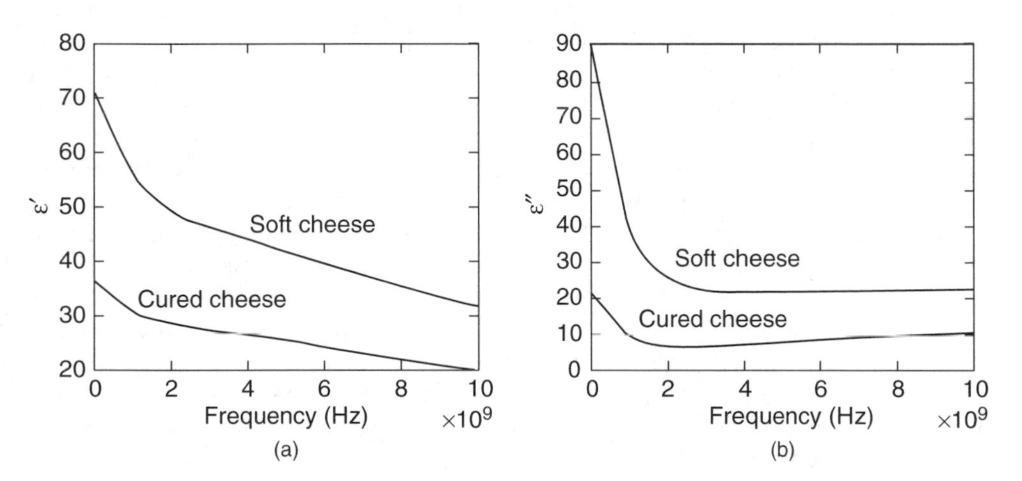

Fig. 3.19. Dielectric constant **(a)** and loss factor **(b)** versus frequency of soft and cured cheese (adapted from Catalá Civera 1999).

increased (Green 1997). Dielectric properties of cottage cheese were studied by Herve et al. (1998) in order to extend the shelf life of the product with microwave treatment. The authors concluded that the cheese with the highest fat content had the lowest dielectric constant. This might be due to the fact that fat content is related to the moisture content in a negative way, and a lower dielectric constant yields a lower dielectric constant. Datta et al. (2005) showed the dielectric constant and loss factor of processed cheese at different compositions for temperatures of 20°C and 70°C. The authors concluded that the dielectric properties of processed cheese are not generally temperature dependent.

Catalá Civera (1999) found different values for the dielectric properties of cured and soft cheese, suggesting that microwave sensors can be a good method of controlling curing processes. In figure 3.19 dielectric constant and loss factor spectra of cured and soft cheese are shown. Dielectric values of soft cheese are higher than cured ones since soft cheese has higher moisture content.

Future Research Topics

As we have seen throughout this chapter, relatively high-potency microwaves can interact with materials and produce changes in them. Therefore, microwave energy can be used as primary energy to produce changes similar to those made by thermal, mechanical, or infrared energy. Obviously microwave energy, alone or combined with other energy sources, has advantages in food technology, such as speed, cleanliness, automation capability, environmental respect (it does not produce CO_2 because there is no combustion process), and, on occasions, energy saving. The processing of food materials by microwave may be designed as a "tailor-made" process that depends on a lot of food characteristics, such as aggregation state (solid, liquid, semiliquid, etc.), size and shape (grains, pastes, whole fruits, cut fruits, etc.), porosity, and so forth. Therefore, the design of an industrial treatment is a multidisciplinary task that requires the use of hybrid engineering equipment.

The first step is to study the dielectric properties of pretreated food versus frequency. For this purpose only the ISM bands (industrial, scientific, and medical bands) can be used—400 MHz, 900 MHz, and 2450 MHz being the most important frequencies. When the microwave treatment band has been chosen, the dynamic studies are performed, both in continuous wave (CW) and in pulsed state. Both the continuous monitoring and the feedback interaction must be performed by microwave sensors, which are the transmitters of the current state of the material, whether it has been treated or not.

In the case of microwave sensors, we are not limited to working in a specific band because such sensors basically behave as receptors, working at a low signal so they do not interfere with the treatment. In most cases scattering parameters are measured, and it is possible to deduce from them the real and imaginary part of the dielectric constant. Figures 3.1 and 3.2 show that the variations between ε' and ε'' with frequency are the response to different phenomena; these phenomena are sometimes detachable. In conclusion, it is clear that future work will lead to multisensors, each one selective to one variable, and as a whole multivariable.

Multivariable knowledge of the process yields a modeling of the material so that its behavior under different treatments can be simulated. There is a lot of work to be done on this issue. Nevertheless, the main guidelines have been traced out:

- Multitreatment
- Multisensor
- Modeling
- Feedback to find the optimum processing

References

Alandes, C., De los Reyes, R., Betoret, N., Romero, J., Andrés, A., Fito, P. 2004a. "Propiedades dieléctricas de muestras de calabaza (c. *Maxima duchesne*) fresca e impregnada: Influencia de su localización en el fruto." In *Procedins of III Congreso Español De Ingeniería De Alimentos,* Pamplona, Spain.

Alandes, C., Betoret, N., De los Reyes, R., Romero, J., Andrés, A., Fito, P. 2004b. "Características microestructurales y capacidad de impregnación de cilindros de calabaza (c. *Maxima duchesne*): Efecto de la localización y del tiempo de vacío." In *Procedins of III Congreso Español De Ingeniería De Alimentos,* Pamplona, Spain.

Bengtsson, N.E., Risman, P.O. 1971. Dielectric properties of food at 3 GHz as determined by a cavity perturbation technique: II. Measurements on food materials. *Journal of Microwave Power* 6:107–23.

Bengtsson, N.E., Melin, J., Remi, K., Soderlind, S. 1963. Measurements of the dielectric properties of frozen and defrosted meat and fish in the frequency range 10–200 MHz. *Journal of the Science of Food and Agriculture* 14:593–604.

Betoret, N., De los Reyes, R., Romero, J., Andrés, A., Fito, P. 2004. "Dielectric and structural properties of squash (C. *maxima Duchesne*)." In *Proceedings of EFFOST,* Warsaw, Poland.

Bilbao Sáinz, C. 2002. "Estudio del secado combinado aire/microondas en manzana Granny Smith." Tesis doctoral de la Universidad Politécnica de Valencia.

Bircan, C., Barringer, S.A. 2002. Determination of protein denaturation of muscle foods using dielectric properties. *Journal of Food Science* 67(1):202–5.

Borgaard, C., Christensen, L.B., Jespersen, Bo L. 2003. "Reflection mode microwave spectroscopy for on-line measurement of fat in trimmings." In *49th ICoMST,* August 31–September 5, Campinas, Brazil.

Catalá Civera, J.M. 1999. "Estudio de estructuras guiadas monomodo para aplicaciones de caracterización dieléctrica de materiales y curado de compuestos elastoméricos a frecuencias de microondas." Tesis doctoral de la Universidad Politécnica de Valencia.

Chamchong, M., Datta, A.K. 1999. Thawing of foods in a microwave oven: I. Effect of power levels and power cycling. *Journal of Microwave Power and Electromagnetic Energy* 34:9–21.

Clerjon, S., Damez, J.L. 2005. "Microwave sensing for food structure evaluation." In *Proceedings of ISEMA 2005, 6th International Conference on Electromagnetic Wave Interaction with Water and Moist Substances,* edited by Kupfer, K., pp. 357–64. MFPA an der Bauhaus—Universität Weimar, May 29 to June 1, Weimar, Germany.

Clerjon, S., Daudin, J.D., Damez, J.L. 2003. Water activity and dielectric properties of gels in the frequency range 200 MHz–6 GHz. *Food Chemistry* 82:87–97.

Datta, A.K., Ananthcswaran, R.C. 2001. *Handbook of Microwave Technology for Food Applications,* edited by Datta, A.K., and Anantheswaran, R.C., series of Food Science and Technology. Marcel Dekker, New York.

Datta, A.K., Sumnu, G., Raghavan, G.S.V. 2005. "Dielectric properties of foods." In *Engineering Properties of Foods,* edited by Rao, M.A., Rizvi, S.S.H., and Datta, A.K., 3rd edition, series of Food Science and Technology, pp.501–65. CRC Press, New York.

Debye, P. 1929. *Polar Molecules.* Chemical Catalogue Co., New York.

Decareau, R.V. 1985. *Microwaves in the Food Processing Industry,* series of Food Science and Technology. Academic Press, New York.

Decareau, R.V., Peterson, R.A. 1986. *Microwave Processing and Engineering.* Ellis Horwood Series in Food Science and Technology. Ellis Horwood, Chichester, UK.

de Loor, G.P., Meijboom, F.W. 1966. The dielectric constant of foods and other materials with high water contents at microwave frequencies. *Journal of Food Technology* 1:313–22.

De los Reyes, R., Romero, J., Betoret, N., Andrés, A., De los Reyes, E., Fito, P. 2005a. "Dielectric properties of fresh and vacuum impregnated squash (*C. moschata Duchesne*): The role of fruit microstructure." In *Proceedings of EFCE,* Glasgow, Scotland.

De los Reyes, R., Fito, P.J., De los Reyes, E., Fito, P. 2005b. "Dielectric properties determination in citric fruits in relation with process variables," In *Proceedings of EFFOST,* Valencia, Spain.

De los Reyes, R., Haas, C., Andrés, A. 2005c. "Changes in the dielectric properties of 'salted cod–water' system during the desalting process and their relation with other physical properties." In *Proceedings of EFFOST,* Valencia, Spain.

De los Reyes, R., Andrés, A., Heredia, A., Khan, M., Betoret, N., De los Reyes, E., Fito, P. 2005d. "Measurements of dielectric properties of fresh and osmotized fruits: Implications on combined hot air-microwave drying." In *Proceedings of EFCE,* Glasgow, Scotland.

Doi, T., Kanzaki, M., Watanabe, T., Nakamuma, H., Shibuya, M., Matsumoto, K. 1991. Non-destructive continuous and simultaneous estimation of moisture and salts contents in butter by the combination of specific gravity and dielectric constant. *Nippon Shokuhin Kogyo Gakkaishi* 38(10):904–9.

Fear, E.C., Hagness, S.C., Meaney, P.M., Okoniewski, M., Stuchly, M.A. 2002. Enhancing breast tumor detection with near-field imaging. *Microwave Magazine,* pp. 48–56.

Fito, P., Chiralt, A., Martín, M.E. 2004. "Current state of microwave applications to food processing." In *Novel Food Processing Technologies,* edited by Barbosa-Canovas, G.V., Tapia, M.S., and Cano, M.P., pp. 1–16. CRC Press, Boca Raton.

Funk, D.B., Gillay, Z., Meszaros, P. 2005. "A unified moisture algorithm for improved RF dielectric grain moisture measurements." In *Proceedings of ISEMA 2005, 6th International Conference on Electromagnetic Wave Interaction with Water and Moist Substances*, edited by Kupfer, K., pp. 99–106. MFPA an der Bauhaus-Universität Weimar. May 29–June 1, Weimar, Germany.

Gabriel, C., Gabriel, S., Corthout, E. 1996. The dielectric properties of biological tissues: I. Literature survey. *Physics in Medicine and Biology* 41:2231–49.

Green, A.D. 1997. Measurements of the dielectric properties of cheddar cheese. *Journal of Microwave Power and Electromagnetic Energy* 32:16–27.

Herve, A.G., Tang, J., Luedecke, L., Feng, H. 1998. Dielectric properties of cottage cheese and surface treatment using microwaves. *Journal of Food Engineering* 37:389–410.

Hewlett-Packard. 1992a. "Basics of measuring the dielectric properties of materials: Application note 1217-1."

———. 1992b. "HP dielectric materials measurements, solutions catalogue of fixtures and software: Complete solutions for dielectric materials measurements: Application note."

Hullberg, A. 2004. "Quality of processed pork: Influence of RN genotype and processing conditions." Doctoral thesis, Swedish University of Agricultural Sciences, Uppsala, Sweden.

Joshi, K. 2005. "High resolution, non-destructive and in-process time domain aquametry for FMCG and other products using microstrip sensors." In *Proceedings of ISEMA 2005, 6th International Conference on Electromagnetic Wave Interaction with Water and Moist Substances*, edited by Kupfer, K., pp. 384–390. MFPA an der Bauhaus-Universität Weimar. May 29–June 1, Weimar, Germany.

Kaatze, U. 1989. Complex permittivity of water as a function of frequency and temperature. *Journal of Chemical and Engineering Data* 34:371–74.

Kent, M. 1972. Microwave dielectric properties of fish meal. *Journal of Microwave Power* 7:109–16.

———. 1977. Complex permittivity of fish meal: A general discussion of temperature, density, and moisture dependence. *Journal of Microwave Power* 12:341–45.

———. 1990. Hand-held instrument for fat/water determination in whole fish. *Food Control* 1:47–53.

———. 2001. "Microwave measurements of product variables." In *Instrumentation and Sensors for the Food Industry,* 2nd edition, edited by Kress-Rogers, E., and Brimelow, C.J.B., pp.233–279. Woodhead Publishing in Food Science and Technology, CRC Press, New York.

Kent, M., Stroud, G. 1999. A new method for the measurement of added glaze on frozen foods. *Journal of Food Engineering* 39:313–21.

Kent, M., Lees, A., Christie, R.H. 1993. Estimation of the fat content of minced meat using a portable microwave fat meter. *Food Control* 4:222–25.

Kent, M., MacKenzie, K., Berger, U.K., Knöchel, R., Daschner, F. 2000. Determination of prior treatment of fish and fish products using microwave dielectric spectra. *European Food Research and Technology* 210:427–33.

Kent, M., Knöchel, R., Daschner, F., Berger, U.K. 2001. Composition of foods including added water using microwave dielectric spectra. *Food Control* 12:467–82.

Kent, M., Peymann, A., Gabriel, C., Knight, A. 2002. Determination of added water in pork products using microwave dielectric spectroscopy. *Food Control* 13:143–49.

Kent, M., Oehlenschlager, J., Mierke-Klemeyer, S., Manthey-Karl, M., Knöchel, R., Daschner, F., Schimmer, O. 2004a. A new multivariate approach to the problem of fish quality estimation. *Food Chemistry* 87:531–35.

Kent, M., Oehlenschlager, J., Mierke-Klemeyer, S., Knöchel, R., Daschner, F., Schimmer, O. 2004b. Estimation of the quality of frozen cod using a new instrumental method. *European Food Research and Technology* 219:540–44.

Kent, M., Knöchel, R., Daschner, F., Schimmer, O., Tejada, M., Huidobro, A., Nunes, L., Batista, I., Martins, A. 2005a. Determination of the quality of frozen hake using its microwave dielectric properties. *International Journal of Food Science and Technology* 40:55–65.

Kent, M., Knöchel, R., Daschner, F., Schimmer, O., Albrechts, C., Oehlenschläger, J., Mierke- Klemeyer, S., Kroeger, M., Barr, U., Floberg, P., Tejada, M., Huidobro, A., Nunes, L., Martins, A., Batista, I., Cardoso, C. 2005b. "Intangible but not intractable: The prediction of food 'quality' variables using dielectric spectroscopy." In *Proceedings of ISEMA 2005, 6th International Conference on Electromagnetic Wave Interaction with Water and Moist Substances*, edited by Kupfer, K., pp. 347–56. MFPA an der Bauhaus-Universität Weimar, May 29–June 1, Weimar, Germany.

Knöchel, R., Barr, U.K., Tejada, M., Nunes, M.L., Oehlenschläger, J., Bennink, D. 2004. Newsletter of the SEQUID (Seafood Quality Identification) project. European Commission Framework Programme V, Quality of Life and Management of Living Resources, RTD Project QLK 1-2001-01643.

Kraszewski, A. 1980. Microwave aquametry. *Journal of Microwave Power* 15: 209–20.
Kuang, W., Nelson, S.O. 1997. Dielectric relaxation characteristic of fresh fruits and vegetables from 3 to 20 GHz. *Journal of Microwave Power and Electromagnetic Energy* 32:114–22.
Kudra, T., Raghavan, G.S.V., Akyel, C., Bosisio, R., van de Voort, F.R. 1992. Electromagnetic properties of milk and its constituents at 2.45 GHz. *International Microwave Power Institute Journal* 27(4):199–204.
Lisovsky, V.V. 2005. "Automatic control of moisture in agricultural products by methods of microwave aquametry." In *Proceedings of ISEMA 2005, 6th International Conference on Electromagnetic Wave Interaction with Water and Moist Substances*, edited by Kupfer, K., pp. 375–83. MFPA an der Bauhaus-Universität Weimar, May 29–June 1, Weimar, Germany.
Lyng, J.G., Zhang, L., Bruton, N.P. 2005. A survey of the dielectric properties of meats and ingredients used in meat product manufacture. *Meat Science* 69:589–602.
Martín Esparza, M.E. 2002. "Utilización de microondas en el secado por aire caliente de manzana (var. Granny Smith): Influencia del pretratamiento por impregnación a vacío." Tesis doctoral de la Universidad Politécnica de Valencia.
Martinez, V., Hinarejos, M.J., Romero, J.B., De los Reyes, R., De los Reyes, E., Fito, P. 2002. "Determinación de propiedades físicas En cítricos a partir de medidas de propiedades electromagnéticas." In *Actas Del 2° CESIA,* Lérida, Spain.
Martisen, O.G., Grimnes, S., Mirtaheri, P. 2000. Noninvasive measurements of post-mortem changes in dielectric properties of haddock muscle—a pilot study. *Journal of Food Engineering* 43:189–92.
Metaxas, A.C., Meredith, R.J. 1993. *Industrial Microwave Heating,* IEE Power Engineering series 4. Peter Peregrinus LTD, London.
Miura, N., Yagihara, S., Mashimo, S. 2003. Microwave dielectric properties of solid and liquid foods investigated by time-domain reflectometry. *Journal of Food Science* 68(4):1396–1403.
Mudgett, R.E. 1995. "Electrical properties of foods." In *Engineering Properties of Foods,* edited by Rao, M.A., and Rizvi, S.S.H., 2nd edition, pp. 389–455. Marcel Decker, New York.
Mudgett, R.E., Smith, A.C., Wang, D.I.C., Goldblith, S.A. 1974. Prediction of dielectric properties in nonfat milk at frequencies and temperatures of interest in microwave processing. *Journal of Food Science* 39: 52–54.
Mudgett, R.E., Goldblith, S.A., Wang, D.I.C., Westphal, W.B. 1980. Dielectric behaviour of a semisolid food at low, intermediate, and high moisture content. *Journal of Microwave Power* 15:27–36.
Ndife, M., Sumnu, G., Bayindrli, L. 1998. Dielectric properties of six different species of starch at 2450 MHz. *Food Research International* 31:43–52.
Nelson, S.O. 1977. Use of electrical properties for grain-moisture measurement. *Journal of Microwave Power* 12(1):67–72.
———. 1978. Radio frequency and microwave dielectric properties of shelled corn. *Journal of Microwave Power* 13:213–18.
Nelson, S.O. 1982. Dielectric properties of some fresh fruits and vegetables at frequencies of 2.45 to 22 GHz. *ASAE Paper 82-3053.*
———. 1991. Dielectric properties of agricultural products—measurements and applications. *Digest of Literature on Dielectrics,* edited by de Reggie, A. *IEEE Trans. Electr. Insul.* 26(5):845–69.
———. 1998. "Dielectric properties measurement techniques and applications." ASAE Annual Int. Meeting, Florida. Paper 98-3067.
———. 1999. Dielectric properties measurement techniques and applications. *Transactions of the ASAE* 42(2):523–29.
———. 2005. "Frequency and temperature dependence of the permittivity of fresh fruits and vegetables." In *Proceedings of ISEMA 2005, 6th International Conference on Electromagnetic Wave Interaction with Water and Moist Substances*, edited by Kupfer, K., pp. 391–398. MFPA an der Bauhaus-Universität Weimar, May 29–June 1, Weimar, Germany.
Nelson, S.O., Prakash, A., Lawrence, K. 1991. Moisture and temperature dependence of the permittivities of some hydrocolloids at 2.45 GHz. *Journal of Microwave Power and Electromagnetic Energy* 26:178–85.
Nelson, S.O., Forbus Jr., W.R., Lawrence, K.C. 1993. Microwave permittivities of fresh fruits and vegetables from 0.2 to 20 GHz. *Transactions of the ASAE* 37(1):183–89.
Nelson, S.O., Forbus Jr., W.R., Lawrence, K.C. 1994. Permittivity of fresh fruits and vegetables from 0.2 to 20 GHz. *Journal of Microwave Power and Electromagnetic Energy* 29:81–93.
Nelson, S.O., Trabelsi, S., Kraszewski, A.W. 1998. Advances in sensing grain moisture content by microwave measurements. *Transactions of the ASAE* 41(2):483–87.

Ohlsson, T., Henriques, M., Bengtsson, N.E. 1974. Dielectric properties of model meat emulsions at 900 and 2800 MHz in relation to their composition. *Journal of Food Science* 39:1153–56.

Okamura, S., Tsukamoto, S. 2005. "New sensor for high moist leaves in green tea roduction." In *Proceedings of ISEMA 2005, 6th International Conference on Electromagnetic Wave Interaction with Water and Moist Substances*, edited by Kupfer, K., pp. 340–46. MFPA an der Bauhaus-Universität Weimar, May 29–June 1, Weimar, Germany.

Pace, W.E., Westphal, W.B., Goldblith, S.A. 1968. Dielectric properties of commercial cooking oils. *Journal of Food Science* 33:30–36.

Prakash, S., Armstrong, J.G. 1970. Measurement of the dielectric constant of butter. *Dairy Industries* 35(10):688–89.

Roebuck, B.D., Goldblith, S.A. 1972. Dielectric properties of carbohydrate-water mixtures at microwave frequencies. *Journal of Food Science* 37:199–204.

Romero, J.B., De los Reyes, R., Betoret, N., Barrera, C., De los Reyes, E., Fito, P. 2004. Dielectric properties of vacuum-impregnated apple (Granny Smith). In 9th International Congress on Engineering and Food (ICEF 9), Montpellier, France.

Rzepecka, M.A., Pereira, R.R. 1974. Permittivity of some dairy products at 2450 MHz. *Journal of Microwave Power* 9:277–88.

Seaman, R., Seals, J. 1991. Fruit pulp and skin dielectric properties for 150 MHz to 6400 MHz. *Journal of Microwave Power and Electromagnetic Energy* 26:72–81.

Shiinoki, Y., Motouri, Y., Ito, K. 1998. On-line monitoring of moisture and salt contents by the microwave transmission method in a continuous salted butter-making process. *Journal of Food Engineering* 38:153–67.

Sipahioglu, O., Barringer, S.A. 2003. Dielectric properties of vegetables and fruits as a function of temperature, ash, and moisture content. *Journal of Food Science* 68(1):234–39.

Sipahioglu, O., Barringer, S.A., Taub, I., Yang, A.P.P. 2003a. Characterization and modeling of dielectric properties of turkey meat. *Journal of Food Science* 68(2):521–27.

Sipahioglu, O., Barringer, S.A., Taub, I., Prakash, A. 2003b. Modeling the dielectric properties of ham as a function of temperature and composition. *Journal of Food Science* 68(3):904–9.

Sone, T., Taneya, S., Handa, M. 1970. Dielectric properties of butter and their application for measuring moisture content during continuous processing. 18th International Dairy Congress. IE:221 (Food Science and Technology Abstract 12P1690 2 [12]).

Sucher, M., Fox, J. 1963. *Handbook of Microwave Measurements.* Polytechnic Press of the Polytechnic Institute of Brooklyn, Brooklyn, New York.

To, E.C., Mudgett, R.E., Wang, D.I.C., Goldblith, S.A. 1974. Dielectric properties of food materials. *Journal of Microwave Power* 9:303–15.

Trabelsi, S., Nelson, S.O. 2005a. "Microwave dielectric properties of cereal grain and oilseed." In *Proceedings of the American Society of Agricultural Engineers,* St. Joseph, Michigan. Paper no. 056165.

———. 2005b. "Microwave dielectric methods for rapid, nondestructive moisture sensing in unshelled and shelled peanuts." In *Proceedings of the American Society of Agricultural Engineers,* St. Joseph, Michigan. Paper no. 056162.

———. 2005c. "Universal microwave moisture sensor." In *Proceedings of ISEMA 2005, 6th International Conference on Electromagnetic Wave Interaction with Water and Moist Substances*, edited by Kupfer, K., pp. 232–35. MFPA an der Bauhaus-Universität Weimar, May 29–June 1, Weimar, Germany.

Trabelsi, S., Kraszewski, A.W., Nelson, S.O. 1998. Nondestructive microwave characterization for bulk density and moisture content determination in shelled corn. *Measurements Science and Technology* 9:1548–56.

Tran, V.N., Stuchly, S.S., Kraszewski, A. 1984. Dielectric properties of selected vegetables and fruits 0.1–10.0 GHz. *Journal of Microwave Power* 19:251–58.

Venkatesh, M.S., Raghavan, G.S.V. 2004. An overview of microwave processing and dielectric properties of agri-food materials. *Biosystems Engineering* 88(1):1–18.

Vicedo, V., Barrera, C., Betoret, N., De los Reyes, R., Romero, J.B., Fito, P. 2002. "Impregnación a vacío de manzana Granny Smith con disoluciones de sacarosa enriquecidas en Ca y Fe: Efecto sobre las propiedades dieléctricas." In *Proceedings of IberDESH,* Valencia, Spain.

Wei, Y., Chiang, P., Sridhar, S. 1992. Ion size effects on the dynamic and dielectric properties of aqueous alkali solutions. *Journal of Chemical Physics* 96(6):4569–73.

Wu, H., Kolbe, E., Flugstad, B., Park, J.W., Yongsawatdigul, J. 1988. Electrical properties of fish mince during multifrequency ohmic heating. *Journal of Food Science* 63:1028–32.

Zhang, L., Lyng, J.G., Brunton, N., Morgan, D., McKenna, B. 2004. Dielectric and thermophysical properties of meat batters over a temperature range of 5–85°C. *Meat Science* 68:173–84.

Zheng, M., Huang, Y.W., Nelson, S.O., Bartley, P., Gates, K.W. 1998. Dielectric properties and thermal conductivity of marinated shrimp and channel catfish. *Journal of Food Science* 63:668–72.

Web Sites

http://www.kha.co.nz/
http://microradar.com/
http://www.thermo.com/

Acknowledgments

We would like to thank the "Consellería de Agricultura, Pesca y Alimentación," of the Valencia government for its economic support and to the R + D + i Linguistic Assistance Office at the Universidad Politécnica of Valencia for their help in revising this chapter.

4 Ultrasounds for Quality Assurance

Bosen Zhao, Otman A. Basir, and
Gauri S. Mittal

Introduction

The ultrasound is energy generated by sound waves of ≥20 kHz. In practice, the frequency used in ultrasonic techniques varies from dozens of kHz to dozens of MHz depending on the objective. Ultrasound is an elastic wave propagating in mediums (gas, liquid, solid, or multiphased complex systems). Ultrasound propagation (velocity and amplitude) is influenced by the medium's characteristics. This attribute is used for material characterization and for food quality assurance, including evaluations of texture, viscosity, density, and particle size. Being elastic waves, ultrasound can be longitudinal waves or shear waves according to the movement of particles during the propagation of the waves. However, not all of the materials experience shear waves: for example, gas and water are considered non-shear-wave bearing. There are at least two representation methods for ultrasound signals: A-scan (amplitude-based, one-dimensional representation) and B-scan (brightness-based, two-dimensional gray-level imaging).

Ultrasound can be reflected and refracted following Snell's law if its wavelength is smaller than the characteristic dimension of the object it encounters in propagation. This attribute is often used for crack, void, and enclosure detection in product inspection. These flaws have different characteristic impedance than the product reflecting ultrasound energy. Characteristic impedance is defined as a product of sound speed and the density of the material (Pierce 1981). To detect those defects of size in the order of millimeters, the ultrasound frequency used for inspection is in the order of megahertz, considering the sound velocity varies from about 1500 to 6000 m/s in liquids and solids. Higher frequency has better resolution in defect detection. In the past few years, there has been a growing interest in, and an increasing amount of research using, the reflection signal for foreign bodies (FBs) or contaminant detection in foods (Zhao et al. 2003a). FB stands for a piece of undesirable solid matter present in a food product, such as a glass fragment in a beverage bottle and metal or bone fragments in meat products. Table 4.1 provides a list of ultrasound applications and findings in food quality assurance detection since the late 1990s.

An ultrasound is applied by using low or high intensity. Low-intensity ultrasound contains <1 W/cm^2, while high intensity ultrasound typically ranges from 10 to 1000 W/cm^2 (McClements 1995). The high-intensity ultrasound is generally used in equipment cleaning, homogenizers, and emulsifiers in food processing by virtue of its sonication effect (Kress-Rogers 2001). The low-intensity ultrasound is mostly used for food quality assurance, including food characterization and FB detection in packed foods. This chapter discusses advances made in food quality assurance using ultrasound.

Table 4.1. Overview of ultrasonic applications in food quality assurance.

Quality Parameters	Food	US Properties	Mode	Frequency	References
Firmness	Avocado	α, V	P/E, contact	50 kHz	Mizrach and Flitsanov 1999; Flitsanov et al. 2000; Mizrach 2000; Mizrach et al. 2000
	Plum	α	P/E, contact	50 kHz	Mizrach 2004
	Tomato	α	P/E, contact	50 kHz	Verlinden et al. 2004
	Mango	*f*, A	C, intrusive	46–52 kHz	Valente and Ferrandis 2003
	Bread, wafer, crackers	A	C, noncontact	5–35 kHz	Juodeikiene and Basinskiene 2004
Rib-eye area	Beef		B-scan	3.5 MHz	Fernandes et al. 2001; Greiner 2003; BIF 2004
Backfat thickness	Beef, pork		B-scan	3.5 MHz	Fernandes et al. 2001; Greiner 2003; BIF 2004; Ishmael 2003; Brondum et al. 1998; Gresham 2005; Youssao et al. 2002; Morlein et al. 2005
Marbling	Beef, pork		B-scan	3.5 MHz	Fernandes et al. 2001; Greiner 2003; BIF 2004; Newcom et al. 2003; Brondum et al. 1998; Gresham 2005; Youssao et al. 2002; Morlein et al. 2005
	Beef		Elastography	2.25 MHz	Ishmael 2003
Elastic modulus	Cheese	V	T, contact	1 MHz	Benedito et al. 2000, 2002; Mulet et al. 2002
	Dough	V	T, contact	37 kHz	Kidmose et al. 2001

Table 4.1. *Continued.*

Quality Parameters	Food	US Properties	Mode	Frequency	References
Loss modulus	Dough	α, V	P/E, T, contact	37 kHz	Kidmose et al. 2001; Letang et al. 2001
Storage time	Cheese	V	T, contact	1 MHz	Benedito et al. 2001
Moisture	Cheese	V	T, contact	1 MHz	Benedito et al. 2000, 2002; Mulet et al. 2002
Composition	Meat product	V	T, contact	1 MHz	Simal et al. 2003
Concentrations	Avocado (oil)	α, V	P/E, contact	50 kHz	Mizrach and Flitsanov 1999
	Fermentation liquid	V	T, contact	2 MHz	Resa et al. 2004
	Sucrose, glycerol, NaCl	α, V	T, contact	2.25 MHz	Dukhin et al. 2005
	Lipid in fish muscle	V	T, contact	2.25 MHz	Shannon et al. 2004
	$NaPO_4$, $NaNO_2$, kaolin	α, V, R	P/E, contact	NA	Bamberger and Greenwood 2004
	Water	α, V	P/E, contact	2–10 MHz	Letang et al. 2001
	Milk, salt water		P/E, contact	1.1 MHz	Mathieu and Schweitzer 2004
Acidity	Mango	α	P/E, contact	50 kHz	Mizrach 2000
Crack	Cheese	α, V	P/E, T, contact	1 MHz	Benedito 2001
Particle size	Corn oil	α, V	P/E, contact	5 MHz	Coupland and McClements 2001
	Air bubble	R	P/E, contact	3. 5 MHz	Kulmyrzaev et al. 2000
	Vegetable oil	V	T, contact	5.35, 5.75 MHz	Ammann and Galaz 2003
	Sugar, lactose	R	P/E, contact	2.25 MHz	Saggin and Coupland 2002
	Emulsions, suspensions	α, V	T, contact	1–100 MHz	Dispersion Technology 2005
Viscosity	Chocolate	DV-PD	P/E, contact	4 MHz	Ouriev 2000
	Tomato juice	DV-PD	T, contact	5 MHz	Dogan et al. 2002, 2003
	Tomato, orange juices	V	P/E, contact	4 MHz	Alig and Lellinger 2000

Continues

Table 4.1. *Continued.*

Quality Parameters	Food	US Properties	Mode	Frequency	References
	Sugar water	SR	P/E, contact	2.25, 10 MHz	Alig and Lellinger 2000; Greenwood and Bamberger 2002; Saggin and Coupland 2004
	Xanthan solution	SR	P/E, contact	10 MHz	Saggin and Coupland 2004
FB detection	In glass bottle		P/E, contact	4 MHz	Zhao et al. 2003a
	In polymer package material		T, noncontact	2–5 MHz	Gan et al. 2002; Bhardwaj 2002
	In polymer package material		P/E, contact	5 MHz	Haeggestrom and Luukkala 2001
	Peach flesh		B-scan	3.5–7.5 MHz	Chivers et al. 1995

Note: C = continuous acoustic excitation method, P/E = pulse/echo method, T = through-transmission method, A = amplitude of resonance, V = velocity of sound, f = frequency of resonance, R = reflectance method, α = attenuation coefficient, DV-PD = Doppler velocimetry pressure drop method for viscosity measurement, and SR = shear reflection for viscosity measurement. Water coupling is grouped into "contact" due to its inspection speed restriction.

The characteristic impedance is often called *acoustic impedance* in much nonacoustic research literature, which does not follow acoustic terminology. Acoustic impedance is defined as the local pressure divided by local particle velocity in an acoustic field (Pierce 1981). These two definitions lead to some confusion.

Ultrasound Measurement and Equipment

In ultrasonic testing, ultrasound velocity and the attenuation coefficient in materials of interest are measured. These are the two most important parameters for material characterization. There are two modes to perform this measurement: pulse/echo (P/E) mode and through-transmission (T) mode.

In P/E mode, a single transducer transmits impulsive ultrasound signals to the material and receives the echoes from it. The ultrasound velocity in the material of interest is calculated by the echo's round-trip distance divided by the time elapsed. The amplitude of the ultrasound is assumed to decay exponentially with the propagation distance due to irreversible energy diffusive processes occurring in the material. The attenuation coefficient (α) is defined as $A = A_0 \cdot e^{-\alpha L}$, where A_0 and A are the ultrasound pressure amplitudes at the origin and the ultrasound pressure at distance L, respectively. For P/E mode, the attenuation coefficient can be obtained by comparing the measured signal amplitudes at given distances and taking into account the characteristic impedances of the materials on the reflection interface.

T mode employs two transducers to execute the same task: one is the transmitter, and the other the receiver. The material of interest is placed in between the two. In addition to velocity and attenuation coefficient, reflectance (R) is also used for material characterization. Reflectance is defined as the ratio of the intensities of the returning and incident acoustic pulses, and it is related to the specific impedances on the interface between the sample and the material contacting the sample (Saggin and Coupland 2004). From the standpoint of signal propagation, reflectance measures the acoustic events occurring on the interface between a given medium and the medium under test. This feature allows R measurement for characterizing materials with high damping to ultrasound signals where sound velocity and attenuation are difficult to measure.

Ultrasound can be generated in an impulsive and also in a continuous way. Continuous ultrasound is used more in high intensity for cavitations than in low intensity for measurements. However, it has an advantage over the impulsive mode in detecting small time intervals when it is used in measurements (Birks and Green 1991).

The ultrasonic waves are generated by a transducer. Traditionally, transducers are made of piezoelectric materials. As the name indicates, electricity is developed when pressure is applied to the material. Reversely, when an electric field is applied to the material, the material rapidly changes shape. Ultrasound can be focused in a manner analogous to focusing light, since ray acoustics analysis applies when the wave length is less than the object dimension. Acoustic lenses are made of solids or liquids. They are normally integrated in transducers (Birks and Green 1991).

Operationally, ultrasound sensors are grouped as contact and noncontact. The main disadvantage for the traditional piezoelectric transducer is that it needs to contact the object being tested to transmit and receive the acoustic energy. The contact mode limits its scope of applications and inspection speed. Transducers are easily worn in contact mode. Using water or gel as a coupling medium can overcome this problem. However, in many cases, use of water is not acceptable. On the other hand, high-speed relative movement between the transducer and the parts can induce serious water turbulence when it is used for scanning purposes. This turbulence deteriorates the signal-to-noise ratio. Air-coupled transducers were developed for noncontact mode to address this issue.

There are two types of air-coupled ultrasonic transducers according to their mechanisms: one using the piezoelectric mechanism and the other using the electrostatic capacitance. In piezoelectric designs, the first problem is to overcome the large mismatch of the characteristic impedance between the piezoelectric ceramic and air. Big impedance mismatch prevents ultrasound energy transmission to the air from the transducer. This has been overcome by manipulating the acoustic impedance transitional layers in front of the piezoelectric element. Using compressed fibers as the final matching layer, a piezoelectric air-coupled transducer can propagate up to 5 MHz ultrasound through nearly all materials in noncontact mode (Bhardwaj 2002). Electrostatic ultrasound transducers are composed of a thin membrane film and a rigid conducting back-plate to form a capacitor. Applied voltages cause the membrane to vibrate, and hence generate ultrasound, whereas a change in charge across the membrane can be used to detect ultrasound. It is recognized that the surface feature of a back-plate plays an important role in sensitivity improvement. Recent electrostatic ultrasound transducers have been made using micromachining techniques to form optimized features on silicon wafers (Gan et al. 2001). This type of transducer has been used to detect defects in Teflon, aluminum, brass, and carbon-fiber-reinforced polymer composites (Hutchins et al. 1998; Gan et al. 2001).

The drawback of air-coupled transducers is their operation mode. They are operated in either T mode or as a separate transmitter and receiver on the same side of an object. T mode is not suitable for the detection of FBs settled at the bottom of containers due to their complexities—for instance, a converged/extruded bottle neck. Same-side-separate mode is suitable for detecting defects on the surface or close to the surface of an object but is not suitable for inner defects such as an FB in a container. Moreover very high attenuation in air, due to thermal dissipation, retains same-side-separate mode's operation at a lower-frequency band in industrial inspections: ≤2 MHz for electrostatic models and ≤5 MHz for air piezoelectric models (Ultran Group 2004). This limits its spatial resolutions. New progress has been reported—piezoelectric models with frequency up to 10 MHz are also now available (Bhardwaj 2004). Another concern is that air-coupled transducers may need more accurate orientation than water-coupled transducers due to a big difference in sound speed in air (330 m/s) and in the test materials such as metal and glass (≤5000 m/s). In such cases, a small deviation in incident angle in the air will lead to a big refraction angle in the test materials, according to Snell's law. Thus, air-coupled transducers might fail to deliver ultrasound energy to the spot to be inspected. In practice, accurate transducer orientation cannot be guaranteed in a production line. This will lead to more failure or false detection during inspection (Basir et al. 2003). For transducer alignment automation, an algorithm was developed using the amplitude information extracted from the ultrasound signals (Zhao et al. 2003b). This allows automatic following of a sensor to curved surfaces such as bottles.

Electromagnetic acoustic transducers (EMATs) and laser-ultrasound systems also provide noncontact operation alternatives. In addition to noncontact, EMATs are able to conveniently produce either compressive or shear waves. Another advantage of EMATs is that the requirement for transducer orientation is not critical. The drawback of EMATs is that they are suitable only for electrically conductive materials. Using the thermoelastic or ablation effect of material for ultrasound excitation, laser-ultrasound systems also provide a number of advantages over conventional ultrasonic techniques. In addition to their noncontact nature, the advantages are high spatial resolution, curved surface applications, reaching hard-to-access areas, and large tolerance of transducer orientation. Disadvantages include low ultrasound wave amplitude when the laser intensity is in the thermoelastic region, expense, and bulky equipment. Food quality assurance applications have not been reported for EMATs and laser-ultrasound systems.

Temperature affects ultrasound measurements. For example, in the temperature range of 20°C–30°C, the sound speed of water increases at the rate of 2.7 m/°C. The attenuation coefficient also increases with the increase in temperature (McCarthy et al. 2005).

Texture Evaluation

Texture is one of the most important attributes used by consumers to assess food quality (Kilcast 2004). The International Standards Organization (ISO) defined food texture as sensory characteristics perceived largely by the sense of touch. Inputs from other senses, especially vision and taste, contribute many times also (Mizrach and Flitsanov 1999). Based on this definition, texture tests often refer to firmness, hardness, tenderness, and elasticity evaluations, depending on food types and their expected nature. Usually, food firmness is measured by a penetrometer. The firmness is scored by the penetration distance of a penetrant placed inside the sample under a controlled force for a fixed period of time. This method is destructive and cannot be used for every-piece testing. Since the late 1990s, ultrasound techniques have been attempted for the texture testing of fruits (Mizrach and Flitsanov 1999; Flitsanov et al. 2000; Mizrach 2004; Verlinden et al. 2004; Mizrach 2000; Mizrach et al. 2000), cheeses (Cho et al. 2001; Benedito et al. 2000, 2001, 2002; Mulet et al. 2002), and meat or meat products (Allen et al. 2001; Abouelkaram et al. 2000; Ophir et al. 1994; Llull et al. 2002). Most of the studies were carried out using two transducers in contact mode. The transducers were in direct contact with the sample or coupled using a water tank. In the work of Cho et al. (2001), two air-coupled transducers were used to characterize the physical properties of cheddar cheese. The sample was placed in between the transmitting and receiving transducers at a distance of 3 cm from each other. The case of using a single transducer was presented by Ophir et al. (1994), who proposed the ultrasound elastography principle to directly calculate the elastic modulus of beef.

Whether to select the ultrasound velocity or the attenuation coefficient for material characterization is debatable. The attenuation coefficient had good correlations with the firmness, dry matter, and oil content for avocados and mangos (Mizrach and Flitsanov 1999; Flitsanov et al. 2000; Mizrach 2000; Mizrach et al. 2000), tomatoes (Verlinden et al. 2004), and plums (Mizrach 2004). Ultrasound velocity showed a nonmonotonic complex relationship that is not adequate for an index of firmness (Mizrach 2000). In the case of cheeses, ultrasound velocity was found better correlated with elastic modulus (Cho et al. 2001; Benedito et al. 2000, 2001, 2002; Mulet et al. 2002). For limited samples, ultrasound velocity was well correlated with the firmness of meat or meat products (Allen et al. 2001; Abouelkaram et al. 2000; Llull et al. 2002). By reviewing the investigations on dairy products, some researchers concluded that the sound speed is a better parameter for investigating effects on a molecular scale. The attenuation is more suitable for small colloidal sizes that characterize effects related to the heterogeneity and phase composition of the particular system (Resa et al. 2004). In a study by Elvira et al. (2005) using ultrasound to detect microbial contamination in packed ultraheat treatment (UHT) milk, the attenuation coefficient seemed better than velocity to correlate the *Bacillus pumilus* contamination. It is premature to conclude if their results support this claim.

The ultrasound elastography method computes local strains of a test material subjected to a given stress. In testing, the first signal is picked up by placing the sensor on the sample with slight pressure. Then the sample is compressed more by enforcing the sensor to pick up the second signal on the same spot. The local strain is interpreted as the time shift of the signal pair between pre- and postcompression of the material. This method has the advantages of simplicity in the principle for understanding and using only one transducer. However, since this method requires a constant stress for compressing the sample, the published data so far are noncalibrated and for the purpose of image feature enhancing (Ophir et al. 1994). This issue may be addressed by a "stress meter" embedded into the ultrasound sensor, as suggested by Ophir et al. (2002).

Carcass Grading

Beef carcass grading includes backfat thickness (BF), rib-eye area (REA), and percentage of intramuscular fat (%IMF, or marbling). As the BF increases, the percentage of retail cuts decreases, and when rib-eye area increases, retail cut yield increases. Marbling contributes slightly to meat tenderness and is associated with the palatability traits of "juiciness" and "flavor." These indexes are evaluated on the carcass after slaughter: BF is measured at the 12th rib perpendicular to the outside fat at a point three-fourths the length of the rib-eye muscle. The rib-eye muscle area is measured at the 12th rib by using a grid or a rib-eye tracing with a compensating polar planimeter. The marbling score is assessed visually by a grader. For the purpose of genetic improvement of beef cattle, and also for meat products planning during processing, these indexes are expected to be known for live cattle. Between the 1980s and 1990s, ultrasound imaging was introduced for the evaluation of these indexes (Cross and Belk 1994). The ultrasound evaluation is first performed by well-trained and certified technicians using B-scan equipment and then compared with the carcass examination after slaughter. In an early report, ultrasound backfat depth had the highest correlations (>0.7), while for the ultrasound-measured rib-eye area and intramuscular fat, the correlations were between 0.4 and 0.6 (Fernandes et al. 2001). Further, the Annual Proficiency Testing and Certification Program (Greiner 2003) confirmed that the ultrasound is an accurate predictor of carcass traits. Average correlations for the 43 participating technicians were 0.89, 0.86, and 0.70 for backfat thickness, rib-eye area, and intramuscular fat, respectively. At present, real-time ultrasound technology has advanced to the state whereby accurate measurements of these three traits can be made on live beef animals. However, marbling is still the most difficult of all ultrasound traits to measure accurately. Equipment calibration, animal preparation, electrical power signal noise, existence of atmospheric radio waves, and transducer-animal contact are some of the factors that can influence measurement accuracy. A minimum of four independent images were collected and the resulting %IMF predictions averaged for this trait (Beef Improvement Federation 2004). The difficulties in the marbling measurement may be improved using elastography. An initial study showed that 88 percent of the variability in Warner-Bratzler (W.B.) shear force tests (used to evaluate tenderness) was accounted for by the elastogram information (Ishmael 2003).

Currently in the United States, most beef cattle scanning is done with an Aloka 500 V with a 17 cm linear array 3.5 MHz transducer or with a Classic Scanner 200 with a 18 cm linear array 3.5 MHz transducer (BIF 2004). Human errors in operation and image interpretation remain the main source of errors for live animal inspection. To ensure unbiased testing, strict protocol should be followed by certified and highly skilled technicians.

An ultrasound-based quality evaluation technique is also applied to live pig and pork carcasses (Newcom et al. 2003; Brondum et al. 1998; Gresham 2005). The equipment used for pig evaluation is the same as that used for cattle, except for the use of a smaller (13 cm) linear array (Youssao et al. 2002). A lightweight (<2 lb), wrist-mounted unit is also available (Gresham 2005). In abattoirs, the fat and meat thicknesses of carcasses are measured using semicircular transducers installed on conveyors running at 1150 carcasses/h speed (Brondum et al. 1998). In a recent study, nearly 80 percent correct classification of roughly "low" and "high" marbling for pork was achieved (Morlein et al. 2005). However, marbling by ultrasound imaging still needs more work to improve its accuracy for more refined classification.

Density Measurement

On-line measurement of density is important for quality control during food processing. For acoustic attenuative materials, many density-measuring devices are based on the reflection coefficient measurement:

$$R = (Z_2 - Z_1)/(Z_2 + Z_1) \tag{4.1}$$

where $Z = \rho c$ is acoustic impedance, ρ and c are respectively the sound velocity and density of the material, and subscripts 1 and 2 indicate respectively the two materials at interface: the known material and the material under test.

Since R can be readily measured at the interface, the test material density (ρ_2) can be calculated as well, provided that the sound velocity (c_2) is known:

$$\rho_2 = \rho_1 \frac{c_1}{c_2} \frac{1+R}{1-R} \tag{4.2}$$

Fox et al. (2004) designed a V-shaped sensor system to measure food batters' specific gravity based on P/E mode. This V-shaped sensor is easy to use as it can be inserted in batters without air bubbles caging on the sensor surface. For regular flat head sensors, it is difficult to avoid air bubbles caging on the sensor surface. The V-shaped sensor measures a "probe gain," a normalized reflection amplitude. The normalized reflection amplitude is defined as the ratio between the ultrasound reflection at the sensor/sample interface and the reflection between the water delay line and Perspex, a V-shaped casing material. In the V-shaped sensor's design, the sound velocity is not obtained directly. Instead, it is assumed that the sound velocity is linearly correlated with the volumetric fraction of air bubbles in the batter. In the final correlation with the probe gain, the specific gravity is proportional to the attenuation coefficient of the batter and assumed "stable." However, the attenuation coefficient is believed to change with the batter aging time.

Greenwood et al. (1999) developed an on-line densitometer for liquids and slurries using equation 4.1. Their design consists of using at least three sensors: one performs a normal incident P/E mode reflection coefficient measurement; the other two constitute a transmitter/receiver pair to perform the reflection coefficient measurement at a large incident angle (60 degrees) at the interface between Z_1 and Z_2. Combining the two measured reflection coefficients, the diffraction angle can be calculated using Snell's law. The sound velocity in the test materials is therefore obtained knowing the incident angle and sound velocity in the known material.

When using equation 4.1 to measure density, a large difference in the acoustic impedances between the solid and the liquid results in a lower sensitivity for changes in the density. Performing a sensitivity analysis on equation 4.1, one can get

$$\frac{\Delta R}{R} = \frac{2}{z - 1/z} \cdot \frac{\Delta z}{z} \tag{4.3}$$

where $z = Z_1/Z_2$ is the impedance ratio.

Equation 4.3 indicates that if z is large, the relative variation in impedance ratio ($\Delta z/z$) is scaled down in producing a measured reflection coefficient variation ($\Delta R/R$). For example, for steel and water, $z = 30$; if $\Delta z/z = 15$ percent, then $\Delta R/R$ is only 1 percent.

However, if we choose z as close to 1, the sensitivity will be greatly increased. This is the reason why in many liquid-density-measuring devices, low-impedance materials are used at the liquid interface, such as Rexolite and Perspex (impedance ratio to water equals about 2).

To address the low-sensitivity issue, there are some other approaches. Hirnschrodt et al. (2000) developed a resonance antireflection method. In this method, the ultrasound sensor is coupled with a liquid delay line whose impedance is close to that of the test liquid. The two liquids are separated by a solid material whose impedance may be very high compared to that of the test liquid. However, the thickness of the separation layer is chosen as half of the wave length. This provides no reflection at the interface between the liquid delay line and the solid layer, and the reflection is only dependent on the impedance ratio between the liquid delay line and the test liquid, which is closer to 1.

Mathieu and Schweitzer (2004) proposed a method using backscattering analysis to measure liquid density. In this method, two wires of different materials are used as the ultrasound interfacing areas to test liquids. The densities of the two wires are chosen so that one is slightly higher and the other is much larger than the density of the test liquids. The two wires' diameters are identical but are small compared to the ultrasonic wave length. In this case, the diffraction effect is dominant. The backscattered signals are used to calculate the reflection coefficient. The higher-density wire is used to calibrate the instrument, whereas the lighter one allows effective measurement for small density changes of test liquids.

A common criterion in selection or design of an ultrasound density sensor for slurries is that the ultrasound wave length should be larger than the particle size.

Rheology Measurement

Viscosity is an important parameter in food processing for equipment and process design, ingredient functionality determination, intermediate or final product quality control, and shelf life evaluation. There are three approaches for applying ultrasound to food rheology measurements.

The first is using an ultrasound Doppler velocimeter to measure the velocity profile of a fluid in a cross section of a pipe, from which the shear rate is computed. The stress on the wall is obtained by using a pressure sensor to measure the pressure drop across the pipe. The fluid viscosity or rheological parameters are determined by fitting in an appropriate rheological model. Equipment is commercially available (Battelle, Pacific Northwest Division). Some thick fluids were tested such as chocolate (Ouriev 2000) and tomato juice (Dogan et al. 2002, 2003) using this method. Choi et al. (2002) measured the velocity profile of fluids (corn syrup and tomato juice) in pipes using the Doppler effect. Thus, this provides an in-line measurement of fluid viscosity. This method is suitable for food quality monitoring and assurance during processing where the food is flowing. For packed foods, the following two methods are used.

The second approach is ultrasonic shear reflection working with a shear wave transducer (Greenwood and Bamberger 2002; Alig and Lellinger 2000). The shear wave transducer is coupled to the fluid and measures the reflection coefficient on the interface between the coupling material and the fluid, instead of measuring the velocity or attenuation in the fluid. The viscosity (μ) of the fluid is obtained by (Saggin and Coupland 2004)

$$\mu = 2RX/\rho\omega \tag{4.4}$$

$$R + jX = \frac{1 - r^2 + j2r\sin\vartheta}{1 + r^2 + 2r\cos\vartheta} \tag{4.5}$$

where R and X are respectively the real and imaginary parts of an acoustic impedance, $j = \sqrt{-1}$ is unit of imaginary number, ρ and ω are the fluid density and angular frequency respectively, r is the measured normalized reflection coefficient (i.e., measured reflected amplitude of the sample divided by the reflected amplitude of a calibrating material), and ϑ is the difference of phase angles between the sample and calibrating material.

Shear ultrasonic reflection provides the food industry with a useful on-line measurement of fluid rheology at high-frequency range (MHz), while the conventional oscillatory rheometer works at low frequency (a few Hz). It allows for better understanding of the rheology of complex fluids at the molecular scale. However, in the shear ultrasound method, the measured rheological parameters at the frequency of MHz can not be validated with conventional methods (rotary cylinder, oscillatory, etc.) due to the small frequency of this equipment. For example, different values are obtained from shear ultrasound and the oscillatory viscometer for the rheological parameters of sugar syrup and xanthan gum (Saggin and Coupland 2004). In another experiment, the viscosity of a sugar water solution obtained by the shear ultrasound method was nonlinearly correlated with rotary cylinder viscometer values (Greenwood and Bamberger 2002).

The third approach simply correlates the ultrasonic properties with the rheological properties obtained by the conventional measuring methods. Potential advantages of this is that velocity measurement is the simplest, most widely used method and probably the most accurate in ultrasonic techniques (Benedito et al. 2000). Certain governing laws or relationships between them have been developed. Experimental work on diluted tomato and orange juices demonstrated that their apparent viscosities are proportional to the quadratic of the longitudinal sound speed shown in figure 4.1 (Zhao et al. 2003c). In another work for developing a "dipstick" viscosity sensor for fluids in vessels, Cegla et al. (2004) tried to correlate the group velocity and attenuation of a surface wave (quasi-Scholte mode) on an immersed plate with the viscosity of glycerol. Research on semisolid materials such

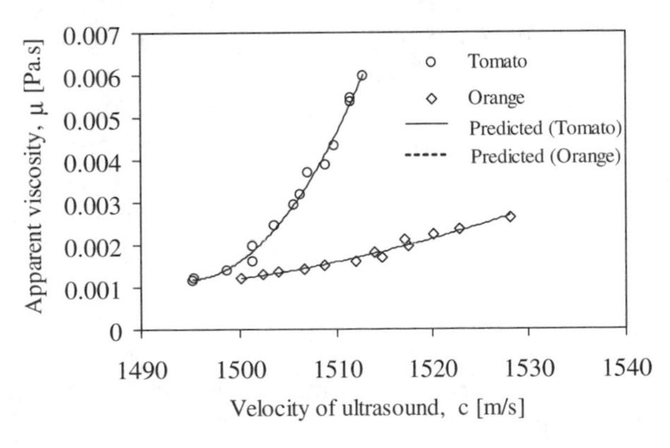

Fig. 4.1. Apparent viscosity of juice as a function of the velocity of sound (predicted and experimental data) (Zhao et al. 2003c, reprinted with permission of Marcel & Dekker).

as dough was conducted using this technique. It was reported that the storage modulus of dough was highly correlated with the longitudinal sound speed during 40 min of aging (Kidmose et al. 2001). Velocity–viscosity correlation seems a promising approach for monitoring a specified product processing.

Particle Size Measurement

The measurement of particle sizes is important in dispersed systems, such as emulsions, suspensions, and foams, due to their effect on stability, appearance, taste, and microbiological status (McClements 1995). Currently there are some techniques for particle size measurement, such as electron microscopy, laser light scattering, and nuclear magnetic resonance. Each of these methods has some drawbacks. For example, electron microscopy needs sample preparation that is disruptive and may generate artifacts. Laser light scattering is only suitable for very diluted solutions. The nuclear magnetic resonance device is expensive.

Ultrasound can be used to determine particle sizes in emulsions or suspensions in a manner analogous to the light-scattering method. It has advantages over the other methods in terms of equipment cost, sample preparation, and concentration range (up to 45 percent by volume, Dukhin and Goetz 2005). When an ultrasound propagates in a complex system, it will interact with the particles and continuous phase. In general, ultrasound for particle size measurement approaches the problem using two-stage modeling: "single particle" and "macroscopic." The first stage attempts to account for all of the ultrasound disturbances surrounding a single particle. At this stage, the microscopic properties of both the continuous phase and particle and the system properties are involved. The single-particle model results usually in a dispersion of ultrasound velocity and attenuation. The second stage relates this "single-particle level" to the macroscopic level at which the experimental data are taken. Principles related to ultrasonic wave interactions with the particles are available (Povey 1997; Dukhin and Goetz 2002). Operationally, ultrasound propagation (velocity and amplitude) is first measured as a function of frequency using P/E mode or T mode. Then the mean particle size is estimated by iteratively adjusting the diameter in the model to minimize the distance between the measured and the theoretical dispersive curves (Coupland and McClements 2001). There are many different types of ultrasound spectrometers, but one widely common apparatus uses a continuous ultrasound signal in the sample whose amplitude is measured by the transducers. After each measurement, at a given frequency, the frequency is changed to make another measurement. A full spectrum and particle size distribution took 2–3 min to acquire in the late 1990s (Povey 2000). This sort of equipment is used for off-line lab analysis. Currently, a model for on-line measurement is available for process monitoring and quality control. This model uses T-mode pulse signals rather than continuous waves. The particle size measured is from 0.005 to 1000 μm (Dispersion Technology 2005).

Casein in water was studied (Povey et al. 1999) using an interferometric device with two pairs of transducers, one for low frequency and the second for high frequency. Mcskimmin techniques can be used to study many inorganic gels due to their relatively larger shear modulus. Fat crystallization has been studied by ultrasound P/E technique (Povey 1998).

Cereal Products Characterization

Cereal products include wafer sheets, crackers, extrudates, and crisp bread. Low-frequency ultrasound is normally used for cereal products' characterization since they are very attenuative to ultrasound propagation due to their cellular structures or porosity. Elmehdi et al. (2003) applied a 54 kHz ultrasound pulse to bread crumb samples. They reported that the ultrasound velocity is proportional to the square root of the bread crumb density. They also found that the cell shapes (spherical or compressed spherical) have a significant effect on the ultrasound propagation velocity. Juodeikiene and Basinskiene (2004) measured the through-transmission signal amplitudes at different frequencies for many cereal products. They proposed a concept of optimal frequency for different samples. The optimal frequency was chosen for each product based on a two-parameter criterion: the minimum standard deviation and the minimum variation coefficient of the through-transmission signal amplitudes. They found an exponential relationship between the mechanical strength and the signal amplitude. The best correlation coefficient was 0.96 for crisp bread at an optimal ultrasound frequency of 17.59 kHz. However, it was not clear how to decide the optimal frequency if the frequency of the minimum standard deviation is not the same as that of the minimum variation coefficient.

Plastic Food Package Inspection

Defects in plastic food packages include incomplete seals, misplaced layers or adhesives, faults along heat seals, channel voids, contaminants, and inclusions, as well as small nicks and cuts. These defects cause pathways for microbial penetration that eventually result in product spoilage or give rise to product deterioration during storage (Ozguler et al. 1999). This concern led to research using water-coupled piezoelectric transducers operated in pulse/echo mode, by which channel defects in plastic scaling were detected as small as 50 μm (Shal et al. 2001).

The bonding strength of packaging seals is another concern for food manufacturers since packages can leak when subjected to vibrations/impacts during transportation. Pascall et al. (2002) investigated, in a water tank, the sealing strength of 335 ml polymeric trays using ultrasound C-scan imaging. The statistical scatter of reflected signals decreased as the sealing strength increased when the polymeric was uniformly fused. However, there is no quantitative measurement for the peel strength in their investigation. Later Ayhan and Zhang (2003) and Ayhan (2004) C-scanned multilayered 200 ml food cups constructed of polystyrene, polyvinylidene chloride, and polyethylene. Using the reflection amplitude of ultrasound pulse signals could pinpoint seal defects but could not characterize the bonding condition, though there was no quantitative measurement on the bonding strength either.

Currently, air-coupled ultrasound packaging seal inspection systems are commercially available. The systems can scan seals of the pouches, trays, and packages made of plastic, foil, film, or paper. They are used to verify seal integrity and quickly pinpoint common defects such as incomplete seals, misplaced layers or adhesives, faults along heat seals, channel voids, contaminants, and inclusions, as well as small nicks, cuts, and other defects. The inspection speed can reach as high as 250 pouches/min (Packaging Technologies and Inspection 2005).

Microbial Contamination Detection

Being noninvasive and nondestructive, ultrasound can be a good tool for detecting microbial contaminations in packaged foods. The research was initiated in the late 1980s for packaged milk quality assurance. Ultrasonic propagation velocity and amplitude remain the principal indices for detecting microorganism growth in milk. Recently ultrasonic detection results were compared with total viable bacterial count and physical–chemical parameters (pH, acidity, and stability to ethanol). It was shown that ultrasonic detection could take place even before detectable physical–chemical changes appear. According to Elvira et al. (2005), contamination is detectable as early as 13 h after the bacteria incubation in milk. A different detection approach, ultrasound streaming along with a Doppler monitoring technique, was also proposed for contamination inspection in the early 1990s (Ross et al. 2004).

Foreign Body (FB) Detection

Using ultrasound to detect FBs in foods has drawn considerable attention since the work of Chivers and coworkers (1995) in which a clinical linear B-scanner was tried to image stone pits embedded in peach flesh. Their work confirmed the feasibility of using low-intensity ultrasound to detect FBs in fruits. Haeggstrom and Luukkala (2001) detected and identified some FBs in plastic-packaged cheeses using a piezoelectric transducer. Experiments were conducted in a water tank with P/E mode. FBs included stones and spheres of bone, wood, glass, plastic, and steel. The detection depth was up to 75 mm in cheeses, which is good enough for the inspection of a standard package with 400 g mass.

Glass fragments finding their way into packed glass bottles has long been a big concern of beverages manufacturers. Challenges for ultrasound detection lie in the strong echo signal from the glass bottle wall that masks the scattered signals from a small glass fragment. This is a case of very bad signal-to-noise ratio (SNR) for signal processing in the time domain. Zhao et al. (2003a, 2005) developed a signal-processing method using short time Fourier transform (STFT) to single out the FB-scattered signal. In their method, the amplitude of a certain frequency component of the echoes' signal was computed along the time domain. Contributions by FB scatterings were separated by improving the SNR on the chosen frequency, provided that the noise was white. Figure 4.2 is an example of this method. Figure 4.2a shows three signals sampled at the same location of a glass bottle filled with water. The first is for "no object," indicating no glass fragment inside; the second and third are those from glass fragments with sizes of about $2 \times 3 \times 1$ mm and $3 \times 3 \times 1.5$ mm, respectively, placed in the bottle. In figure 4.2a, the echo labeled as 1 is the outer surface echo, 2 is the inner surface echo, 3 through 6 are the echo reverberations in the glass bottle wall, and label 7 is the outer surface echo delayed and attenuated by the transducer delay line (which is outside the interest of the study). According to the theoretical analysis, the fragment reflection signals start from the second reflection at the same moment as the inner wall reflection. However, its existence is not visually perceivable in figure 4.2a due to the high-intensity reflection of the inner surface of the wall. Figure 4.2b illustrates the computation results of the center frequency components of the spectrum for the three signals in figure 4.2a. The differences between the three processed signals become visible starting from the second peak. The very high energy interference of the inner surface echo was demasked. The

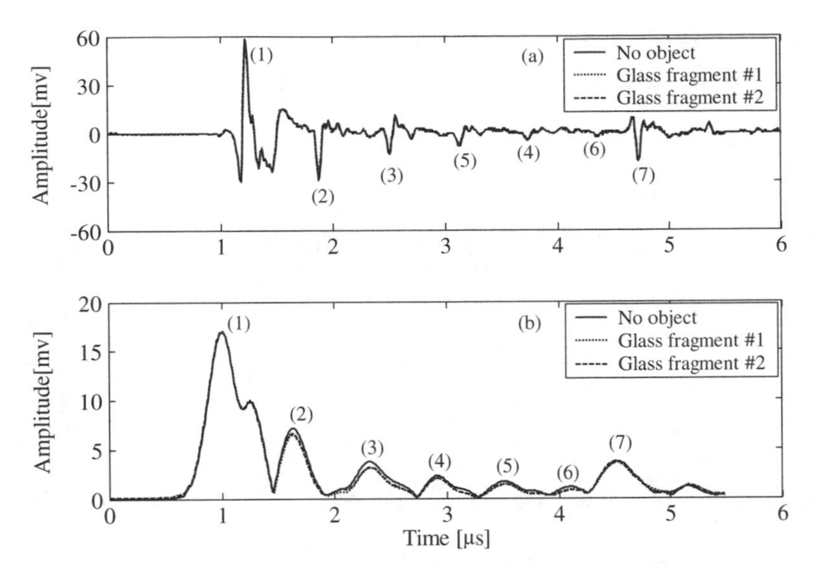

Fig. 4.2. **(a)** Ultrasonic echo signals sampled from outside of a glass bottle in the conditions of "no object" inside; "glass fragment 1 (≈ 2 × 3 × 1 mm)" and "glass fragment 2 (≈ 3 × 3 × 1.5 mm)" inside. **(b)** Envelope of the center frequency component of the three signals in a computation using short-time Fourier transform.

first peak remained the same, since the presence of an object in the bottle did not affect the outer surface echo signals.

The center frequency of a signal is defined as

$$f_c = \sum_{i=1}^{m/2} A_i \cdot f_i \Big/ \sum_{i=1}^{m/2} A_i \qquad (4.6)$$

where A_i and f_i ($= i \times f_s/m$) are respectively the magnitude and the frequency of the ith component in the spectrum of width $m/2$, m is the signal length in the time domain, and the f_s is the sampling frequency in hertz.

Figure 4.3 shows the glass fragments detected and the experiment setup where a piezoelectric transducer was mounted on an X-Y table and coupled with water. Similarly, Knorr et al. (2004) discussed their work detecting glass and plastic pieces in a tin can, glass bottle, and plastic beaker. In their research, conventional medical B-scan equipment was used.

As mentioned previously, air coupling is more efficient than water or solid contact for realizing high-speed ultrasound FB detection. Using an air-coupled electrostatic transducer through-transmission mode, Gan et al. (2002) detected a small piece of stainless steel plate with a dimension of 1.5 × 7 mm suspended in a polymer water bottle. They also detected corn starches in water-filled microwaveable polymer food containers of wall thickness 0.45 mm. Parallel to this progress, Bhardwaj (2002) detected almond nuts in milk chocolate using an air-coupled piezoelectric transducer. He also reported the feasibility of using the air-coupled transducer to detect FBs in containers filled with liquid. Currently, the application of the air-coupled transducer is in FB detection for food products before packing or packaged in low-impedance materials such as polymer.

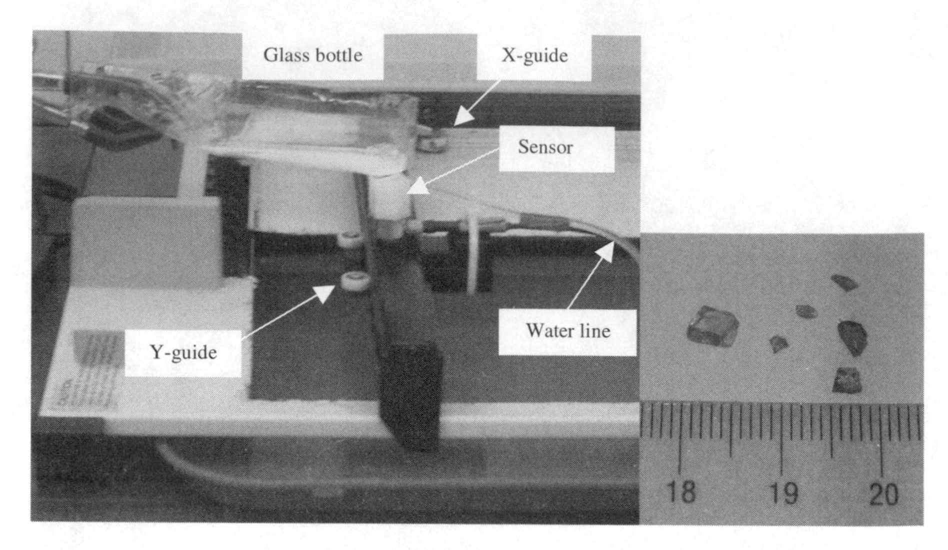

Fig. 4.3. Experimental setup to detect glass fragments in packed foods using the ultrasonic method.

FB detection in high-impedance materials such as glass and metal has not been reported. Again, the main reason is believed to be the big impedance mismatch between the air and the container.

Conclusions

Inspection or detection speed is a very important factor that will decide if an ultrasound technique can be transferred from the lab to the production floor. Food-processing production speed is about 2.5–7.5 pieces/s for fruits (Peleg 1999). Beverage and beer production is assumed to be approximately the same speed. It is difficult for contact coupling or water coupling to match up with this speed. Noncontact ultrasound transducer techniques will play an important role in the near future for quality characterization and foreign body detection in large-volume food production and inspection. Air-coupled transducers will find more applications in addition to their successful application in package quality inspection. Transducer combinations could be more cost effective than developing other ultrasound techniques or higher-performance transducers. For example, a laser transmitter in conjunction with an air-coupled receiver may be applied to many high-impedance packaged (metal, glass, and porcelain) foods in inspections for properties measurement and foreign body detection. This is due to the fact that air-coupled transducers are better at receiving than transmitting (Green 2004), while optical fibers provide laser projection in a flexible way.

Extensive work on material modeling will be needed to better understand and correlate ultrasound parameters with food texture and rheology. On the other hand, more information can be extracted from the ultrasound signals. Only the time domain signal processing method was used in almost all of the cited works on texture and rheology characterization, while the dispersive characteristics of ultrasound have been widely used in particle size measurements. This property—particle size—has not been fully investigated yet.

References

Abouelkaram, S., Suchorski, K., Buquet, B., Berge, P., Culioli, J., Delachartre, P., Basset, O. 2000. Effects of muscle texture on ultrasonic measurements. *Food Chem* 69:447–55.

Alig, I., Lellinger, D. 2000. Ultrasonic methods for characterizing polymeric material. *Chem Innovation* 302:14–20.

Allen, P., Dwyer, C., Mullen, A.M., Buckin, V., Smyth, C., Morrissey, S. 2001. Using Ultrasound to Measure Beef Tenderness and Fat Content, April, http://www.teagasc.ie/research/reports/foodprocessing/4532/eopr-4532.pdf.

Ammann, J.J., Galaz, B. 2003. Sound velocity determination in gel-based emulsions. *Ultrasonics* 41:569–79.

Ayhan, Z. 2004. Seal bond characterization of laminated plastic food cups by electron scanning and optic microscopes. *Packag Technol Sci* 17:205–11.

Ayhan, Z., Zhang, H. 2003. Evaluation of heat seal quality of aseptic food containers by ultrasonic and optical microscopic imaging. *Eur Food Res Technol* 217:365–68.

Bamberger, J.A., Greenwood, M.S. 2004. Non-invasive characterization of fluid foodstuffs based on ultrasonic measurements. *Food Res Int* 37:621–25.

Basir, O.A., Zhao, B., Mittal, G.S. 2003. "Detection of foreign bodies in foods using ultrasound." In *Detecting Foreign Bodies in Food*, Woodhead Publishing, London, pp. 204–25.

Benedito, J., Carcel, J.A., Sanjuan, N., Mulet, A. 2000. Use of ultrasound to assess cheddar cheese characteristics. *Ultrasonics* 38:727–30.

Benedito, J., Carcel, J., Gerbert, M., Mulet, A. 2001. Quality control of cheese maturation and defects using ultrasonics. *J Food Sci* 661:100–104.

Benedito, J., Carcel, J.A., Gonzalez, R., Mulet, A. 2002. Application of low intensity ultrasonics to cheese manufacturing processes. *Ultrasonics* 40:19–23.

Bhardwaj, M.C. 2002. "Non-contact ultrasound: the last frontier in non-destructive testing and evaluation." In *Encyclopaedia of Smart Materials*, John Wiley and Sons, New York, pp. 690–714.

———. 2004. Evaluation of piezoelectic transducers to full scale non-contact ultrasonic analysis mode, World Conference on Non-Destructive Testing—2004, Montreal, Canada. Ultran Group, http://www.ultrangroup.com/docs.php.

BIF. 2004. "Animal evaluation guidelines," chap. 3, Beef Improvement Federation, http://www.beefimprovement.org/guidelines/Chap3.PDF.

Birks, A.S., Green, R.E. 1991. *Ultrasonic Testing Nondestructive Testing Handbook,* 2nd edition, American Society for Nondestructive Testing, pp. 29, 258–62.

Brondum, J., Egebo, M., Agerskov, C., Busk, H. 1998. On-line pork carcass grading with the Autofom ultrasound system. *J Animal Sci* 76:1859–68.

Cegla, F., Cawley, P., Lowe, M. 2004. Fluid characterization using the quasi-Scholte mode, NDT 2004—the 43rd Annual British Conference on NDT, Session 6C—modelling, Torquay, UK, http://www.bindt.org/Mk1Site/NDT2004Abstr5.html.

Chivers, R.C., Russel, H., Anson, L.W. 1995. Ultrasonic studies of preserved peaches. *Ultrasonics* 33:75–77.

Cho, B. Irudayaraj, J. Bhardwaj, M.C. 2001. Rapid measurement of physical properties of cheddar cheese using a non-contact ultrasound technique. *Trans ASAE* 446:1759–62.

Choi, Y.J., McCarthy, K.L., McCarthy, M.J. 2002. Thermographic techniques for measuring flow properties. *J Food Sci* 67:2718–24.

Coupland, J.N., McClements, D.J. 2001. Droplet size determination in food emulsions: comparison of ultrasonic and light scattering methods. *J Food Eng* 50:117–20.

Cross, H.R., Belk, K.E. 1994. Objective measurements of carcass and meat quality. *Meat Sci* 36:191–202.

Dispersion Technology. 2005. Catalogue, Dispersion Technology, http://www.dispersion.com/pages/products/products.htm.

Dogan, N., McCarthy, M.J., Powell, R.L. 2002. In-line measurement of rheological parameters and modeling of apparent wall slip in diced tomato suspensions using ultrasonics. *J Food Sci* 676:2235–40.

———. 2003. Comparison of in-line consistency measurement of tomato concentrates using ultrasonics and capillary methods. *J Food Process Eng* 25:571–87.

Dukhin, A.S., Goetz, P.J. 2002. *Ultrasound for Characterizing Colloids Particle Sizing, Zeta Potential Rheology*, 15, Elsevier Publ., New York.

———. 2005. Characterization of chemical polishing materials by means of acoustic spectroscopy. http://www.dispersion.com/.

Dukhin, A.S., Goetz, P.J., Travers, B. 2005. Use of ultrasound for characterizing dairy products. *J Dairy Sci* 88:1320–34.
Elmehdi, H.M., Page, J.H., Scanlon, M.G. 2003. Using ultrasound to investigate the cellular structure of bread crumb. *J Cereal Sci* 38:33–42.
Elvira, L., Sampedro, L., Matesanz, J., Gomez-Ullate, Y., Resa, P., Iglesias, J.R., Echevarrıa, F.J., Montero de Espinosa, F. 2005. Non-invasive and non-destructive ultrasonic technique for the detection of microbial contamination in packed UHT milk. *Food Res Int* 38:631–38.
Fernandes, T., Miller, S.P., Devitt, C.J.B., Wilton, J.W. 2001. Serial real-time ultrasound to predict carcass quality in commercial beef cattle. http://64.233.161.104/search?q=cache:Z0HllBgfwREJ:bru.aps.uoguelph.ca/Articles01/2001-pg1.pdf+ultrasound+beef+grading&hl=en.
Flitsanov, U., Mizrach, A., Liberzon, A., Akerman, M., Zauberman, G. 2000. Measurement of avocado softening at various temperatures using ultrasound. *Postharvest Biol Technol* 20:279–86.
Fox, P., Smith, P.P., Sahi, S. 2004. Ultrasound measurements to monitor the specific gravity of food batters. *J Food Eng* 65:317–24.
Gan, T.H., Hutchines, D.A., Billson, D.R., Schindel, D.W. 2001. The use of broadband acoustic transducers and pulse-compression technique for air-coupled ultrasonic imaging. *Ultrasonics* 39:181–94.
Gan, T.H., Hutchines, D.A., Billson, D.R. 2002. Preliminary studies of a novel air-coupled ultrasonic inspection system for food containers. *J Food Eng* 53:315–23.
Gestrelius, H., Hertz, T.G., Nuamu, M., Persson, H.W., Lindstrom, K. 1993. A non-destructive method for microbial quality control of aseptically packaged milk. *LWT—Food Sci Technol* 26:334–39.
Green, R.E. Jr. 2004. Non-contact ultrasonic techniques. *Ultrasonics* 42:9–16.
Greenwood, M.S., Bamberger, J.A. 2002. Measurement of viscosity and shear wave velocity of liquid or slurry for on-line process control. *Ultrasonics* 39:623–30.
Greenwood, M.S., Skorpik, J.R., Bamberger, J.A., Harris, R.V. 1999. Online ultrasonic density sensor for process control of liquids and slurries. *Ultrasonics* 37:159–71.
Greiner, S.P. 2003. Ultrasound applications for the beef industry livestock update, http://www.ext.vt.edu/news/periodicals/livestock/aps-03_04/aps-218.html.
Gresham, J.D. 2005. The ultrasound review: determination of backfat thickness and loin eye muscle depth in live swine with the 50S Tringa, Pie Medical and Classic Ultrasound, http://www.utm.edu/departments/caas/anr/faculty/gresham.php.
Haeggstrom, E., Luukkala, M. 2001. Ultrasound detection and identification of foreign bodies in food products. *Food Control* 12:37–45.
Hirnschrodt, M., Jena, A.V., Vontz, T., Fischer, B., Lerch, R. 2000. Ultrasonic characterization of liquid using resonance antireflection. *Ultrasonics* 38:200–205.
Hutchins, D.A., Schindel, D.W., Bashford, A.G., Wright, W.M.D. 1998. Advances in ultrasonic electrostatic transduction. *Ultrasonics* 36:1–6.
Ishmael, S. 2003. Breast cancer research may lead to better beef quality, Cattle Today, http://www.cattletoday.com/archive/2003/October/CT296.shtml.
Jayasooriya, S.D., Bhandari, B.R., Torley, P., D'Arcy, B.R. 2004. Effect of high power ultrasound waves on properties of meat: a review. *Int J Food Prop* 72:301–19.
Juodeikiene, G., Basinskiene, L. 2004. Non-destructive texture analysis of cereal products. *Food Res Int* 37:603–10.
Kidmose, U., Petersen, L., Nielsen, M. 2001. Ultrasonics in evaluation rheological properties of dough from different wheat varieties and during ageing. *J Texture Stud* 32:321–34.
Kilcast, David (ed.). 2004. *Texture in Foods, Volume 2: Solid Foods.* Leatherhead Food International, UK.
Knorr, D., Zenker, M., Heinz, V., Lee, D.U. 2004. Applications and potential of ultrasonics in food processing. *Trends Food Sci and Technol* 15:261–66.
Kress-Rogers, E. 2001. "Ultrasound propagation in foods and ambient gas: principles and applications." In *Instrumentation and Sensors for the Food Industry*, edited by Kress-Rogers, E., Brimelow, C.J.B., Woodhead, Cambridge, UK, pp. 364–65.
Kulmyrzaev, A., Cancelliere, C., McClements, D.J. 2000. Characterization of aerated foods using ultrasonic reflectance spectroscopy. *J Food Eng* 46:235–41.
Letang, C., Piau, M., Verdier, C., Lefebvre, L. 2001. Characterization of wheat-flour-water doughs: a new method using ultrasound. *Ultrasonics* 39:133–41.
Llull, P., Simal, S., Feminia, A., Benedito, J., Rossello, C. 2002. The use of ultrasound velocity measurement to evaluate the textural properties of sobrassada from Mallorca. *J Food Eng* 52:323–30.

Mathieu, J., Schweitzer, P. 2004. Measurement of liquid density by ultrasound backscattering analysis. *Meas Sci Technol* 15:869–76.

McCarthy, M.J., Wang, L., McCarthy, K.L. 2005. "Ultrasound properties." In *Engineering Properties of Foods,* edited by Rao, M.A., Rizvi, S.S.H., Datta, A.K., Taylor and Francis, Boca Raton, pp. 567–609.

McClements, D.J. 1995. Advances in the application of ultrasound in food analysis and processing. *Trends Food Sci Technol* 6:293–99.

Mizrach, A. 2000. Determination of avocado and mango fruit properties by ultrasonic technique. *Ultrasonics* 38:717–22.

———. 2004. Assessing plum fruit quality attributes with an ultrasonic method. *Food Res Int* 37:627–31.

Mizrach, A., Flitsanov, U. 1999. Nondestructive ultrasonic determination of avocado softening process. *J Food Eng* 40:139–44.

Mizrach, A., Flitsanov, U., Akerman, M., Zauberman, G. 2000. Monitoring avocado softening in low-temperature storage using ultrasonic measurements. *Comput Electron Agric* 26:199–207.

Morlein, D., Rosner, F., Brand, S., Jenderka, K.V., Wicke, M. 2005. Non-destructive estimation of the intramuscular fat content of the longissimus muscle of pigs by means of spectral analysis of ultrasound echo signals. *Meat Sci* 69:187–99.

Mulet, A., Benedito, J., Golas, Y., Carcel, J.A. 2002. Noninvasive ultrasonic measurements in the food industry. *Food Rev Int* 18 (2 and 3):123–32.

Newcom, D.W., Baas, T.J., Stalder, K.J. 2003. Genetics of pork quality, National Swine Improvement Federation Conference, Des Moines, Iowa, http://www.nsif.com/Conferences/2003/contents.html.

Ophir, J., Miller, R.K., Ponnekanti, H., Cespedes, I., Whittaker, A.D. 1994. Elastography of beef muscle. *Meat Sci* 36:239–50.

Ophir, J., Kaisar Alam, S., Garra, B.S., Kallel, F., Konofagou, E.E., Krouskop, T., Merritt, C.R.B., Righettr, R., Souchon, R., Srinivasan, S., Varghese, T. 2002. Elastography: imaging the elastic properties of soft tissues with ultrasound. *J Med Ultrason* 29:155–71.

Ouriev, B. 2000. Ultrasound Doppler Based In-line Rheometry of Highly Concentrated Suspensions, Ph.D thesis, Institute of Food Science, Swiss Federal Institute of Technology, Zurich.

Ozguler, A., Morris, S.A., O'Brien, W.D. Jr. 1999. Evaluation of defects in the seal region of food packages using the ultrasonic contrast descriptor. *Packaging Technol Sci* 12:161–71.

Pacific Northwest Division. 2005. Ultrasonic Rheometer, Brochures and Marketing Material, Pacific Northwest Division, Battelle. http://www.battelle.org/agrifood/rheometer.stm.

Packaging Technologies and Inspection. 2005. *SEAL-SCAN*™ Model Pti-550, Packaging Technologies and Inspection LLC. http://www.ptiusa.com.

Pascall, M.A., Richtsmeier, J., Riemer, J., Farahbakhsh, B. 2002. Non-destructive packaging seal strength analysis and leak detection using ultrasonic imaging. *Packaging Technol Sci* 15:275–85.

Peleg, K. 1999. Development of a commercial fruit firmness sorter. *J Agric Eng Res* 72:231–38.

Pierce, A.D. 1981. *Acoustics: An Introduction to Its Physical Principles and Applications.* Mcgraw-Hall, New York, pp. 14–16, 516–21.

Povey, M.J.W. 1997. *Ultrasonic Techniques for Fluids Characterization*, Academic Press, San Diego, pp. 141–57.

———. 1998. Ultrasonics in food. *Contemp Phys* 3(6):467–78.

———. 2000. Particulate characterization by ultrasound. *PSTT* 311:373–80.

Povey, M.J.W., Golding, M., Higgs, D., Wang, Y. 1999. Ultrasonic spectroscopy studies of casein in water. *Int Dairy J* 9:299–303.

Resa, P., Elvira, L., de Espinosa, F.M. 2004. Concentration control in alcoholic fermentation processes from ultrasonic velocity measurements. *Food Res Int* 37:587–94.

Ross, K.A., Pyrak-Nolte, L.J., Campanella, O.H. 2004. The use of ultrasound and shear oscillatory tests to characterize the effect of mixing time on the rheological properties of dough. *Food Res Int* 37:567–77.

Saggin, R., Coupland, J.N. 2001. Concentration measurement by acoustic reflectance. *J Food Sci* 665: 681–85.

———. 2002. Ultrasonic monitoring of powder dissolution. *J Food Sci* 674:1473–77.

———. 2004. Rheology of xanthan/sucrose mixtures at ultrasonic frequencies. *J Food Eng* 65:49–53.

Shal, N.N., Rooney, P.K., Ozguler, A., Morris, S.A., O'Brien, W.D. Jr. 2001. A real-time approach to detect seal defects in food packages using ultrasonic imaging. *J Food Prot* 649:1392–98.

Shannon, R.A., Probert-Smith, P.J., Lines, J., Mayia, F. 2004. Ultrasound velocity measurement to determine lipid content in salmon muscle; effects of myosepta. *Food Res Int* 37:611–20.

Simal, S., Benedito, J., Clemente, G., Femenia, A., Rossello, C. 2003. Ultrasonic determination of the composition of a meat-based product. *J Food Eng* 58:253–57.

Ultran Group. 2004. Products Catalogue, http://www.ultrangroup.com/docs.php.

Valente, M., Ferrandis, J.Y. 2003. Evaluation of textural properties of mango tissue by a near-field acoustic method. *Postharvest Biol Technol* 29:219–28.

Verlinden, B.E., De Smedt, V., Nicolai, B.M. 2004. Evaluation of ultrasonic wave propagation to measure chilling injury in tomatoes. *Postharvest Biol Technol* 32:109–13.

Youssao, I., Verleyen, V., Michaux, C., Leroy, P.L. 2002. Choice of probing site for estimation of carcass lean percentage in Piétrain pig using the real-time ultrasound. *Biotechnol Agron Soc Environ* 64:195–200.

Zhao, B., Basir, O.A., Mittal, G.S. 2003a. Detection of metal, glass and plastic pieces in bottled beverages using ultrasound. *Food Res Int* 36:513–21.

———. 2003b. A self-aligning ultrasound sensor for detecting foreign bodies in glass containers. *Ultrasonics* 413:217–22.

———. 2003c. Correlation analysis between beverage viscosity and sound velocity. *Int J Food Prop* 63:443–48.

———. 2005. Estimation of ultrasound attenuation and dispersion using short time Fourier transform. *Ultrasonics* 43:375–81.

5 NMR for Food Quality and Traceability

Raffaele Sacchi and Livio Paolillo

Introduction

Food quality and traceability are becoming ever more important aspects in the food chain and market. Safety, nutrition, and sensory quality are important for both fresh and processed food products, and recent issues are related to the assurance of the geographic origin of food, the raw material used (botanical variety or cultivar, animal species), and the processes applied to produce the food product.

The need for advanced and specific analytical techniques to assess food quality and verify in an objective way the "food's history" is linked to this developing scene. In this context chromatographic and spectroscopic methods are widely applied, and new methods are tested each year to improve the ability to verify quality, genuineness, and authenticity of foods.

The application of nuclear magnetic resonance (NMR) spectroscopy to the analysis and quality control of foods has also shown great development in the last few years. The increase of new applications and the attention to this technique by scientists, official control institutions, and food industries can be attributed both to the high specificity and versatility of the NMR technique and to the improvement of NMR instrument performances and availability. In the last 10 years, research centers and industries purchased a lot of new NMR equipment, thus stimulating new applications, and a similar trend can be expected in the near future.

The scientific literature on NMR in food science is increasing. In addition to specialized books and a high number of papers published in scientific journals, since 1992, every two years in Europe an International Conference on Applications of Magnetic Resonance in Food Science takes place in which scientists worldwide present new applications of the technique in the field of food science and technology (Webb et al. 2001; Belton et al. 2003).

The aim of this chapter is not a full review of all NMR applications in food science (over 1000 papers are available in the literature, as well as several books) but to show some examples of recent applications of this versatile and powerful technique. In particular, examples of applications of NMR in the fields of food quality (nutritional, sensory, freshness) and traceability (geographic origin, botanical origin, animal species, process technology applied to foods) have been selected to show different NMR approaches to different analytical problems.

Before describing selected applications of NMR to different food systems, a brief (very basic) general presentation of NMR spectroscopy—of its principles, instrumentations, and advantages and limits compared with other conventional analytical techniques—will be given.

NMR Spectroscopy

The Very Basic Principles of NMR

The principles of NMR spectroscopy are nowadays well-known in the scientific community, and it is not necessary to discuss here what can be easily found in many excellent textbooks. For the purpose of this review, it is only important to stress the main parameters that are used in treating the NMR data in food analysis.

Relevant parameters are primarily chemical shifts and relaxation times. Chemical shifts generally referred to in terms of δ (ppm) describe the properties of nuclei embedded in different chemical environments. For protons the usual range falls between 0 and 12 ppm, as referred to the TMS (tetramethylsilane), or equivalent molecules, signal set equal to $\delta = 0$ ppm. Other nuclei such as C13, P31, and N15 have distinct advantages in terms of chemical shift range in the order of more than 100 ppm but also have disadvantages due to their much weaker sensitivity. Instruments equipped with high magnetic fields (9 tesla or higher) overcome the sensitivity problem, and in fact these nuclei are now easily accessible.

Relaxation times describe the physical pathway by which perturbed nuclei return to their original equilibrium state. Relaxation times are then somewhat connected to the ease by which nuclei exchange information with the environment. In a three-dimensional space two types of relaxation times can exist: the spin-lattice relaxation time T_1 that brings the nuclei aligned along the z axis (where the magnetic field is applied) and the spin-spin relaxation time T_2 that is concerned with the physical phenomena occurring within the xy plane.

Relaxation times depend on the molecular weight of the compounds under investigation. Low-molecular-weight material generally yields nice, sharp spectra due to long relaxation times. On the contrary high-molecular-weight material yields broad signals and causes a quite short value of the relaxation times. These differences are used, for example, to discriminate water and fat in food or, in general, different components of different molecular weights.

Different NMR Techniques: Applicability, Advantages, and Disadvantages of Each One in Food Control

High-Resolution NMR (HR-NMR)

HR-NMR is a spectroscopic methodology that has a very high sensitivity and that makes use of high-magnetic fields that permit the observation of very detailed spectral parameters. These two factors are indeed quite important in assigning components and measuring their intensity ratios. In general qualitative and quantitative criteria are considered in HR-NMR. Qualitative criteria are discussed in terms of linearity and selectivity. NMR is definitely the best analytical method from the linearity point of view since the intensity of resonances is strictly proportional to the number of nuclei resonating at a certain frequency. Selectivity is also extremely good because NMR differentiates all the isotopes of the elements and even for a given isotope is able to yield measurable differences in chemical shift for different chemical environments.

From a quantitative point of view sensitivity, precision, and accuracy are crucial criteria. Sensitivity depends on the signal-to-noise ratio, which can be considered acceptable when it is higher than 10. Modern NMR spectrometers easily meet this requirement. Precision and accuracy can be determined from mean standard deviations on replicates. It has been shown that by using high-molecular-field spectrometers precision and accuracy are comparable to the GC (gas chromatography) techniques.

Site-Specific Natural Isotope Fractionation by NMR (SNIF-NMR)

This technique is based on the determination of the natural isotope contents of the main components of a genuine substance. The identification of the origin of a product is generally based on the identification of some minor components that are in many cases hard to detect and quantify. SNIF-NMR, instead, measures the isotopic ratio in a given system and in particular that of the main components, thus providing evidence of possible adulterations. For example, the C13/C12 ratio for vanillin extracted from vanilla beans is appreciably different from the measured ratio in the presence of adulteration with synthetic vanillin. Moreover SNIF-NMR enables the simultaneous estimate of more than one isotope ratio in the same substance, such as C13/C12 and D/H ratios, and contributes to a better definition of product authenticity.

Solid-State NMR

While in liquids, molecular motion permits a rather complete averaging of line-broadening magnetic interactions normally present in all materials. In solids these interactions are indeed present and contribute to a vast line-broadening phenomenon. For this reason solid-state NMR equipment requires considerable modifications with respect to solution NMR instruments. Dipolar broadening has to be eliminated by more sophisticated accessories in order to observe narrow lines. Strategies such as magic angle spinning (MAS) and extensive decoupling of protons when C13 nuclei are investigated permit the observation of components in the solid material. In addition relaxation times of components with different internal mobility can be measured, yielding complementary information on, for instance, tissue composition.

NMR Imaging (MRI)

Magnetic resonance imaging, which earned the Nobel Prize for Paul Lauterbur and Peter Mansfield in 2003, is a technique designed to reconstruct an image of an object, of a tissue, or of a whole human body from its NMR parameters. The relevant parameters are the relaxation times of the most abundant nuclei that possess higher sensitivity, namely hydrogen nuclei. Relaxation times of proton signals in different tissues are measured across the tissue itself in an inhomogeneous magnetic field, thus permitting the construction of an image where different tissues show different NMR-behaved protons. This is the case, for instance, for hydrogen nuclei in water and in fat that are by far the most populated proton species. Results obtained for living tissues can also be extended to water-rich foods, where protons in water and protons in fats can be easily discriminated.

The relevant aspect of MRI is that slices of tissues can be worked out by computer-assisted data processing. Each slice will show the many diverse objects (portions that differ in their chemically different proton content) in the tissue, leading to a clear observation of the tissue morphology.

Low-Field NMR and Relaxometry

Low-field NMR is a technique primarily used in industries for control purposes of plant processes. It is based on the use of inexpensive low-field apparatuses (0.5–0.7 T magnets) that give information on abundant components with different relaxation times. Thus relaxometry is considered in this technique, and an example is the classical evaluation of water and fat content in different foods. Water being a more mobile molecule has a longer relaxation time and is easily monitored against less mobile larger molecules such as fats. The technique is rather important in food processing and aging, where it gives fast responses on the water/fat relative ratios.

NMR in Food Science

The use of the different magnetic resonance techniques outlined in the preceding depends on different analytical purposes and on different food matrices. Five types of NMR experiments have been found particularly useful in food science (Gidley et al. 2003).

"Solution-state" quantitative high-resolution NMR methods are particularly suited to defining molecular structures of single molecules (the classical approach of NMR and mass spectrometry [MS] as tools to identify natural or synthetic molecules) and also to analyze increasingly complex mixtures in solution. Starting from the basic application to wine, honey, oils, and fruit juices, a good possibility is actually offered by NMR spectrometers at medium-high fields (7–12 tesla corresponding to proton frequencies of 300–600 MHz) for the analysis of foods characterized by high complexity. Several examples of NMR multicomponent analysis performed in a direct and nondestructive way directly on the raw food in water or organic solution will be discussed in the next paragraphs of this chapter.

"Solid-state" high-resolution NMR methods have been applied to many foods in order to characterize different molecular fractions (carbohydrate, proteins, lipids) present inside the solid food matrix (cereals, meats, cheese). The recent developments in solid-state NMR equipment allow the identification of low-molecular-weight components directly in the solid food sample.

Relaxation time NMR measurements are suitable as a probe of oil and water present in many foods and to assess the relative abundance of oil and water. These studies, of great importance for food industry process control and on-line studies of complex foods, have a minor application in the field of food traceability but a high practice value for simple quality control of food matrices in which fats (total fats, solid/liquid fat ratio) and water (total content and water activity and forms) can be related to sensory quality and shelf life (solid fats in margarines, starch retrodegradation, bread storage, food emulsion, etc).

Diffusion NMR methods take advantage of magnetic field gradients and are useful in defining both bulk diffusion and molecular self-diffusion rates noninvasively.

NMR imaging methods are also applied increasingly for the noninvasive visualization of food and can be of aid in the nondestructive control of fish, meat, and vegetable products.

The next paragraphs briefly provide an overview of the main applications that yield relevant information as far as the assessment of quality and traceability in foods is concerned. Some examples have been selected, mostly in the field of food lipid science. A short review dealing with a field as large as "NMR in food quality and traceability" has to be rather selective, and the choice of topics, examples, and figures of importance is inevitably subjective. In this review for readers unfamiliar with NMR, most attention is paid to high-resolution NMR applications rather than to the quantitative measurements of component mixtures that represent straightforward applications of low-resolution NMR.

NMR Applications for Food Quality and Traceability

With respect to the objective definition of food quality parameters, all NMR techniques can give potential information (directly or indirectly) related to the molecular composition, the physical status of water and fat, the starch and protein in emulsions, the internal structure of solid foods, and so on. In other words, all those characteristics of a food that satisfy the law and consumer expectations.

Some of the information available by NMR is not easily obtained by other methodologies. Important parameters can also be measured with chromatographic, sensory, or analytical tests, and a fairly large number of recent studies show good correlations between NMR and other traditional methods applied to food quality control.

The ability of high-resolution NMR to monitor in a noninvasive way all abundant molecules present in a raw material or in a complex system is a major driver for NMR applications in food science. This does not need to involve identification of each signal present in the NMR spectra, as pattern recognition and related techniques have been rapidly developed in the last few years. A lot of applications use the NMR spectra as a "fingerprint" of foods, useful in defining quality.

With the aim of developing applications for "food traceability" (an analytical test performed to assess the history and origin of a food), different NMR techniques and instruments have been employed, but the best results have come from the application of high-resolution NMR. Recent literature offers examples of interesting, rapid, and specific applications of high-resolution NMR (1) in differentiating foods obtained using different processing conditions or technologies and (2) to assess adulterations of low-grade foods in mixtures.

The assessment of the biological or geographic origin of foods has been performed using high-resolution NMR, both by recording the quantitative spectra of ^{1}H, ^{13}C, ^{17}O, or ^{31}P nuclei in combination with multivariate statistics and by determining the isotopic enrichment of selected nuclei. The natural distribution of isotopes, in fact, is a powerful tool for food authentication since the distribution of isotopes is dependent on many factors and the enrichment of selected isotopes can be of aid in assessing botanical species, variety, and place of production/growing.

The isotope ratio $^{2}H/^{1}H$ was the first parameter recorded in the 1980s by NMR and was used to identify the geographic origin of wine and the presence of sugars coming from sources other than grapes. The method developed at the University of Nantes (France) by Martin and Martin (1981), which defined site-specific natural isotope fractionation (SNIF)-NMR, is probably one of the best answers to the analytical problem of botanical and geographic traceability, and it was widely applied in the last years to several liquid foods (wine, fruit juices, oils).

As previously stated, the advantage of this technique, as compared to chromatography, is that all the molecules present in a mixture are detected without manipulation, separation, or treatment on a molar basis. Each different carbon or proton family gives rise to signals with well-defined chemical shifts, thus allowing the simultaneous identification and the quantification of the components by using one or more diagnostic peaks.

The following paragraphs will show some recent examples of NMR applications to quality control and traceability, grouped in relation to the studied foods. Finally, at the end of the chapter the future horizons of this powerful technique will be briefly discussed.

Vegetable Oils and Fats

One of the main fields in which high-resolution NMR has been applied is food lipids. The use of NMR in lipid science began some 40 years ago with the pioneering applications of proton NMR to oils and fats with the aim of assessing global unsaturation (Johnson and Shoolery 1962). Over 400 scientific papers have since been published on this topic, and also a "remake" of the classical work of Johnson and Shoolery has been recently proposed as a "new method" using modern NMR instruments (Miyake et al. 1998).

In the last few years carbon-13 and proton NMR have been successfully applied to oils, fats, and vegetable food lipids to assess their fatty acid composition (Sacchi et al. 1989, 1990a), the distribution of acyl chain on triacylglycerols (Sacchi et al. 1991, 1992), and the presence of minor "tracing" compounds (Sacchi et al. 1996), in order to assess oil quality (Sacchi et al. 1994) and authenticity (Sacchi et al. 1998).

All this has been based on the basic work done by several researchers on the identification of acyl structures and chemical shift assignment (Gunstone 1991; Wollenberg 1990).

The applications of proton and carbon-13 NMR to oil and fat analysis have already been reviewed many times (Pollard 1986; Sacchi et al. 1997; Aparicio et al. 1999; Sacchi 2001; Zamora et al. 1994).

Olive Oils

The applications of high-resolution (^{1}H and ^{13}C) NMR spectroscopy to virgin olive oil analysis have already been reviewed (Sacchi 2001; Sacchi et al. 1997; Zamora et al. 1994). Virgin olive oil can be easily analyzed in terms of freshness, quality, positional distribution of saturated and unsaturated fatty acids, and adulteration with foreign (seed) oils or with esterified/refined olive and olive-pomace oils. Recently, the possible contributions of high-resolution NMR to the definition of geographic origin of virgin olive oil and the process traceability have also been explored (Sacchi et al. 1998; Mannina et al. 2000, 2001, 2003).

Quality Assessment

As far as basic quality assessment is concerned, ^{13}C-NMR allows a rapid and direct analysis of total diacylglycerols and of the relative amounts of *sn*-1,2 and *sn*-1,3-diacylglycerols in olive oil (Sacchi et al. 1992). The same information can be also obtained by using proton NMR (Sacchi et al. 1991). The NMR determination of diacylglycerols in virgin olive oils can be a useful index in detecting the presence of partially neutralized oils added to virgin ones.

Process Traceability and Authenticity of Extra Virgin Olive Oil
Diacylglycerols naturally present in virgin olive oil do not exceed 2–3 percent (mainly *sn*-1,2-diacylglycerols) arising from the triacylglycerol biosynthesis. Larger amounts of diacylglycerols (mainly *sn*-1,3-diacylglycerols) are found in neutralized olive oils ("refined olive oil" and "refined olive-pomace oil") produced from a starting material (olives or olive-pomace) with a high level of free fatty acids. Refining processes (involving treatments with chemical or physical neutralization, bleaching, deodorization, and winterization) remove free fatty acids but cause only a partial loss of diacylglycerols (absorbed in the soapy phase during oil neutralization and washing). Therefore, the total content of diacylglycerols and the ratio *sn*-1,3-diacylglycerols/total diacylglycerols (*sn*-1,2 + *sn*-1,3-diacylglycerols) can provide a good discrimination between virgin olive oils and refined oils ("olive oils" and "olive-pomace oils") (Sacchi et al. 1989, 1990a, 1991).

Positional Distribution of Saturated Fatty Acids: Detection of Adulteration with Esterified Oils
The analysis of esterified oils (considered nonedible in the European Union) mixed with olive oils is normally carried out using standard methods based on the quantitative analysis of saturated fatty acids in the *sn*-2 position of triacylglycerols. The EU official analytical procedure requires several steps based on (1) hydrolysis of triacylglycerols by pancreatic lipase, (2) extraction of lipids by diethylether, (3) thin layer chromatography (TLC) fractionation of lipids and recovery of the 2-monoacylglycerol band, and (4) transmethylation of 2-monoacylglycerols (2-MGs) and gas chromatographic analysis of the fatty acid composition of 2-MGs. The NMR method based on the analysis of the ^{13}C spectrum yields immediate results, operating directly on the oil samples without any chemical manipulation, and it seems to represent the only direct instrumental method by which the positional distribution of fatty acids on glycerols can be identified (Sacchi et al. 1992). Determinations made on virgin olive oils spiked with a known amount of esterified oils showed a good quantitative accuracy of the NMR method (Sacchi et al. 1992). From the experimental data, it was observed that the method outlined in the preceding can be easily applied to detect a level of 2–3 percent saturated components in the *sn*-2 position. This limit can be lowered by using high magnetic fields (11–14T) or by increasing spectral line intensities with overnight accumulations. In order to obtain the best resolution, good probe temperature control during accumulation and appropriate proton decoupling are required.

Further information about virgin olive oil purity can be obtained from the analysis of the unsaponifiable fraction, which, in virgin olive oil, is constituted mainly of squalene, β-sitosterol, and aliphatic alcohols. ^{13}C-NMR analysis of the unsaponifiable matter has been recently used in combination with multivariate statistical analysis for the discrimination of virgin olive oil from olive-pomace oil and refined olive oils (Zamora et al. 2002b).

Geographic Origin and Varietal Traceability of Extra Virgin Olive Oil
Multivariate statistical analysis of high-field proton NMR spectral data, obtained from 55 extra-virgin olive oil samples coming from four Italian regions, indicates a possible contribution of NMR in the authentication of virgin olive oil geographic origin (Sacchi et al. 1998; Mannina et al. 2000, 2001, 2003). Apart from the obvious need of collecting reference spectra for each year of production (sampling from the representative varieties of each region and data-bank preparation), the application of multivariate statistics to NMR

spectral data offers an interesting developing area for the certification of the geographic origin of extra-virgin olive oils.

Fish Oils and Fish Lipids

In the last 10 years, high-resolution carbon-13 and proton-NMR spectroscopies (Aursand and Alexon 2001; Aursand and Grasdalen 1992; Aursand et al. 2000) have been successfully applied to the analysis of fish oils and lipids. These analyses focused not only on the rapid determination of lipid classes and fatty acid composition but also on the assessment of the positional distribution of fatty acids along the glycerol moiety, on the fish quality, as well as on answers to issues like lipid oxidation and lipolysis, species and geographic traceability (Medina et al. 1994a, 1994b, 1998, 2000).

As previously stated, ^{13}C- and ^{1}H-NMR are complementary techniques that define lipid composition and chemical alterations. ^{13}C-NMR offers the advantage of a higher resolution in terms of chemical shift scattering but lacks in sensitivity and requires longer experimental times in order to obtain quantitative spectra. Proton NMR, on the other hand, is very rapid and quantitative, but proton signals of lipid molecules are present in a narrow spectral range, thus reducing resolution. Therefore, different applications have been carried out taking into account advantages and disadvantages of the two techniques.

An overview of recent applications of high-resolution proton NMR to fish oils and lipids has recently been written (Sacchi et al., in press). The detection and quantification of ω-3 (*n*-3) polyunsaturated fatty acids (PUFAs) in fish oils and lipids make up one of the main applications. The use of this modern technique allows the rapid, nondestructive, and structure-specific analysis of fish oils and of lipids extracted from different fish muscles and seafoods. The nutritional quality and the measure of total ω-3 PUFAs can be determined by the careful integration of the *n*-3 methyl protons. Furthermore, the way to quantify cholesterol, phospholipids, and docosahexaenoic acid (DHA) (22:6*n*-3) in complex fish lipid mixtures is also described. The analysis of oxidation products can also be performed using high-field instruments (500–600 MHz). Finally, several applications to fish lipid quality control (i.e., oxidation and composition of commercially available oils enriched in *n*-3 PUFA, effect of thermal processing during tuna canning, evaluation of ripening in salted anchovies) are discussed.

Quantitative Determination of *n*-3 PUFAs

The identification and quantification of the ω-3 (*n*-3) PUFAs for their intrinsic dietary benefits in human health are important because ω-3 (*n*-3) PUFAs make up one of the major quality attributes of fish oils. Proton NMR spectroscopy offers a rapid and structure-specific method for the global quantification of *n*-3 PUFAs in fish lipids by taking advantage of only two selected proton NMR methyl resonances (Sacchi et al. 1993a, 1993b; Igarashi et al. 2000, 2003). A simple one-dimensional TOCSY experiment made on a sample of tuna lipids allows the assignment of *n*-3 methyls at 0.95 ppm, which are coupled to the ω-2 (methylene), ω-3,4 (olefinic), and finally diallylic (ω-5) protons. Therefore, the careful integration of the clearly resolved methyl resonances of *n*-3 PUFAs and of the saturated *n*-6, *n*-7, and *n*-9 monounsaturated fatty acids (MUFAs) and PUFAs allows the quantitative measurement of the total *n*-3 components. For this application spectrometers working at reasonably high magnetic fields (10 tesla or higher) are used.

In a more recent work, the results outlined in the preceding have been applied more extensively and proposed as an International Union of Pure and Applied Chemistry (IUPAC) official method for the determination of DHA and of *n*-3 fatty acids in fish oils (Igarashi 2003). In this research several samples of different origins (tuna oil, bonito oil, salmon oil, and sardine oil) were examined in different laboratories and using spectrometers at different magnetic fields to determine the accuracy of the measured peak intensities and, therefore, the quantification of fish oil composition in terms of *n*-3 components and DHA. The relative composition (mol%) of the *n*-3 fatty acids was obtained by simply comparing the relative intensities of the methyl resonances at 0.95 ppm (*n*-3 fatty acids) with those at 0.8 ppm (all other fatty acids). Moreover, the DHA relative composition (mol%) was shown to be measured from the assigned C2 and C3 proton resonances at 2.38 ppm well separated from the C2 methylene protons of the other fatty acids at 2.28 ppm (Aursand and Grasdalen 1992; Aursand et al. 2000). The DHA weight concentration in mg/g can also be easily measured by comparing its C2, C3 methylene resonance intensities with those of known amounts of ethyleneglycol dimethyl ether (EGDM) as an internal standard, whose resonances occur at 3.35 ppm (methyls) and at 3.5 ppm (methylenes) (Igarashi et al. 2003).

The data obtained in 13 different NMR laboratories from five different countries (Japan, Norway, Italy, France, and Denmark) were subjected to statistical analyses and demonstrated that both repeatability and reproducibility fall in a quite low range of relative standard deviations and in particular vary between 1.73 and 4.27 percent for DHA concentrations in mg/g. These values are acceptable for quality control in industrial laboratories, and therefore this methodology was proposed as an official method of analysis, considering that its major advantage over other possible analytical methods is its operational ease and the nondestructive NMR analysis. The family of *n*-3 fatty acids is assessed specifically, using proton resonances not affected by molecular weight of fatty acids and lipid classes. In fact, common transesterification methods used for gas chromatographic analysis do not esterify free fatty acids that are not measured by high-resolution gas chromatography (HRGC) in highly lipolyzed oils. In these cases, GC/MS is needed for careful quantification; also needed is a specific derivative (treating the sample using diazomethane) or time-consuming total esterification methods.

1H-NMR Spectra of Fish Oils and Lipids Extracted from Fish Muscles

Proton NMR spectra of commercial fish oil, tuna lipids, salmon lipids (Aursand et al. 2000; Sacchi et al. 1993), and bonito and sardine oils permit the evaluation of both fatty acid profiles and lipid classes composition. NMR spectra of natural lipid extracts from fish muscles differ from those of refined fish oils due to the relevant presence not only of triacylglycerols as major components but also of a large amount of polar lipids (phospholipids, diacylglycerols, free fatty acids) or other lipid-soluble unsaponifiable components (cholesterol).

The main resonances, commonly found in all fish oils and lipids, occur at 0.85 ppm for methyl groups of saturated *n*-6 and *n*-9 acyl chains, at 0.95 ppm for methyl groups of ω-3 components, at 1.65 ppm for the methylene protons in C3 position, at 2.05 ppm for the allylic protons, at 2.31 for methylenes in C2, at 2.76 ppm for the doubly allylic protons, and at 5.39 ppm for olefinic protons superimposed to glyceryl resonances (Sacchi et al. 1993b). In addition to these resonances another signal appearing at 3.37 ppm can be assigned to phosphatidylcholine methyls. Some of these signals, therefore, may be used for assessing lipid composition and fatty acid composition. In addition, the careful study of the minor resonances detectable in diagnostic spectral areas, the use of high-field NMR equipments and two-dimensional experiments, as well as the combination of chromatographic techniques in concentrating minor lipid fractions with proton NMR allow obtaining, in a short time, very useful information. Interesting information, with respect to the quality assessment and process control of fish products, is reported in the following paragraphs.

Lipolysis and Fish Processing

Sacchi et al. (1993a) and Medina et al. (1994b, 1995, 1998, 2000) applied high-resolution ^{13}C-NMR in fish-processing quality control, in particular monitoring the effect of tuna canning on ω-3 PUFAs, lipolysis, and oxidation.

Lipid classes and their evolution, due to enzymes and thermal processing, measured by NMR were shown to be a good quality parameter of raw, processed fish and fish oils. Proton NMR, on the contrary, does not allow a direct quantification of carboxyl protons of free fatty acids, and in addition methylene protons bound to the carbon C2 overlap with those of esterified acyl chains.

Oxidation Products

Proton spectroscopy can also be applied to studying the oxidative degradation of fish lipids by the same approach developed for other oils and lipids (Saito and Nakamura 1990; Claxson et al. 1994; Falch et al. 2004, 2005). This technique is able to detect, simultaneously, primary and secondary oxidation products in oils and lipids, and to study the effect of industrial processing (salting, thermal treatments, etc.) on lipids. This can be achieved by two analytical strategies. The simplest, but of minor applicative interest, is the study of the relative intensity changes of the main components, particularly PUFAs. This approach was followed by Saito and Nakamura (1990), who demonstrated that the oxidative deterioration of fish oils could be monitored by estimating the ratio of olefinic protons (4.9–5.8 ppm) to aliphatic protons (0.5–3 ppm). This ratio decreases as the oxidative deterioration proceeds. This procedure is applicable also at low magnetic fields and could be of importance in monitoring industrial processing or in the rapid control of global unsaturation and fatty acid composition.

A second and more interesting application has to be considered, although more difficult and requiring more skilled procedures. It concerns the study and the identification of new signals in the spectra due to components generated in the oxidation process. In an interesting paper, Falch et al. (2004) have shown, in the proton spectra of oxidized fish lipids, signals at approximately 6.5, 6, 5.6, 4.4, and 2.9 ppm originated from the protons of hydroperoxides and conjugated *trans-cis* double bonds formed during oxidation of polyunsaturated fatty acids, mainly from oxidized *n*-3 double bonds.

Secondary oxidation lipid products such as aldehydes can be observed in the proton spectrum. Signals of saturated and unsaturated aldehyde protons between 9.3 and 9.9 ppm have also been detected for unsalted salmon fillet lipids submitted to the oxidation process (Guillén and Ruiz 2004). Guillén and Ruiz monitored the lipid oxidation by measuring changes in relative ratios of acyl groups belonging to ω-3 and saturated and unsaturated acyl groups.

Meat and Fish Products (Liquid HR, Solid HR, and Imaging)

High-resolution proton NMR has been applied by Al-Jowder et al. (2001) in combination with chemometrics to the problem of meat products authentication. Samples of meat products were homogenized in 1 M HCl for one minute; then the suspension was centrifuged, and the supernatant, mixed with D_2O, directly transferred into an NMR tube and analyzed. From the chemometric analysis (PCA—principal component analysis, PLS—partial least squares, LDA—linear discriminant analysis) of spectral data, the adulteration by undeclared mixing of offal (beef liver and kidney) to pure beef muscle was detected, and a quantitative calibration of the content of liver and kidney was made. The work demonstrated that even in meat products a simple extraction of meat tissues and the use of high-field (500 MHz) proton NMR lead to the rapid identification and quantification of several diagnostic metabolites (glucose, lactic acid, several amino acids, creatine, creatinine, carnosine, carnitine, anserine, glycylbetaine, myo-inositol).

High-resolution carbon-13 has been applied to the characterization of meat lipids (Di Luccia et al. 1994). Among different aspects (fatty acid composition, lipid hydrolysis, oxidation) related to the quality and freshness of meat, interesting information has been gained by the study of NMR on the positional distribution of fatty acids. One of the factors strongly affecting the nutritional quality of meat is the fatty acid positional distribution. The ability of NMR to offer a global view of the stereodistribution of fatty acids on triacylglycerols is a nice contribution to meat quality assessments.

The application of NMR imaging (MRI) techniques to the characterization of meat structure has been recently reviewed by Bonny et al. (2001). This technique allows an excellent view of *marbling* (intramuscular fat distribution), which is one of the most important parameters for meat quality and for its reaction to some processing (i.e., salting, jam production, etc.). With respect to the visual evaluation of meat cuts, MRI at high field (4.7 T) offers the advantages of a nondestructive procedure and a three-dimensional view of marbling.

A similar approach was developed by Cernadas et al. (2001) to classify loin from the Iberian pig, in relation to marbling, and allowed Cernadas and co-workers to stress differences in meat quality in two different families of Iberian pork. The use of a medical MRI apparatus and an appropriate algorithm in the MRI data evaluation was shown to be a valid and nondestructive alternative for loin quality evaluation. MRI has also been applied as a tool to evaluate the adiposity distribution in fish (Collewet et al. 2001).

The potential contribution of NMR to the definition of the geographic origin of fish products and fish species has also been recently explored (Aursand et al. 2000; Aursand and Alexon 2001). Deuterium SNIF-NMR, in combination with data from gas chromatography, isotope ratio mass spectrometry and statistical analysis, allowed differentiation between wild and farmed salmon from Norway and Scotland (Aursand et al. 2000). The possibility of using carbon-13 NMR as a "fingerprint" in origin recognition of wild and farmed fish and to discriminate different fish species was confirmed by Aursand and Alexon (2001). Carbon-13 NMR quantitative data (282 carbon-13 resonances detected in the spectra) analyzed using Kohonen neural networks and self-organizing feature maps allowed a quantitative classification of fish species (halibut, mackerel, and salmon).

Milk and Dairy Products

Belloque and Ramos (1999) reviewed the applications of NMR spectroscopy to milk and dairy products. Several applications are focused on the study of the physical state of milk and cheese and the structural characterization and conformational/aggregational state of milk proteins. High-resolution NMR spectroscopy for qualitative and quantitative analyses has been applied primarily to milk analysis, its application to dairy products being limited by the complexity of the systems. The lipid fraction of milk has been studied in order to define both the fatty acid composition and the fatty acid distribution on the *sn*-1,3 and *sn*-2 position in milk fat triacylglycerols (Kalo et al. 1996).

An interesting application devoted to quality and traceability assessment has been developed in order to distinguish natural milk caseins from commercial caseinates (Ward and Bastian 1998). By using ^{31}P-NMR, the resonance of phosphoserine was used to evaluate the degree of phosphorylation in super- and dephosphorylated caseins. The ^{31}P-NMR spectra of commercial caseinates has shown some phosphoserine-depleted β-casein fractions, which may provide a way to distinguish natural caseins from commercial caseinates.

The application of SNIF-NMR to dairy science is quite limited, since the process involved in cheese making is often complex and can involve the use of components of different sources, which can alter the isotopic ratio.

Wine, Beer, and Fruit Juices

The study of fermented beverages like wine and beer, as well as of fruit-based products and fruit juice containing sugars and other water-soluble components, has been carried out extensively by high-resolution NMR and SNIF-NMR in order to assess quality, authenticity, and geographic traceability.

Besides simple applications such as the determination of methanol content in wines (Sacchi et al. 1990b), high-field proton NMR has been applied as a screening tool to determine the authenticity of orange or other juices. The ^{1}H-NMR spectrum is very complex and contains quantitative information on sugars, flavonoids, organic acids, amino acids, and also minor components (Vogels et al. 1996; Belton et al. 1996). These data, when subjected to a main component analysis, are able to permit a rapid discrimination between authentic and adulterated samples.

The first application of SNIF-NMR to food analysis was made just to quantify the deuterium/hydrogen content (D/H) in specific sites of ethanol obtained by fermentation of natural sugars from different botanical sources. The D/H ratio on the methyl site of ethanol allowed the distinguishing of ethanol arising from grape, beet, and other botanical sources (Martin and Martin 1981, 1990, 1995). The first application of this finding was controlling the addition of sugar in wine. The same information was then applied to the geographic discrimination of wines.

Following basic research done by the University of Nantes in the 1980s, Day et al. (1995) determined the geographic origin in French wines, using the joint analysis of elemental and isotopic composition.

Martin et al. (1996) also demonstrated the ability of deuterium SNIF-NMR to detect the addition of beet sugar in concentrated and single-strength fruit juices.

Fresh Vegetables (Solid NMR and Imaging)

An approach recently used for vegetables concerns the NMR identification of those molecules that are related to quality, shelf life, and the expression of different genetic materials. The opportunity to use NMR for this purpose, not just for the classical (liquid-state) high-resolution approach, arises from the progress made in NMR equipment able to analyze in high-resolution mode complex solid-liquid or solid systems (Harris 2001). The magic angle spinning (MAS) NMR probes allow observing, with good spectral resolution, signals arising from low-molecular-weight solutes within biological tissue without any pretreatment of the sample. In addition, whole-water or organic extracts can be analyzed by high-field magnetic liquid NMR.

Elegant examples of this strategy can be found using some fresh fruits (Duarte et al. 2001, 2005) and green tea leaves (Gidley et al. 2003).

Honey

The deuterium/hydrogen ratio derived from SNIF-NMR applied to honey arising from temperate regions was found to be very close to that found in beet sugar; the addition of beet sugar in tropical honey, however, can be easily performed. SNIF-NMR has also been able to assess the authenticity of citrus honeys (Lindner et al. 1996).

Latest Developments and Future Horizons

NMR spectroscopy offers advantages and disadvantages over other analytical techniques. The advantages are its nondestructiveness and its ease in determining the relative amounts of the main components. The disadvantages are due to the limited sensitivity of the NMR method, which requires not less than fractions of milligrams for a rapid analysis. This makes the determination of minor components time-consuming and difficult, although modern developments in equipment, such as cryoprobes, can increase the sensitivity range by a factor of 100, thus extending sensitivity to the micromolar range.

A horizon in which NMR already has shown its potential contribution is the area of the relatively new discipline of "metabolomics," which has the target of simultaneously describing all molecular components present in a system. As genomic, proteomic, and lipidomic analyses of crop plants and farm animals become more common, NMR spectroscopy will play a major role in this field of food science.

NMR will remain the single most versatile and informative spectroscopic method for food analysis (Gidley et al. 2003). Improvements can be expected in

- NMR hardware and electronics (dynamic range, higher sensitivity with respect to minor components detected at very low micromolar levels, cryoprobes, etc.), with increased sensitivity and resolution of spectra;
- the development of hyphenated techniques (liquid chromatography [LC]-NMR and LC-NMR-MS for automatized approaches in complex food systems);
- robustness and reduced costs;
- the ingenuity of experiments and the ease of use of NMR spectrometers.

References

Al-Jowder, O., Casuccelli, F., Defernez, M., Kemsley, E.K., Wilson, R.H., Colquhoun, I.J. 2001. High resolution NMR studies of meat composition and authenticity. In *Magnetic resonance in food science: a view to the future,* edited by Webb, G.A., Belton, B.S., Gil, A.M., Delgadillo I., pp. 232–38, Royal Society of Chemistry, Cambridge, UK.

Aparicio, R., Mcintyre, P., Aursand, M., Eveleigh, L., Marighetto, N., Rossell, B., Sacchi, R., Wilson, R., Woolfe, M. 1999. Oils and fats. In *Food Authenticity: Issues and Methodologies,* European Community, Concerted Action n. AIR3-CT94–2452, pp. 209–69, Eurofins Scientific, Nantes, France.

Aursand, M., Alexon, D.E. 2001. Origin recognition of wild and farmed salmon (Norway and Scotland) using ^{13}C-NMR spectroscopy in combination with pattern recognition techniques. In *Magnetic resonance in food science: a view to the future,* edited by Webb, G.A., Belton, B.S., Gil, A.M., Delgadillo I., pp. 227–31, Royal Society of Chemistry, Cambridge, UK.

Aursand, M., Grasdalen, H. 1992. Interpretation of the ^{13}C-NMR spectra of omega-3 fatty acids extracted from the white muscle of Atlantic salmon (*Salmo salar*). *Chem Phys Lipids,* 62:239–51.

Aursand, M., Mabon, F., Martin, G.J. 2000. Characterization of farmed and wild salmon (*Salmo salar*). *JAOCS,* 77:659–66.

Belloque, J., Ramos, M. 1999. Application of NMR spectroscopy to milk and dairy products. *Trends Food Sci Technol,* 10:313–20.

Belton, B.S., Gil, A.M., Webb, G.A., Rutledge, D. 2003. Magnetic resonance in food science: latest development. In *Proceedings of the Sixth International Conference on Applications of Magnetic Resonance in Food Science* (Paris, France, September 4–6, 2002), pp. 1–272, Royal Society of Chemistry, London.

Belton, M.S., Delgadillo, I., Holmes, E., Nicholls, A., Nicholson, J.K., Spraul, M. 1996. Use of high-field ^{1}H-NMR spectroscopy for the analysis of liquid foods. *J Agric Food Chem,* 44:1483–87.

Bonny, J.M., Laurent, W., Renou, J.P. 2001. Characterisation of meat structure by NMR imaging at high field. In *Magnetic resonance in food science: a view to the future,* ed. Webb, G.A., Belton, B.S., Gil, A.M., Delgadillo, I., pp. 17–21, Royal Society of Chemistry, Cambridge, UK.

Cernadas, E., Antequera, T., Rodrìguez, P.G., Duran, M.L., Gallardo, R., Villa, D. 2001. Magnetic resonance imaging to classify loin from Iberian pig. In *Magnetic resonance in food science: a view to the future,* edited by Webb, G.A., Belton, B.S., Gil, A.M., Delgadillo, I., pp. 239–45, Royal Society of Chemistry, Cambridge, UK.

Claxson, A.W.D., Hawkes, G.E., Richardson, D.P., Naughton, D.P., Haywood, R.M., Chander, C.L., Atherton, M., Lynch, E.J., Grootveld, M.C. 1994. Generation of lipid peroxidation products in culinary oils and fats during episodes of thermal stressing: a high field ^{1}H NMR study. *Febs Lett,* 355:81–90.

Collewet, G., Toussaint, C., Davenel, A., Akoka, S., Médale, F., Fauconneau, B., Haffray, P. 2001. Magnetic resonance imaging as a tool to quantify the adiposity distribution in fish. In *Magnetic resonance in food science: a view to the future,* edited by Webb, G.A., Belton, B.S., Gil, A.M., Delgadillo, I., pp. 252–58, Royal Society of Chemistry, Cambridge, UK.

Day, M.P., Ben-Li, Z., Martin, G.J. 1995. Determination of the geographical origin of wine using joint analysis of elemental and isotopic composition: II. Differentiation of the principal production zone in France for the 1990 vintage. *J Sci Food Agric,* 67:113–23.

Di Luccia, A., Satriani, A., Sacchi, R., Gigli, S., Ferrara, L. 1994. A study of neutral lipids from river buffalo (*Bubalus bubalus L.*) meat. *Proceedings of the IVth World Buffalo Congress* (São Paulo, Brazil, June 23–30, 1993), 2:118–20.

Duarte, I.F., Delgadillo, I., Spraul, M., Humpfer, E., Gil, A.M. 2001. An NMR study of the biochemistry of mango: the effect of ripening, processing and microbial growth. In *Magnetic resonance in food science: a view to the future,* edited by Webb, G.A., Belton, B.S., Gil, A.M., Delgadillo I., pp. 259–66, Royal Society of Chemistry, Cambridge, UK.

Duarte, I.F, Goodfellow, B.J., Gil, A.M., Delgadillo, I. 2005. Characterization of mango juice by high resolution NMR, hyphenated NMR and diffusion ordered spectroscopy. *Spectroscopy Lett,* 38(3):319–42.

Falch, E. 1994. Generation of lipid peroxidation products in culinary oils and fats during episodes of thermal stressing: a high field 1H NMR study. *FEBS Lett,* 355(1):81–90.

Falch, E., Anthonsen, H., Axelson, D., and Aursand, M. 2004. Correlation between ^{1}H NMR and traditional analytical methods for determining lipid oxidation in ethylesters of docosahexaenoic acid. *JAOCS,* 81(12): 1105–9.

Falch, E., Størseth, T.R., Aursand, M. 2005. HR NMR to study quality changes in marine by-products. In *Magnetic resonance in food science: the multivariate challenge,* edited by Engelsen, S.B., Belton, P.S., Jakobsen, H.J., pp. 238–52, Royal Society of Chemistry, Cambridge.

Falch, E., Rustad, T., Aursand, M. In press. By-products from gadiform species as raw material for production of marine lipids as ingredients in food or feed. *Process Biochem.*

Gidley, M.J., Ablett, S., Martin, D.R. 2003. The food supply chain: the present role and future potential of NMR. In *Magnetic resonance in food science, latest developments,* edited by Belton, P.S., Gil, A.M., Webb, G.A., Rutledge, D., pp. 3–16, Royal Society of Chemistry, Cambridge, UK.

Guillén, M.D., Ruiz, A. 2004. Study of the oxidative stability of salted and unsalted salmon fillets by 1H nuclear magnetic resonance. *Food Chem,* 86:297–304.

Gunstone, F.D. 1991. High resolution NMR studies of fish oils, *Chem Phys Lipids,* 59:83–89.

Harris, R.K. 2001. Recent advances in solid-state NMR. In *Magnetic resonance in food science: a view to the future,* edited by Webb, G.A., Belton, B.S., Gil, A.M., Delgadillo, I., pp. 3–16, Royal Society of Chemistry, Cambridge, UK.

Igarashi, T., Aursand, M., Hirata, Y., Gribbestad, I.S., Wada, S., Nonaka, M. 2000. Nondestructive quantitative determination of docosahexaenoic acid and n-3 fatty acids in fish oils by high-resolution ^{1}H nuclear magnetic resonance spectroscopy. *J Am Oil Chem Soc,* 77:737.

Igarashi, T., Aursand, M., Sacchi, R., Paolillo, L., Nonaka, M., Wada, S. 2003. Determination of docosahexaenoic acid and n-3 fatty acids in refined fish oils by ^{1}H-NMR spectroscopy: IUPAC interlaboratory study. *J AOAC Int,* 85(6):1341–54.

Johnson, L.F., Shoolery, J.N. 1962. Determination of unsaturation and average molecular weight of natural fats by NMR. *Anal Chem,* 34:1136.

Kalo, P., Kempinnen, A., Kilpelainen, I. 1996. Determination of positional distribution of butyril groups in milkfat triacylglycerols, triacylglycerol mixtures, and isolated positional isomers of triacylglycerols by gas chromatography and ^{1}H nuclear magnetic resonance. *Lipids,* 31:331–36.

Lindner, P., Bermann, E., Gamarnik, B. 1996. Characterisation of citrus honey by deuterium NMR. *J Agric Food Chem,* 44:139–40.

Mannina, L., Patumi, M., Proietti, N., Segre, A.L. 2000. D.O.P. (denomination of protected origin) geographical characterization of Tuscan extra virgin olive oils using high-field ^{1}H NMR spectroscopy. *Ital J Food Sci,* 21:15–24.

Mannina, L., Patumi, M., Proietti, N., Bassi, D., Segre, A.L. 2001. Geographical characterization of Italian extra virgin olive oils using high-field ^{1}H NMR spectroscopy. *J Agric Food Chem,* 49:2687–96.

Mannina, L., Dugo, G., Salvo, F., Cicero, L., Ansanelli, G., Calcagni, C., Segre, A. 2003. Study of the cultivar-composition relationship in Sicilian olive oils by GC, NMR, and statistical methods. *J Agric Food Chem,* 51:120–27.

Martin, G.J., Martin, M.L. 1981. Deuterium labelling at the natural abundance level as studied by high field quantitative 2H NMR. *Tetrahedron Lett,* 22(36):3525–28.

Martin, G.J., Martin, M.L. 1995. Stable isotope analysis of food and beverages by nuclear magnetic resonance. *Annu Rep NMR Spectrosc,* 31:81–104.

Martin, M.L., Martin, G.J. 1990. Deuterium NMR in the study of site-specific natural isotope fractionation (SNIF-NMR). In *NMR basic principles and progress,* vol. 23, edited by Diehl, P., Fluck, E., Günther, H., Kosfield, R., Seelig, J., pp. 1–61, Springer-Verlag, Berlin.

Martin, G.J., Zhang, B.L., Martin, M.L., Dupuy, P. 1983. Application of quantitative deuterium NMR to the study of isotope fractionation in the conversion of saccharides to ethanols. *Biochem Biophys Res Commun,* 111(3):890–96.

Martin, G.G., Wood, R., Martin, G.J. 1996. Detection of added beet sugar in concentrated and single strength fruit juices by deuterium nuclear magnetic resonance (SNIF-NMR method): collaborative study. *J AOAC Int,* 79(4):917–28.

Medina, I., Sacchi, R., Aubourg, S. 1994a. ^{13}C nuclear magnetic resonance monitoring of free fatty acid release after fish thermal processing. *J Am Oil Chem Soc,* 71(5):479–82.

Medina, I., Aubourg, S., Sacchi, R., Addeo, F., Giudicianni, I., Paolillo, L. 1994b. Quality assurance of processed and stored fish by high resolution NMR analysis of lipids. *Proceedings of the Second International Conference on Applications of Magnetic Resonance in Food Science* (Aveiro, Portugal, September 19–21), p. 43.

Medina, I., Sacchi, R., Aubourg, S. 1995. A ^{13}C-NMR study of lipid alteration during fish canning: effect of filling medium. *J Sci Food Agric,* 69:445–50.

Medina, I., Sacchi, R., Giudicianni, I., Aubourg, S.P. 1998. Oxidation of fish lipids during thermal stress as studied by ^{13}C nuclear magnetic resonance spectroscopy. *J Am Oil Chem Soc,* 75:147–54.

Medina, I., Sacchi, R., Aubourg, S. 2000. Application of ^{13}C-NMR to the selection of the thermal processing conditions of canned fatty fish. *Eur Food Res Technol,* 210:176–78.

Miyake, Y., Yokomizo, K., Matsuzaki, N. 1998. Determination of unsaturated fatty acid composition by high resolution nuclear magnetic resonance spectroscopy. *J Am Oil Chem Soc,* 75(7):1091–94.

Pollard, M. 1986. Nuclear magnetic resonace spectroscopy (high resolution). In *Analysis of oils and fats,* edited by Hamilton, R.J., Rossell, J.B., pp. 401–34, Elevier Applied Science Publishers, Barking, UK.

Sacchi, R. 2001. High resolution NMR of virgin olive oil. In *Magnetic resonance in food science—a view to the next century,* pp. 213–26, Royal Society of Chemistry, Cambridge, UK.

Sacchi, R., Addeo, F., Giudicianni, I., Paolillo, L. 1989. La spettroscopia di risonanza magnetica nucleare nell'analisi degli oli di oliva. *Rivista Italiana delle Sostanze Grasse,* 56(4):171–78.

Sacchi, R., Addeo, F., Giudicianni, I., Paolillo, L. 1990a. Applicazione della spettroscopia ^{13}C-NMR alla determinazione di mono-digliceridi ed acidi grassi liberi nell'olio di oliva di pressione e nei rettificati. *Rivista Italiana delle Sostanze Grasse,* 57(5):245–52.

Sacchi, R., Maione, V., Giudicianni, I., Nota, G., Paolillo, L., Addeo, F. 1990b. The determination of methanol in wines by proton NMR spectroscopy. *Ital J Food Sci,* 2(2):113–22.

Sacchi, R., Paolillo, L., Giudicianni, I., Addeo, F. 1991. Rapid ^{1}H-NMR determination of 1,2 and 1,3 diglycerides in virgin olive oils. *Ital J Food Sci,* 3(4):253–62.

Sacchi, R., Addeo, F., Giudicianni, I., Paolillo, L. 1992. Analysis of the positional distribution of fatty acids in olive oil triacylglycerols by high resolution ^{13}C-NMR of the carbonyl region. *Ital J Food Sci,* 4(2): 117–23.

Sacchi, R., Medina, I., Aubourg, S.P., Giudicianni, I., Paolillo, L., Addeo, F. 1993a. Quantitative high-resolution ^{13}C-NMR analysis of lipid extracted from the white muscle of Atlantic tuna (*Thunnus Alalunga*). *J Agric Food Chem,* 41:1247–53.

Sacchi, R., Medina, I., Aubourg, S.P., Paolillo, L., Addeo, F. 1993b. Proton nuclear magnetic resonance rapid and structure-specific determination of ω-3 polyunsaturated fatty acids in fish lipids. *J Am Oil Chem Soc,* 70(3):225–28.

Sacchi, R., Chemin, S., Paolillo, L., Addeo F. 1993c. Evolution of virgin olive oil diacylglycerols during olive ripening, oil extraction and storage as studied by ^{13}C-NMR spectroscopy. In *EEC-FLAIR Symposium on the Sensory and Nutritional Quality of Virgin Olive Oil* (Milano, November 23–24).

Sacchi, R., Addeo, F., Segre, A.L., Rossi, E., Mannina, L., Paolillo, L. 1994. Direct quality control of edible oils and fats using high field ^{1}H-NMR spectroscopy. In *Proceedings of the Second International Conference on Applications of Magnetic Resonance in Food Science* (Aveiro, Portugal, September 19–21), p. 52.

Sacchi, R., Addeo, F., Spagna Musso, S., Paolillo, L., Giudicianni, I. 1995. High resolution ^{13}C-NMR study of vegetable margarines. *Ital J Food Sci,* 7(1):27–36.

Sacchi, R., Patumi, M., Fontanazza, G., Barone, P., Fiordiponti, P., Mannina, L., Rossi, E., Segre, A.L. 1996. A high field ^{1}H-nuclear magnetic resonance study of the minor components in virgin olive oils. *J Am Oil Chem Soc,* 73(6):747–58.

Sacchi, R., Addeo, F., Paolillo, L. 1997. ^{1}H and ^{13}C-NMR of virgin olive oil: an overview. *Magn Reson Chem,* 35:S133–45.

Sacchi, R., Mannina, L., Fiordiponti, P., Barone, P., Paolillo, L., Patumi, M., Segre, A.L. 1998. Characterization of Italian extra virgin olive oils using ^{1}H-NMR spectroscopy. *J Agric Food Chem,* 46:3947–51.

Sacchi, R., Savarese, M., Falcigno, L., Giudicianni, I., Paolillo, L. In press. Proton NMR of fish oils and lipids. In *Handbook of modern magnetic resonance, marine science,* vol. 2, edited by G.H. Webb, Kluwer Academic Publishers, UK.

Saito, H., Nakamura, K. 1990. Application of the NMR method to evaluate the oxidative deterioration of crude and stored fish oils. *Agric Biol Chem,* 54(2):533–34.

Vogels, J.T.W.E., Tervel, L., Tas, A.C., Van den Berg, F., Dukel, F., Van der Greef, J. 1996. Detection of adulteration of orange juices by a new screening method using proton NMR spectroscopy in combination with pattern recognition techniques. *J Agric Food Chem,* 44:175–78.

Ward, L.S., Bastian, E.D. 1998. Isolation of beta-casein A-1–4P in commercial caseinates. *J Agric Food Chem,* 46:77–83.

Webb, G.A., Belton, B.S., Gil, A.M., Delgadillo, I. 2001. *Magnetic resonance in food science: a view to the future,* proceedings of the Fifth International Conference on Applications of Magnetic Resonance in Food Science (Aveiro, Portugal, September 18–20, 2000), Royal Society of Chemistry, London, pp. 1–271.

Wollenberg, K.F. 1990. Quantitative high resolution ^{13}C nuclear magnetic resonance of the olefinic and carbonyl carbons of edible vegetable oils. *J Am Oil Chem Soc,* 67:487–94.

Zamora, R., Navarro, J.L., Hidalgo, F.J. 1994. Identification and classification of olive oils by high-resolution ^{13}C nuclear magnetic resonance. *J Am Oil Chem Soc,* 71:361–64.

Zamora, R., Gomez, G., Dobarganes, M.C., Hidalgo, F.J. 2002a. Oil fractionation as a preliminary step in the characterization of vegetable oils by high-resolution 13C NMR spectroscopy. *JAOCS,* 79:261–66.

Zamora, R., Gomez, G., Hidalgo, F.J. 2002b. Classification of vegetable oils by high-resolution ^{13}C NMR spectroscopy using chromatographically obtained oil fractions. *JAOCS,* 79:267–72.

6 Electronic Nose for Quality and Safety Control

Naresh Magan and Natasha Sahgal

Introduction

Today, quality assurance (QA) systems based on Hazardous Analysis Critical Control Point (HACCP) principles are preferred to quality control (QC) systems, which require careful monitoring and the withholding of contaminated material from the marketplace. QA systems are also becoming the norm in developing effective prevention strategies in food production processes. This requires effective identification of the critical control points (CCPs) and effective monitoring of these points in the food chain. Thus reliable and accurate methods are required for enabling management of the CCPs to be implemented (Tothill and Magan 2003). Thus the rapid detection of spoilage microorganisms in the food production and processing chain is critical regardless of the system being used. There are, however, different levels of detection that may be required in a food production process. The diagnostic tool required depends very much on the type and amount of information required. For example, is qualitative yes or no information required? Is a quantitative measurement needed? What is the sensitivity required? For example, does the monitoring require detection of 10^2 or 10^4 cells in a food raw material? How often does a measurement need to be made and does it need to be on line or at line? The economics of implementation in the food industry are determined by all these factors.

Generally, two factors determine the use of a particular technique: the time required to complete a specific test and that required for the result. For example, traditional serial dilution or washing techniques that depend on agar media and an incubation period require 1.5–2h of labor and perhaps 48–72h for an answer. Enzyme-linked immunosorbent assay (ELISA) techniques may require less time and give an answer much more rapidly (<30min). Lateral flow devices require simple sample preparation and give an answer in a few minutes (Danks et al. 2003). Today, the pressure is for relatively cheap diagnostic tools that can be used routinely and that give results within a few minutes at most to enable informed and effective management of the food production chain.

There are a number of indicators of microbial spoilage that have been employed in food quality assurance programs. These have included analyses of microbial populations; changes in enzyme activity; respiratory losses; changes in chemical components such as chitin, ergosterol, or ATP; immunofluorescence; immunosensors and DNA probes; lateral flow devices; and microbial volatiles (Tothill and Magan 2003). This chapter is predominantly concerned with the potential of using the characteristic volatile fingerprints produced by microbial contaminants as a tool for determining microbial quality of food raw materials and processed food stuffs as an indicator of shelf life.

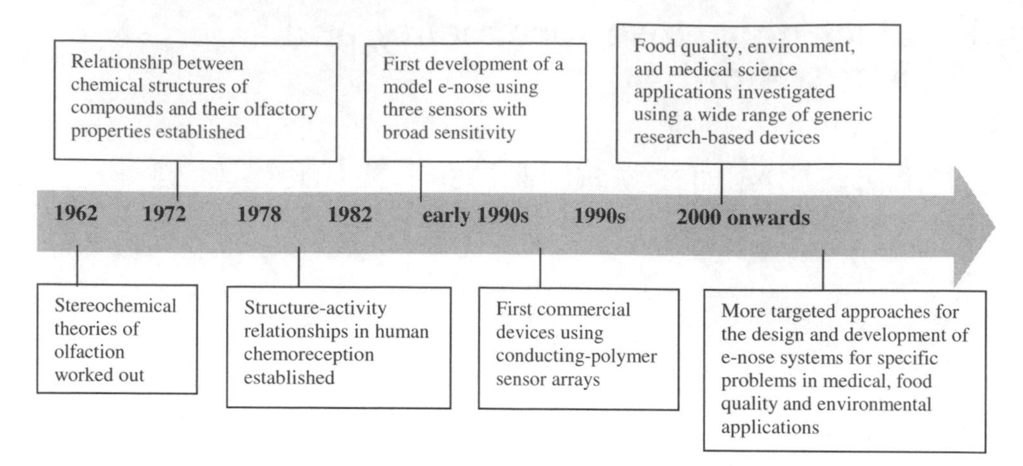

Fig. 6.1. Historical perspective of sensor-based electronic devices (adapted from Turner and Magan 2004).

Gardener and Bartlett (1999) defined an electronic nose as "an instrument which comprises an array of electronic chemical sensors with partial specificity and an appropriate pattern recognition system, capable of recognising simple or complex odours." Figure 6.1 summarizes the development of such sensor arrays from a historical perspective. This chapter will consider (1) measurement and data analyses, (2) types of sensor formats used, (3) examples of applications for raw material QA, and (4) examples for processed foods. Information relevant to medical applications has been recently reviewed (Turner and Magan 2004).

Measurement and Data Analyses

An electronic nose (e.nose) system consists of three basic building blocks: the vapor phase flows over a sensor array, the interaction with the sensor surfaces occurs, and this response is analyzed using software for output and interpretation of the raw or normalized data. There are a number of different types of sensor arrays that have been used in this type of technology. Indeed, sensor technology is a rapidly developing field and over the past decade has resulted in a range of different sensor formats and the development of complex microarray-type sensor devices. In the specific area of e.nose systems, several different physicochemical techniques have been used to produce sensor arrays for odor characterization. Each of the different sensor formats is described briefly here.

Conducting polymer sensors: Conducting polymer sensor arrays consist of unique polymers with different reversible physicochemical properties and sensitivity to groups of volatile compounds. These compounds interact with and attach to the polymer surface, changing the resistance under ambient temperature conditions. This in turn changes the signal, which is monitored for each sensor type, enabling an array to be constructed that has overlapping detection ranges for different groups of volatile compounds. Sample presentation is crucial for this type of sensor to avoid humidity and drift problems. Conducting polymer sensors operate at ambient temperature conditions, which is conducive to many food applications.

Metal oxide sensors: The oxide materials in these sensors contain chemically adsorbed oxygen species, which can interact with the volatile molecules, thus altering the conductivity of the oxide. The selectivity of these sensors can be changed by using different amounts of noble metals or by changing the operating temperature. They are very sensitive, robust, and resistant to humidity and aging effects, although they can suffer from drift over time.

Metal oxide silicon field effect sensors: These sensors are related to metal oxide sensors, but the output signal is derived from a change in potential when the volatile molecules react at a catalytic surface. The operating temperature for these sensors is in the range of 100°C–200°C. They are sensitive to many organic compounds.

Piezoelectric crystals: Sensors containing piezoelectric crystals use the radio frequency resonance of quartz materials coated with acetyl cellulose or lecithin membranes. The adsorption of volatile molecules onto the membrane produces a change in the magnitude of the resonance frequency that is related to the mass of the volatile analyte. The selectivity of these sensors is dictated by the thickness of the coatings.

Surface acoustic wave devices: These devices are an alternative to the preceding sensors that are based on waves emitted along the surface of a crystal by the electric field of surface-deposited aluminum electrodes.

Optical sensors: These sensors are based on a light source that excites the volatile analyte, and the signal can be measured in the resulting absorbance, reflectance, fluorescence, or chemiluminescence.

Electrochemical sensors: These sensors contain electrodes and an electrolyte. The responses generated are dependent on the electrochemical characteristics of the volatile molecules that are oxidized or reduced at the working electrode, with the opposite occurring at the counter electrode. The voltage generated by the reactions between the electrodes is measured and has been used to measure CO, SO_2, and H_2S.

There is a large body of literature on all these sensor technologies (Turner 2000), but the key to the devices under consideration here is the fact that total specificity is not required. Using multifactorial approaches, it is simply enough for the elements of the array to react differently to various analytes, enabling discrimination to be made between samples based on the volatile fingerprints.

New sensor technologies: New approaches are being tested to provide better sensor shelf life, greater sensitivity, and more consistency under variable conditions of temperature and humidity for commercial applications. Surface plasmon resonance can be used. Changes in the optical properties of the polymer materials used in sensors or the resonance of associated cantilevers can be monitored based on changes in the mass of the volatile compound being analyzed. Discotic liquid crystals, which consist of an aromatic core surrounded by hydrocarbon side chains, are very sensitive to the presence of volatile molecules and insensitive to humidity; these crystals could provide advances. Materials based on metalloporphyrins can detect a broader range of fingerprints because of the diversity of metal ions and substituted porphyrins available. Recently, an expanded colorimetric sensor array system based on metallated tetraphenylporphyrins and chemoresponsive dyes has been developed that has good sensitivity, with thresholds of detection for amines, carboxylic acids, and thiols that are better than those of the human nose. Preliminary data using this technique with bacteria have been promising (Suslick et al. 2002).

The application and benefits of e.nose-type devices depend on two critical components. First, the sample must be presented in the correct format to optimize the interaction of volatiles in the headspace with the sensor array being used. Sample size must be standardized and should be presented in a consistent manner under steady state environmental conditions. Thus, humidity, temperature, and sample size must all be standardized to ensure that data sets can be compared and analyzed with confidence. This also minimizes inherent drift problems over time, although these can be overcome by using a set of appropriate standards for regular recalibration. Second, pattern recognition must enable large data sets to be analyzed rapidly to obtain results and more importantly interpret them (Pearce et al. 2003). Normally, volatile odor pattern data are received in the form of normalized data sets based on the divergence, area, adsorption, or desorption components of the individual sensor responses. This generates a significant amount of data and requires effective and appropriate data management. The techniques used to analyze such data sets have included simple supervised techniques such as discriminant function analysis (DFA), which can parametrically classify an "unknown" or "random" sample from a population or group. DFA has been successfully used to detect bacteria, yeasts, and some filamentous fungi (Needham et al. 2005). A simple unsupervised multivariate method such as cluster analysis has also been used to identify a volatile odor class without prior information on the nature of the volatile fingerprint. Principal component analysis (PCA) is a popular nonparametric technique that makes no assumptions about the data and that groups related data in the three-dimensional space of a multivariate system where the parameters are partially correlated and the results are displayed in a two-dimensional pattern recognition map.

Where real-time data analyses are required, neural networks (NNs) that consist of a series of algorithms that are more appropriate for nonlinear sensor systems are used. This enables a specific system to be developed for a specific target at the required level of sensitivity. By having enough background (control) data and by using so-called backpropagation approaches, sensor drift and nonlinear data sets can be taken into account and used for effective prediction of the group into which a real sample falls (Magan et al. 2001). Of course, a large number of training sets are often required to develop appropriate NN systems. Where there is still an overlap between groups, the potential exists for the use of "fuzzy logic" NNs. These are more flexible and can be trained rapidly with large amounts of sensor array data from samples to provide a foundation of healthy background volatile fingerprints. This subsequently makes differentiation of contaminated samples easier and more rapid, often in a matter of minutes. However, it is clear that for different microbial contaminations specific NN analysis systems might need to be developed. Some could be qualitative only and useful for screening, whereas others could be semiquantitative and give more information for treatment of the disease (Turner and Magan 2004). Metal oxide, conducting polymer, and discotic crystals have all been utilized in different array formats to try to qualitatively and semiquantitatively obtain information on, and differentiate between, volatile production patterns produced by spoilage microorganisms in food matrices. However, it is important that the results are combined with multivariate data analyses systems to enable rapid interpretation of volatile patterns and for user-friendly answers to be obtained.

Food Applications

This approach has received much attention because it requires no sample preparation and is thus noninvasive. It can use a larger representative sample, and the sample presentation

Table 6.1. Summary of the range of food matrices and types of spoilage that have been examined using e.nose systems.

Food Product	Types of Spoilage
Grain/flour	Molds/insects/toxins
Bread	Bacteria/yeasts/molds
Milk	Bacteria/yeasts
Cheese	Molds
Coffee	Bacteria/yeasts/molds
Fish	Microbial taints/freshness
Tea	Cultivars/specialty coffees
Beers	Tainting
Wines	Tainting/yeasts
Nuts	Quality/toxins
Fruit	Quality/disease
Meat	Microbial taints/freshness

can be standardized easily. The disadvantage is that it is qualitative and at most semiquantitaive and some sensor systems can be prone to drift. However, new generation sensors may significantly reduce these problems. Table 6.1 shows the range of food products for which this approach has been used, either for research or in practical applications. This demonstrates the broad range of food products that have been examined.

Commercial Products

There is a range of different e.nose instruments, based on different sensor arrays. Table 6.2 gives some examples of the broad spectrum of approaches used in commercial instruments. The price of this technology varies significantly, and takeup may be slow because of this factor. However, as the development of sensor arrays becomes cheaper, the economics of the production of commercial devices and exploitation of this technology should improve rapidly.

Detection of Contamination in Food Raw Materials

It is well-known that microbial contaminants produce a range of volatiles. These volatile fingerprints vary with individual microorganisms. For example, using gas chromatography–mass spectrometry (GC-MS), it has been shown that the volatile profiles produced by *Pseudomonas fragi*, *Saccharomyces cerevisiae,* and *Penicillium verrucosum* are very different (Needham et al. 2005). It has been shown that raw materials such as grain that is poorly stored have off-odors due to alcohols, esters, ketones, mono and sesquiterpenes, and aldehydes. The dominant volatiles in grain were found to be 3-methyl-1-butanol, 1-octen-3-ol, and 3-octanone (Magan and Evans 2000). Studies with grain having different levels of molding showed that by using a standardized grain amount, a conducting polymer array of sensors, and a radial-based neural network it was possible to carry out real-time analyses of grain samples in 10 min (Evans et al. 2000). This enabled a decision to be made on whether grain was acceptable or not. Interestingly, results were very comparable with those produced by an odor panel classification. This approach was based on

Table 6.2. Commercially available electronic nose instruments.

Manufacturer	Place of Origin	Sensor Type	Number of Sensors	Size of Instrument
Agilent Technologies (Hewlett-Packard Co.)	Germany	QMS[a]	—	Desktop
Airsense Analytics GmbH	Germany	MOS[b]	10	Laptop
Alpha MOS-Multi Organoleptic Systems	France	CP,[c] MOS, QCM,[d] SAW[e]	6–24	Desktop
Applied Senso (Nordic Sensor Technologies)	Sweden	MOS, MOSFET,[f] QCM	22	Laptop
Scensive Technologies (Bloodhound Sensors Ltd.)	UK	CP, DLC[g]	14	Laptop
Cyrano Science Inc.	USA	CP	32	Palmtop
Electronic Sensor Technology Inc.	USA	SAW	1	Desktop
HKR-Sensorsysteme GmbH	Germany	QCM, MS[h]	6	Desktop
Lennartz Electronic GmbH	Germany	MOS, QCM, electrochemical	16–40	Desktop
EEV Ltd. Chemical Sensor Systems	UK	CP, MOS, QCM	8–28	Desktop
Microsensor Systems (Sawtek) Inc.	USA	SAW	2	Palmtop
OSMETEC	USA	CP	32	Desktop
RST Rostock (Daimler Chrysler Aerospace)	Germany	QCM, SAW, MOS	6–10	Desktop
SMart Nose	Switzerland	MS	—	Desktop
Technobiochip	Italy	QCM	8	Laptop
Chemsensing Inc	USA	MP[i]	2	—
Element Ltd.	Iceland	MOS	—	Desktop
Environics Industry	Finland	IMCELL[j]	—	—

Source: Adapted from Nagle et al. 1998; Pearce et al. 2003; http://www.nose-network.org/review/.
[a]Quadrupole mass spectrometry.
[b]Metal oxide semiconductor.
[c]Conducting polymer.
[d]Quartz crystal microbalance.
[e]Surface acoustic wave.
[f]Metal oxide semiconductor field-effect transistor.
[g]Discotic liquid crystal.
[h]Mass spectrometry based.
[i]Metalloporphyrins.
[j]Ion mobility.

acceptable or unacceptable criteria and showed a small group of samples in a so-called zone of uncertainty, which suggested that they were in the process of going moldy and thus needed further investigation. This study also investigated the potential for differentiation between mold contamination and mite contamination by using a conducting polymer sensor array combined with a metal oxide array for the pest infestation. This suggests that the opportunity exists to combine such systems for real-time analyses of different types of contaminants from a single sampling.

Other studies using metal oxide sensors have suggested that the presence of 1 percent contaminated grain could be discriminated (de Lacy Costello et al. 2003). However, in trials with harvested naturally contaminated grain some false-positives were obtained when compared to other criteria used at intake facilities for processing. The results were, importantly, not found to be linked to grain moisture content. These studies all suggest that the quality of grain can be effectively managed using this approach.

Processed Foods and Detection of Contamination

Bakery Products

Some studies have also been carried out to examine the potential for using this approach for managing cereal grain quality prior to processing for bakery products. Although most e.nose systems are qualitative for QA, the level of detail required determines the use of the technological approach. Recent studies have demonstrated that discrimination between mold-contaminated and noncontaminated bread was possible within 24–30 h after inoculation, prior to any visible growth and more sensitive than enzyme assays or CFU population measurements (Keshri et al. 2002; Needham et al. 2005). Studies have also employed a mass-spectrometry-based e.nose system to examine the mold contamination of Spanish bakery products, using PCA, DFA, and fuzzy ARTMAP approaches and solid phase microextraction to concentrate the volatiles produced by the molds (Vinaixa et al. 2004). They successfully (98 percent) discriminated mold growth from controls in 48 h, again prior to any visible growth. After 96 h 88 percent discrimination between a range of spoilage species was possible. Potential thus exists for applications in bakery production plants to monitor batches. Each measurement using this type of system would require 20–25 min. This could quite easily be included in a QA system, provided the initial investment gives a relatively rapid return.

Potential also exists for examining contamination with mycotoxins for which regulatory limits exist. Some studies have suggested that since the biochemical pathways for mycotoxin-producing strains of a species may differ from those of nonproducing strains, the volatile fingerprint from strains of the same species may thus differ. Keshri and Magan (2000) examined this and demonstrated that mycotoxigenic and nonmycotoxigenic strains of *Fusarium* section Liseola species that produce fumonisins could be discriminated using volatile production patterns with an e.nose based on conducting-polymer-based sensor arrays. Studies with *Penicillium verrucosum* and ochratoxin/citrinin production have also shown some promise (Needham and Magan 2003). Recent studies have suggested that changes in bacterial populations in milk and water can be detected at between 10^3–10^4 CFUs per ml (Magan et al. 2001; Canhoto and Magan 2003).

Cheese Maturity and Contamination

Recent detailed studies have been carried out to examine whether e.nose technology can be used for quality assessment and monitoring the ripening process of Danish blue cheese (Trihaas and Nielsen 2005; Trihaas et al. 2005). Using a metal-oxide-based commercial e.nose, researchers were able to demonstrate that between 5 and 33 weeks of storage it was possible to effectively monitor and predict maturity stages in different dairy units using the volatile fingerprints as a qualitative measurement. They were also able to correlate the key volatiles produced during ripening by comparing them with GC-MS data on different types of volatiles and the amounts produced during cheese ripening. This was coupled with multivariate analyses using PCA and soft independent modeling of class analogy (SIMCA). They developed a model that could effectively estimate and classify the age of batches of cheeses from different dairy units and between cheeses from the same unit. The results also compared very favorably with a trained sensory panel. Studies by these researchers have also demonstrated that it is possible to discriminate between maturing blue cheeses and the presence of contaminants such as *Geotrichum candidum* and *Penicillium roqueforti* present on the surface of ripening cheeses (P. V. Nielsen, personal communication). There is thus interest in the cheese industry in Denmark for the application of e.nose technology as part of QA management of the production process.

Sensitivity of E.nose Systems for Microbial Detection in Food

Recent studies have attempted to differentiate between nonmicrobial and microbial spoilage. Studies by Needham et al. (2005) used a bacterium, yeast, and filamentous fungus and inoculated bread for examination after different time periods (24, 48, 72 h) at 25°C. This was compared with uninoculated control bread and that containing lipoxygenase to simulate nonmicrobial spoilage. This comparison showed that it was possible to discriminate between nonmicrobial volatiles and volatiles produced by a *Penicillium* species, but not between the yeast and bacterial contamination (after 48 h) prior to any visible growth being present. Figure 6.2 shows the cluster analysis dendrogram that illustrates this. In the study by Needham et al. (2005) a conducting polymer sensor array was used.

Studies with milk-based media have been carried out to try to differentiate between different bacterial and yeast contaminants. Studies were carried out with *Bacillus cereus, Kluyveromyces lactis, Staphylococcus aureus,* and *Candida pseudotropicalis*. Using initial levels of 10^4 CFUs/ml the contaminants were analyzed after 5 h incubation using a conducting polymer sensor array. The data were subject to cross-validation using individual samples of each treatment as unknowns, and the results showed that this could be successfully done, accounting for 83 percent of the data (Magan et al. 2001). Figure 6.3 shows the cross-validation results from this study.

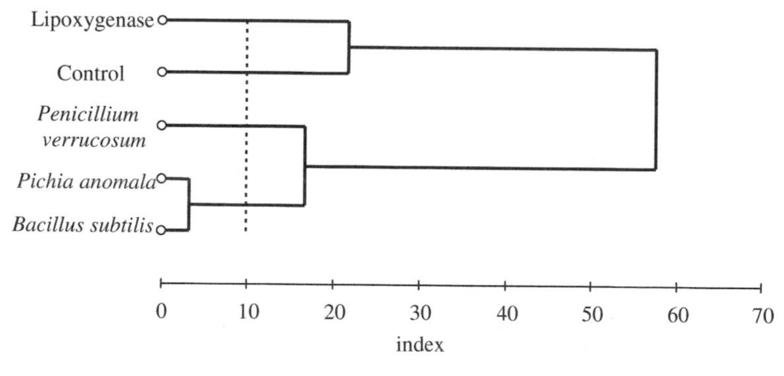

Fig. 6.2. Cluster analysis of microbial/enzymic spoilage on bread based on volatile production patterns after incubation for 48 h at 25°C (from Needham et al. 2005).

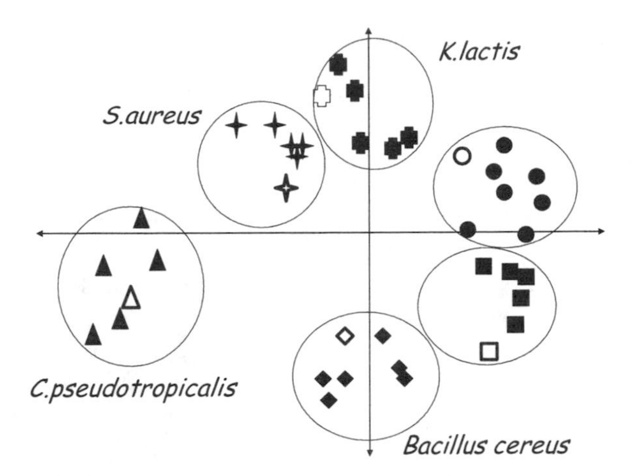

Fig. 6.3. Cross-validation of unknowns of six groups in milk-based media after 5h incubation at 25°C. Initial concentration was 10^4 CFUs/ml. The open symbols were correctly validated in each of the groups. Butanol and the milk medium represent the uninoculated controls. The circles and squares represent control medium and butanol blanks respectively (adapted from Magan et al. 2001).

Studies with different microorganisms and food products suggest that the minimum number of cells that can be detected in food matrices within 24–48h of incubation is 10^3–10^4 for grain and milk and 10^3 for bread and cheese. Studies by Keshri et al. (2002) showed that it was possible to detect molds in bread matrices within 48h and much earlier than when using enzymatic assays or traditional serial dilution methods. Where detection of 10–100 CFUs is needed, then it may be possible to do this by the addition of specific enzymes to enhance volatile headspace generation. This was successfully achieved in medical applications for detection of *Mycobacterium tuberculosis* (Pavlou et al. 2004).

Potential of Using Electronic Tongue Systems

There has been recent interest in using arrays of electrodes with different selectivities and sensitivities by deposition of different electroactive surfaces. Usually electrochemical detection by comparison with reference electrodes or cyclic or square wave voltammetry is used. This approach has been used in relation to monitoring liquids and slurries and thus may have both food and feed applications. This approach has been recently examined for analysis of port wine age, discrimination of orange juices, Spanish red wine, mineral water, mold discrimination in media and food matrices, and olive oil. This approach has potential for examining adulteration of high-value food products rapidly.

In recent studies port wine from 10 to 40 years old were successfully discriminated using an electronic tongue approach (Runitskaya et al. 2005). Mold growth in media has also been examined in studies. These studies showed that measurements could be used to discriminate between 1, 3, 7, and 10 days using an electronic tongue system (Soderstrom et al. 2005). Varying the medium content with different nutrients also influenced the results. However, these studies do show that some potential exists for effective utilization of electronic tongue systems for food and feed applications.

Conclusions

Overall, e.nose systems and the development of electronic tongue systems have been demonstrated to be very useful for monitoring and QA, especially where a yes/no answer is required rapidly. For applications in the food industry, where margins are narrow, the economics of the systems are critical. However, sensor technologies are evolving very rapidly, and we can expect better sensitivity, stability, and shelf life for food and feed applications. Generic systems will only be applied if appropriate NNs and associated software need to be developed in parallel for specific applications.

The opportunity now exists to take data gathered remotely at different sites and use advanced information-technology approaches, satellite communication, and Web-based knowledge systems to analyze these data and give results from a central point within minutes. This would open up the potential for using electric nose and tongue systems throughout a food chain—also to monitor, perhaps more effectively, food-borne pathogens. However, from our experience with the bakery and dairy industry, where noninvasive and shelf-life testing is of particular interest, the following factors need to be considered. Flexibility, in an at-line environment; ease of operation and maintenance; appropriate software; time; and labor savings must be clear. From an economic point of view, because margins are very narrow in these industries, the price should be in the region of 10,000 to 15,000 euros. The technology does have advantages for effective management systems where rapid diagnostics techniques are advantageous for conserving and/or improving QA.

References

Canhoto, O., and Magan, N. 2003. Potential for the detection of microorganisms and heavy metals in potable water using electronic nose technology. *Biosensors and Bioelectronics* 18(5–6):751–54.

Danks, C., Ostoja-Starzewska, S., Flint, J., and Banks, J. 2003. The development of a lateral flow device for the discrimination of OTA producing and non-producing fungi. *Aspects of Applied Biology* 68:21–28.

de Lacey Costello, B.J.P., Ewne, R.J., Gunson, H., Ratcliffe, N.M., Sivanand, P.S., and Spencer-Phillips, P.T.N. 2003. A prototype sensor system for the early detection of microbially linked spoilage in stored wheat grain. *Measurement Science and Technology* 14:397–409.

Evans, P., Persaud, K.C., McNeish, A.S., Sneath, R.W., Hobson, N., and Magan, N. 2000. Evaluation of a radial basis function neural network for determination of wheat quality from electronic nose data. *Sensors and Actuators B* 69:348–58.

Gardener, J.W., and Bartlett, P.N. 1999. *Electronic Noses—Principles and Applications.* Oxford University Press.

Keshri, G., and Magan, N. 2000. Detection and differentiation between mycotoxigenic and nonmycotoxigenic strains of *Fusarium* spp. using volatile production profiles and hydrolytic enzymes. *Journal of Applied Microbiology* 89:825–33.

Keshri, G., Vosey, P., and Magan, N. 2002. Early detection of spoilage moulds in bread using volatile production patterns and quantitative enzyme assays. *Journal of Applied Microbiology* 92:165–72.

Magan, N., and Evans, P. 2000. Volatiles in grain as an indicator of fungal spoilage, odour descriptors for classifying spoiled grain and the potential for early detection using electronic nose technology: a review. *Journal of Stored Product Protection* 36:319–40.

Magan, N., Pavlou, A., and Chrysanthakis, I. 2001. Milk-sense: a volatile sensing system recognises spoilage bacteria and yeasts in milk. *Sensors and Actuators B* 72:28–34.

Nagle, H.T., Gutierrez-Osuna, R., and Schiffman, S.S. 1998. The how and why of electronic noses. *Spectrum, IEEE* 35(9): 22–31.

Needham, R., and Magan, N. 2003. Detection and differentiation of toxigenic and non-toxigenic *Penicillium verrucosum* strains on bakery products using an electronic nose. *Aspects of Applied Biology* 68:217–22.

Needham, R., Williams, J., Beales, N., Voysey, P., and Magan, N. 2005. Early detection and differentiation of spoilage of bakery products. *Sensors and Actuators B* 106:20–23.

Olssen, J., Borjesson, T., Lundstedt, T., and Schnuerer, J. 2002. Detection and quantification of ochratoxin A and deoxynivalenol in barley grains by GC-MS and electronic nose. *International Journal of Food Microbiology* 72:203–14.

Pavlou, A., Turner, A.P.F., and Magan, N. 2002. Recognition of anaerobic bacterial isolates in vitro using electronic nose technology. *Letters in Applied Microbiology* 35:366–69.

Pavlou, A., Magan, N., Meecham-Jones, J., Brown, J., Klatser, P., and Turner, A.P.F. 2004. Detection of *Mycobacterium tuberculosis* (TB) in vitro and in situ using electronic nose in combination with a neural network system. *Biosensors and Bioelectronics* 20:538–44.

Pearce, T.C., Schiffmann, S.S., Nagle, H.T., and Gardner, J.W. 2003. *Handbook of Machine Olfaction: Electronic Nose Technology.* Wiley-VCH Verlag, Weinheim.

Runitskaya, A., Delgadillo, I., Legin, A., Rocha, S., Da Costa, A.-M., Simoes, T. 2005. Analysis of port wines using an electronic tongue. In *Proceedings of 11th ISOEN,* ed. S. Marco and I. Montoliu, pp. 178–79. Barcelona University.

Soderstrom, C., Boren, H., Krantz-Rulcker, C. 2005. Use of an electronic tongue and HPLC with electrochemical detection to differentiate molds in culture media. *International Journal of Food Microbiology* 97:247–57.

Suslick, K.S., Kosal, M.A., McNamara, W.B., and Sen, A. 2002. Smellseeing: a colorimetric electronic nose. In *Technical Digest, Proceedings of ISOEN,* pp. 27–28. Aracne, Italy.

Tothill, I.E., and Magan, N. 2003. Rapid detection methods for microbial contamination. In *Rapid and On-Line Instrumentation for Food Quality Assurance,* ed. I.E. Tothill, pp. 136–60. Woodhead Publishing, Cambridge, UK.

Trihaas, J., and Nielsen, P.V. 2005. Electronic nose technology in quality assessment: monitoring the ripening process of Danish blue cheese. *Journal of Food Science* 70:44–49.

Trihaas, J., Vognsen, L., and Nielsen, P.V. 2005. Electronic nose: new tool in modelling the ripening of Danish blue cheese. *International Dairy Journal* 15:679–91.

Turner, A.P.F., and Magan, N. 2004. Electronic noses and disease diagnostics. *Nature Reviews Microbiology* 2:161–66.

Turner, A.P.F. 2000. Biosensors—sense and sensitivity. *Science* 290(5495):1315–17.

Vinaixa, M., Marín, S., Brezmes, J., Llobet, E., Vilanova, X., Correig, X., Ramos, A.J., and Sanchis, V. 2004. Early detection of fungal growth in bakery products using an e-nose based on mass spectrometry. *Journal of Agriculture and Food Chemistry* 52 (60):68–74.

7 Rapid Microbiological Methods in Food Diagnostics

Daniel Y. C. Fung

Introduction

Rapid microbiological methods is a dynamic area in applied microbiology dealing with the study of improved methods in the isolation, early detection, characterization, and enumeration of microorganisms and their products in clinical, food, industrial, and environmental samples. In the past 15 years this field has emerged into an important subdivision of the general field of applied microbiology and is gaining momentum nationally and internationally as an area of research and application to monitor the numbers, kinds, and metabolites of microorganisms related to food spoilage, food preservation, food fermentation, food safety, and food-borne pathogens.

Fung (2002a) conducted a comprehensive review of the historical development of rapid microbiological methods and updated a variety of methods in the field. Many methods and procedures currently in use in food microbiology laboratories were developed more than 100 years ago. The "conventional" methods used by many regulatory laboratories around the world are based on laborious, large-volume usage of liquid and solid media and reagents, as well as time-consuming procedures both in operation and data collection. However, these methods remain the gold standards in applied microbiology and in the books of regulatory agencies nationally and internationally to the present time.

Many conferences, workshops, symposia, and meetings were held in the past 20 years around the world to address improved methods in applied microbiology. Also many "hands-on" workshops have been developed to test these new methods in laboratory settings. The most comprehensive workshop was developed by the author in 1981 at Kansas State University. This eight-day workshop combines lectures by outstanding, internationally known scientists and laboratory exercises under the supervision of experts from academia as well as industrial company representatives to demonstrate the latest methods, techniques, diagnostics kits, instruments, systems, and new concepts to the participants. Since 1981 more than 3500 scientists from 60 countries and 46 states have come to the intensive workshop. In 2005 the workshop celebrated its quarter-century anniversary with a silver medallion presented to all participants.

Following are important developments related to rapid methods in food diagnostics. The information should be of interests to food science and microbiology students, microbiologists, food scientists, consultants, quality assurance and control managers, laboratory directors, and researchers.

Advances in Sample Preparation and Treatments

One of the most important steps for successful microbiological analysis of any material is sample preparation. Without proper sampling procedures the data obtained will have limited meaning and usefulness. With the advancement of microbiological techniques and miniaturization of kits and test systems to ever-smaller sizes, proper sample preparation becomes critical. Microbiological samples can be grouped as solid samples, liquid samples, surface samples, and air samples.

Solid Samples

Common laboratory procedures for solid samples include aseptic techniques to collect samples and rapid transport (less than 24 h) to a laboratory site in a frozen state for frozen foods, and chilled state for most other foods. The purpose is to minimize growth or death of the microorganisms in the food to be analyzed. The next step is to aseptically remove a subsample such as 5 g, 10 g, 25 g, or more for testing. Sometimes samples are obtained from different lots and composited for analysis. In food microbiology, almost always the food is diluted to 1 : 10 dilution (i.e., 1 part of food in 9 parts of sterile diluent) and then homogenized by a variety of instruments. To make a 1 : 10 dilution, the procedure is simple, but when an analyst has to make 10 or more samples, this becomes laborious and time-consuming. An instrument called Gravimetric Diluter, marketed by Spiral Biotech (Bethesda, MD), can automatically perform this function. The analyst simply puts an amount of food (e.g., 10.5 g) into a sterile Stomacher bag, sets the desired dilution (1 : 10), and the instrument will deliver the appropriate amount of sterile diluent (e.g., 94.5 g). Thus, the dilution operation can be done automatically and efficiently. The dilution factor can be programmed to deliver other factors, such as 1 : 25, 1 : 50, and so on. Manninen and Fung (1992) found this system to be efficient and accurate over a wide range of dilutions.

After dilution, the sample needs to be homogenized. Traditionally, a sterile blender or Osterizer is used to homogenize the food suspension for 1 to 2 min before further diluting the sample for microbiological analysis. The disadvantages of using a blender include (1) the blender must be cleansed and resterilized between each use, (2) aerosols may be generated and contaminate the environment, and (3) heat may be generated mechanically and may kill some bacteria. In the past 25 years the Stomacher invented by Anthony Sharpe has become standard equipment in food analysis laboratories. About 40,000 Stomacher units are in use worldwide. The sample is placed in a sterile plastic bag, and an appropriate amount of sterile diluent is added. The sample in the bag is then "massaged" by two paddles of the instrument for 1 to 2 min, and then the contents can be analyzed with or without further dilution. The advantage of the Stomacher include (1) no need to resterilize the instrument between samples because the sample (housed in a sterile plastic bag) does not come in contact with the instrument, (2) disposable bags allow analysis of large number of samples efficiently, (3) no heat or aerosols will be generated, and (4) the bag with the sample can serve as a container for time course studies.

Recently Anthony Sharpe invented the Pulsifier for dislodging microorganisms from foods without excessively breaking the food structure. The Pulsifier has an oval ring that can house a plastic bag with sample and diluent. When the instrument is activated, the ring will vibrate vigorously for a predetermined time (30 to 60 s). During this time microorganisms on the food surface or in the food will be dislodged into the diluent with minimum destruction of the food. Fung et al. (1998) evaluated the Pulsifier against the Stomacher with 96 food items (including beef, pork, veal, fish, shrimp, cheese, peas, a variety of vegetables, cereal, and fruits) and found that the systems gave essentially the same viable cell count in the food but the "Pulsified" samples were much clearer than the "Stomached" samples. A more recent report by Kang et al. (2001) found that the Pulsifier and Stomacher had a correlation coefficient of 0.971 and 0.959 for total aerobic count and coliform count, respectively, with 50 samples of lean meat tissues. The "Pulsified" samples, however, contained much less meat debris than "Stomached" samples. In the case of Stomached samples, much meat debris occurred, which interfered with plating samples on agar. More recently, Wu et al. (2003) conducted a comprehensive study of the Pulsifier versus the Stomacher on 30 vegetables and reported no difference in total count and coliform between the two methods. However, there were distinct differences in the liquid between the methods, with pulsified samples having less turbidity, less total solids, and higher pH than the stomached samples. The superior quality of microbial suspensions with minimum food particles from the Pulsifier has positive implications for general analysis as well as for techniques such as adenosine triphosphate (ATP) bioluminescence tests, DNA/RNA hybridization, polymerase chain reaction (PCR) amplifications, enzymatic assays, and so on.

Liquid Samples

Liquid samples are easier to manipulate than solid samples. After appropriate mixing (by vigorous hand shaking or by instrument), one only needs to aseptically introduce a known volume of liquid sample into a container and then add a desired volume of sterile diluent to obtain the desired dilution ratio (1 : 10, 1 : 100, etc.). Further dilutions can be made as necessary. There are now many automated pipetting instruments available for sample dilutions marketed in the field. Viscous and semisolid samples need special consideration such as the use of large-mouth pipettes during operation. Regardless of the consistency of the semisolid sample, 1 ml of sample is considered as 1 ml of liquid for ease of making dilution calculations. It should also be noted that in a dilution series there are dilution errors involved; thus, the more dilutions one makes the more errors one will introduce.

Surface Samples

Sampling of surfaces of food or the environment presents a different set of concerns. The analyst needs to decide on the proper unit to report the findings, such as number of bacteria per centimeter square, per inch square, or other unit. One can analyze different shapes of the surface such as a square, rectangle, triangle, or circle. A sterile template will be useful for this purpose. Occasionally, one has to analyze unusual shapes such as the surface of an egg, an apple, or a hair net or the entire surface of a chicken. The calculation of these areas becomes quite complex. For intact meat or other soft tissues, one can excise an area of the food by using a sterile knife, assuming that all the organisms are on the surface and that the meat itself is sterile. Often a sterile moistened cotton swab is used to obtain microbes from the surface of a known area, and then the swab is placed into a diluent of known volume (e.g., 5 ml), shaken, and then plated on a general purpose agar or a selective agar. Instead of a cotton swab, one can use contact materials to sample surfaces. This, for example, includes selective and nonselective agar in a Rodac plate, adhesive tape, sterile gauge, and a sterile sponge.

The nature and characteristics of the surfaces are also very important. Surface-sampling techniques for dry surfaces, wet surfaces, oily surfaces, slimy surfaces, meat, chicken skin, orange skin, stainless steel, concrete, rocks, hair nets, and so forth are very different. A lot of microorganisms will remain on the surface even after the same area is repeatedly sampled. Biofilms are very hard to completely remove from any surface. This, however, should not be a deterrent to using surface-sampling techniques if one can relate the numbers obtained to another parameter such as cleanliness of the surface or quality of a food product. Lee and Fung (1986) conducted a comprehensive review on surface-sampling techniques for bacteriology.

Recently Fung et al. (2000) developed a convenient method to obtain surface samples called "hands-free, 'Pop-up' adhesive tape method for microbial sampling of meat surfaces." In this procedure the 3M Pop-up tape unit is placed on the wrist of an analyst so that both hands can be free for manipulating experimental materials—for obtaining the meat sample, arranging agar plates, labeling samples, and so on. When the time is ready, the analyst simply pulls one piece of tape out of the unit from the wrist and uses the tape to obtain a microbial sample from the meat surface (15 s), then transfers the tape to an agar surface (15 s), and finally incubates the plate for a viable cell count of the meat surface. The correlation coefficient of the Pop-up tape method and the more cumbersome conventional swab/rinse method for obtaining viable cell counts was 0.91. Thus the simple Pop-up tape method is a viable alternative to other methods for estimating microbial surface contamination.

Air Samples

Air sampling in food microbiology has received much less attention compared with other sample techniques already discussed. Due to recent concerns about environmental air pollution, indoor air quality, public health, and the threat of bioterrorism, there is a renewed interest in rapid techniques to monitor microbes and their toxins in the air. The most common way to estimate air quality is the use of "air plates," where the lid of an agar plate is removed and the agar surface exposed to the air of the environment for a determined time, such as 10 min, 30 min, or a couple of hours. The plate is then covered and incubated, and later colonies are counted. If the colony numbers exceed a certain value, for example 15 per plate, the air quality may be considered unacceptable. However this simple method is "passive," and the information is not too quantitative. A much better method is (1) to "actively" pass a known volume of air through an instrument to measure biological particles over an agar surface (impaction) to obtain viable cell numbers after incubation of the agar or (2) to trap microorganisms in a liquid sample (impingement) and then analyze the liquid for various viable cells.

There are a variety of commercially available air samplers. Some of them are quite sophisticated, such as the Anderson Air Sampler, which can separate particle sizes from the environment in six stages, from large particles (more than 5 μm in diameter) to small particles (0.2 μm). The author has used the SAS sampler (Bioscience International, Bethesda, MD) for many years with good results. With this instrument, a Rodac plate or an ordinary plate with a suitable agar is clipped in place. A cover with a precision pattern of holes (to direct air flow precisely) is then screwed on. After activating the instrument, a known volume of air is sucked through the holes, and the particles hit and are lodged onto the surface of the agar. After operation (e.g., 60 liters of air in 20 s) the air sampler cover is removed, the lid of the agar plate is replaced, and the plate is incubated. The number of colonies developed on the agar can be converted to colony-forming units (CFUs) per cubic meter. A similar system named MAS 100 Air Sampler is marketed by EM Science, Darmstadt, Germany. Al-Dagal and Fung (1993) suggested that for food-processing plants 0–100 CFUs/cubic meter is considered clean air, 100–300 CFUs/cubic meter is acceptable air, and over 300 CFUs/cubic meter is considered not acceptable. Applied food microbiologists are constantly searching for better sample preparation methods to improve recovery of microbes from foods and the environment.

This section and the preceding ones only dealt with improvements related to solid and liquid foods, surfaces of food and food contact areas, and air samples. A great variety of physical, chemical, physicochemical, and biological sampling methods used in clinical sampling, industrial sampling, and environmental sampling can also be explored by food microbiologists to make sampling of microorganisms in foods more precise and accurate.

Advances in Total Viable Cell Count Methodologies

One of the most important pieces of information concerning food quality, food spoilage, food safety, and the potential implications of food-borne pathogens is the total viable cell count of food, water, food contact surfaces, and air of the food plants. The conventional "standard plate count" method has been in use for the past 100 years. The method involves preparing the sample, diluting the sample, plating the sample with a general nonselective agar, incubating the plates at 35°C, and counting the colonies after 48 h. The operation of the conventional standard plate count method, although simple, is time-consuming both in terms of operation and data collection. Also this method utilizes a large number of test tubes, pipettes, dilution bottles, dilution buffer, sterile plates, and incubation space and requires the cleanup of reusable materials and resterilizing them for further use.

Several methods have been developed, tested, and used effectively in the past 20 years as alternative methods for viable cell count. Most of these methods were first designed to perform viable cell counts and relate the counts to standard plate counts. Later, coliform count, fecal coliform count, and yeast and mold counts were introduced in these systems. Further developments in these systems include differential counts, pathogen counts, and even pathogen detection after further manipulations. Many of these methods have been extensively tested in many laboratories throughout the world and have gone through AOAC International collaborative study approvals. The aim of these methods is to provide reliable viable cell counts of food and water in more convenient, rapid, simple, and cost-effective alternative formats compared to the cumbersome standard plate count method.

The Spiral Plating Method is an automated system to obtain a viable cell count (Spiral Biotech, Bethesda, MD). A stylus is used to spread a liquid sample on the surface of a pre-poured agar plate (selective or nonselective) in a spiral shape (the Archimedes spiral) with a concentration gradient starting from the center and decreasing as the spiral progresses outward on the rotating plate. The volume of the liquid deposited at any segment of the agar plate is known. After the liquid containing microorganisms is spread, the agar plate is incubated overnight at an appropriate temperature for the colonies to develop. The colonies appearing along the spiral pathway can be counted either manually or electronically. The time for plating a sample is only several seconds compared to minutes used in the conventional method. Also, using a laser counter, an analyst can obtain an accurate count in a few seconds, as compared with a few minutes in the tiring procedure of counting colonies by the naked eye. The system has been used extensively in the past 20 years with satisfactory microbiological results from meat, poultry, seafood, vegetables, fruits, dairy products, spices, and so forth. Manninen et al. (1991) evaluated the spiral plating system against the conventional pour plate method using both manual count and laser count and found that the counts were essentially the same for bacteria and yeast. Newer versions of the spiral plater are Autoplater (Spiral Biotech, Bethesda, MD) and Whitley Automatic Spiral Plater (Microbiology International, Rockville, MD). With these automatic instruments an analyst need only present the liquid sample and the instrument completely and automatically processes the sample, including resterilizing the unit for the next sample.

The ISOGRID system (Neogen, Lansing, MI) consists of a square filter with hydrophobic grids printed on the filter to form 1600 squares for each filter. A food sample is first weighed, homogenized, diluted, and enzyme treated, then passed through the filter assisted by a vacuum. Microbes are trapped on the filter and into the squares. The filter is then placed on prepoured nonselective or selective agar and then incubated for a specific time and temperature. Since a growing microbial colony cannot migrate over the hydrophobic material, all colonies have a square shape. The analyst can then count the squares as individual colonies. Automatic instruments are also available to count these square colonies in seconds. Again, this method has been used to test a great variety of foods in the past 20 years.

Petrifilm (3M, St. Paul, MN) is a system with appropriate rehydratable nutrients embedded in a series of films in the unit. To obtain a viable cell count, the protective top layer is lifted, 1 ml of liquid sample is introduced to the center of the unit, and then the cover is replaced. A plastic template is placed on the cover to make a round mold. The rehydrated medium will support the growth of microorganisms after suitable incubation time and temperature. The colonies are directly counted in the unit. This system has a shelf life of over one year in cold storage. The attractiveness of this system is that it is simple to use, is small in size, has a long shelf life, doesn't require preparing agar, and has results that are easy to read. Recently the company also introduced a Petrifilm counter so that an analyst only needs to place the Petrifilm with colonies into the unit and the unit will automatically count and record the viable cell count in the computer. The manual form of the Petrifilm has been used for many food systems and is gaining international acceptance as an alterative method for a viable cell count.

Redigel system (3M, St. Paul, MN) consists of tubes of sterile nutrient with a pectin gel in the tube but no conventional agar. This liquid system is ready for use, and no heat is needed to "melt" the system since there is no agar in the liquid. After an analyst mixes 1 ml of liquid sample with the liquid in the tube, the resultant contents are poured into a special petri dish coated with calcium. The pectin and calcium will react and form a gel that will solidify in about 20 min. The plate is then incubated at the proper time and temperature, and the colonies are counted the same way as the conventional standard plate count method.

The four methods described in the preceding have been in used for almost 20 years. Chain and Fung (1991) conducted a comprehensive evaluation of all four methods against the conventional standard plate count method using seven different foods, 20 samples each, and found that the alternative systems and the conventional method were highly comparable at an agreement of $r = 0.95$. In the same study they also found that the alternative systems cost less than the conventional system for making viable cell counts.

A newer alternative method, the SimPlate system, (BioControl, Bellevue, WA) has 84 wells imprinted in a round plastic plate. After the lid is removed, a diluted food sample (1 ml) is dispensed onto the center landing pad, and 10 ml of rehydrated nutrient liquid provided by the manufacturer is poured onto the landing pad. The mixture (food and nutrient liquid) is distributed evenly into the wells by swirling the SimPlate in a gentle, circular motion. Excessive liquid is absorbed by a pad housed in the unit. After 24 h of incubation at 35°C, the plate is placed under UV light. Positive fluorescent wells are counted, and the number is converted in the most probable number (MPN) table to determine the number of bacteria present in the SimPlate. The method is simple to use with minimum amount of preparation. A 198-well unit is also available for samples with high counts. Using different media, the unit can also make counts of total coliforms and *Escherichia coli* counts, yeast and mold counts, and even *Campylobacter*.

The preceding methods are designed to count aerobic microorganisms. To count anaerobic microorganisms, one has to introduce the sample into the melted agar, and after solidification the plates need to be incubated in an enclosed anaerobic jar. In the anaerobic jar, oxygen is removed by the hydrogen generated by the Gas Pack in the jar to create an anaerobic environment. After incubation, the colonies can be counted and reported as an anaerobic count of the food. The method is simple but requires expensive anaerobic jars and disposable Gas Packs. Also it takes about 1 h before the interior of the jar becomes anaerobic. Some strict anaerobic microorganisms may die during this 1 h period of reduction of oxygen. The author developed a simple anaerobic double-tube system that is easy to use and provides an instant anaerobic condition for the cultivation of anaerobes from foods (Fung and Lee 1981). In this system, the desired agar (ca. 23 ml) is first autoclaved in a large test tube (OD 25 × 150 mm). When needed, the agar is melted and tempered at 48°C. A liquid food sample (1 ml) is added into the liquid. A smaller sterile test tube (OD 16 × 150 mm) is inserted into the large tube with the food sample and the melted agar. By so doing, a thin film is formed between the two test tubes. The unit is tightly closed by a screw cap. The entire unit is placed into an incubator for the colonies to develop. No anaerobic jar is needed for this simple anaerobic system. After incubation, the colonies developing in the agar film can be counted and provide an anaerobic count of the food being tested. This simple method has been used extensively for applied anaerobic microbiology in the author's laboratory for about 20 years. (Ali and Fung 1991; Schmidt et al. 2000). The preceding methods are designed to grow colonies to visible sizes for enumeration and to report the data as CFUs per gram, milliliter, or centimeter squares of the food being tested.

A few "real-time" viable cell count methods have been developed and tested in recent years. These methods rely on using "vital" stains to stain "live" cells or adenosine triphosphate (ATP) detection of live cells. All these methods need careful sample preparation, filtration, careful selection of dyes and reagents, and instrumentation. Usually the entire system is quite costly. However, they can provide one-shift results and can handle large numbers of samples.

The direct epifluorescent filter techniques (DEFT) method has been tested for many years and is in use in the United Kingdom for raw milk quality assurance programs. In this method, the microorganisms are first trapped on a filter and then the filter is stained with acridine orange dye. The slide is observed under UV microscopy. "Live" cells usually fluoresce orange-red, orange-yellow, or orange-brown, whereas "dead" cells fluoresce green. The slide can be read by the eye or by a semiautomated counting system marketed by Bio-Foss. A viable cell count can be made in less than an hour.

The Chemunex Scan RDI system (Monmouth Junction, NJ) involves filtering cells on a membrane and staining cells with vital dyes (Fluorassure), and after about 90 min incubation (for bacteria), the membrane with stained cells is read in a scanning chamber that can scan and count fluorescing viable cells. This system has been used to test disinfecting solutions against such organisms as *Pseudomonas aeruginosa*, *Serratia marcescens*, *E. coli*, and *Staphylococcus aureus* with satisfactory results.

The MicroStar system developed by the Millipore Corporation utilizes ATP bioluminescence technology by trapping bacteria in a specialized membrane (Milliflex). Individual live cells are trapped in the matrix of the filter and grow into microcolonies. The filter is then sprayed with permeablizing reagent in a reaction chamber to release ATP. The bioluminescence reagent is then sprayed onto the filter. Live cells give off light due to the presence of ATP, the light is measured by a CCD camera, and fluorescent particles (live cells) are counted.

These are new developments in staining technology, ATP technology, and instrumentation for viable cell counts. The application of these methods for the food industry is still in the evaluation stage. The future looks promising.

Advances in Miniaturization and Diagnostic Kits

Identification of normal flora, spoilage organisms, food-borne pathogens, starter cultures, and so on in food microbiology is an important part of microbiological manipulations. Conventional methods, dating back to more than 100 years ago, utilize large volumes of medium (10 ml or more) to test for a particular characteristic of a bacterium (e.g., lactose broth for lactose fermentation by *E. coli*). Inoculating a test culture into individual tubes one at a time is also very cumbersome. Through the years many microbiologists have devised vessels and smaller tubes to reduce the volumes used for these tests (Hartman 1968).

This author (Fung and Hartman 1975) has systematically developed many miniaturized methods to reduce the volume of reagents and media (from 5 to 10 ml to about 0.2 ml) for microbiological testing in a convenient microtiter plate that has 96 wells arranged in a 8 × 12 format. The basic components of the miniaturized system are the commercially sterilized microtiter plates for housing the test cultures, a multiple inoculation device, and containers to house solid media (large petri dishes) and liquid media (in another series of microtiter plates with 0.2 ml of liquid per well). The procedure involves placing liquid cultures (pure cultures) to be studied into sterile wells of a microtiter plate (ca. 0.2 ml for each well) to form a master plate. Each microtiter plate can hold up to 96 different cultures, 48 duplicate cultures, or various combinations as desired. The cultures are then transferred by a sterile multipoint inoculator (96 pins protruding from a template) to solid or liquid media. Sterilization of the inoculator is performed by alcohol flaming. Each transfer represents 96 separate inoculations in the conventional method. After incubation at an appropriate temperature, the growth of cultures on solid media or liquid media can be observed and recorded, and the data can be analyzed. These methods are ideal for studying large numbers of isolates or for research involving challenging large numbers of microbes against a host of test compounds. Through the years using the miniaturized systems, the author has characterized thousands of bacterial cultures isolated from meat and other foods, studied the effect of organic dyes against bacteria and yeasts, and performed challenge studies of various compounds against microbes with excellent results.

Many useful microbiological media (such as the *Candida albicans* isolation medium of Goldschmidt et al. 1991 and the *Klebsiella pneumoniae* isolation medium of Chein and Fung 1990) were discovered through this line of research.

Other scientists also have miniaturized many systems and developed them into diagnostic kits around the late 1960s to 1970s. Currently, API, Enterotube, Minitek, Crystal ID, MicroID, RapID, Biolog, and Vitek systems are available. Most of these systems were first developed for identification of enterics (*Salmonella, Shigella, Proteus, Enterobacter*, etc.). Later, many of the companies expanded the systems' capacity to identify nonfermentors, anaerobes, gram-positive organisms, and even yeast and molds. Miniaturized systems are accurate, efficient, labor saving, space saving, and cheaper than the conventional methods. Originally, an analyst needed to read the color reaction of each well in the diagnostic kit and then use a manual identification code to "key" out the organisms. Recently diagnostic companies have developed automatic readers phasing in with computers to provide rapid and accurate identification of the unknown cultures.

The most successful and sophisticated miniaturized automated identification system is the Vitek system (bioMerieux, Hazelwood, MO), which utilizes a plastic card that contains 30 tiny wells in which each has a different reagent. The unknown culture in a liquid form is "pressurized" into the wells in a vacuum chamber, and then the cards are placed in an incubator for a period of time ranging from 4 to 12 h. The instrument periodically scans each card and compares the color changes or gas production of each tiny well with the data base of known cultures. Vitek can identify a typical *E. coli* culture in 2 to 4 h. Each Vitek unit can handle 120 cards or more automatically. There are a few thousand Vitek units in use currently in the world. The data base is especially good for clinical isolates.

The Biolog system (Hayward, CA) is also a miniaturized system using the microtiter format for growth and reaction information similar to the systems developed by the author in 1970s. Pure cultures are first isolated on agar and then suspended in a liquid to the appropriate density (ca. 6 log cells/ml). The culture is then dispensed into a microtiter plate containing different carbon sources in 95 wells and 1 nutrient control well. The plate with the pure cultures is then incubated overnight, after which the microtiter plate is removed and the color pattern of the wells with carbon utilization is observed and compared with profiles of typical patterns of microbes. This system is very ambitious and tries to identify more than 1400 genera and species of environmental, food, and medical isolates from major groups of gram-positive, gram-negative, and other organisms. There is no question that miniaturization of microbiological methods has saved much material and operational time and has provided needed efficiency and convenience in diagnostic microbiology. The flexible systems developed by the author and others can be used in many research and development laboratories for studying large numbers of cultures. The commercial systems have played key roles in diagnostic microbiology and have saved many lives due to rapid and accurate characterization of pathogenic bacteria. These miniaturized systems and diagnostic kits will continue to be very useful and important in the medical and food microbiology arenas.

Miniaturization can also be used to improve the viable cell count procedure. This is possible in two areas. The first area is to actually miniaturize the conventional viable cell count procedure, which involves growing bacteria on agar after dilution of the sample. The second area is miniaturization of the entire most probable number (3- or 5-tube MPN) procedure used extensively for water testing for almost 100 years in public health laboratories.

More than 30 years ago the author (Fung and Kraft 1968; Fung and LaGrange 1969) miniaturized the viable cell count procedure by diluting the samples in the microtiter plate using 0.025 ml size calibrated loops in a 1 : 10 dilution series. One can simultaneously dilute 12 samples to 8 series of 1 : 10 dilutions in a matter of minutes. After dilution, the samples can be transported by a calibrated pipette and spot plating 0.025 ml on agar. One conventional agar plate can accommodate 4 to 8 spots. After incubation, colonies in the spots can be counted, and the number of viable cells in the original sample can be calculated since all the dilution factors are known. The accepted range of colonies to be counted in one spot is 10 to 100. The conventional agar plate standard is from 25 to 250 colonies per plate. In a similar vein, the author also miniaturized the most probable number method in the microtiter plate by diluting a sample in a 3-tube miniaturized series (Fung and Kraft 1969). In one microtiter plate one can dilute 4 samples each in triplicate (3-tube MPN) to 8 series of 1 : 10 dilution. After incubation, the turbidity of the wells are recorded, and a modified 3-tube MPN table can be used to calculate the MPN of the original sample. This procedure recently received renewed interest in the scientific community.

Walser (2000) in Switzerland reported the use of an automated system for a microtiter plate assay to perform classical MPN of drinking water. He used a pipetting robot equipped with sterile pipetting tips for automatic dilution of the samples and after incubation placed the plate in a microtiter plate reader and obtained MPN results with the use of a computer. The system can cope with low or high bacterial load from 0 to 20,000 colonies per milliliter. Irwin et al. (2000) in the United States also worked on a similar system by using a modified Gauss-Newton algorithm and 96-well microtechnique for calculating MPN using Microsoft Excel spreadsheets.

Advances in Immunological Testings

Antigen and antibody reaction has been used for decades for detecting and characterizing microorganisms and their components in medical and diagnostic microbiology. Antibodies are produced in animal systems when a foreign particle (antigen) is injected into the system. By collecting the antibodies and purifying them, one can use these antibodies to detect the corresponding antigens. These antibodies can be polyclonal (a mixture of several antibodies in the antiserums that can react with different sites of the antigens) or monoclonal (there is only one pure antibody in the antiserum that will react with only one epitope of the antigens). Both polyclonal antibodies and monoclonal antibodies have been used extensively in applied food microbiology. There are many ways to perform antigen-antibody reactions, but the most popular format in recent years is the "sandwiched" enzyme-linked immunosorbent assay, popularly known as the ELISA.

Briefly, antibodies (e.g., anti-*Salmonella* antibody) are fixed on a solid support (e.g., wells of a microtiter plate). A solution containing a suspect target antigen (e.g., *Salmonella*) is introduced to the microtiter well. If the solution has *Salmonella,* the antibodies will capture the *Salmonella*.

After washing away food debris and excess materials, another anti-*Salmonella* antibody complex is added to the solution. The second anti-*Salmonella* antibody will react with another part of the trapped *Salmonella*. This second antibody is linked with an enzyme such as horse radish peroxidase. After another washing to remove debris, a chromagen complex such as tetramethylbenzidine and hydrogen peroxide is added. The enzyme will react with the chromagen and will produce a color compound that will indicate that the first antibody has captured *Salmonella*. If all the reaction procedures are done properly and the liquid in a microtiter well exhibits a color reaction, then the sample is considered positive for *Salmonella*.

This procedure is simple to operate and has been used for decades with excellent results. It should be mentioned that the ELISA needs about 1 million cells to be reactive, and therefore before the ELISA is performed, the food sample has to go through an overnight incubation so that the target organism has reached a detectable level. The total time to detect a pathogen by this system should include the enrichment time of the target pathogens.

Many diagnostic companies (such as BioControl, Organon Teknika, and Tecra) have marketed ELISA test kits for food-borne pathogens and toxins such as *Salmonella*, *E. coli*, and staphylococcal enterotoxins. However, the time involved in samples addition, incubating, washing and discarding of liquids, adding another antibody complex, washing, and finally adding reagents for color reaction all contribute to the inconvenience of the manual operation of the ELISA. Recently some companies have completely automated the entire ELISA procedure.

The VIDAS system (bioMerieux, Hazelwood, MO) is an automated system that can perform the entire ELISA procedure automatically and can complete an assay in from 45 min to 2 h depending on the test kit. Since VIDAS utilizes a more sensitive fluorescent immunoassay for reporting the results, its system is named ELFA. All the analyst needs to do is to present to the reagent strip a liquid sample of an overnight enriched sample. The reagent strip contains all the necessary reagents in a ready-to-use format. The instrument will automatically transfer the sample into a plastic tube called the SPR (solid phase receptacle), which contains antibodies to capture the target pathogen or toxin. The SPR will be automatically transferred to a series of wells in succession to perform the ELFA test. After the final reaction, the result can be read and interpretation of a positive or negative test will be automatically determined by the instrument. Presently, VIDAS can detect *Listeria*, *Listeria monocytogenes*, *Salmonella*, *E. coli* O157, staphylococcal enterotoxin, and *Campylobacter*. More than 14,000 VIDAS units are in use in 2006. BioControl (Bellevue, WA) markets an Assurance EIA system, which can be adapted to automation for high-volume testing. Assurance EIA is available for *Salmonella*, *Listeria*, *E. coli* O157 : H7, and *Campylobacter*. The message of this discussion is that many ELISA test kits are now highly standardized and the test can be performed automatically to increase efficiency and reduce human errors.

Another exciting development in immunology is the use of lateral flow technology to perform antigen-antibody tests. In this system, the unit has three reaction regions. The first well contains antibodies to react with target antigens. These antibodies have color particles attached to them. A liquid sample (after overnight enrichment) is added to this well, and if the target organism (e.g., *E. coli* O157 : H7) is present, it will react with the antibodies. The complex will migrate laterally by capillary action to the second region, which contains a second antibody designed to capture the target organism. If the target organism is present, the complex will be captured, and a blue line will form due to the color particles attached to the first antibody. Excess antibodies will continue to migrate to the third region, which contains another antibody that can react with the first antibody (which has now become an antigen), to the third antibody and will form a blue band. This is a "control" band indicating that the system is functioning properly. The entire procedure takes only about 10 min. This is truly a rapid test!

Neogen (manufacturer of the Reveal system, Lansing, MI) and BioControl (manufacturer of the VIP system, Bellevue, WA) are the two main companies marketing this type of system for *E. coli* O157, *Salmonella,* and *Listeria.* The newest entry to this field is Eichrom Technologies, which markets a similar lateral migration system called Eclipse for the detection of *E. coli* O157 : H7. Merck KgaA (Darmstadt, Germany) is also working on a similar lateral migration system for many common food-borne pathogens; this system uses a more sensitive gold particle system to report the reactions.

A number of interesting methods utilizing growth of the target pathogen are also available to detect antigen-antibody reactions.

The BioControl 1-2 test (BioControl, Bellevue, WA) is designed to detect motile *Salmonella* in foods. In this system, the food sample is first pre-enriched for 24 h in a broth, and then 0.1 ml is inoculated into one of the chambers in an L-shaped system. The chamber contains a selective enrichment liquid medium for *Salmonella.* There is a small hole connecting the liquid chamber with a soft agar chamber through which *Salmonella* can migrate. An opening on the top of the soft agar chamber allows the analyst to deposit a drop of polyvalent anti-H antibodies against flagella of *Salmonella.* The antibodies move downward in the soft agar due to gravity and diffusion. If *Salmonella* is present, it will migrate throughout the soft agar. As the *Salmonella* and the anti-H antibodies meet, they will react and form a visible V-shaped "immunoband." The presence of the immunoband indicates the presumptive positive for *Salmonella* in the food sample. This reaction occurs after overnight incubation of the unit. This system is easy to use and interpret and has gained popularity because of its simplicity.

Tecra (Roseville, Australia) developed a unique *Salmonella* detection system that combines immunocapturing and growth of the target pathogen and the ELISA in a simple-to-use self-contained unit. The food is first pre-enriched in a liquid medium overnight, and an aliquot is added into the first tube of the unit. Into this tube a dip stick coated with *Salmonella* antibodies is introduced and left in place for 20 min, at which time the antibodies will capture the *Salmonella,* if present. The dip stick with *Salmonella* attached will then be washed and placed into a tube containing growth medium. The dip stick is left in this tube for 4 h. During this time if *Salmonella* is present, it will start to replicate, and the newly produced *Salmonella* will automatically be trapped by the coated antibodies. Thus, after 4 h of replication, the dip stick will be saturated with trapped *Salmonella.* The dip stick will be transferred to another tube containing a second antibody conjugated to enzyme and allowed to react for 20 min. After this second antigen-antibody reaction the dip stick is washed in the fifth tube and then placed into the last tube for color development similar to other ELISAs. A purple color developed on the dip stick indicates the presence of *Salmonella* in the food. The entire process, from incubation of food sample to reading of the test results, is about 22 h, making it an attractive system for detection of *Salmonella.* The system can now also detect *Listeria.*

A truly innovative development in applied microbiology is the immunomagnetic separation system. Vicam (Somerville, MA) pioneered this concept by coating antibodies against *Listeria* on metallic particles. Large numbers of these particles (in the millions) are added to a liquid suspected to contain *Listeria* cells. The antibodies on the particles will capture the *Listeria* cells after the mixture is rotated for about an hour. After the reaction, the tube is placed next to a powerful magnet, which will immobilize all the metallic particles to the side of the glass test tube regardless of whether the particles have or have not captured the *Listeria* cells. The rest of the liquid will be decanted. By removing the magnet from the tube, the metallic particles can again be suspended in a liquid. At this point, the only cells in the solution will be the captured *Listeria*. By introducing a smaller volume of liquid (10 percent of the original volume), the cells are now concentrated by a factor of 10. Cells from this liquid can be detected by direct plating on selective agar, ELISA, PCR reaction, or other microbiological procedures in almost pure culture state. Immunomagnetic capture can save at least one day in the total protocol of pre-enrichment and enrichment steps of pathogen detection in food.

Dynal (Oslo, Norway) developed this concept further by use of very homogeneous paramagnetic beads that can carry a variety of molecules such as antibodies, antigens, and DNA. Dynal has developed beads to capture, for example, *E. coli* O157, *Listeria*, *Cryptosporidium*, and *Giardia*. Furthermore, the beads can be supplied without any coating materials, and scientists can tailor them to their own needs by coating the necessary antibodies or other capturing molecules for detection of target organisms. Currently many diagnostic systems (ELISA, PCR, etc.) are including the immunomagnetic capture step to reduce incubation and increase sensitivity of the entire protocol.

Fluorescent antibody techniques have been used for decades for the detection of *Salmonella* and other pathogens. Similar to the DEFT test designed for viable cell count, fluorescent antibodies can be used to detect a great variety of target microorganisms. Tortorello and Gendel (1993) used this technique to detect *E. coli* O157:H7 in milk and juice. Wu et al. (2004) combined immunomagnetic separation with a continuous circulation system (Pathathrix system) and an effective ELISA (Colortrix system) to detect 1 to 10 CFUs/25 g of *E. coli* O157:H7 in raw ground beef in 5.25 h.

Antigen-antibody reactions make up a powerful system for rapid detection of all kinds of pathogens and molecules. This section describes some of the more useful methods developed for applied food microbiology. Some systems are highly automated, and some systems are exceedingly simple to operate. It should be emphasized that many of the immunological tests described in this section provide presumptive positive or presumptive negative screening test results. For negative screening results, the food in question is allowed to be shipped for commerce. For presumptive positive test results, the food will not be allowed for shipment until confirmation of the positive is made by the conventional microbiological methods.

This field of immunological testing will continue to evolve as detection methodologies are explored.

Advances in Instrumentation and Biomass Measurements

Instruments are needed to monitor changes in a population, such as adenosine triphosphate (ATP) levels; specific enzymes; pH; electrical impedance, conductance, and capacitance; generation of heat; and radioactive carbon dioxide. It is important to note that for the information to be useful, these parameters must be related to viable cell count of the same sample series. In general, the larger the number of viable cells in the sample, the shorter the detection time of these systems. A scatter gram is then plotted and used for further comparison of unknown samples. The assumption is that as the number of microorganisms increases in the sample, these physical, biophysical, and biochemical events will also increase accordingly. When a sample has 5 log or 6 log organisms/ml, detection time can be achieved in about 4 h.

All living things utilize ATP. In the presence of a firefly enzyme system (luciferase and luciferin system), oxygen, and magnesium ions, ATP will facilitate the reaction to generate light. The amount of light generated by this reaction is proportional to the amount of ATP in the sample. Thus, the light units can be used to estimate the biomass of cells in a sample. The light emitted by this process can be monitored by sensitive and automated fluorimeters. Some of the instruments can detect as little as 100 to 1000 femtograms (one femtogram, fg, is −15 log in grams). The amount of ATP in 1 CFU has been reported as 0.47 fg, with a range of 0.22 to 1.03 fg. Using this principle, many researchers have used ATP to estimate microbial cells in solid and liquid foods.

There has been a paradigm shift in the field of ATP detection in recent years. Instead of detecting the ATP of microorganisms, the systems are now designed to detect ATP from any source for hygiene monitoring. The idea is that a dirty food-processing environment will have a high ATP level and a properly cleansed environment will have a low ATP level regardless of what contributed to the ATP in these environments. Once this concept is accepted by the food industry, there will be an explosion of ATP systems on the market.

In all of these systems, the key is to be able to obtain an ATP reading in the form of relative light units (RLUs) and to relate these units to the cleanliness of the food-processing surfaces. Most systems identify an acceptable RLU, unacceptable RLU, and marginal RLU for different surfaces in food plants. Since there is no standard in what constitutes an absolutely acceptable ATP level on any given environment, these RLUs are quite arbitrary. In general, a dirty environment will have high RLUs, and after proper cleaning the RLUs will decrease. Besides the sensitivity of the instruments, for an analyst to select a particular system, the following attributes are considered: simplicity of operation, compactness of the unit, computer adaptability, cost of the unit, support from the company, and documentation of the usefulness of the system.

Currently the following ATP instruments are available: Lumac (Landgraaf, the Netherlands), BioTrace (Plainsboro, NJ), Lightning (BioControl, Bellevue, WA), Hy-Lite (EM Science, Darmstadt, Germany), Charm 4000 (Charm Sciences, Malden, MA), Celsis system SURE (Cambridge, UK), Zylux (Maryville, TN), Profile 1 (New Horizon, Columbia, MD), and others.

As microorganisms grow and metabolize nutrients, large molecules change to smaller molecules in a liquid system and cause a change in electrical conductivity and resistance in the liquid as well as at the interphase of electrodes. These changes can be expressed as impedance, conductance, and capacitance changes. When a population of cells reaches about 5 log/ml, it will cause a change of these parameters. Thus, when a food has a large initial population, the time to make this change will be shorter than when a food has a smaller initial population. The time for the curve to change from the baseline and accelerate upward is the detection time of the test sample, which is inversely proportional to the initial concentration of microorganisms in the food. In order to use these methods, a series of standard curves must be constructed by making viable cell counts of a series of food with different initial concentrations of cells and then measuring the resultant detection time. A scattergram can then be plotted. Thereafter, in the same food system, the number of the initial population of the food can be estimated by the detection time on the scattergram.

The Bactometer (bioMerieux, Hazelwood, MO) has been in use for many years to measure impedance changes in foods, water, cosmetics, and so forth by microorganisms.

Samples are placed in the wells of a 16-well module, which is then plugged into the incubator to start the monitoring sequence. As the cells reach the critical number (5 log to 6 log/ml), the change in impedance increases sharply, and the monitor screen shows a slope similar to the log phase of a growth curve. The detection time can then be obtained to determine the initial population of the sample. If one sets a cut off point of log 6 organisms/g of food for acceptance or rejection of the product and the detection time is 4 h ± 15 min, then one can use the detection time as a criterion for quality assurance of the product. Food that exhibits no change of impedance curve after more than 4 h 15 min in the instrument is acceptable, while food that exhibits a change of impedance curve before 3 h 45 min is not acceptable. For convenience the instrument is designed such that the sample bar for a food on the screen will flash red for an unacceptable sample, green for an acceptable sample, and yellow for a marginally acceptable sample.

A similar system called RABIT (Rapid Automated Bacterial Impedance Technique), marketed by Bioscience International (Bethesda, MD), is available for monitoring microbial activities in food and beverages. Instead of the 16-well module used in the Bactometer, individual tubes containing electrodes are used to house the food samples.

The Malthus system (Crawley, UK) uses conductance changes of the fluid to indicate microbial growth. It generates conductance curves similar to impedance curves used in the Bactometer. It uses individual tubes for food samples. Water heated to a desirable temperature (e.g., 35°C) is used as the temperature control (instead of heated air, which is used in the previous two systems).

All these systems have been evaluated by various scientists in the past 10 to 15 years with satisfactory results. All have their advantages and disadvantages depending on the type of food being analyzed. These systems can also be used to monitor target organisms such as coliform, yeast, and mold by specially designed culture media. In fact, the Malthus system has a *Salmonella* detection protocol that has AOAC International approval.

Through sophisticated computer algorithms and instrumentation, BacT/Alert microbial detection system (Organon Teknika, Durham, NC) utilizes colorimetric detection of carbon dioxide production by microorganisms in a liquid system. Food samples are diluted and placed in special bottles with appropriate nutrients for growth of microorganisms and production of carbon dioxide. At the bottom of the bottle there is a sensor that is responsive to the amount of carbon dioxide in the liquid. When a critical amount of the gas is produced, the sensor changes from dark green to yellow, and this change is detected by reflectance colorimetry automatically. The units can accommodate 120 or 240 culture bottles. Detection time of a typical culture of *E. coli* is about 6 to 8 h.

Soleris (Centrus, Ann Arbor, MI), a new automatic system, utilizes color changes of media during the growth of cultures to detect and estimate organisms in foods and liquid systems. The uniqueness of the system is that the color compounds developed during microbial growth are diffused into an agar column in the unit and the changes are measured automatically without the interference of food particles. Depending on the initial microbial load in the food, same-shift microbial information can be obtained. The system is easy to use and can accommodate 32 samples for one incubation temperature or 128 samples for four independent incubation temperatures in different models. The system is designed for bioburden testing and Hazardous Analysis Critical Control Point (HACCP) control; can test for indirect total viable cells, coliform, *E. coli*, yeast, mold, and lactic acid bacteria counts; and can be used for swab samples and environmental samples.

Basically, any type of instrument that can continuously and automatically monitor turbidity and color changes of a liquid in the presence of microbial growth can be used for rapid detection of the presence of microorganisms. There will definitely be more systems of this nature on the market in years to come.

Advances in Genetic Testings

So far, all the rapid tests discussed for detection and characterization of microorganisms were based on phenotypic expressions of genotypic characteristics of microorganisms. Phenotypic expression of cells is subject to growth conditions such as temperature, pH, nutrient availability, oxidation-reduction potentials, environmental and chemical stresses, toxins, water activities, and so on. Even immunological tests depend on phenotypic expression of cells to produce the target antigens to be detected by the available antibodies or vice versa. The conventional "gold standards" of diagnostic microbiology rely on phenotypic expression of cells and are inherently subject to variation.

Genotypic characteristics of a cell are far more stable. The natural mutation rate of a bacterial culture is about 1 in 100 million cells. Thus, there is a push in recent years to make genetic test results the confirmative and definitive identification step in diagnostic microbiology. The debate is still continuing, and the final decision has not been reached by governmental and regulatory bodies for microbiological testing. Genetic-based diagnostic and identification systems are discussed in this section.

Hybridization of the DNA sequence of an unknown bacterium by a known DNA probe is the first stage of genetic testings. The Genetrak system (Framingham, MA) is a sensitive and convenient system to detect pathogens such as *Salmonella, Listeria*, *Campylobacter,* and *E. coli* O157 in foods. In the beginning, the system utilized radioactive compounds bound to DNA probes to detect DNAs of unknown cultures. Currently RNA probes are used to hybridize target RNAs (such as from *Salmonella*) and use enzymatic reactions to detect the presence of the pathogens. After enrichment of cells (e.g., *Salmonella*) in a food sample for about 18 h, the cells (target cells as well as other microbes) are lysed by a detergent to release cellular materials (DNA, RNA, and other molecules) into the enrichment solution. Two RNA probes (designed to react with one piece of target *Salmonella* RNA) are added to the solution. The capture probe with a long tail of a nucleotide (e.g., adenine, AAAAA) is designed to capture the RNA onto a dip stick with a long tail of thymine (TTTTT). The reporter probe with an enzyme attached will react with any part of the RNA fragment. If *Salmonella* RNA molecules are present, the capture probes will attach to one end of the RNA, and the reporter probes will attach to the other end. A dip stick coated with many copies of a chain of complementary nucleotide (e.g., thymine, TTTTT) will be placed into the solution. Since adenine (A) will hybridize with thymine (T), the chain (TTTTT) on the dip stick will react with the AAAAA and thus capture the target RNA complex onto the stick. After washing away debris and other molecules in the liquid, a chromagen is added. If the target RNA is captured, then the enzyme present in the second probe will react with the chromagen and will produce a color reaction indicating the presence of the pathogen in the food. In this case, the food is positive for *Salmonella*. The Genetrak system has been evaluated and tested for many years and has AOAC International approval for the procedure for many food types.

Polymerase chain reaction (PCR) is now an accepted method for detecting pathogens by amplification of the target DNA and detecting the target PCR products. Basically, a

DNA molecule (double helix) of a target pathogen (e.g., *Salmonella*) is first denatured at about 95°C to form single strands, and then the temperature is lowered to about 55°C for two primers (small oligonucleotides specific for *Salmonella*) to anneal to specific regions of the single-stranded DNA. The temperature is increased to about 70°C for a special heat-stable polymerase, the TAQ enzyme from *Thermus aquaticus*, to add complementary bases (A, T, G, or C) to the single-stranded DNA and complete the extension to form a new double strand of DNA. This is called a thermal cycle. After this cycle, the tube will be heated to 95°C again for the next cycle. After one thermal cycle one copy of DNA will become two couples. After about 21 cycles and 31 cycles, 1 million and 1 billion copies of the DNA will be formed, respectively. This entire process can be accomplished in less than an hour in an automatic thermal cycler. Theoretically, if a food contains one copy of *Salmonella* DNA, the PCR method can detect the presence of this pathogen in a very short time. After PCR reactions, one still needs to detect the presence of the PCR products to indicate the presence of the pathogen. In the classical PCR procedure the time-consuming electrophoresis method was used to separate the PCR products from other molecules on the gel. Currently much-improved "real-time PCR systems" are being developed and used.

The following are brief discussions of four commercial kits for PCR reactions and detection of PCR products.

The BAX® for Screening family of PCR assays for food-borne pathogens (Qualicon, Wilmington, DE) combines DNA amplification and automated homogeneous detection to determine the presence or absence of a specific target. All primers, polymerase, and deoxynucleotides necessary for PCR as well as a positive control and an intercalating dye are incorporated into a single tablet. The system works directly from an overnight enrichment of the target organisms. No DNA extraction is required. Assays are available for *Salmonella* (Mrozinski et al. 1998), *E. coli* 0157:H7 (Johnson et al. 1998; Hochberg et al. 2000), *Listeria* genus, and *Listeria monocytogenes* (Steward and Gendel 1998; Norton et al. 2000, 2001). The system uses an array of 96 blue LEDs as the excitation source and a photomultiplier tube to detect the emitted fluorescent signal. This integrated system improves the ease of use of the assay. In addition to simplifying the detection process, the new method converts the system to a homogeneous PCR test. The homogenous detection process monitors the decrease in fluorescence of a double-stranded DNA (dsDNA) intercalating dye in solution with dsDNA as a function of temperature. Following amplification, melting curves are generated by slowly ramping the temperature of the sample to a denaturing level (95°C). As the dsDNA denatures, the dye becomes unbound from the DNA duplex, and the fluorescent signal decreases. This change in fluorescence can be plotted against temperature to yield a melting curve waveform. This assay thus eliminates the need for gel-based detection and yields data amenable to storage and retrieval in an electronic database. In addition, this method reduces the hands-on time of the assay and reduces the subjectivity of the reported results. Further, melting curve analysis makes possible the ability to detect multiple PCR products in a single tube. The inclusivity and exclusivity of the BAX® system assays reach almost 100 percent, meaning that false-positive and false-negative rates are almost zero. Currently the USDA is using the BAX® system to detect *Salmonella* in raw ground beef. The automated BAX® system can now be used with assays for the detection of *Cryptosporidium parvum* and *Campylobacter jejuni/coli* and for the quantitative and qualitative detection of genetically modified organisms in soy and corn.

The TaqMan system of Applied Biosystems (Foster City, CA) also amplifies DNA by PCR protocol. However, during the amplification step a special molecule is annealed to the single-stranded DNA to report the linear amplification. The molecule has the appropriate sequence for the target DNA. It also has two attached particles. One is a fluorescent particle, and another one is a quencher particle. When the two particles are close to each other, no fluorescence occurs. However when the Taq polymerase is adding bases to the linear single strand of DNA, it will break this molecule away from the strand (like the Pac-Man in computer games). As this occurs, the two particles will separate from each other, and fluorescence will occur. By measuring fluorescence in the tube, a successful PCR reaction can be determined. In the author's laboratory this system has been successfully used to detect *Yersinia enterocolitica* in raw meat and tofu samples (Vishnubhatla et al. 2001).

A new system called *molecular beacon technology* (Stratagene, La Jolla, CA) has been developed and can be used for food microbiology in the future (Robinson et al. 2000). In this technology, all reactions are again in the same tube. A molecular beacon is a tailor-made hairpin-shaped hybridization probe. The probe is used to attach to target PCR products. On one end of the probe a fluorophore is attached, and on the other end a quencher of the fluorophore. In the absence of the target PCR products the beacon is in a hairpin shape, and there is no fluorescence. However, during PCR reactions and the generation of target PCR products, the beacons will attach to the PCR products and cause the hairpin molecule to unfold. As the quencher moves away from the fluorophore, fluorescence will occur, and this can be measured. The measurement can be done as the PCR reaction is in progression, thus allowing real-time detection of target PCR products and thus the presence of the target pathogen in the sample. This system has the same efficiency as the TaqMan system, but the difference is that the beacons detect the PCR products themselves, while the TaqMan system only reports the occurrence of a linear PCR reaction and not the presence of the PCR product directly. By using molecular beacons containing different fluorophores, one can detect different PCR products in the same reaction tubes, and thus one is be able to perform "multiplex" tests of several target pathogens or molecules. The use of this technology is very new and not well-known in food microbiology areas.

One of the major problems of PCR systems is the problem of contamination of PCR products from one test to another. Thus, if any PCR products from a positive sample (e.g., *Salmonella* PCR products in a previous run) enter the reaction system of the next analysis, it may cause a false-positive result. Probelia system, developed by Institut Pasteur (Paris, France) attempts to eliminate PCR product contamination by substituting the base uracil for the base thymine in the entire PCR protocol. Thus, in the reaction tube there are adenine, uracil, guanine, and cytosine and no thymine. During PCR reaction the resultant Probelia PCR products will be AUGC pairings and not the natural ATGC pairings.

The PCR products are read by hybridization of known sequences in a microtiter plate. The report of the hybridization is by color reaction similar to an ELSIA test in the microtiter system.

After one experiment is completed, a new sample is added into another tube for the next experiment. In the tube there is an enzyme called *uracil-D-glycosylase,* which will hydrolyze any DNA molecules that contain a uracil. Therefore, if there are contaminants from a previous run, they will be destroyed before the beginning of the new run. Before a new PCR reaction, the tube with all reagents is heated to 56°C for 15 min for UDG to hydrolyze any contaminants. During the DNA denaturization step the UDG will be inactivated and will not act on the new PCR products containing uracil. Currently, Probelia can detect *Salmonella* and *Listeria monocytogenes* from foods. Other kits under development include *E. coli* O157 : H7, *Campylobacter,* and *Clostridium botulinum.*

Theoretically, the PCR system can detect one copy of a target pathogen from a food sample (e.g., *Salmonella* DNA). In practice, about 200 cells are needed to be detected by current PCR methods. Thus, even in a PCR protocol the food must be enriched for a period of time (e.g., overnight or at least 8 h incubation of food in a suitable enrichment liquid) so that there are enough cells for the PCR process to be reliable. Besides the technical manipulations of the system, which can be complicated for many food microbiology laboratories, two major issues need to be addressed: inhibitors of PCR reactions and the question of live and dead cells. In food, there are many enzymes, proteins, and other compounds that can interfere with the PCR reaction and result in false-negatives. These inhibitors must be removed or diluted. Since PCR reaction amplifies target DNA molecules, even DNA from dead cells can be amplified, and thus food with dead *Salmonella* can be declared as *Salmonella* positive by PCR results. Thus, food that was properly cooked but contained DNA of dead cells may be unnecessarily destroyed because of a positive PCR test.

PCR can be a powerful tool for food microbiology once all the problems are solved and analysts are convinced of the applicability in routine analysis of foods.

The RiboPrinter microbial characterization system (Du Pont Qualicon, Wilmington, DE) characterizes and identifies organisms to genus, species, and subspecies levels automatically. To obtain a RiboPrint of an organism, the following steps are followed:

1. From an agar plate a pure colony of bacteria suspected to be the target organism (e.g., *Salmonella*) is picked by a sterile plastic stick.
2. Cells from the stick are suspended in a buffer solution by mechanical agitation.
3. An aliquot of the cell suspension is loaded into the sample carrier to be placed into the instrument. Each sample carrier has space for eight individual colony picks.
4. The instrument will automatically prepare the DNA for analysis by restriction enzyme and lysis buffer to open the bacteria, release, and cut DNA molecules. The DNA fragments will go through an electrophoresis gel to separate DNA fragments into discrete bands. Last, the DNA probes, conjugate, and substrate will react with the separated DNA fragments, and light emission from the hybridized fragments will then be photographed. The data are stored and compared with known patterns of the particular organism.

The entire process takes 8 h for eight samples. However, at 2 h intervals, another eight samples can be loaded for analysis.

Different bacteria will exhibit different patterns (e.g., *Salmonella* versus *E. coli*) and even the same species can exhibit different patterns (e.g., *Listeria monocytogenes* has 49 distinct patterns). Some examples of numbers of RiboPrint patterns for some important food pathogens are *Salmonella*, 145; *Listeria*, 89; *E. coli*, 134; *Staphylococcus*, 406; and *Vibrio*, 63. Additionally, the data base includes 300 *Lactobacillus*, 43 *Lactococcus,* 11 *Leuconostoc,* and 34 *Pediococcus*. The current identification database provides 3267 RiboPrint patterns representing 98 genera and 695 species (Fung 2005).

Another important system is the pulsed-field gel electrophoresis patterns of pathogens. In this system, pure cultures of pathogens are isolated and digested with restriction enzymes, and the DNA fragments are subjected to a system known as *pulsed-field gel electrophoresis,* which effectively separates DNA fragments on the gel (DNA fingerprinting). For example in a food-borne outbreak of *E. coli* O157 : H7, biochemically identical

E. coli O157 : H7 cultures can exhibit different patterns. By comparing the gel patterns from different sources, one can trace the origin of the infection or search for the spread of the disease and thereby control the problem.

In order to compare data from various laboratories across the country, the PulseNet system was established under the National Molecular Subtyping Network for Food-borne Disease Surveillance at the Centers for Disease Control and Prevention (CDC). There are many other genetic base methods, but they are not directly related to food microbiology and are beyond the scope of this review. It is safe to say that many genetic base methods are slowly but surely finding their ways into food microbiology laboratories, and they will provide valuable information for quality assurance, quality control, and food safety programs in the future.

Advances in Biosensors

The biosensor is an exciting field in applied microbiology. The basic idea is simple, but the actual operation is quite complex and involves much instrumentation. Basically, a biosensor is a molecule or a group of molecules of biological origin attached to a signal recognition material.

When an analyte comes in contact with the biosensor, the interaction will initiate a recognition signal that can be reported in an instrument.

Many types of biosensors have been developed, such as enzymes (a great variety of enzymes have been used), antibodies (polyclonal and monoclonal), nucleic acids, and cellular materials.

Sometime whole cells can also be used as biosensors. Analytes detected include toxins (e.g., staphylococcal enterotoxins, tetrodotoxins, saxitoxin, and botulinum toxin), specific pathogens (e.g., *Salmonella*, *Staphylococcus*, and *E. coli* O157 : H7), carbohydrates (e.g., fructose, lactose, and galactose), insecticides and herbicides, ATP, antibiotics (e.g., penicillins), and others. The recognition signals used include electrochemical (e.g., potentiometry, voltage changes, conductance and impedance, light addressable, etc.), optical (e.g., UV, bioluminescence and chemiluminescence, fluorescence, laser scattering, reflection and refraction of light, surface plasmon resonance, and polarized light), and miscellaneous transducers (e.g., piezoelectric crystals, thermistor, acoustic waves, and quartz crystal).

An example of a simple enzyme biosensor is the sensor for glucose. The reaction involves the oxidation of glucose (the analyte) by glucose oxidase (the biosensor) with the end products of gluconic acid and hydrogen peroxide. The reaction was reported by a Clark oxygen electrode, which monitors the decrease in oxygen concentration amperometrically. The range of measurement is from 1 to 30 mM, with a response time of 1–1.5 min and recovery time of 30 s. The lifetime of the unit is several months. Some of the advantages of enzyme biosensors are that they are binding to the subject, highly selective, and rapid acting. Some of the disadvantages are their expense, loss of activity when they are immobilized on a transducer, and loss of activity due to deactivation. Other enzymes used include galactosidase, glucoamylase, acetylcholinesterase, invertase, lactate oxidase, and so on. Excellent review articles and books on biosensors are presented by Eggins (1997), Cunningham (1998), Goldschmidt (1999), and others.

Recently, much attention has been directed to biochip and microchip developments to detect a great variety of molecules including food-borne pathogens.

Due to the advancement in miniaturization technology, as many as 50,000 individual spots (e.g., DNA microarrays), with each spot containing millions of copies of a specific DNA probe, can be immobilized on a specialized microscope slide. Fluorescent labeled targets can be hybridized to these spots and be detected. An excellent article by Deyholos et al. (2001) described the application of microarrays to discover genes associated with a particular biological process, such as the response of a plant (*Arabidopsis*) to NaCl stress, and provided a detailed analysis of a specific biological pathway such as one-carbon metabolism in maize.

Biochips can also be designed to detect all kinds of food-borne pathogens when a variety of antibodies or DNA molecules against specific pathogens are imprinted on the chip for the simultaneous detection of pathogens such as *Salmonella, Listeria, E. coli*, and *Staphylococcus aureus* by the same chip. According to Elaine Heron of Applied Biosystems of Foster City, California (Heron 2000), biochips are an exceedingly important technology in life sciences, and the market value is estimated to be as high as £5 billion by the middle of this decade. This technology is especially important in the rapidly developing field of proteomics, which requires massive amounts of data that generate valuable information.

Certainly, the developments of these biochips and microarray chips are impressive for obtaining a large amount of information for biological sciences. As for food-borne pathogen detection, there are several important issues to consider. These biochips are designed to detect minute quantities of target molecules. The target molecules must be free from contaminants before being applied to the biochips. In food microbiology, the minimum requirement for pathogen detection is one viable target cell in 25 g of a food such as ground beef. A biochip will not be able to seek out such a cell from the food matrix without extensive cell amplification (either by growth or PCR) or sample preparation by filtration, separation, absorption, centrifugation, and so on. Any food particle in the sample will easily clot the channels used in biochips. These preparations will not allow the biochips to provide "real-time" detection of pathogens in foods.

Another concern is viability of the pathogens to be detected by biochips. Monitoring the presence of some target molecule will only provide the presence or absence of the target pathogen and will not provide the viability of the pathogen in question. Some form of culture enrichment to ensure growth is still needed in order to obtain meaningful results. It is conceivable that the biomass of microbes can be monitored by biochips, but instantaneous detection of specific pathogens such as *Salmonella*, *Listeria*, and *Campylobacter* in the food matrix during food processing is still not possible. The potential of biochips and microarrays for food pathogen detection is great, but at this moment much more research is needed to make this technology a reality in applied food microbiology.

Fung's 10 Predictions for the Future

1. **Viable cell counts will still be used**. It is the firm belief of the author that viable cell counts (total aerobic count, anaerobic count, coliform/*E. coli* count, differential count, and pathogenic count) will remain an important parameter to assess the potential safety and hygiene quality of food supplies.
2. **Real-time monitoring of hygiene will be in place**. Several exciting developments in this area have occurred, such as ATP bioluminescence, catalase measurement,

and instant protein detection kits. Recently, BioControl (Bellevue, WA) introduced a protein testing kit called FLASH, which can detect the presence of protein on food contact surfaces almost instantaneously. Positive surfaces change the color of the swab from yellow to green/blue in 5 s. This system has been validated in the author's laboratory (Olds et al. 2005). This type of real-time monitoring system will be developed for other compounds in the future.

3. **PCR, ribotyping, and genetic tests will become a reality in food laboratories.**
4. **ELISA and immunological tests will be completely automated and widely used**. After pre-enrichment of food samples (overnight incubation or 8 h incubation), an analyst can place the sample into an automated system and monitor the presence of the target pathogen in a matter of 1 to 2 h. Automated systems will continue to be developed and used in the future.
5. **Dip stick technology will provide rapid answers**. Many forms of dip sticks are available for screening of pathogens by "lateral" migration of an antigen-antibody complex. These kits can detect target organisms in about 10 min after enrichment of the cultures overnight. This type of technology will continue to be developed and used in the future.
6. **Biosensors will be in place for HACCP programs**. A variety of biosensors are now available on the market to monitor microbes, but they are not yet suitable to use in routine monitoring of pathogens in the food industry. More research and development will be needed to have this technology in place for the food industry.
7. **Instant detection of target pathogens will be possible by a computer-generated matrix in response to particular characteristics of pathogens**. Microarrays, microchips, and nano technology will be gaining importance in food microbiology in the near future.
8. **Effective separation and concentration of target cells will greatly assist in rapid identification**. A variety of approaches have been mentioned in this chapter. These developments will continue to improve detection sensitivity and speed of detection of target pathogens
9. **A microbiological alert system will be in food packages**. It is conceivable that a series of reagents in the form of "bar codes" will be placed inside packaging materials and will change color due to the development of gas (ammonia, hydrogen sulfide, hydrogen, carbon dioxide, etc.), acid, or extremes in temperature to indicate a potential spoilage problem.
10. **Consumers will have rapid alert kits for pathogens at home**. Nowadays there are urine, blood glucose, pregnancy, and even AIDS test kits available for the consumer to use at home. It is possible that rapid alert kits for food spoilage and even food pathogen detection will be developed for home use (Fung 2002b).

In conclusion, the future looks very bright for the field of rapid methods and automation in microbiology. The potential is great, and many exciting developments will certainly unfold in the near and far future.

References

Al-Dagal, M. M. and D. Y. C. Fung. 1993. Aeromicrobiology: An assessment of a new meat research complex. J. Environ. Health 56(1):7–14.

Ali, M. S. and D. Y. C. Fung. 1991. Occurrence of *Clostridium perfringens* in ground beef and turkey evaluated by three methods. J. Food Prot. 11:197–203.

Chain, V. S. and D. Y. C. Fung. 1991. Comparison of Redigel, Petrifilm, Spiral Plate System, Isogrid, and aerobic plate count for determining the numbers of aerobic bacteria in selected food. J. Food Prot. 54: 208–11.

Chein, S. P. and D. Y. C. Fung. 1990. Acriflavine violet red bile agar for the isolation and enumeration of *Klebsiella pneumoniae*. Food Microbiol. 7:73–79.

Cunningham, A. J. 1998. Bioanalytical Sensors. John Wiley and Sons, New York.

Deyholos, M., H. Wang, and D. Galbraith. 2001. Microarrays for gene discovery and metabolic pathway analysis in plants. Life Sci. 2(1):2–4.

Eggins, B. 1997. Biosensors: An Introduction. John Wiley and Sons, New York.

Fung, D. Y. C. 2002a. Rapid method and automation in microbiology. Crit. Rev. Food Sci. Food Saf. 1(1):3–22.

———. 2002b. Where are we now? On the fast track with rapid and automated methods. Food Saf. 8(3):18–25.

———. 2005. Handbook for Rapid Methods and Automation in Microbiology Workshop. Kansas State University, Manhattan, p. 750.

Fung, D. Y. C. and P. A. Hartman. 1975. Miniaturized microbiological techniques for rapid characterization of bacteria. Heden, C. G. and T. Illeni, ed., New Approaches to the Identification of Microorganisms. John Wiley and Sons, New York, chap. 21.

Fung, D. Y. C. and A. A. Kraft. 1968. Microtiter method for the evacuation of viable cells in bacterial cultures. Appl. Microbiol. 16:1036–39.

———. 1969. Rapid evaluation of viable cell counts using the microtiter system and MPN technique. J. Milk Food Technol. 322:408–9.

Fung, D. Y. C. and W. S. LaGrange. 1969. Microtiter method for bacterial evaluation of milk. J. Milk Food Technol. 32:144–46.

Fung, D. Y. C. and C. M. Lee. 1981. Double-tube anaerobic bacteria cultivation system. Food Sci. 7:209–13.

Fung, D. Y. C., A. N. Sharpe, B. C. Hart, and Y. Liu. 1998. The Pulsifier: A new instrument for preparing food suspensions for microbiological analysis. J. Rapid Methods Autom. Microbiol. 6:43–49.

Fung, D. Y. C., L. K. Thomspon, B. A. Crozier-Dodson, and C. L. Kastner. 2000. Hands-free "Pop-up" adhesive type method for microbial sampling of meat surfaces. J. Rapid Methods Autom. Microbiol. 8(3):209–17.

Goldschmidt, M. C. 1999. Biosensors: Scope in microbiological analysis. In Robinson, R., C. Batt, and P. Patel., eds., Encyclopedia of Food Microbiology. Academic Press, New York, pp. 268–78.

Goldschmidt, M. C., D. Y. C. Fung, R. Grant, J. White, and T. Brown. 1991. New aniline blue dyes medium for rapid identification and isolation of *Candida albicans*. J. Clin. Microbiol. 29(6):1095–99.

Hartman, P. A. 1968. Miniaturized Microbiological Methods. Academic Press, New York.

Heron, E. 2000. Applied biosystem: Innovative technology for the life sciences. Am. Lab. 32:(24)35–38.

Hochberg, A. M., P. N. Gerhardt, T. K. Cao, W. Ocasio, W. M. Barbour, and P. M. Morinski. 2000. Sensitivity and specificity of the test kit BAX for screening/*E. coli* 0157:H7 in ground beef: Independent laboratory study. J. AOAC Int. 83(6):1349–56.

Irwin, P., S. Tu, W. Damert, and J. Phillips. 2000. A modified Gauss-Newton algorithm and ninety-six well micro-technique for calculating MPN using EXCEL spread sheets. J. Rapid Methods Automat. 8(3): 171–92.

Johnson, J. L., C. L. Brooke, and S. J. Fritschel. 1998. Comparison of BAX for screening/*E. coli* 0157:H7 vs. conventional methods for detection of extremely low levels of *Escherichia coli* 0157:H7 in ground beef. Appl. Environ. Microbiol. 64:4390–95.

Kang, D. H., R. H. Dougherty, and D. Y. C. Fung. 2001. Comparison of Pulsifier and Stomacher to detach microorganisms from lean meat tissues. J. Rapid Methods Autom. Microbiol. 9(1):27–32.

Lee, J. Y. and D. Y. C. Fung. 1986. Surface sampling technique for bacteriology. J. Environ. Health 48:200–205.

Manninen, M. T. and D. Y. C. Fung. 1992. Use of the gravimetric diluter in microbiological work. J. Food Prot. 55:59–61.

Manninen, M. T., D. Y. C. Fung, and R. A. Hart. 1991. Spiral system and laser counter for enumeration of microorganisms. J. Food Prot. 11:177–87.

Mrozinski, P. M., R. P. Betts, and S. Coates. 1998. Performance tested methods: Certification process for BAX for screening/*Salmonella*; A case study. J AOAC Int. 81:1147–54.

Norton, D. M., M. McCamey, K. J. Boor, and M. Wiedmann. 2000. Application of the BAX for screening/genus *Listeria* polymerase chain reaction system for monitoring *Listeria* species in cold-smoked fish and in the smoked fish processing environment. J. Food Prot. 63:343–46.

Norton. D. M., M. McCamey, K. L. Gall, J. M. Scarlett, K. J. Boor, and M. Wiedmann. 2001. Molecular studies on the ecology of *Listeria monocytogenes* in the smoked fish processing industry. App. Environ. Microbiol. 67:198–205.

Olds, D. A., D. Y. C. Fung, and C. W. Shanklin. 2005. Semiquantitative evaluation of protein residues in foods using the flash rapid cleaning validation method. J. Rapid Methods Autom. Microbiol. 13(3):135–47.

Robinson, J. K., R. Mueller, and L. Filippone. 2000. New molecular beacon technology. Am. Lab. 32(24):30–34.

Schmidt, K. A., R. H. Thakur, G. Jiang, and D. Y. C. Fung. 2000. Application of a double tube system for the enervation of *Clostridium tyrobutyricum*. J. Rapid Methods Autom. Microbiol. 8(1):21–30.

Steward, D. and S. M. Gendel. 1998. Specificity of the BAX polymerase chain reaction system for detection of the foodborne pathogen *Listeria monocytogenes*. J. AOAC Int. 81:817–22.

Tortorello, M. and S. M. Gendel. 1993. Fluorescent antibodies applied to direct epifluorescent filter techniques for microscopic enumeration of *Escherichia coli* O157:H7 in milk and juice. J. Food Prot. 56:672.

Vishnubhatla, A., D. Y. C. Fung, W. Wonglumsom, R. D. Oberst, M. P. Hays, and T. G. Nagaraja. 2001. Evaluation of a 5 nuclease (TaqMan) assay for the detection of virulent strains for *Yersinia enterocolitica* in raw meat and tofu samples. J. Food Prot. 64(3):354–60.

Walser, P. E. 2000. Using conventional microtiter plate technology for the automation of microbiology testing of drinking water. J. Rapid Methods Autom. Microbiol. 8(3):193–208.

Wu, V. C. H., P. Jitareerat, and D. Y. C. Fung. 2003. Comparison of the Pulsifier and the Stomacher for recovery of microorganisms in vegetables. J. Rapid Methods Autom. Microbiol. 11(2):145–52.

Wu, V. C. H., V. Gill, R. D. Oberst, R. K. Phebus, and D. Y. C. Fung. 2004. Rapid protocol (5.25 h) for the detection of *Escherichia coli* O157:H7 in raw ground beef by an immuno-capture system (Pathatrix) in combination with Colortrix and CT-SMAC. J. Rapid Methods Autom. Microbiol. 12(1):57–68.

Acknowledgments

This material is based upon work supported by the Cooperative State Research, Education, and Extension Service, U.S. Department of Agriculture, under agreement No. 93-34211-8362. Contribution No. 06-2-14, Kansas Agricultural Experimental Station, Manhattan, Kansas.

8 Molecular Technologies for Detecting and Characterizing Pathogens

Geraldine Duffy and Terese Catarame

Introduction

Worldwide food-borne microbial infections continue to cause a substantial public health and economic burden. Ensuring the microbiological safety of food is achieved both by the implementation of food safety management systems such as Hazard Analysis Critical Control Points (HACCPs) and by testing of foods to ensure they conform to set microbiological criteria. These criteria may be set by EU or national regulations, by customers (retail, fast food sector, etc.), or by the manufacturers themselves as part of a food safety management system to ensure that process controls and performance criteria have been achieved.

There are many challenges in the detection of food pathogens, which include the fact that they are generally present in very low numbers in food (often <100 cfu g^{-1}), and sometimes in the midst of up to 1 million other microorganisms, and they may be on the surface or imbedded in a complex food matrix. As many foods contain agents inhibitory to microorganisms and may be stored at chill or frozen temperatures, the microorganisms on the food are often in an injured/stressed condition and indeed may be in a viable but not culturable state. Under the right conditions, these sublethally injured organisms do have the potential to recover and cause illness, and therefore their detection is of public health importance.

Traditional methods for the detection of bacterial pathogens from foods rely on culturing of the organisms on agar plates (1 to 2 days) and usually necessitate an initial liquid enrichment step (1 to 2 days) to increase the numbers of the target organism and to enable recovery and detection of the organism from the food. The suspect colony on the agar plate can then be identified by morphological, immunological, or biochemical means (De Boer and Beumer 1999).

These methods are thus very time-consuming, taking 5 to 7 days to detect specific pathogenic microorganisms. In food microbiological analysis, while rapid methods are becoming more widely available and accepted, many still lack sufficient sensitivity, robustness, and specificity for use by the food industry. Major international advances in the 1990s in the field of biotechnology rapidly progressed the state of the art in the field of microbial detection. In particular, there were major developments in the level of genomic information available for food-borne pathogens that could be exploited to develop methods to detect and genetically characterize microorganisms.

One of the most practical and useful applications of molecular tools is their specificity as they target genetic regions unique to the organism. And depending on the gene target, they can yield valuable information about virulence properties of the organism. They are also invaluable in detecting and identifying infectious agents for which routine growth-based culture and microscopy methods are not adequate, such as viruses, parasites, and bacteria that are difficult to culture. Although molecular biological assays are more specific, sensitive, and faster than conventional microbiological methods, the complexities of food matrices continue to pose unique challenges that may preclude the direct application of molecular biological methods. Consequently, a cultural enrichment period (24 to 48 h) is generally still required for food samples prior to analysis.

Molecular diagnostic techniques are based on the detection of a fragment of genetic material (nucleic acids, i.e., DNA or RNA) that is unique to the target organism, and as such they are highly specific. DNA is a large molecule composed of a series of subunits called *nucleotides*. Each nucleotide consists of a phosphate group, a 5-carbon sugar, deoxyribose, and one of four different nitrogenous bases (adenine, thymine, guanine, and cytosine [A, T, G, C]). These bases bond specifically according to the number of hydrogen bonds that they form, and as a result A always pairs with T, and G with C. The nucleotides are linked together to form a polynucleotide chain, and two strands of polynucleotides are twisted together to form a double helix. The base pairs can be arranged along the length of the helix in any order, and it is the sequence of base pairs along the length of the molecule that is important and that forms the genetic "fingerprint." RNA is similar to DNA in that it is made up of nucleotides. The only difference being that the 5-carbon sugar is ribose instead of deoxyribose and the base uracil is found instead of thymine. A single gene might be a segment of DNA or RNA consisting of hundreds or thousands of nucleotides. Depending on the level of specificity required (genus, species, strain), different regions and lengths of the genome can be used as targets (De Boer and Beumer 1999).

Polynucleotides with a sequence complementary to the single-stranded DNA or RNA can be used as probes. DNA sequences that are homologous to all members of the family of interest, or unique sequences that are particular to a species, can be chosen for the construction of the probe. Probes that target different genera, species, or serovars can thereby be constructed. Genes associated with virulence/pathogenicity are often also used for detection of pathogens. Nucleic-acid-based assays can either be performed directly or after amplification of the target sequence.

Hybridization-Based Methods

Hybridization is the process by which a DNA probe binds to a complementary region of single-stranded DNA or RNA (Entis et al. 2001). Target nucleic acids are denatured by high temperature or high pH, and then a probe labeled with a radioisotope or a chemical is added. If the target nucleic acid in the sample contains the same nucleotide sequence as that of the gene probe, it will bind and hybridize with the target DNA or RNA (Hill et al. 2001; Liebana 2002). Immobilization of either the target DNA or a capture probe prior to hybridization allows for simple removal of the unbound reporter probe. Solution hybridization assays in which the probe and target nucleic acids are free in solution have been developed; however, there are difficulties posed in removing the probe hybrids after solution hybridization (Barbour and Tice 1997). Liquid-phase hybridization is commercially available as GENE-TRAK® assays, which have been developed for various food pathogens, including *Salmonella*, *Listeria*, *Escherichia coli*, and *Staphylococcus aureus*.

DNA Hybridization Methods

DNA hybridization tests may be performed in many ways. The Southern blot method, which was developed by Southern (1975), is based on the transfer and immobilization of DNA fragments and separation by gel electrophoresis, all conducted on a solid support, such as nylon or nitrocellulose membrane. DNA transfer to solid support is generally accomplished by capillary methods, but electroblotting, positive pressure, and vacuum transfer procedures can also be used (Surzycki 2000a). Once immobilized, the DNA is available for hybridization with a labeled DNA or RNA probe. Solid-phase hybridization can also be performed as colony blots or dot blots. In colony blots, an aliquot of sample (i.e., homogenized food) is spread-plated onto an appropriate agar. After incubation, an imprint of the colony/colonies is transferred to a solid support such as a membrane or paper filter by pressing the support onto the agar surface. Next the cells are lysed by a combination of high pH and temperature, which also denatures and affixes the DNA to the support. DNA attached to the solid support is then detected by a labeled probe, and the excess probe is removed by washing (Hill et al. 2001).

In the dot blot technique, microorganisms grown in liquid or solid culture are transferred with a loop or other device, as a "dot" onto a localized area of a membrane. The colony blot is particularly useful for identification of bacteria directly from agar plates.

RNA Hybridization Methods

RNA gel blots, or northern hybridization analysis, was introduced shortly after the DNA blotting technique (Alwine et al. 1977). In this technique, RNA molecules are fractioned by size, using denaturing agarose gel electrophoresis. The fractionated RNA is transferred to a solid support, such as nylon or a nitrocellulose membrane, and then, similar to the Southern blot analysis, the membrane is hybridized with a specific labeled probe. Northern blot analysis is more difficult to perform than Southern blot because RNA is sensitive to degradation by RNAases, which are difficult to inactivate and have a universal presence (Surzycki 2000c).

Fluorescent In Situ Hybridization (FISH)

In situ hybridization (ISH) with radiolabeled DNA was first reported by Pardue and Gall (1969) and John et al. (1969) for examination of cells. It was applied to bacteria for the first time in 1988 (Giovannoni et al.), and with the advent of fluorescent labels became more widely used (Amann et al. 1990). Fluorescent in situ hybridization (FISH) is a technique that detects nucleic acid sequences by a fluorescently labeled probe that hybridizes specifically to its complementary target gene within the intact cell. The target gene is the intercellular ribosomal RNA (rRNA) of a particular organism. The rRNA genes contain both highly conversed and highly variable sequence regions and are transcribed at very high numbers in viable cells (Amann et al. 1995). Cells are treated with a single-stranded oligonucleotide probe complementary to the rRNA sequence of the target organism. Labeling of the probe with a fluorescent molecule enables the organisms to be detected by fluorescent microscopy. Fang et al. (2003) demonstrated that a *Salmonella*-specific oligonucleotide probe complementary to the 23S rRNA (Sal-3), labeled with the fluorochrome Cy3, detected *Salmonella* in naturally contaminated food samples after 16h in a pre-enrichment broth. However, to date the dependence of FISH on microscopic detection has limited its potential, as microscope methods require considerable operator training and in general cannot be fully automated.

Nucleic Acid Amplification

Nucleic acid methods that include an amplification step for the target DNA/RNA are now routinely employed in molecular biology. These methods, as outlined in the following, increase the target nucleic acid material by up to a millionfold and are particularly important in the arena of food microbiology, where one of the major hurdles is the recovery and detection of very low numbers of a particular pathogen. However, even with the incorporation of an amplification step, a liquid enrichment of the food sample (24 to 48 h) is generally still required to yield sufficient nucleic acid material from the target organism for the reaction.

Polymerase Chain Reaction (PCR)

The most popular method of amplification is the polymerase chain reaction (PCR) technique. Since its development in 1983 (Mullis 1990) and the publication of the first experimental data on PCR (Saiki et al. 1985), this reaction has become an essential tool in molecular biology. In this technique, the DNA is extracted from the organism, and the double strands are denatured into single-stranded DNA. Short sequence DNA primers are annealed to the complementary DNA target in the organism. The primers are then extended across the target sequence using a heat-stable DNA polymerase (usually *Taq* polymerase, a thermostable and thermoactive enzyme from *Thermus aquaticus*) in the presence of free deoxynucleoside triphosphates (dNTPs), resulting in a double replication of the starting target material. Multiple repeats of the denaturation, annealing, and extension steps result in an exponential increase in the levels of the initial target DNA, thus greatly increasing the sensitivity of the method (Entis et al. 2001). Theoretically, a single gene target can be amplified a millionfold to allow detection in only a few hours. The sensitivity of PCR is limited, in part by the small sample volumes used in PCR and by the presence of inhibitory substances such as proteinases and collagen present in many foods (Powell et al. 1994; Kim et al. 2001). Sample preparation, including an enrichment step to increase the number of target cells, effective extraction of the pathogen from the enriched food, extraction of DNA, and the use of the appropriate DNA polymerases and primers are critical to achieve the desired sensitivity in PCR reactions (Rådström et al. 2003).

In conventional PCR, the steps in the PCR cycle are carried out in an automated, programmable block heater. The PCR products are separated by gel electrophoresis, stained with ethidium bromide, and visualized using ultraviolet light (Olsen et al. 1995). A range of gene targets have been used in the PCR detection of pathogens. For *Salmonella* targets include species-specific genes (Kumar et al. 2003; Trkov and Avguštin 2003) and virulence genes (*invA*) (Rahn et al. 1992; C.-H. Chiu and Ou 1996). A number of suitable gene targets have been suggested for the detection and/or differentiation of enterohemorrhagic *E. coli*. Species-specific targets include the rfb_{O157} gene of the O157 antigen (Paton and Paton 1998), *wzx* (O-antigen flippase) and *wzy* (O-antigen polymerase) for serogroups O26 and O113 (DebRoy et al. 2004), and virulence genes *Vt1, Vt2, eae*A, and *hlyA*)

Table 8.1. Conventional PCR assays applied to food pathogens.

Pathogen	Gene Target	Food	Reference
Salmonella	*16S-23S rRNA*	Chicken and milk	T. H. Chiu et al. 2005
Salmonella	*ogdH*	Chicken	Jin et al. 2004
E. coli O157	*eaeA*	Ground beef	Uyttendaele et al. 1999
E. coli O157	*eaeA, hlyA, vt1, vt2*	Minced beef	Fitzmaurice et al. 2004
L. monocytogenes	*actA*	Soft cheese	Longhi et al. 2003
L. monocytogenes	*inlAB*	Frankfurters	Jung et al. 2003
Campylobacter	*flaA*	Chicken	Oyofo et al. 1997
Campylobacter	*16S rRNA*	Poultry	Mateo et al. 2005
Mycobacterium paratuberculosis	*16S rRNA*	Milk	Tasara et al. 2005

(Paton and Paton 1998; Fagan et al. 1999). Targets for *L. monocytogenes* and *Yersinia enterocolitica* include the listeriolysin *O* gene (Thomas et al. 1991) and the *16 S rRNA* gene (Wannet et al. 2001; Jaradat et al. 2002; Longhi et al. 2003; Kot and Trafny 2004). These targets have been widely employed for confirmation of pathogens from suspect colonies on agar plates and for direct detection from fecal samples. PCR methods have now been developed for detection of pathogens from foods, and table 8.1 outlines some recently reported conventional PCR methods for a range of food pathogens and identifies the target gene and food.

One of the main disadvantages of conventional PCR is that laboratories must take specific precautions to avoid amplicon carryover and consequently false-positive results. In addition, this technique is generally time-consuming and labor-intensive (Persing 1991). A limiting factor in the uptake of PCR methods, and indeed many other rapid pathogen methods by the food industry, relates to a lack of full validation against the accepted cultural method in accordance with the International Standards Organization (ISO). It is generally essential, when testing against microbial criteria set by regulatory authorities, to use a recognized ISO method or an equivalent rapid method. For *Salmonella* the procedure for determining equivalence is outlined in EN/ISO 16140: 2003. The lack of validation and ring trials is now being recognized as a shortfall, and research to fully validate methods is now being undertaken. D'Agostino et al. 2004 report a PCR assay for detection of *L. monocytogenes* in raw milk that was evaluated in a collaborative trial involving 13 European laboratories. Similarly, Josefsen et al. 2004 have reported on the validation of a PCR-based method for detection of food-borne thermotolerant campylobacters from chicken in a multicenter collaborative trial. Malorny et al. 2004 have reported a multicenter validation of a PCR-based method for detection of *Salmonella* in chicken and pig samples, and Abdulmawjood et al. 2004 have reported on a multicenter interlaboratory trial for PCR-based detection of food-borne *E. coli* O157. PCR is commercially available as the BAX system (Dupont Qualicon) for a range of food-borne pathogens.

Real-Time PCR

Real-time PCR is now increasingly replacing conventional PCR as a rapid, sensitive, and specific molecular diagnostic technique (Bellin et al. 2001). Since its introduction, real-time PCR has revolutionized the field of molecular diagnostics, and the technique is

Table 8.2. Some real-time PCR instruments currently available.

Instrument	Manufacturer	Location
ABI 7000	Applied Biosystems	Foster City, California
ABI 7900HT	Applied Biosystems	Foster City, California
i-Cycler iQ	Bio-Rad	Hercules, California
LightCycler	Roche Applied Science	Indianapolis, Indiana
SmartCycler	Cepheid	Sunnyvale, California
Mx3000P	Stratagene	Cedar Creek, Texas
RotorGene	Corbett Research	Queensland, Australia
Apollo ATC 901	Apollo Instrumentation	San Diego, California
BAX System	Dupont/Qualicon	Wilmington, Delaware
DNA Engine Opticon 2	MJ Research	Waltham, Massachusetts
R.A.P.I.D.	Idaho Technology	Salt Lake City, Utah

Source: Adapted from McKillip and Drake 2004.

being used in a rapidly expanding number of applications (Arya et al. 2005). Real-time PCR allows continuous monitoring of amplification through the use of fluorescent double-stranded (ds) DNA intercalating dyes or sequence-specific probes (Wittner et al. 1997). Real-time PCR assays offer many advantages over traditional PCR methods. They are much quicker to perform, with the enhancements in speed attributed to reduced amplification time and the elimination of an additional step(s) needed for product detection. Moreover, the use of a closed system for amplification and detection minimizes the potential for amplicon carryover contamination (Bankowski and Anderson 2004). Although expensive in capital terms, real-time PCR-based strategies are becoming more popular in research and public health laboratories and are being increasingly used for specific diagnostic applications and pathogen detection (Bellin et al. 2001; Taylor et al. 2001). A number of real-time PCR instruments are commercially available for use in PCR (table 8.2). Real-time PCR methods can be divided into those that are not sequence-specific, such as DNA minor groove binding dyes, and those that are sequence-specific and might even afford simultaneous detection and confirmation of the target amplicon during the PCR reaction (McKillip and Drake 2004). Currently, the method for nonspecific real-time detection of PCR amplicons employs fluorescent double-stranded DNA intercalating dyes such as SYBR Green I (McKillip and Drake 2004). SYBR Green I binds to the minor groove of ds DNA during the extension step of the PCR and falls off during the denaturation step (Bustin 2000; Lekanne Deprez et al. 2002). The specificity and sensitivity of SYBR Green I is limited as it binds to all ds DNA, including nonspecific PCR products and PCR primer dimers (Wittner et al. 1997; Bustin 2000). SYBR Green I has been successfully used in combination with melting curve analysis for mutation screening and allele discrimination (Lyon et al. 1998; Bennet et al. 2003), but it is also useful for food pathogen detection (table 8.3).

A diverse array of fluorescently labeled probes are in use for sequence-specific detection of target DNA or RNA (McKillip and Drake 2004). They involve fluorescence resonance energy transfer (FRET) between fluorogenic labels or between one fluorophore and a quencher group (Didenko 2001).

FRET is a process by which energy is passed between molecules separated by 10–100 Å that have overlapping emission and absorption spectra (Stryer and Haugland 1967; Clegg

Table 8.3. Real-time PCR assays applied to food pathogens.

Real-Time PCR Chemistry	Pathogen	Gene Target	Food	Reference
SYBR Green I	*E. coli* O157	*stx1, stx2*	—	Jothikumar and Griffiths 2002
	Salmonella	*fimI*	—	Jothikumar et al. 2003
	Salmonella	*16S rRNA*	Retail beef, pork, turkey, chicken	Catarame et al. 2006
	L. monocytogenes, Salmonella	*HlyA, invA*	Sausage	Wang et al. 2004
HybProbes	*E. coli* O157	*eaeA*	Beef products	Ellingson et al. 2005
	Campylobacter	*16S rRNA*	Chicken	Abu-Halaweh et al. 2005
	Salmonella	*sipB, sipC*	Meat products	Ellingson et al. 2004
	Salmonella	*hylA*	Ground beef, beef hotdogs	Nguyen et al. 2004
	E. coli O157	*rfbE*	Ground beef, beef hotdogs	Nguyen et al. 2004
	E. coli O157, O26, O111	*per, wzy, fliC-fliA, vt1, vt2*	Minced beef	O'Hanlon et al. 2004
TaqMan	*E. coli* O157	*eaeA, stx1, stx2*	Beef	Sharma 2002
	Salmonella		Beef, shrimp	Kimura et al. 1999
	L. monocytogenes	*hlyA*	—	Lunge et al. 2002
Molecular beacons	*E. coli* O157	*rfbE*	Raw milk, apple juice	Fortin et al. 2001
	Salmonella	*himA*	—	Chen et al. 2000
	L. monocytogenes	*hlyA, iap*	Dried nonfat milk	Koo and Jaykus 2003

1992). Regardless of the specific means in which the fluorophore-quenching pair is applied, these methods have the advantage of sequence specificity that ds DNA intercalating dyes do not offer (McKillip and Drake 2004). Sequence-specific detection can be performed using the linear oligoprobes such as the hybridization probes (HybProbes), dual-labeled oligoprobes (TaqMan probes), or hairpin oligonucleotides (molecular beacons) (Mackay, 2004). Several variations of the basic FRET chemistry exist, although many of these remain unproven in food systems (McKillip and Drake 2004). Real-time PCR technology has been used to detect a range of food-borne pathogens (table 8.3).

An advantage of real-time PCR assays over conventional PCR methods is that they can be used for quantification of initial target DNA. Accurate calculation of the initial amount of DNA and elimination of false-negative results can be obtained with the inclusion of an internal amplification control, which is now becoming mandatory (Hoorfar et al. 2004). It consists of an internal control amplified at the same time as the target gene but detected by a second fluorophore. Future applications of the real-time PCR technology may include the development of quantitative real-time PCR methods able to quantify the number of bacteria directly in complex materials such as food.

RNA-Based Amplification Assays

While DNA is generally selected as a target molecule in designing PCR assays for food pathogens, a limitation of this approach is that it is not possible to distinguish between viable and nonviable bacteria, though this is somewhat overcome by sample enrichment that increases the numbers of viable cells and target DNA. mRNA, which has a short half-life, is a better target for the determination of viability (Sheridan et al. 1998). The isolation of RNA is an essential step in the analysis of patterns and mechanisms of gene expression (Surzycki 2000a). Moreover, the viruses of importance in food-borne illness have RNA, rather than DNA, as genomic material.

Reverse Transcriptase PCR (RT-PCR)

Reverse transcriptase polymerase chain reaction (RT-PCR) is a variation of the PCR reaction and employs the enzyme reverse transcriptase to convert messenger RNA (mRNA) into complementary DNA (cDNA), which is subsequently amplified by DNA PCR (Sambrook and Russell 2001). Thus, this sensitive and powerful technique allows an exponential increase in the amount of mRNA in the form of cDNA copies. The benefits of this procedure include its sensitivity, its large dynamic range, the potential for high throughout, as well as accurate quantification. To achieve this, however, appropriate normalization strategies are required to control for experimental error introduced during the multistage process required to extract and process the RNA (Huggett et al. 2005). In fact, one of the main difficulties with this technique is that isolation of RNA is technically more difficult than DNA and is also less stable. RNA samples also can be contaminated with residual DNA, which makes it impossible to correctly determine RNA concentration or perform RT-PCR, and thus removal of DNA is critical in this technique (Surzycki 2000b). RT-PCR has been used to monitor cell viability in bacteria of relevance for the food industry, such as VTEC (McIngvale et al. 2002), *L. monocytogenes* (Klein and Juneja 1997), and *Salmonella* (Szabo and Mackey 1999) and also parasites important in food-borne transmission such as *Cryptosporidium* and *Giardia* (Caccio 2004). RT-PCR has been used to identify viruses implicated in food-borne outbreaks in different countries (Sair et al. 2002; Di Pinto et al. 2003; Kobayashi et al. 2004). RT-PCR assays have also been used as part of a large surveillance study on the importance of enteric viruses as causes of illness across Europe (Koopmans et al. 2003).

Nucleic Acid Sequence-Based Amplification (NASBA)

An alternative to RT-PCR is the nucleic acid sequence-based amplification (NASBA). It is a sensitive, isothermal, transcription-based amplification system specifically designed for the detection of RNA targets (Deiman et al. 2002). This method is reported to specifically amplify RNA but not DNA (Heim et al. 1998). This system selectively amplifies RNA through the concerted action of three enzymes: reverse transcriptase, RNAaseH, and RNA polymerase (Cook 2003). Since NASBA amplifies RNA using an RNA T7-polymerase promoter to generate multiple RNA products, ds DNA is not denatured and consequently not amplified (Chan and Fox 1999; Simpkins et al. 2000). NASBA has traditionally been used for the amplification of blood-borne viruses (Kievits et al. 1991; Van Gemen et al. 1993), but it was also optimized for detection of pathogenic bacteria such as *L. monocytogenes* (Uyttendaele et al. 1995) and *Salmonella* (Cook et al. 2002). D'souza and Jaykus

(2003) report on the use of NASBA for the detection of *Salmonella enterica* serovar Enteritidis (*S.* Enteritidis) in representative inoculated foods (fresh meats, poultry, fish, ready-to-eat salads, and bakery products) following an 18 h pre-enrichment. The primer and probe set were based on mRNA sequences of the *dnaK* gene of *Salmonella*. NASBA has also been shown to be most useful for detection of organisms that are impossible or difficult to culture, such as viruses from sliced turkey and lettuce (Jean et al. 2004) and *Mycobacterium paratuberculosis* from milk (Rodriguez-Lazaro et al. 2004).

Subtyping

Accurate typing and subtyping of pathogenic bacteria are essential if human cases of infection are to be linked within epidemiological investigations, and sources of infection traced (Thomson-Carter 2001). Typing methods can be classified as phenotypic (detecting characteristics expressed by the organism) or genotypic (directly examining the organism's genetic content) (Maslow and Mulligan 1996). Phenotypic methods include phage typing, serotyping, and biotyping. A range of genotypic techniques for the detection of bacteria in food have been developed and applied as outlined in the following.

Restriction Fragment Length Polymorphism (RFLP)

Restriction fragment length polymorphism (RFLP) analysis investigates certain types of sequence polymorphisms, so-called point mutations that can be base exchanges, base deletions, or insertions (Meyer and Birchmeier 1995). The basic mechanism of RFLP analysis relies on the ability of restriction enzymes, the "endonucleases," to cut double-stranded DNA according to a certain succession of bases in a process called *digestion* (Hummel 2003). RFLP can be performed by digestion of DNA samples followed by analysis using standard gel-transfer hybridization procedures. Another method is the restriction digestion of a PCR-amplified DNA segment that contains a variably present restriction site. The technique thus requires some knowledge of the DNA sequence flanking that restriction site (Dietrich et al. 1999).

Pulse Field Gel Electrophoresis (PFGE)

Pulse field gel electrophoresis (PFGE) employs restriction enzymes (endonucleases) to make a limited number of cuts in bacterial chromosomes to provide chromosomal restriction patterns that form a "fingerprint" for each organism (Méndez-Álvarez et al. 1995). In this technique, DNA is first prepared and encapsulated in agarose plugs, digested with appropriate restriction enzymes depending on the target organism, and subjected to electrophoresis in which an electric field periodically changes direction and/or intensity relative to the agarose gel (Maule 1998). It is important that the test organisms are embedded in agarose plugs and that the DNA is released in situ because this minimizes shearing of the DNA before it is digested with restriction enzymes (Maslow and Mulligan 1996). The periodical reorientation of the electric field allows the separation of large DNA fragments, which would not be adequately separated by conventional agarose gel electrophoresis using a constant (direction) electric field (Maslow and Mulligan 1996). In PFGE, each time the field is switched, larger molecules take longer to change direction and have less time to move during each pulse, so they migrate slower than smaller molecules, leading to overall separation (Basim and Basim 2001). In this way, while conventional agarose gel electrophoresis may achieve a maximum resolution of 50 kb, PFGE is able to separate molecules as large as 12 Mb (Maule 1998). The concept of a contour-clamped homogeneous electric field to separate large DNA molecules was introduced by Chu and coworkers (1986). PFGE requires highly technically skilled personnel but is very reproducible and has the advantage of generating genetic data that can be statistically analyzed. Using standardized protocols, PFGE is used to share genetic information on food pathogens in national and international networks such as PulseNet (a U.S. national network of public health and food regulatory agency laboratories coordinated by the Centers for Disease Control and Prevention) and the Europe International surveillance network (ENTER-NET) for the enteric infections.

PFGE has been used extensively in the investigation of sporadic cases and outbreaks of food-borne microbial infection and has also proved invaluable in molecular characterization and epidemiology of infection related to verotoxigenic *E. coli* (Thomson-Carter 2001), *Salmonella* (Ross and Heuzenroeder 2005), *L. monocytogenes* (Okwumabua et al. 2005), and *Clostridium botulinum* (E. A. Johnson et al. 2005).

Amplification-Based Typing Methods

Various typing systems that use PCR-amplification-based methods have been described. One of the main disadvantages of amplification-based methods is that many factors can affect reliability and reproducibility (Tyler et al. 1997).

Random Amplification of Polymorphic DNA (RAPD)

Random amplification of polymorphic DNA (RAPD) is a PCR-based method in which the primers are chosen arbitrarily rather than based on knowledge of the sequence to be amplified. The stringency of primer annealing is low to allow priming of imperfectly matched sequences. PCR performed under these conditions generates a complex pattern of PCR products that is, at least in theory, unique to a particular bacterial strain. This method is distinct from the classic PCR in its use of a single primer instead of two and a low-stringency annealing temperature (Williams et al. 1990). Cocolin et al. (2005) have reported on the benefits of a RADP-PCR as an epidemiological tool for *L. monocytogenes.*

Amplified Fragment Length Polymorphism (AFLP)

The amplified fragment length polymorphism (AFLP) technique is based on the selective PCR amplification of restriction fragments from a total digest of genomic DNA. The technique involves the digestion of DNA with restriction enzymes and then the ligation of the fragments' ends to nucleotide adapters that are designed in such a way that the initial restriction site is not restored after ligation. This allows simultaneous restriction and ligation, while religated original fragments are cleaved again. Finally PCR amplification of restriction fragments is achieved by using the adapter and restriction site sequence as target sites for primer annealing. Visualization of the amplified fragments is then performed by gel electrophoresis (Vos et al. 1995). Two restriction enzymes are used, one with an average cutting frequency (like *Eco*RI) and a second one with a higher cutting frequency (like *Mse*I) (Sharbel 1999). The vast majority of bands detected on AFLP gels are fragments flanked by both enzyme recognition sites. Like RAPD, AFLP analysis is applicable to DNA of any origin and complexity without prior sequence information (Masiga et al. 2000) and combines the reliability of the RFLP technique with the power of the PCR technique (Vos et al. 1995). AFLP has been shown to be a suitable tool for discrimination between *Salmonella enterica* serovar Typhimurium DT126 isolates from separate food-related outbreaks in Australia (Ross and Heuzenroeder 2005).

Repetitive Extragenic Palindromic PCR (Rep-PCR)

Other amplification-based typing methods use PCR amplification with primers complementary to regions of a bacterial genome that contain repetitive sequences (Versalovic et al. 1991, 1994). After PCR the amplified fragments can be resolved in a gel matrix, yielding a profile referred to as Rep-PCR (repetitive extragenic palindromic PCR) genomic fingerprinting (Versalovic et al. 1994). In the Enterobacteriaceae family, several families of repetitive sequences have been found; some examples are Rep, ERIC (enterobacterial repetitive intergenic consensus), and boxC (module box) (Baldy-Chudzik 2001). The corresponding protocols are referred to as Rep-PCR, ERIC-PCR, and box-PCR (Rademarker and De Bruijn 1998). The Rep sequences have been found in *E. coli* and are 35 to 40 bp long and include an inverted repeat (Gilson et al. 1984). ERIC sequences are also found in *E. coli* and are larger elements of 124 to 127 bp and contain a highly conserved central inverted repeat (Hulton et al. 1991). The 154 bp box element, described in 1992 (Martin et al. 1992), was discovered in the genome of *Streptococcus pneumoniae* and is present in approximately 25 copies per genome. Reading from 5′-3′, the box elements are composed of three sub-units: box a (59 nucleotides), box b (45 nucleotides), and box c (50 nucleotides) (Martin et al. 1992). The sequences within the boxA subunit appear to be conserved among diverse bacterial species (Koeuth et al. 1995). The boxA1R primer, based on the boxA sequence, has been used in Rep-PCR analysis of pathogenic and nonpathogenic *E. coli* isolates (J. R. Johnson and O'Bryan 2000; Seurinck et al. 2003). The repetitive elements are thought to be highly conserved because rep sites are essential protein-DNA interaction sites and/or these sequences may propagate themselves as "selfish" DNA by gene conversion (Versalovic et al. 1991). Catarame et al. (in press) have reported on a rep-PCR protocol for the characterization of *E. coli* O157.

Multilocus Sequence Typing (MLST)

Multilocus sequence typing (MLST) has recently come to the forefront as a method for strain characterization and for epidemiological tracking of bacterial infections. MLST exploits the unambiguous nature and electronic portability of nucleotide sequence data in microorganisms.

This tool can be applied to a wide rang of bacteria (Urwin and Maiden 2003; Maiden et al. 1998), and a public database is maintained as well as primers and protocols for species for which methods have been established (http://www.mlst.net). This technique has been shown to be useful for typing *Campylobacter* spp. (Miller et al. 2005).

Developmental Methods

Microarrays

Microarrays consist of immobilized biomolecules on planar surfaces, microchannels, microwells, or beads. Biomolecules commonly immobilized on microarrays include oligonucleotides, PCR products, proteins, lipids, peptides, and carbohydrates (Venkatasubbarao 2004). As for northern and Southern blotting, the underlying principle of a DNA microarray analysis is the hybridization of complementary nucleic acid strands. Single-stranded DNA arrayed on a solid substrate are designed to act as a probe to selectively capture complementary labeled targets from a biological sample. In a DNA microarray experiment, the first step is the extraction of RNA from cells. The mRNA is isolated and then reverse transcribed into cDNA and converted to a form of labeled polynucleotides, called *targets*. The labeled polynucleotides are then applied to the array and captured by immobilized polynucleotides (probes) (Simon et al. 2003). Different microarray platforms exist including glass slides, silicon chips (Affymetrix GeneChip™ Arrays), and nylon membranes, and the labeling process may vary depending on the microarray platform (Simon et al. 2003). Unlike other hybridization formats, DNA microarrays allow significant miniaturization and also speed, automation, and high throughput (Gupta et al. 1999).

The global nature of the DNA microarray technique holds tremendous promise for the unfolding of complex genetic and metabolic networks. The technology can be used to study gene expression in complex microbial populations, such as those found in food and gastrointestinal tracts (Al-Khaldi et al. 2002). DNA microarrays used to study gene expression have allowed the simultaneous analysis of thousands of different genes. In this way, the use of DNA microarrays to study gene expression results in large datasets, which are frequently difficult to analyze and could lead to confusing hypotheses and conclusions. Adequate experimental design and statistical tools are therefore critical in this technique. In an effort to establish standards for microarray experiment annotation and data interpretation, the Microarray Gene Expression Data Group (MGED) was established in November 1999 (http://www.mged.org). A set of guidelines has been suggested to outline the minimum information required for microarray experiment design and data representation that allows the uniformity for reproduction and verification by other researchers.

Although DNA microchips have been used mostly for gene expression studies, the technique has a great potential for use in diagnostic microbiology. Studies have been conducted using DNA microarrays spotted with genetic markers for bacteria genomes that are associated with various pathogens and virulence factors. In one such study six genes were chosen to identify 15 strains of *Salmonella, Shigella,* and *E. coli* (Chizhikov et al. 2001). In another study, *E. coli* O157:H7 was detected using spotted DNA microarray labeled with oligonucleotide probes that were complementary to four virulence loci (Call et al. 2001). DNA microarray technology has also been applied for the rapid detection of *Campylobacter* spp. and the differentiation of two closely related *Campylobacter* species taken directly from chicken samples (Keramas et al. 2003). Hing et al. (2004) report on a method for simultaneous detection of a number of common food-borne pathogenic bacteria using oligonucleotide array technology and a pair of universal primers designed for PCR amplification of the *23S rRNA* gene.

Ahn and Walt (2005) report on a highly novel method using fiber-optic DNA microarrays for rapid and sensitive detection of *Salmonella* spp. A fiber-optic DNA microarray was prepared using microsphere-immobilized oligonucleotides (in microwells created by etching optical fiber bundles) with probes specific for the *Salmonella invA* and *spvB* genes.

With further development, microarrays have the greatest potential to enable instant and simultaneous detection of multiple pathogens in foods in a single user-friendly step.

Conclusion

As increasing numbers of microorganisms are fully sequenced, numerous possibilities increase to exploit this information for detection and characterization of microorganisms. Recent advances in the area of DNA/RNA microarray technology offer real potential for detection of multiple microorganisms in a single rapid step. Molecular tools now also offer huge opportunities for full traceability of pathogens along the food chain. However despite all these exciting scientific advances, it must be borne in mind that for the food industry to adapt methods, there are a number of important criteria that must be met. To demonstrate compliance with microbiological criteria, rapid methods employed by the food industry must be fully validated and shown to be equivalent to gold standard cultural methods. For example for *Salmonella* there is now an ISO procedure that outlines the necessary steps for demonstrating the equivalence of the rapid method with the gold standard culture methods. In addition, for the industry to adapt a method, it must be user-friendly, should not require highly specialized technical staff, and should be robust enough for use in a busy laboratory with high sample throughput. Equally, initial capital cost and running cost are considerations, though they may be offset by savings from obtaining results earlier, thus leading to earlier product release. If all these issues are borne in mind when developing and validating rapid methods, these methods will no doubt be widely employed by the food industry.

References

Abdulmawjood, A., Bulte, M., Roth, S., Schonenbrucher, H., Cook, N., D'Agostino, M., Burkhard, M., Jordan, K., Pelkonen, S., and Hoorfar, J. 2004. Toward an international standard for PCR-based detection of foodborne *Escherichia coli* O157: validation of the PCR-based method in a multicenter interlaboratory trial. *Journal of Association of Analytical Communities International* 87(4):856–60.

Abu-Halaweh, M., Bates, J., and Patel, B. K. 2005. Rapid detection and differentiation of pathogenic *Campylobacter jejuni* and *Campylobacter coli* by real-time PCR. *Research Microbiology* 156(1):107–14.

Ahn, S., and Walt, D. R. 2005. Detection of *Salmonella* spp. using microsphere-based, fiber-optic DNA microarrays. *Analytical Chemistry* 77(15):5041–47.

Al-Khaldi, S. F., Martin, S. A., Rasooly, A., and Evans, J. D. 2002. DNA microarray technology used for studying foodborne pathogens and microbial habitats: minireview. *Journal of Association of Analytical Communities International* 85:906–10.

Alwine, J. C., Kemp, D. J., and Stark, G. R. 1977. Method for detection of specific RNAs in agarose gels by transfer to diazobenzyloxymethyl—paper and hybridization with DNA probes. *Proceedings of the National Academy of Sciences of the United States of America* 74:5350–54.

Amann, R. I., Krumholz, L., and Stahl, D. A. 1990. Fluorescent-oligonucleotide probing of whole cells for determinative, phylogenetic, and environmental studies in microbiology. *Journal of Bacteriology* 172:762–70.

Amann, R. I., Ludwig, W., and Schleifer, K. H. 1995. Phylogenetic identification and *in situ* detection of individual microbial cells without cultivation. *Microbiological Reviews* 59:143–69.

Arya, M., Shergill, I. S., Williamson, M., Gommersall, L., Arya, N., and Patel, H. R. 2005. Basic principles of real-time quantitative PCR. *Expert Review of Molecular Diagnosis* 5:209–19.

Baldy-Chudzik, K. 2001. Rep-PCR—a variant to RAPD or an independent technique of bacteria genotyping? A comparison of the typing properties of Rep-PCR with other recognised methods of genotyping of microorganisms. *Acta Microbiologica Polonica* 50(3–4):189–204.

Bankowski, M. J., and Anderson, S. M. 2004. Real-time nucleic acid amplification in clinical microbiology. *Clinical Microbiology Newsletter* 26(2):9–15.

Barbour, W. M., and Tice, G. 1997. Genetic and immunologic techniques for detecting foodborne pathogens and toxins. *In* Food Microbiology: Fundamentals and Frontiers, Doyle, M. P., Beuchat, L. R., and Monteville, T. J. (eds.), pp. 710–27. ASM Press, Washington.

Basim, E., and Basim, H. 2001. Pulsed field gel electrophoresis (PFGE) technique and its use in molecular biology. *Turkish Journal of Biology* 25:405–18.

Bellin, T., Pulz, M., Matussek, A., Hempen, H. G., and Gunzer, F. 2001. Rapid detection of enterohemorrhagic *Escherichia coli* by real-time PCR with fluorescent hybridisation probes. *Journal of Clinical Microbiology* 39:370–74.

Bennet, C. D., Campbell, M. N., Cook, C. J., Dyre, D. J., Nay, L. M., Nielsen, D. R., Rasmussen, R. P., and Bernard, P. S. 2003. The LightTyper™ high-throughput genotyping using fluorescent melting curve analysis. *Biotechniques* 34:1288–95.

Bustin, S. A. 2000. Absolute quantification of mRNA using real-time reverse transcription polymerase chain reaction assays. *Journal of Molecular Endocrinology* 25:169–93.

Caccio, S. M. 2004. New methods for the diagnosis of *Cryptosporidium* and *Giardia*. *Parassitologia* 46:151–55.

Call, D. R., Brockman, F. K., and Chandler, D. P. 2001. Detecting and genotyping *Escherichia coli* O157:H7 using multiplexed PCR and nucleic acid microarrays. *International Journal of Food Microbiology* 67: 71–80.

Catarame, T., O'Hanlon, K. A., Blair, I. S., McDowell, D. A., and Duffy, G. 2006. Comparison of a real time PCR assay with a culture method for the detection of *Salmonella* in retail meat samples. *Journal of Food Safety* 26(1):1–15.

Catarame, T., Duffy, G., O'Hanlon, K., Downer, R., Sheridan, J. J., Blair, I. S., and McDowell, D. A. In press. Comparison of PFGE and Rep-PCR in differentiation of verocytotoxigenic *Escherichia coli*. *Journal of Microbiological Methods.*

Chan, A. B., and Fox, J. D. 1999. NASBA and other transcriptions-based amplification methods for research and diagnostic microbiology. *Reviews in Medical Microbiology* 10:185–96.

Chen, W., Martinez, G., and Mulchandani, A. 2000. Molecular beacons: a real-time polymerase chain reaction assay for detecting *Salmonella*. *Analytical Biochemistry* 280:166–72.

Chiu, C.-H., and Ou, J. T. 1996. Rapid identification of *Salmonella* serovars in faeces by specific detection of virulence genes, *invA* and *spvC*, by an enrichment broth culture-multiplex PCR combination assay. *Journal of Clinical Microbiology* 34(10):2619–22.

Chiu, T. H., Chen, T. R., Hwang, W. Z., and Tsen H. Y. 2005. Sequencing of an internal transcribed spacer region of 16S-23S rRNA gene and designing of PCR primers for the detection of *Salmonella* spp. in food. *International Journal of Food Microbiology* 97(3):259–65.

Chizhikov, V., Rasooly, A., Chumakov, K., and Levy, D. 2001. Microarray analysis of microbial virulence factors. *Applied and Environmental Microbiology* 67:3258–63.

Chu, G., Vollrath, D., and Davis, R. W. 1986. Separation of large DNA molecules by contour-clamped homogeneous electric fields. *Science* 234:1582–85.

Clegg, R. M. 1992. Fluorescence resonance energy transfer and nucleic acids. *Methods in Enzymology* 211:353–88.

Cocolin, L., Stella, S., Nappi, R., Bozzetta, E., Cantoni, C., and Comi, G. 2005. Analysis of PCR-based methods for characterization of *Listeria monocytogenes* strains isolated from different sources. *International Journal of Food Microbiology* 103(2):167–78.

Cook, N. 2003. The use of NASBA for the detection of microbial pathogens in food and environmental samples. *Journal of Microbiological Methods* 53:165–74.

Cook, N., Ellison, J., Kurdziel, A. S., Simpkins, S., and Hays, J. P. 2002. A NASBA-based method to detect *Salmonella enterica* serotype Enteriditis strain PT4 in liquid whole egg. *Journal of Food Protection* 65: 1177–78.

D'Agostino, M., Wagner, M., Vazquez-Boland, J. A., Kuchta, T., Karpiskova, R., Hoorfar, J., Novella, S., Scortti, M., Ellison, J., Murray, A., Fernandes, I., Kuhn, M., Pazlarova, J., Heuvelink, A., and Cook, N. 2004. A validated PCR-based method to detect *Listeria monocytogenes* using raw milk as a food model—towards an international standard. *Journal of Food Protection* 67(8):1646–55.

De Boer, E., and Beumer, R. R. 1999. Methodology for detection and typing of foodborne microorganisms. *International Journal of Food Microbiology* 50:119–30.

DebRoy, C., Roberts, E., Kundrat, J., Davis, M. A., Briggs, C. E., and Fratamico, P. M. 2004. Detection of *Escherichia coli* serogroups O26 and O113 by PCR amplification of the wzx and wzy genes. *Applied and Environmental Microbiology* 70(3):1830–32.

Deiman, B., Van Aarle, P., and Sillekens, P. 2002. Characteristics and applications of nucleic acid sequence-based amplification (NASBA). *Molecular Biotechnology* 20(2):163–79.

Didenko, V. V. 2001. DNA probes using fluorescence resonance energy transfer (FRET): designs and applications. *Biotechniques*. 31:1106–21.

Dietrich, W. F., Weber, J. L., Nickerson, D. A., and Kwok, Pui-Yan. 1999. Identification and analysis of DNA polymorphisms. *In* Genome Analysis: A Laboratory Manual. Mapping Genomes, vol. 4, Birren, B., Green, E. D., Hieter, P., Klapholz, S., Myers, R. M., Riethman, H., and Roskams J. (eds.). Cold Spring Harbor Laboratory Press, New York.

Di Pinto, A., Forte, V. T., Tantillo, G. M., Terio, V., and Buonavoglia, C. 2003. Detection of hepatitis A virus in shellfish (*Mytilus galloprovincialis*) with RT-PCR. *Journal of Food Protection* 9:1681–85.

D'souza, D. H., and Jaykus, L. A. 2003. Nucleic acid sequence based amplification for the rapid and sensitive detection of *Salmonella enterica* from foods. *Journal of Applied Microbiology* 95(6):1343–50.

Ellingson, J. L., Anderson, J. L., Carlson, S. A., and Sharma, V. K. 2004. Twelve hour real-time PCR technique for the sensitive and specific detection of *Salmonella* in raw and ready-to-eat meat products. *Molecular and Cellular Probes* 18(1):51–57.

Ellingson, J. L., Koziczkowski, J. J., Anderson, J. L., Carlson, S. A., and Sharma, V. K. 2005. Rapid PCR detection of enterohemorrhagic *Escherichia coli* (EHEC) in bovine food products and feces. *Molecular and Cellular Probes* 19(3):213–17.

Entis, P., Fung, D. Y. C., Griffiths, M. W., McIntyre, L., Russell, S., Sharpe, A. N., and Tortorello, M. L. 2001. Rapid methods for detection, identification and enumeration. *In* Compendium of Methods for the Microbiological Examination of Foods, Downes, F. P., and Ito, K. (eds.), pp. 89–126. American Public Health Association Publications, Washington, D.C.

Fagan, P. K., Hornitzky, M. A., Bettelheim, K. A., and Djordjevic, S. P. 1999. Detection of shiga-like toxin (*stx1* and *stx2*), intimin (*eaeA*) and enterohemorrhagic *Escherichia coli* (EHEC) hemolysin (EHEC *hlyA*) genes in animal faeces by multiplex PCR. *Applied and Environmental Microbiology* 65(2):868–72.

Fang, Q., Brockmann, S., Botzenhart, K., and Wiedenmann, A. 2003. Improved detection of *Salmonella* spp. in foods by fluorescent *in situ* hybridization with 23S rRNA probes: a comparison with conventional culture methods. *Journal of Food Protection* 66:723–31.

Fitzmaurice, J., Duffy, G., Kilbride, B., Sheridan, J. J., Carroll, C., and Maher, M. 2004. Comparison of a membrane surface adhesion recovery method with an IMS method for use in a polymerase chain reaction method to detect *Escherichia coli* O157:H7 in minced beef. *Journal of Microbiological Methods* 59(2): 243–52.

Fortin, N. Y., Mulchandani, A., and Chen, W. 2001. Use of real-time polymerase chain reaction and molecular beacons for the detection of *Escherichia coli* O157:H7. *Analytical Biochemistry* 289:281–88.

Gilson, E., Clément, J. M., Brutlag, D., and Hofnung, M. 1984. A family of dispersed repetitive extragenic palindromic DNA sequences in *E. coli*. *European Molecular Biology Organisation Journal* 3(6):1417–21.

Giovannoni, S. J., Delong, E. F., Olsen, G. J., and Pace, N. R. 1988. Phylogenetic group specific oligodeoxynucleotide probes for identification of single microbial cells. *Journal of Bacteriology* 170:720–26.

Gupta, P. K., Roy, R. J., and Prasad, M. 1999. DNA chips, microarrays and genomics. *Current Science* 77:875–84.

Heim, A., Grumbach, I. M., Zeuke, S., and Top, B. 1998. Highly sensitive detection of gene expression of an intronless gene: amplification of mRNA, but not genomic DNA by nucleic acid sequence based amplification (NASBA). *Nucleic Acids Research* 26(9):2250–51.

Hill, W. E., Datta, A. R., Feng., P., Lampel, K. A., and Payne, W. L. 2001. Bacteriological analytical manual online. http://www.cfsan.fda.gov/~ebam/bam-24.html.

Hing, B. X., Jiang, L. F., Hu, Y. S., Fang, D. Y., and Guo, H. Y. 2004. Application of oligonucleotide array technology for the rapid detection of pathogenic bacteria of foodborne infections. *Journal of Microbiological Methods* 58(3):403–11

Hoorfar, J., Cook, N., Malorny, B., Wagner, M., De Medici, D., Abdulmawjood, A., and Fach, P. 2004. Diagnostic PCR: making internal amplification control mandatory. *Journal of Applied Microbiology* 96(2): 221–22.

Huggett, J., Dheda, K., Bustin, S., and Zumla, A. 2005. Real-time RT-PCR normalisation: strategies and considerations. *Genes and Immunity* 6(4):279–84.

Hulton, C. S., Higgins, C. F., and Sharp, P. M. 1991. ERIC sequences: a novel family of repetitive elements in the genomes of *Escherichia coli*, *Salmonella* Typhimurium and other Enterobacteriaceae. *Molecular Microbiology* 5:825–34.

Hummel, S. 2003. Ancient DNA Typing: Methods, Strategies and Applications, pp. 111–30. Springer-Verlag, Berlin.

Jaradat, Z. W., Schutze, G. E., and Bhunia, A. K. 2002. Genetic homogeneity among *Listeria monocytogenes* strains from infected patients and meat products from two geographic locations determined by phenotyping, ribotyping and PCR analysis of virulence genes. *International Journal of Food Microbiology* 76:1–10.

Jean, J., D'Souza, D. H., and Jaykus, L. A. 2004. Multiplex nucleic acid sequence-based amplification for simultaneous detection of several enteric viruses in model ready-to-eat foods. *Applied and Environmental Microbiology* 70(11):6603–10

Jin, U. H., Cho, S. H., Kim, M. G., Ha, S. D., Kim, K. S., Lee, K. H., Kim, K. Y., Chung, D. H., Lee, Y. C., and Kim, C. H. 2004. PCR method based on the ogdH gene for the detection of *Salmonella* spp. from chicken meat samples. *Journal of Microbiology* 42(3):216–22.

John, H., Birnstiel, M., and Jones, K. 1969. RNA:DNA hybrids at the cytogenetic level. *Nature* 223:582–87.

Johnson, E. A., Tepp, W. H., Bradshaw, M., Gilbert, R. J., Cook, P. E., McIntosh, E. D. 2005. Characterization of *Clostridium botulinum* strains associated with an infant botulism case in the United Kingdom. *Journal Clinical Microbiology* 43(6):2602–7.

Johnson, J. R., and O'Bryan, T. T. 2000. Improved repetitive-element PCR fingerprinting for resolving pathogenic and non pathogenic phylogenetic groups within *Escherichia coli*. *Clinical and Diagnostic Laboratory Immunology* 7(2):265–73.

Josefsen, M. H., Cook, N., D'Agostino, M., Hansen, F., Wagner, M., Demnerova, K., Heuvelink, A. E., Tassios, P. T., Lindmark, H., Kmet, V., Barbanera, M., Fach, P., Loncarevic, S., and Hoorfar, J. 2004. Validation of a PCR-based method for detection of food-borne thermotolerant campylobacters in a multicenter collaborative trial. *Applied and Environmental Microbiology* 70(7):4379–83.

Jothikumar, N., and Griffiths, M. W. 2002. Rapid detection of *Escherichia coli* O157:H7, with multiplex real-time PCR assays. *Applied and Environmental Microbiology* 68(6):3169–71.

Jothikumar, N., Wang, X., and Griffiths, M. W. 2003. Real-time multiplex SYBR green I-based PCR assay for simultaneous detection of *Salmonella* serovars and *Listeria monocytogenes*. *Journal of Food Protection* 66(11):2141–45.

Jung, Y. S., Frank, J. F., Brackett, R. E., and Chen, J. 2003. Polymerase chain reaction detection of *Listeria monocytogenes* on frankfurters using oligonucleotide primers targeting the genes encoding internalin AB. *Journal of Food Protection* 66(2):237–41.

Keramas, G., Bang, D. D., Lund, M., Madsen, M., Rasmussen, S. E., Bunkenborg, H., Telleman, P., and Christensen, C. B. 2003. Development of a sensitive DNA microarray suitable for rapid detection of *Campylobacter* spp. *Molecular and Cellular Probes* 17:187–96.

Kievits, T., Van Gemen, B., Van Strijp, D., Schukkink, R., Dircks, M., Adriaanse, H., Malek, L., Sooknanan, R., and Lens, P. 1991. NASBA isothermal enzymatic in vitro nucleic acid amplification optimised for the diagnosis of HIV-1 infection. *Journal of Virological Methods* 35:273–86.

Kim, C. H., Khan, M., Morin, D. E., Hurley W. L., Tripathy, D. N., Kehrli, M., Oluoch, A. O., and Kakoma, I. 2001. Optimization of the PCR for detection of *Staphylococcus aureus nuc* gene in bovine milk. *Journal of Diary Science* 84:74–83

Kimura, B., Kawasaki, S., Fugii, T., Kusunoki, J., Itoh, T., and Flood, S. J. A. 1999. Evaluation of TaqMan PCR assay for detecting *Salmonella* in raw meat and shrimp. *Journal of Food Protection* 62:329–35.

Klein, P. G., and Juneja, V. K. 1997. Sensitive detection of viable *Listeria monocytogenes* by reverse transcription-PCR. *Applied and Environmental Microbiology* 63:4441–48.

Kobayashi, S., Natori, K., Takeda, N., and Sakae, K. 2004. Immunomagnetic capture RT-PCR for detection of norovirus from foods implicated in a foodborne outbreak. *Microbiology and Immunology* 48:201–4.

Koeuth, T., Versalovic, J., and Lupski, J. R. 1995. Differential subsequence conservation of interspersed repetitive *Streptococcus pneumoniae* BOX elements in diverse bacteria. *Genome Research* 5(4):408–18.

Koo, K., and Jaykus, L.-A. 2003. Detection of *Listeria monocytogenes* from a model food by fluorescence resonance energy transfer-based PCR with an asymmetric fluorogenic probe set. *Applied and Environmental Microbiology* 69:1082–88.

Koopmans, M., Vennema, H., Heersma, H., Van Strien, E., Van Duynhoven, Y., Brown, D., Reacher, M., Lopman, B., and European Consortium on Foodborne Viruses. 2003. Early identification of common-source foodborne virus outbreaks in Europe. *Emerging Infectious Diseases* 9:1136–42.

Kot, B., and Trafny, E. A. 2004. The application of PCR to the identification of selected virulence markers of *Yersinia* genus. *Poultry Journal of Veterinary Science* 7:27–31.

Kumar, H. S., Sunil, R., Venugopal, M. N., Karanasagar, I., and Karunasagar, I. 2003. Detection of *Salmonella* spp. in tropical seafood by polymerase chain reaction. *International Journal of Food Microbiology* 88:91–95.

Lekanne Deprez, R. H., Fijnvandraat, A. C., Ruijter, J. M., and Moorman, A. F. M. 2002. Sensitivity and accuracy of quantitative real-time polymerase chain reaction using SYBR green I depends on cDNA synthesis conditions. *Analytical Biochemistry* 307:63–69.

Liebana, E. 2002. Molecular tools for epidemiological investigations of *S. enterica* subspecies *enterica* infections. *Research in Veterinary Science* 72:169–75.

Longhi, C., Maffeo, A., Penta, M., Petrone, G., Seganti, L., and Conte, M. P. 2003. Detection of *Listeria monocytogenes* in Italian-style soft cheeses. *Journal of Applied Microbiology* 94(5):879–85.

Lunge, V. R., Miller, B. J., Livak, K. J., and Batt, C. A. 2002. Factors affecting the performance of 5′ nuclease PCR assays for *Listeria monocytogenes* detection. *Journal of Microbiological Methods* 51:361–68.

Lyon, E., Millson, A., Phan, T., and Wittwer, C. T. 1998. Detection and identification of base alterations within the region of factor V Leiden by fluorescent melting curves. *Molecular Diagnostics* 3:203–10.

Maiden, M. C. J., Bygraves, J. A., Feil, E. R., Morelli, G., Russell, J. E., Urwin, R., Zhang, Q., Zhou, J., Zurth, D., Caugant, A., Feavers, I. M., Achtman, M., and Spratt, B. G. 1998. Multilocus sequence typing: a portable approach to the identification of clones within populations of pathogenic microorganisms. *Proceedings of National Academy of Science, USA* 95:3140–45

Malorny, B., Cook, N., D'Agostino, M., De Medici, D., Croci, L., Abdulmawjood, A., Fach, P., Karpiskova, R., Aymerich, T., Kwaitek, K., Hoorfar, J., and Malorny, B. 2004. Multicenter validation of PCR-based method for detection of *Salmonella* in chicken and pig samples. *Journal of Association of Analytical Communities International* 87(4):861–66.

Martin, B., Humbert, O., Camara, M., Guenzi, E., Walker, J., Mitchell, T., Andrew, P., Prudhomme, M., Alloing, G., Hakenbeck, R., Morrison, D. A., Boulnois, G. J., and Claverys, J.-P. 1992. A highly conserved repeated DNA element located in the chromosome of *Streptococcus pneumoniae*. *Nucleic Acids Research* 20(13):3479–83.

Masiga, D. K., Tait, A., and Turner, C. M. R. 2000. Amplified restriction fragment length polymorphism in parasite genetics. *Parasitology Today* 16:350–53.

Maslow, J., and Mulligan, M. E. 1996. Epidemiologic typing systems. *Infection Control and Hospital Epidemiology* 17(9):595–604.

Mateo, E., Carcamo, J., Urquijo, M., Perales, I., and Fernandez-Astorga, A. 2005. Evaluation of a PCR assay for the detection and identification of *Campylobacter jejuni* and *Campylobacter coli* in retail poultry products. *Research Microbiology* 156(4):568–74.

Maule, J. 1998. Pulsed field gel electrophoresis. *Molecular Biotechnology* 9:107–26.

McIngvale, S. C., Elhanafi, D., and Drake, M. A. 2002. Optimisation of reverse transcriptase PCR to detect viable shiga-toxin-producing *Escherichia coli*. *Applied and Environmental Microbiology* 68:799–806.

McKillip, J., and Drake, M. 2004. Real-time nucleic acid-based detection methods for pathogenic bacteria in food. *Journal of Food Protection* 67(4):823–32.

Méndez-Álvarez, S., Pavòn, V., Esteve, I., Guerrero, R., and Gaju, N. 1995. Analysis of bacterial genomes by pulsed field gel electrophoresis. *Microbiologìa SEM* 11:323–36.

Meyer, D. and Birchmeier, C. 1995. Multiple essential functions of neuregulin in development. *Nature* 378:386–90.

Miller, W. G., On, S. L., Wang, G., Fontanoz, S., Lastovica, A. J., and Mandrell, R. E. 2005 Extended multilocus sequence typing system for *Campylobacter coli*, *C. lari*, *C. upsaliensis*, and *C. helveticus*. *Journal of Clinical Microbiology* 43(5):2315–29.

Mullis, K. B. 1990. The unusual origin of the polymerase chain reaction. *Scientific American* 262(4):56–61.

Nguyen, L. T., Gillespie, B. E., Nam, H. M., Murinda, S. E., and Oliver, S. P. 2004. Detection of *Escherichia coli* O157:H7 and *Listeria monocytogenes* in beef products by real-time polymerase chain reaction. *Foodborne Pathogenic Diseases* 1(4):231–40

O'Hanlon, K. A., Catarame, T. M. G., Duffy, G., Sheridan, J. J., Blair, I. S., and McDowell, D. A. 2004. Rapid detection and quantification of *E. coli* O157/O26/O111 in minced beef by real-time PCR. *Journal of Applied Microbiology* 96: 1013–23.

Okwumabua, O., O'Connor, M., Shull, E., Strelow, K., Hamacher, M., Kurzynski, T., Warshauer, D. 2005. Characterization of *Listeria monocytogenes* isolates from food animal clinical cases: PFGE pattern similarity to strains from human listeriosis cases. *FEMS Microbiology Letters* 249(2):275–81.

Olsen, J. E., Aabo, S., Hill, W., Notermans, S., Wernars, K., Granum, P. E., Popovic, T., Rasmussen, H. N., and Olsvik, Ø. 1995. Probes and polymerase chain reaction for detection of foodborne bacterial pathogens. *International Journal of Food Microbiology* 28:1–78.

Oyofo, B. A., Abd el Salam, S. M., Churilla, A. M., and Wasfy, M. O. 1997. Rapid and sensitive detection of *Campylobacter* spp. from chicken using the polymerase chain reaction *Zentralbl Bakteriology* 285(4):480–85.

Pardue, M. L., and Gall, J. G. 1969. Molecular hybridisation of radioactive DNA to the DNA of cytological preparations. *Proceedings National Academy of Science, USA* 64:600–604

Paton, A. W., and Paton, J. C. 1998. Detection and characterisation of shiga toxigenic *Escherichia coli* by using multiplex PCR assays for *stx1*, *stx2*, *eaeA*, enterohemorrhagic *E. coli hlyA*, *rfb*O111 and *rfb*O157. *Journal of Clinical Microbiology* 36:598–602.

Persing, D. H. 1991. Polymerase chain reaction: trenches to benches. *Journal of Clinical Microbiology* 29(7):1281–85.

Powell, H. A., Gooding, C. M., Garret, S. D., Lund, B. M., and McKee, R. A. 1994. Proteinase inhibition of the detection of *Listeria monocytogenes* in milk using polymerase chain reaction. *Journal of Clinical Microbiology* 18:59–61.

Rademarker, J. L. W., and De Bruijn, F. J. 1998. Characterisation and classification of microbes by rep-PCR genomic fingerprinting and computer assisted pattern analysis. *Applied and Environmental Microbiology* 64:2096–2104.

Rådström, P., Knutsson, R., Wolffs, P., Dahlenborg, M., and Löfström, C. 2003. Pre-PCR processing of samples. *In* PCR Detection of Microbial Pathogens, Sachse, K., and Frey, J. (eds.), pp. 31–50. Humana Press, Totowa, New Jersey.

Rahn, K., De Grandis, S. A., Clarke, R. C., McEwen, S. A., Galan, J. E., Ginocchio, C., Curtiss, R. III, and Gyles, C. L. 1992. Amplification of an *invA* gene sequence of *Salmonella* Typhimurium by polymerase chain reaction as a specific method of detection of *Salmonella*. *Molecular and Cellular Probes* 6(4):71–279.

Rodriguez-Lazaro, D., Lloyd, J., Herrewegh, A., Ikonomopoulos, J., D'Agostino, M., Pla, M., Cook, N. 2004 A molecular beacon-based real-time NASBA assay for detection of *Mycobacterium avium* subsp. *paratuberculosis* in water and milk. *FEMS Microbiology Letters* 237(1):119–26.

Ross, I. L., and Heuzenroeder, M. W. 2005. Use of AFLP and PFGE to discriminate between *Salmonella enterica* serovar Typhimurium DT126 isolates from separate food-related outbreaks in Australia. *Epidemiology and Infection* 133(4):635–44.

Saiki, R. H., Scharf, S., Faloona, F., Mullis, K. B., Horn, G. T., Ehrlich, H. A., and Arnheim, N. 1985. Enzymatic amplification of beta-globulin genomic sequences and restriction site analysis for diagnosis of sickle cell anemia. *Science* 230:1350–54.

Sair, A. I., D'Souza, D. H., Moe, C. L., and Jaykus, L. A. 2002. Improved detection of human enteric viruses in foods by RT-PCR. *Journal of Virological Methods* 100:57–69.

Sambrook, J., and Russell, D. W. 2001. Amplification of cDNA generated by reverse transcription of mRNA. *In* Molecular Cloning: A Laboratory Manual vo, l. 2, pp. 8.46–8.53. Cold Spring Harbor Laboratory Press, New York.

Seurinck, S., Verstraete, W., and Siciliano, S. D. 2003. Use of 16S-23S rRNA intergenic spacer region PCR and repetitive extragenic palindromic PCR analyses of *Escherichia coli* isolates to identify nonpoint fecal sources. *Applied and Environmental Microbiology* 69(8):4942–50.

Sharbel, T. F. 1999. Amplified fragment length polymorphism: a non-random PCR-based technique for multilocus sampling. *In* DNA Profiling and DNA Fingerprinting, Epplen, J. T., and Lubjuhn, T. (eds.), pp. 177–94. Birkhäuser Verlag, Basel.

Sharma, V. K. 2002. Detection and quantification of enterohemorrhagic *Escherichia coli* O157, O111 and O26 in beef and bovine faeces by real-time polymerase chain reaction. *Journal of Food Protection* 65:1371–80.

Sheridan, G. E. C., Masters, C. I., Shallcross, J. A., and Mackey, B. M. 1998. Detection of mRNA by reverse transcription-PCR as an indicator of viability in *Escherichia coli*. *Applied and Environmental Microbiology* 64:1313–18.

Simon, R. M., Korn, E. L., McShane, L. M., Wright, G. W., and Zhao, Y. 2003. DNA microarray technology. *In* Design and Analysis of DNA Microarray Investigations, Dietz, K., Gail, M., Krickeberg, K., Samet, J., and Tsiatis, A. (eds.), pp. 5–10. Springer-Verlag, New York.

Simpkins, S. A., Chan, A. B., Hays, J., Popping, B., and Cook, N. 2000. An RNA transcription-based amplification technique (NASBA) for the detection of viable *Salmonella enterica*. *Letters in Applied Microbiology* 30:75–79.

Southern, E. M. 1975. Detection of specific sequences among DNA fragments separated by gel electrophoresis. *Journal of Molecular Biology* 98:5502–17.

Stryer, L., and Haugland, R. P. 1967. Energy transfer: a spectroscopic ruler. *Proceedings of the National Academy of Sciences of the United States of America* 58:719–26.

Surzycki, S. 2000a. DNA transfer and hybridisation. *In* Basic Techniques in Molecular Biology, Surzycki, S. (ed.), pp. 233–62. Springer-Verlag, Berlin.

———. 2000b. Isolation and purification of RNA. *In* Basic Techniques in Molecular Biology, Surzycki, S. (ed.), pp. 119–44. Springer-Verlag, Berlin.

———. 2000c. Northern transfer and hybridisation. *In* Basic Techniques in Molecular Biology, Surzycki, S. (ed.), pp. 263–97. Springer-Verlag, Berlin.

Szabo, E. A., and Mackey, B. M. 1999. Detection of *Salmonella* Enteriditis by transcription-polymerase chain reaction (PCR). *International Journal of Food Microbiology* 51(2–3):113–22.

Tasara, T., Hoelzle, L. E., and Stephan, R. 2005. Development and evaluation of a *Mycobacterium avium* subspecies *paratuberculosis* (MAP) specific multiplex PCR assay. *International Journal of Food Microbiology* 104(3):279–87.

Taylor, M. J., Hughes, M. S., Skuce, R. A., and Neill, S. D. 2001. Detection of *Mycobacterium bovis* in bovine clinical specimens using real-time fluorescence resonance energy transfer probe rapid-cycle PCR. *Journal of Clinical Microbiology* 39:1272–78.

Thomas, E. J., King, R. K., Burchak, J., and Gannon, V. P. 1991. Sensitive and specific detection of *Listeria monocytogenes* in milk and ground beef with the polymerase chain reaction. *Applied and Environmental Microbiology* 57(9):2576–80.

Thomson-Carter, F. 2001. Typing methods for VTEC. *In* Verocytotoxigenic *E. coli,* Duffy, G., Garvey, P., and McDowell, D. A. (eds.), pp. 91–111. Food Science and Nutrition Press, Connecticut.

Trkov, M., and Avguštin, G. 2003. An improved 16S rRNA based PCR method for the specific detection of *Salmonella enterica. International Journal of Food Microbiology* 80(1):67–75.

Tyler, K. D., Wang, G., Tyler, S. D., and Johnson, W. M. 1997. Factors affecting reliability and reproducibility of amplification-based DNA fingerprinting of representative bacterial pathogens. *Journal of Clinical Microbiology* 35:339–46.

Urwin, R., and Maiden, M. C. J. 2003. Multi-locus sequence typing: a tool for global epidemiology. *Trends in Microbiology* 11:479–87

Uyttendaele, M., Schukkink, R., Van Gemen, B., and Debevere, J. 1995. Development of NASBA®, a nucleic acid amplification system, for identification of *Listeria monocytogenes* and comparison to ELISA and a modified FDA method. *International Journal of Food Microbiology* 27:77–89.

Uyttendaele, M., van Boxstael, S., and Debevere, J. 1999. PCR assay for detection of the *E. coli* O157:H7 eae-gene and effect of the sample preparation method on PCR detection of heat-killed *E. coli* O157:H7 in ground beef. *International Journal of Food Microbiology* 52(1–2):85–95.

Van Gemen, B., Kievits, T., Schukkink, R., Van Strijp, D., Malek, L. T., Sooknanan, R., Huisman, H. G., and Lens, P. 1993. Quantification of HIV-1 RNA in plasma using NASBA during HIV-1 primary infection. *Journal of Virological Methods* 43:177–87.

Venkatasubbarao, S. 2004. Microarrays-status and prospects. *Trends in Biotechnology* 22:630–37.

Versalovic, J., Koeuth, T., and Lupski, J. R. 1991. Distribution of repetitive DNA sequences in eubacteria and application to fingerprinting of bacterial genomes. *Nucleic Acids Research* 24:6823–83.

Versalovic, J., Schneider, M., De Bruijn, F. J., and Lupski, J. R. 1994. Genomic fingerprinting of bacteria using repetitive sequence-based polymerase chain reaction. *Methods in Molecular and Cellular Biology* 5:25–40.

Vos, P., Hogers, R., Bleeker, M., Reijans, M., van de Lee, T., Hornes, M., Frijters, A., Pot, J., Peleman, J., Kuiper, M., and Zabeau, M. 1995. AFLP: a new technique for DNA fingerprinting. *Nucleic Acids Research* 23:4407–14.

Wang, X., Jothikumar, N., and Griffiths, M. W. 2004. Enrichment and DNA extraction protocols for the simultaneous detection of *Salmonella* and *Listeria monocytogenes* in raw sausage meat with multiplex real-time PCR. *Journal of Food Protection* 67(1):189–92.

Wannet, W. J., Reessink, M., Brunings, H. A., and Maas, H. M. 2001. Detection of pathogenic *Yersinia enterocolitica* by a rapid and sensitive duplex PCR assay. *Journal of Clinical Microbiology* 12:4483–86.

Williams, J. G. K., Kubelik, A. R., Livak, K. J., Rafalski, J. A., and Scott, V. T. 1990. DNA polymorphisms amplified by arbitrary primers are useful as genetic markers. *Nucleic Acids Research* 18:6531–35.

Wittner, C. T., Herrmann, M. G., Moss, A. A., and Rasmussen, R. P. 1997. Continuous fluorescence monitoring of rapid cycle DNA amplification. *Biotechniques* 22:130–38.

9 DNA-Based Detection of GM Ingredients

Alexandra Ehlert, Francisco Moreano,
Ulrich Busch, and Karl-Heinz Engel

Introduction

Deoxyribonucleic acid (DNA) is a ubiquitously occurring cell constituent providing highly specific information on organisms at every taxonomic level. Therefore, it is, per se, an ideal target for diagnostic approaches. The major breakthrough for DNA analysis was the development of the polymerase chain reaction (PCR) [1, 2]. This technique enabling the exponential amplification of selected DNA sequences with high degrees of sensitivity and specificity paved the way for the incorporation of DNA analysis as a standard tool in the portfolio of modern analytical methods. Meanwhile the technique has been widely established in many fields, such as clinical diagnostics, forensic sciences, or histopathology. This chapter reviews applications of DNA-based methods for the analysis of foods. In the last decade the need for methods to detect and to quantify DNA from genetically modified organisms (GMOs) has been a major driver for the development and optimization of PCR-based techniques. Therefore, this field of application will be used here to outline principles, challenges, and current developments of DNA analysis in foods. In addition, DNA-based approaches in the areas of food authentication, detection of food-borne pathogenic microorganisms, and screening for food allergens will be reviewed.

Analysis of GMOs

Many countries have established regulatory frameworks regarding the use of recombinant DNA techniques in the course of the production of foods and food ingredients [3]. In addition to safety aspects, labeling of foods derived from GMOs is a central issue of the public debate.

In the European Union, legal requirements for labeling of GMO-derived foods were first provided by the so-called Novel Foods Regulation [4] and its amendments [5–7]. The need for labeling was triggered by the detection of either protein or DNA from the GMOs. This initiated the first wave of analytical approaches, mainly focusing on the detection of DNA from GMOs via qualitative PCR. Recently, new regulations have been enforced containing harmonized provisions for the risk assessment and authorization of GMOs, as well as traceability, labeling, and postmarketing surveillance of the use of GMOs in the food and feed chain [8, 9]. Regulation (EC) 1830/2003 defined the establishment of a traceability system allowing the documentation and the monitoring of the flow of GMOs and GMO-derived products at all stages along the food and feed chain. According to regulation (EC) 65/2004, information on traceability includes a unique code identifier for the respective transformation event, in the case of products consisting of or containing

GMOs [10]. With this novel strategy labeling requirements have been extended to highly processed products and are no longer dependent on a positive testing of recombinant DNA or protein. However, it is acknowledged that in agriculture adventitious contaminations with traces of GMO-derived material cannot be excluded. Therefore, the current regulations determine certain thresholds for adventitious or technically unavoidable presence of GMOs in food and feed. Labeling is not required if the proportion of material containing an authorized GMO is not higher than 0.9 percent (considered individually for each ingredient). A limit of 0.5 percent has been set for products containing material derived from nonauthorized GMOs, which have benefited from a favorable risk evaluation by the European Food Safety Authority. No levels are tolerated for material derived from nonauthorized GMOs.

To be able to control compliance with these legal provisions, appropriate methods for detection, as well as quantification, of GMOs are required. The European Commission's Reference Laboratory provides official methods for the event-specific detection and quantification of material from authorized GMOs. Despite these valuable methods major analytical challenges arise from the increasing number of authorized and unauthorized GMOs, the lack of reference materials, and the need to determine GMO contents in composite and processed foods.

Qualitative PCR

PCR permits the detection of minute amounts of specific DNA sequences. The technique enables exponential amplification of a specific DNA fragment in vitro using short oligonucleotides (primers) flanking the sequence of interest and a thermostable DNA polymerase. In general, one reaction cycle consists of three steps: allowing melting of double-stranded DNA, annealing of the primer, and enzymatic elongation. Running of multiple cycles of this exponential amplification leads to a detectable quantity of the desired DNA fragment. Detailed descriptions and applications of this widely used methodology have been published [11, 12]. Every gene of interest (e.g., a transgene) can be amplified in this way. Following the PCR, the amplicons formed are generally separated according to their lengths via gel electrophoresis. Verification of their identities is achieved by cleavage with restriction enzymes and subsequent separation of the digestion products via gel electrophoresis, by Southern blotting, or by direct sequencing.

Requirements of DNA Preparation

Analogous to other analytical approaches, appropriate sampling is the first requirement to be met when subjecting foods to DNA analysis. Inhomogeneous distribution of GMOs in bulk materials can be a major contributor to overall analytical variance. Errors may occur at the various stages (sampling, subsampling, preparation of aliquots for analytical steps). Sampling strategies are especially important if low GMO concentrations are to be analyzed [13, 14]. One of the objectives of the European Network of GMO Laboratories (ENGL) is to identify and to develop sampling strategies to support EU legislation [15].

The next essential step is extracting from the food matrix sufficient amounts of DNA exhibiting the quality required for successful PCR analysis. A broad spectrum of DNA extraction methods is available [16–19]. In general, plant tissue is ground; a detergent is applied to disrupt the cell membranes and to inactivate endogenous nucleases;

different agents are used to remove proteins (proteinase K), polysaccharides (CTAB), and lipids (chloroform); and finally the DNA is isolated by alcohol/salt precipitation [20, 17]. Various extraction kits are available, in which the purification of DNA is achieved by chaotropic salts on silica columns or by binding the DNA on magnetic particles. At present the CTAB method and DNA-binding silica materials are most commonly used for isolation of DNA from GMO samples.

Various components of the matrix analyzed or chemicals applied during DNA extraction may influence the purity of DNA and inhibit the PCR reaction [21]. To overcome false-negative results, the use of homologous or heterologous internal positive controls in the PCR amplification presents a powerful tool [22].

Specificity of PCR—Choice of Target Sequences

Depending on the sequences selected for PCR amplification, the detection of GMOs can be categorized into four levels of specificity: screening methods, gene-specific methods, construct-specific methods, and event-specific methods [13, 23]. Target sequences resulting in these different specificities of the PCR assays are schematically shown in figure 9.1. Screening methods target regulatory elements commonly used in transformations, such as the cauliflower mosaic virus (CaMV), 35S promoter (P-35S), or the *Agrobacterium tumefaciens* nopaline synthase terminator (T-NOS). The detection of such sequences indicates the presence of GMO-derived DNA, but positive signals may also be due to other factors, such as naturally occurring CaMV [24]. The specificity can be increased by amplifying the sequence coding for the gene of interest. However, such gene-specific methods do not allow a distinction between different GMOs carrying the same transgene. Construct-specific methods target junctions between regulatory sequences and the gene of interest. However, the complete gene construct may have been transformed into different crops. In such cases, targeting the junction at the integration site between the plant genome and the inserted DNA provides the highest level of specificity. It also allows a differentiation between authorized and nonauthorized GMOs containing similar transgenic constructs. In the field of GMO analysis event-specific methods have been described for the detection of

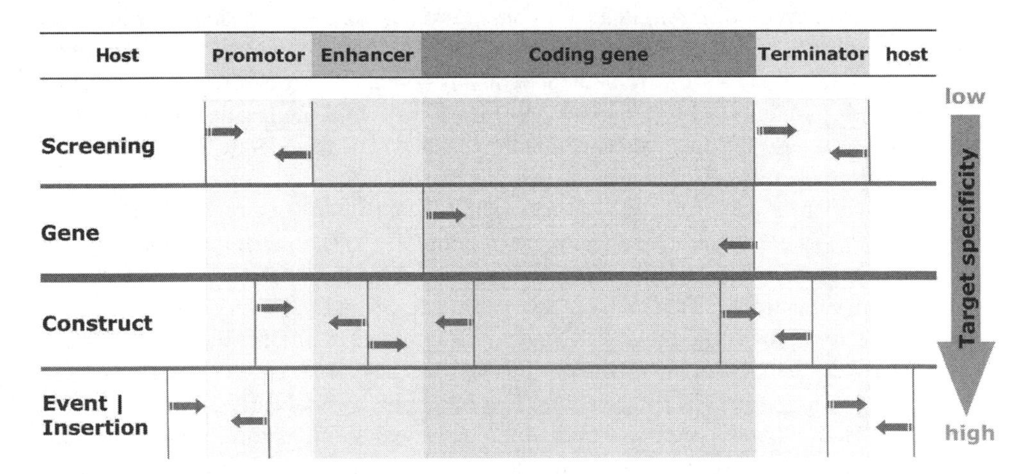

Fig. 9.1. Target specificity of PCR assays (adapted from [23]).

several maize lines (Bt11, Bt176, T25, MON863, MON810, NK603), canola event GT73, and Roundup Ready soya [25–40].

Influence of Food Composition and Processing

The detection of GMOs in food faces a number of challenges arising from the complexity of food compositions and the technological parameters of manufacturing processes. Food manufacturing may affect the quality and quantity of DNA in processed products. Fragmentation of DNA is initiated by shear forces, heat treatment, pH variations, enzymatic activities, and fermentations resulting in reduced average size of DNA [41]. The effects of using degraded DNA as a template in PCR-based detection systems have been investigated by following diverse manufacturing practices. The choice of the size of the target sequence influences the detectability of DNA [42–44]. Using insect-resistant Bt176 maize as an example, it could be shown that the probability of detecting the GMOs decreases rapidly in the course of heat treatment when targeting the complete 1914 bp sequence of the synthetic *cryIA*(*b*) gene. On the other hand, a shorter target sequence (211 bp), covering part of the CDPK promoter and the *cryIA*(*b*) gene, is detectable even after heating for 105 min. In addition, thermal treatment in combination with acidic conditions dramatically increases the DNA degradation and thus the probability of detection (fig. 9.2) [43].

Degradation of DNA in the course of food processing negatively affects the detection efficiency, especially when long sequences are the targets. This results in false-negative results when analyzing samples of processed foods and feeds. Further challenges arising from the quantitative determination of GMOs in composite and processed foods will be discussed later in the chapter.

Quantitative PCR

Competitive PCR

Standard endpoint detection of DNA sequences as performed in qualitative PCR analysis cannot be applied to quantitative determinations due to the discontinuity of the amplification efficiency between different PCR reactions. First approaches of quantitative analysis of DNA were based on the co-amplification of the designated target and an exogenous standard (competitor) [45, 46]. In the course of this so-called competitive PCR, samples containing constant amounts of template DNA are spiked with increasing amounts of the competitor. Both possess identical primer binding sites and nearly identical lengths so that equivalent amplification efficiencies in the course of the PCR reaction are to be expected. The quantification is performed by comparing the signal intensities of both amplicons, measured after gel electrophoresis. The point of equivalence (i.e., where the molar ratios of target and internal standard are equal) is determined at the intersection of the linear regression curve with the abscissa [47].

The applicability of competitive PCR for the quantification of DNA has been demonstrated for several examples [26, 48–53]. However, the labor-intensive approach requires extensive handling with PCR products and involves a high risk of cross-contamination. The time-consuming and extremely material-intensive technique requires several reaction mixes for the measurement of one point of equivalence and visualization of PCR products by gel electrophoresis, in combination with complex gel documentation/evaluation.

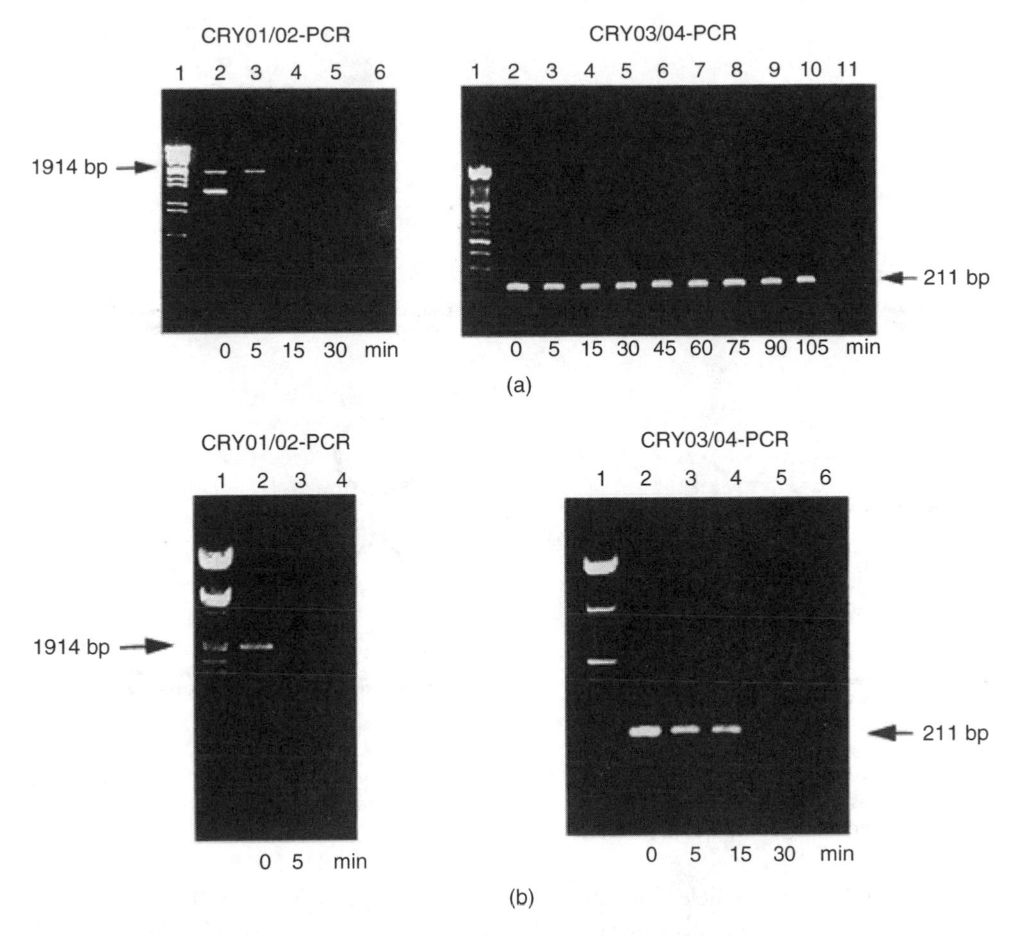

Fig. 9.2. Effect of heat treatment at neutral pH **(a)** and at pH 2 **(b)** on the detection of Bt-maize (for details see [43]).

Real-Time PCR

Real-time PCR is the state-of-the-art technique to detect and to quantify DNA. This technique permits the direct on-line measurement of PCR product amounts at every stage of the reaction by using fluorescence techniques. The fluorescence signals are proportional to the amounts of PCR products generated and can be observed by different approaches. As shown in figure 9.3, double-stranded (ds) DNA-intercalating dyes (SYBR Green), hydrolysis probes (TaqMan Probes), or reversible hybridization probes (HybProbes, Beacons) are used [54–57].

The main disadvantage of double-strand-specific intercalating dyes (fig. 9.3A) is the unavoidable detection of nonspecific PCR products such as primer dimers, besides the specific amplicons. For example, SYBR Green binds independently from the sequence to the minor groove of dsDNA. The possibility of separating the favored amplicon from unspecific background by melting curve analysis and the easy application at relatively low costs are benefits of this detection format. Another technique is TaqMan Probes,

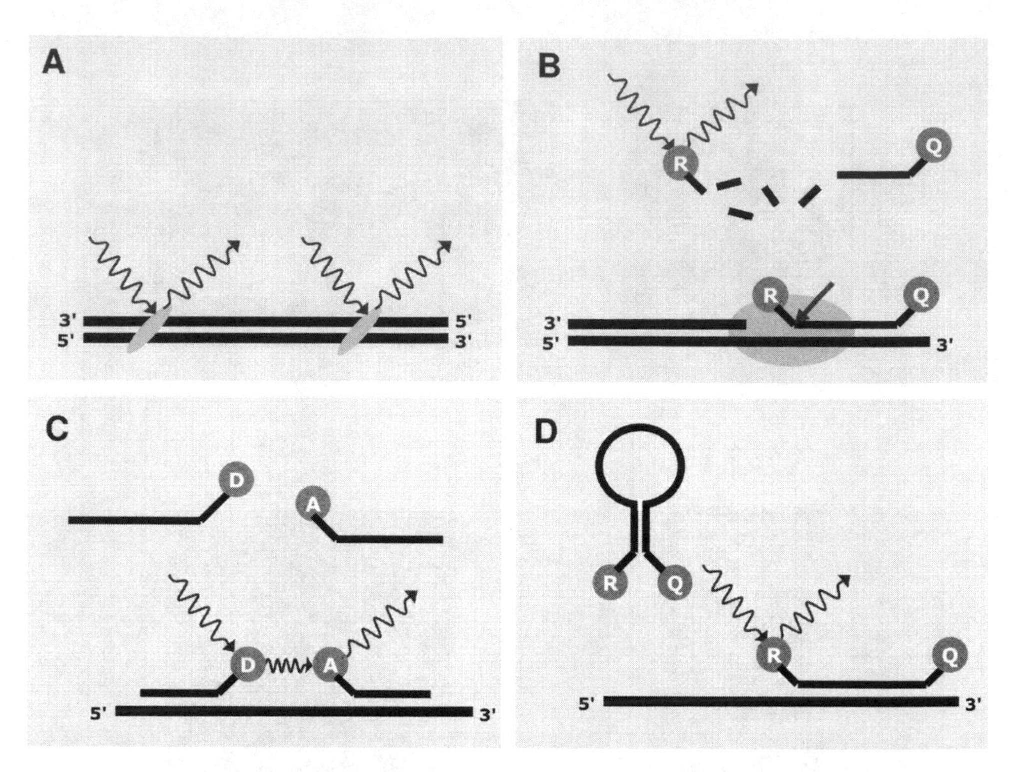

Fig. 9.3. Commonly used detection formats of real-time PCR: **(A)** DNA-intercalating dye, **(B)** hydrolysis probes, **(C)** HybProbes, and **(D)** molecular beacon; A = acceptor, D = donor, Q = quencher, R = reporter.

specific synthetic oligonucleotides hybridizing to the target DNA (fig. 9.3B). A probe, complementary to the target sequence, is labeled with a reporter fluorophor at the 5′-end and with a quencher dye at the 3′-end; thus light emission is suppressed. During elongation the hybridized probe is hydrolyzed by the 5′-3′ exonuclease activity of the *Taq* DNA polymerase, and hence the released reporter emits fluorescence after excitation. The generated signal is proportional to the exponential amplification of templates. In the case of hybridization probes, the fluorescence resonance energy transfer (FRET) is directly measured. Two sequence-specific probes hybridizing closely adjacent to the target, within a distance of one to five bases, are used. As illustrated in figure 9.3C, donor and acceptor probes are 3′- and 5′-terminally labeled, respectively. In the case of successful hybridization of both HybProbes, the excitation of the donor is transferred to the acceptor, and the emitted fluorescence can be detected. The molecular beacon, another hybridization format, is labeled on both ends with reporter and quencher fluorophor, respectively (fig. 9.3D). The central part of the probe is complementary to the target, whereas the terminal part is self-hybridized forming a stem-loop structure. The probe binds to the template during the annealing phase, the dyes are no longer quenched, and a fluorescence signal is obtained. The advantages of internally hybridizing probes are the additional sequence specificity and the monitoring of PCR efficiency.

At the exponential phase of the PCR amplification, the template copy number of target sequences can be extrapolated on the basis of a standard curve. Figure 9.4 displays the

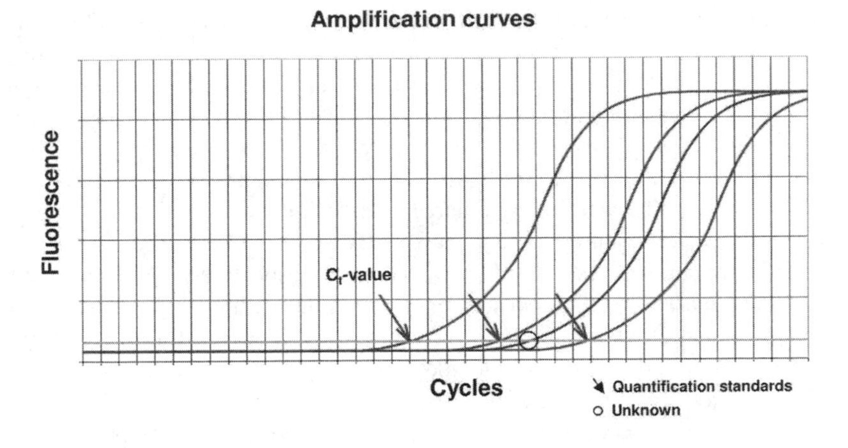

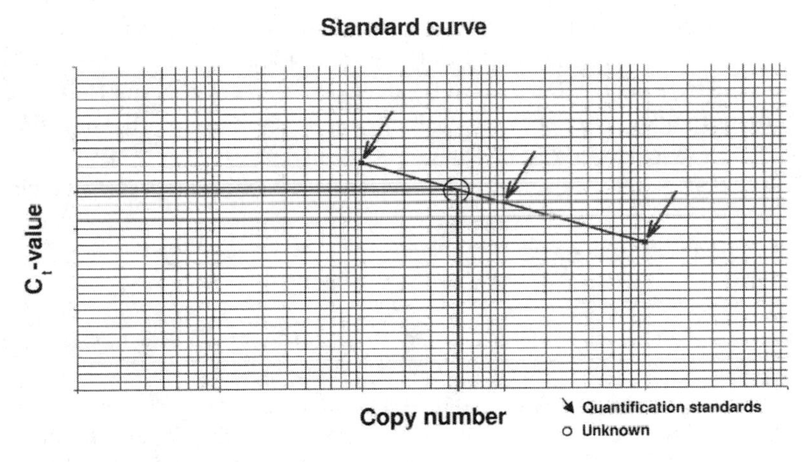

Fig. 9.4. Principle of DNA quantification via real-time PCR. The threshold intersects the amplification curves in the exponential phase of PCR. The respective C_T-values of the quantification standards are correlated with the starting copy numbers in the standard curve.

construction of standard curves by estimating the so-called threshold cycle (C_T) values or crossing points (C_P) from external quantification standards of known target concentration. The generated standard curve describes the logarithmic plotting of starting copy numbers and the determined C_T values. An exact quantification of unknown samples is only ensured if the amplification efficiency is equal to that of the standards used. PCR efficiency can be determined by examining the slope of the linear trend line, which should be ideally –3.32 for 100 percent PCR efficiency [57].

In the field of GMO analysis the quantification in processed products requires simultaneous assessment of the recombinant DNA and of a species- or taxon-specific reference gene. Thus a determination of ingredient-related GMO contents as legally required is possible. Cloned plasmid fragments [58], synthetic hybrid amplicons [59, 60], and certified reference materials (CRM standards) have been described as quantification standards [61].

The first application of real-time PCR to the quantitative analysis of GMOs in foods was described by Wurz et al. [50]. The method was developed to detect a recombinant region in the genome of Roundup Ready soya and a plant-specific sequence within the lectin (*le1*) gene. Special attention was paid to avoid significant differences in amplicon lengths, meeting basic requirements for its application in processed foods. The approach was afterward tested with certified reference materials containing 0.1–2 percent transgenic soya, yielding results that were in good agreement with the expected data.

Approaches for the quantitative detection of Bt176 maize and Roundup Ready soybeans, targeting the transgenes *cryIA(b)* and *cp4-epsps,* respectively, have also been introduced [62]. The maize-specific zein (*ze1*) and the soya-specific lectin (*le1*) genes were used as endogenous reference targets. For the first time, PCR conditions were optimized to allow the quantification of transgenic and isogenic targets in one reaction vessel, thus eliminating tube-to-tube variations.

To date, various methods for the quantification of GMO proportions in raw food materials have been presented and validated in international interlaboratory trials [37, 38, 50, 62–67]. Commercial kits are available for the quantification of transgenic soy (Roundup Ready), maize (Maximizer™ Bt176 and Bt11, Liberty Link™ T25, Yield Guard™ MON810, Roundup Ready NK603 and GA21, StarLink™, and Herkulex™), and canola (Liberty Link™). Interlaboratory testing of obtainable kits for the quantification of Roundup Ready soybeans and Bt176 maize has been performed [67]. Examples of real-time PCR methods based on event-specific detection of GMOs are listed in table 9.1.

Challenges and Developments

Copy Numbers of Genes—Zygosity and Ploidy

The relative quantification strategy applied in GMO analysis determines the ratio of transgene copies to the respective copies of a reference gene, assuming a 1 : 1 relationship between the two genes. In diploid homozygous lines (e.g., achieved by means of self-pollination) this prerequisite is normally met [13, 68]. However, the generation of genetically enhanced lines for commercial purposes involves crossbreeding with optimized conventional varieties. The resulting hybrids with altered levels of ploidy have lost the original correlation between transgene and reference gene. Additionally, unequal levels of ploidy can be found in separate tissues of one organism, as for example in the diploid embryo and the triploid endosperm of maize kernels [68, 69].

The relative quantification of GMO-derived material requires special attention be paid to the selection of suitable reference genes. Preferably, the reference target is a stable gene of known copy number in all varieties and unique to the species. A systematic evaluation of plant-species-specific reference genes has shown that these criteria are not fulfilled by every candidate gene [70].

Lack of Reference Material—Hybrid Molecules

The availability of reference materials plays an essential role in the course of the development and the validation of detection and quantitation systems of GMOs as well as for the implementation of surveillance testing. Commercially available reference standards do not cover the entire spectrum of authorized GMOs, and there is (per se) a complete lack of reference material for the analysis of nonauthorized GM crops.

New approaches using DNA fragments cloned in a plasmid as external calibration standards were first described for the determination of Roundup Ready soybeans [58] and

Table 9.1. Examples of event-specific real-time PCR systems applied in analysis of GMOs.

GMO	Amplicons (length)	Real-Time System Detection Chemistry	Literature
Roundup Ready Soybean	Lectin (102 bp), RRS (85 bp)	LightCycler, TaqMan	[37]
	Lectin (228 bp), RRS (77 bp)	ABI PRISM 7700, LUX primer (sunrise primer)	[39]
	Lectin (115 bp), RRS (199 bp)	LightCycler, FRET probes	[134]
	Lectin (80 bp), RRS (94 bp)	ABI PRISM 7700, TaqMan	[38]
	Lectin (159 bp), RRS (170 bp)	LightCycler, Scorpion primer	
Maize T25	Maize *adh*1, T25 (155 bp)	ABI PRISM 7700, TaqMan	[28]
Maize MON810	MON810 (106 bp)	ABI PRISM 7700, TaqMan	[32]
	Maize *zein* (69 bp), MON810 (115 bp)	ABI PRISM 7700, TaqMan	[31]
Maize MON810, NK603	Maize *ivr* (79 bp), MON810 (223 bp), NK603 (113 bp)	ABI PRISM 7700, TaqMan	[34]
Maize NK603	Maize *adh*1 (136 bp), NK603 (102 bp)	LightCycler, ABI 7900 HT, TaqMan	[35, 135]
Maize Bt11	Maize *ivr* (111 bp), Bt11 3′ junction (70 bp), Bt11 5′ junction (82 bp)	LightCycler, TaqMan	[25]
Maize Bt11, Bt176, GA21, canola GT73	Bt11 (93 bp), Bt176 (77 bp), GA21 (88 bp), GT73 (103 bp)	ABI PRISM 7000, TaqMan	[27]
Maize CBH351 (StarLink)	Junction A (138 bp), Junction B (100 bp)	ABI PRISM 7700, SYBR Green	[136]
Maize MON863	Maize *SSIIb* (88 bp), MON863 (90 bp)	Rotor-Gene 2000, TaqMan	[30]
Cotton MON1445, MON531	Cotton *SadI* (107 bp), MON1445 (99 bp), MON531 (121 bp)	Rotor-Gene 2000, TaqMan	[137]

are now being applied to the analysis of various GMOs [27, 32, 71, 72]. The synthesis of hybrid amplicon molecules by a novel two-step PCR amplification represents another strategy to obtain quantification standards. The reaction starts with separate amplification of the targeted recombinant and taxon-specific sequences with bipartite primers generating complementary overhangs. In the second PCR the mixed purified amplicons are able to self-prime due to the overhang sequences and generate the complete hybrid molecules containing both targets [59, 60].

The application of this approach to the quantitative screening of genetically modified rapeseed lines with a number of transformation events has been described [60]. Two duplex real-time PCR assays allow the simultaneous detection of the construct-specific junction between the 35S promoter and the *pat* gene in LibertyLink™ lines and between the *bar* gene and the g7 terminator in SeedLink™ lines, as well as the detection of a rapeseed-specific acetyl-CoA carboxylase gene. The moderate level of specificity avoids false-positive results and presents a valuable tool for the purpose of surveillance testing of GMOs in food and feed products.

Quantification of DNA in Composed and Processed Foods
Industrially produced foods usually contain various ingredients. Such ingredients may be derived from the same (GM) crop but may differ significantly in technofunctional properties. For instance, mixtures of corn-milling fractions with different particle sizes are industrially applied to influence the characteristics of bakery products. It was demonstrated that unequal efficiencies in the extraction of DNA from fractions differing in particle size distributions may contribute to distortions of GMO quantification [73, 74]. For corn-milling fractions a strong correlation between the degree of comminution and the DNA yields in the extracts was observed. Real-time PCR quantification of the GMO content in mixtures containing conventional and transgenic corn of different particle size distributions resulted in significant over- or underestimations of GMO contents [75].

As expected from the phenomena described for qualitative PCR, the length of the targeted DNA fragment is also crucial for the quantitative analysis of DNA in processed foods. Application of a method validated in an interlaboratory ring trial for the quantification of Bt176 maize delivered accurate data for unprocessed reference materials but resulted in a significant underestimation of the GMO contents in heat-treated samples [67, 76]. The differences in amplicon lengths of the targeted reference gene (79 bp) and the transgene (129 bp) permitted the assumption that distortions in the results obtained by relative quantification were caused by the increased probability of fragmentation of the longer sequence. This could be confirmed by following the heat-induced DNA degradation in mixtures of conventional and transgenic corn (1 percent) [75, 77]. Two established quantitative assays differing in the lengths of the recombinant and reference target sequences (A: $\Delta l_A = -25$ bp; B: $\Delta l_B = +16$ bp; values related to the amplicon length of the reference gene) were applied. Method A resulted in underestimated recoveries of the GMO contents in heat-treated products, reflecting the favored degradation of the longer target sequence used for the detection of the transgene. In contrast, method B resulted in increasing overestimation of the recoveries of the GMO contents in the course of the heat treatment.

Validation
Comparable to other analytical methods, PCR-based approaches for detection and quantification of DNA also have to be validated [78]. The limits of detection (LOD) and quantification (LOQ) are method specific but do also depend on the sample being analyzed. Three types of detection and quantification limits have been distinguished: (1) the absolute limits (i.e., the lowest number of copies required at the first PCR cycle to obtain a probability of at least 95 percent of detecting/quantifying correctly), (2) the relative limits (i.e., the lowest relative percentage of GM material that can be detected/quantified under optimal conditions), and (3) the practical limits (i.e., limits considering factors such as the actual contents of the DNA sample and the absolute limits of the method) [37]. Both the LOD/LOQ of the method and the practical LOD/LOQ of the test sample should be reported, together with the results [37, 23].

Validation of quantitative assays for GMOs in foods cannot be limited to unprocessed reference materials. Validation procedures must demonstrate that neither food composition nor processing will result in distortions of relative quantification results. Future standard protocols should target recombinant and taxon-specific sequences of nearly equal lengths.

Multiplex Approaches

To ensure compliance with labeling requirements, appropriate analytical approaches are demanded for the detection, identification, and quantification of the steadily increasing number of GMOs. Multiplex assays allowing the simultaneous detection of several GMOs in a single PCR reaction have been developed to meet this challenge [29, 34, 40, 79–85]. Examples of multiplex approaches for the detection of GMOs are given in table 9.2.

The simultaneous detection of screening elements in combination with construct-specific targets provides a profiling-like strategy for GMO analysis, including unauthorized ones [79, 81, 83, 84]. Other approaches partly use similar primers for the detection of different targets to overcome the difficulties of simultaneous amplification with multiple oligonucleotides in one single reaction [29, 80]. The disadvantages of these approaches using gel electrophoresis for separation are the long amplicon lengths and length differences of the targets, which pose a problem when applying the method to the analysis of processed foods [82]. At present the simultaneous amplification using nine primer pairs for the detection of different GM maize events represents the most comprehensive multiplex PCR method (fig. 9.5) [85].

Additionally, qualitative applications for identification of GMOs in food have been introduced using detection via microarray technology [40, 86, 87]. Peptide nucleic acids (PNAs) have been described as useful microarray probes for the analysis of Roundup Ready soybeans and different maize lines [86–87]. PNAs are analogues of DNA with peptides rather than pentose sugar phosphates forming the backbone. This results in a high affinity of PNA oligomers to hybridize with DNA; they are more sensitive to single mismatches and thus provide higher sequence specificity [88]. A combined assay of multiplex polymerase chain reaction and ligation detection reaction coupled with microarray has been developed [89] and applied to the analysis of traces of GMOs in foods [90–92]. This approach involves an additional confirmation step of the PCR products through ligation of sequence-specific probes prior to microarray hybridization. Only the ligated products generate a fluorescence signal when hybridized to the array with their unique zip code sequence [89, 90]. A novel multiplex quantitative DNA array-based PCR has been presented for the quantification of transgenic maize in food and feed [93].

One of the latest developments in multiplex approaches for quantitative analysis is the technique of ligation-dependent probe amplification (LPA). Originally, ligation-dependent PCR was applied to the relative quantification of DNA in the field of medical diagnostics [94]. The suitability of this method for the detection and relative quantification of GMOs in food samples has been demonstrated using commercially available maize standards [95]. A synthetic probe set is used for the detection of target sequences to avoid complex cloning and preparation steps required for the isolation of single-stranded DNA probes. These sequence-specific probes contain a target-specific hybridization site and identical primer-binding sites (PBSs) at their 5′- and 3′-ends, respectively. In case of successful hybridization to adjacent sites of the target sequence, the probes are ligated by a thermostable ligase. The use of spacer sequences between hybridization sites and PBSs ensures ligation products with lengths characteristic for each of the target DNA. In the second step

Table 9.2. Examples of multiplex approaches applied in GMO analysis.

Targeted Transgenes	Endogenous Reference Genes	Technique	Detection	Literature
Maize Bt11, Bt176, MON810, GA21, RR-soybeans	Maize *zein*, soybean *le1*	Multiplex conventional PCR + ligation-detection reaction (LDR)	Microarray	[90–92]
NOS terminator, 35S promoter	Maize *zein*, soybean *le1*	Multiplex conventional PCR	Agarose gel electrophoresis, EtBr staining	[79]
Maize Bt11, Bt176, MON810, T25, GA21	Maize *zein*	Multiplex conventional PCR	CGE-LIF	[96]
Maize Bt11, Bt176, MON810, GA21, RR-soybean	Maize *zein*, soybean *le1*	Multiplex conventional PCR	Agarose gel electrophoresis, EtBr staining	[80]
Maize Bt11, Bt176, MON810, GA21, RR-soybeans	Maize *zein*, soybean *le1*	Multiplex conventional PCR + asymmetric PCR with labeled primer	PNA-microarray	[86, 87]
Maize Bt11, MON810, T25, GA21		Multiplex conventional PCR	Agarose gel electrophoresis, EtBr staining	[29]
P35S, NOS, *cp4-epsps, nptII, cryIA(b), pat, bar, gox, oxy, barnase*	Maize *ivr*, soy *le1*, β-actin, canola cruciferin	Multiplex conventional PCR	Agarose gel electrophoresis, EtBr staining	[81]
P35S, T35S, NOS, *cp4-epsps, nptII, cryIA(b),pat, bar, gox*, Pr-act		Multiplex conventional PCR	Agarose gel electrophoresis, EtBr staining	[82, 83]
Maize Bt11, Bt176, MON810, T25, GA21, NK603, MON863, TC1507	Maize *SSIIb*	Multiplex conventional PCR	Agarose gel electrophoresis, EtBr staining, capillary electrophoresis	[85]
NOS, 35S-*epsps, cryIA(b), pat*	Maize *zein*	Multiplex conventional PCR	Agarose gel electrophoresis, EtBr staining	[84]
P35S, NOS, Amp, Maize Bt11, Bt176, MON810, T25, GA21, DBT418, CBH351	Maize *hmga*	Two-step PCR, labeled probe hybridization	Microarray	[93]
18S rRNA, NOS, NOS/*cp4-epsps*, CP4/CTP, 35S/CTP, 35S/plant, *cp4-epsps, nptII, cryIA(b), bar*, PG	Maize *ivr*, soybean *le1*, rapeseed napin	Asymmetric PCR, labeled primer	Microarray	[40]

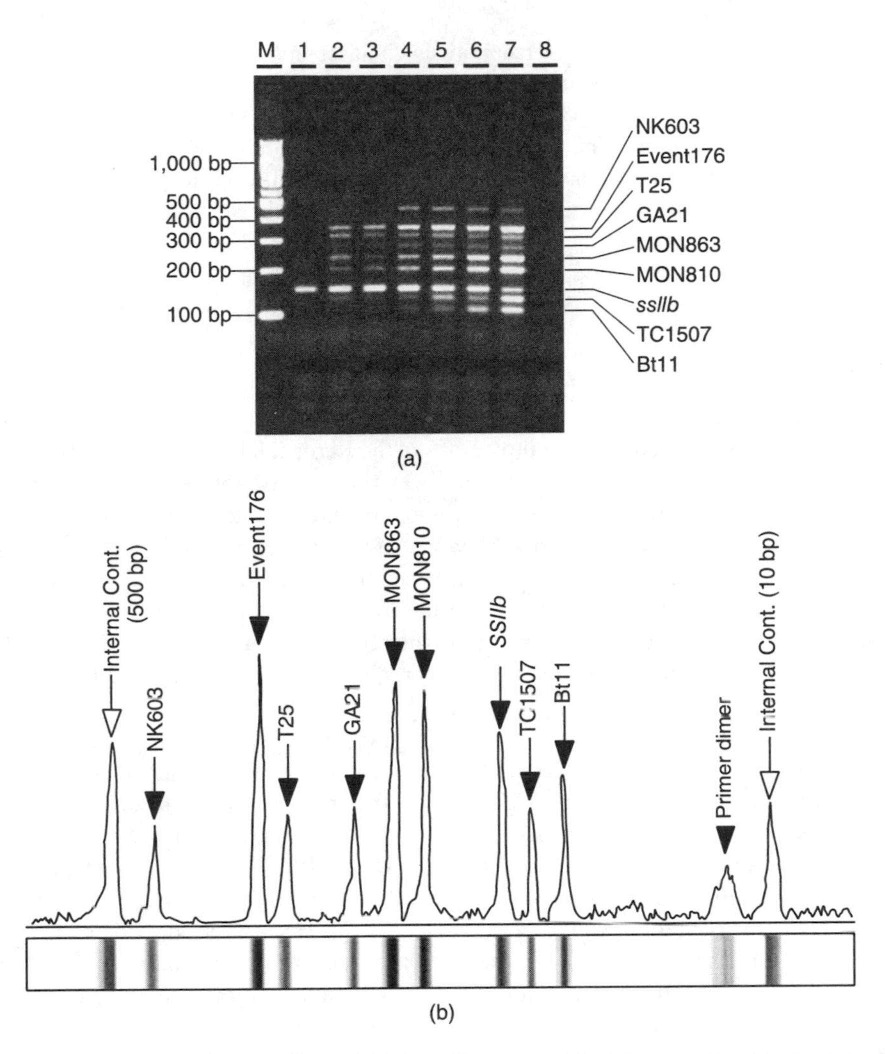

Fig. 9.5. Detection and separation of PCR products amplified from a simulated mixture of eight events of GM maize on agarose gel **(a)** and via capillary electrophoresis **(b)** (for details see [85]).

of the reaction, the ligation products are amplified competitively using one labeled pair of primers. Labeled PCR amplicons are finally separated by capillary electrophoresis and detected via laser-induced fluorescence.

The use of one pair of universal primers in the LPA method avoids one of the major difficulties of multiplex PCR applications, that being the complexity of amplification reaction because of the use of multiple pairs of primers. In contrast to multiplex PCR reactions, the application of this novel approach offers great flexibility due to its modular system that can be complemented with further probes to broaden the range of target sequences. The use of synthetic oligonucleotides as probes and the employment of classical thermocycler and detection methods enable the implementation of the technique in commonly equipped laboratories.

Detection using capillary gel electrophoresis via laser-induced fluorescence (CGE-LIF) represents a very sensitive and rapid separation and detection approach in an automated manner. Different amplicons can be distinguished by size using either labeled PCR primers or DNA-intercalating dyes. The better sensitivity and resolution of CGE-LIF compared to agarose gel electrophoresis have been demonstrated [96]. Capillary electrophoresis is a useful tool for optimization of multiplex PCR reactions because of the possibility of obtaining quantitative data in the form of peak areas/heights [97]. Compared to real-time PCR, quantitative results can be achieved in a cheaper way.

Food Authentication

Proof of authenticity and transparency along the entire food chain are key elements of food regulatory frameworks and labeling provisions in the European Union [98, 99]. In the past, species identification of animals or plants has been routinely performed with protein-based methods such as isoelectric focusing or immunological techniques. These methods can be successfully applied to raw materials. However, food processing and complex matrices make definite species distinction via protein-based methods difficult. Owing to the higher stability of DNA compared to proteins and the high degree of information provided by properly selected DNA sequences, species identification by nucleic-acids-based methods has become an important tool particularly for the authentication of processed foods. DNA-based analysis finds broad applications (e.g., differentiation of dairy products, fish and meat species, potatoes, rice, wheat, and grapes) [100–102]. A classic area is the differentiation of fish and meat species in processed products. Several methods use PCR in combination with restriction fragment length polymorphism analysis to distinguish meat species (beef, pork, ovine, lamb, goat, turkey, and chicken) and innumerable fish species by targeting different genes of the mitochondrial DNA [100, 103–9]. Also real-time PCR applications detecting beef and pork have been described [110–11]. Microarray hybridization of amplified mitochondrial cytochrome *b* gene DNA was applied to determine animal species in cheese and meat samples [112]. An extensive description of DNA-based methods used in authenticity testing and an overview on the various meat and fish species studied have been provided [113].

Nucleic-acid-based detection is also being used to assess the authenticity of premium quality products. The determination of the origin of olives, the distinction between high-price basmati rice and other long grain rice varieties, and the authentication of grape musts have been tackled using microsatellite markers [102, 114–17]. The identification of 50 potato varieties was possible using three different microsatellite primer sets [118]. Differentiation of fruits is possible even in processed products (e.g., the addition of low-price mandarin juice in oranges juices could be detected by analyzing multicopy chloroplast DNA) [119]. Substitution of durum wheat by common wheat (*Triticum aestivum*) in pasta could be detected by targeting the D genome present in the hexaploid wheat but not in durum wheat [120].

Detection of Food-Borne Pathogens

Microbiological analyses are integral elements of food safety assessment procedures. Recently, the European Union specified microbiological criteria for foods in regulation (EC) 2073/2005 to prevent risks for human health through toxins and metabolites of microorganisms and in particular the presence of pathogenic microorganisms [121]. Criteria have been determined for absence/presence of human pathogens like *Salmonella*, *Listeria monocytogenes*, enterohemorrhagic *Escherichia coli*, and toxin-producing organisms such as *Staphylococcus aureus* and *Clostridium botulinum* in different food categories. Traditionally, time-consuming analyses based on growth, isolation, and morphological and physiological characterization of microorganisms are performed. These methods are of limited value if immediate information on potential outbreaks of food-associated infections or data allowing on-line control of modern production processes are required [101]. Once again, polymerase chain reaction offers a promising alternative detection approach.

The first PCR approach described in food analysis dealt with the detection of invasive *Shigella* species on lettuce [122]. In the meantime, standardized methods for the detection of *Salmonella* and enterohemorrhagic *E. coli* (EHEC) via PCR have been used for routine testing by national surveillance authorities [123]. Multiplex PCR assays for the simultaneous detection of up to six bacterial pathogens have been described [124]. As in the analysis of GMOs, the sensitive detection via capillary gel electrophoresis and laser-induced fluorescence has been applied to a multiplex PCR of three food-borne pathogens [125]. Commercial kits for the identification of *Salmonella*, *Listeria*, and *Campylobacter* species and *E. coli*, *Pseudomonas aeruginosa*, and *Staphylococcus aureus* are available and represent useful tools in industrial Hazardous Analysis Critical Control Point (HACCP) concepts and other hygienic control measures.

Detection of Allergens

In industrialized countries 1–2 percent of adults and adolescents and up to 8 percent of children are affected by food allergies. The symptoms may range from skin irritations to severe anaphylactic reactions with fatal consequences. Around 90 percent of the adverse reactions observed have been associated with eight food groups: cow's milk, eggs, fish, crustaceans, peanuts, soybeans, tree nuts, and wheat [126, 127]. Besides these major food allergens a broad spectrum of fruits, vegetables, seeds, spices, meats, and lactose have been reported to possess allergenic potential [126–28].

Taking into account the recommendations of the Codex Alimentarius Commission [129], the European Commission amended the European Food Labeling Directive 2000/13/EC with a list of ingredients to be labeled [98]. Annex IIIa of directive 2003/89/EC includes cereals containing gluten, crustaceans, eggs, fish, peanuts, soybeans, milk and dairy products (including lactose), nuts, celery, mustard, sesame seeds, and products thereof. To protect the health of consumers, the declaration of these ingredients has been made mandatory regardless of their amounts in the final product [130].

Allergens are mainly proteins whose routine food analysis is based on immunological detection using either specific IgE from human serum or antibodies raised in animals. Major challenges are the need to check for the presence of food allergens at extremely low levels and to detect trace amounts of hidden allergens in composite and processed foods [127]. If the allergen itself cannot be targeted, PCR-based methods amplifying specific DNA sequences offer alternative tools [131, 132]. DNA presents a more stable analyte than proteins and is less affected by denaturation, and unambiguous sequences allow specific discrimination of closely related species. Sequences specific for allergenic food proteins are targeted by an increasing number of PCR assays. Examples are listed in table 9.3. No DNA-based method for the detection of egg is described. Methods available

Table 9.3. Examples of PCR methods applied for detection of allergens.

Allergen	Target/Acc. No.	Method	Literature
Celery	Mannitol dehydrogenase (Mtd) gene/ AF067082	Conventional PCR, AGE-PCR	[138]
	api g 1	Real-time PCR, TaqMan	[139]
Cereals containing gluten	Wheat, barley, rye, chloroplast *trnL* gene	Quantitative competitive PCR	[53, 140]
	Wheat (*acc1, RALyase*), barley (*γ-hordein*)	Real-time PCR, TaqMan	[141]
	Wheat *waxy D1* gene/AF113844	Real-time PCR, TaqMan	[142]
	Wheat, barley, PKABA1 gene	Duplex Real-time PCR	[143]
Peanuts	Ara h 2 gene/AY007229	Real-time PCR, TaqMan	[144]
	Peanut Ara h 2 gene /L77197, hazelnut *Cor a 1*/ Z72440	Duplex PCR, PNA-array detection	[145]
	Ara h 2 gene/L77197	Two-step PCR, DNA-PNA duplex detection with $DiSC_2$ dye	[146]
	Ara h 2 gene/L77197	Real-time PCR, TaqMan	[147]
Tree nuts	Walnut *jug r 2*/AF066055	Real-time PCR, TaqMan	[148]
	Hazelnut *Cor a 1*/AF136945	Conventional PCR, AGE-PCR	[149]
	Hazelnut *Cor a 1*/ Z72440	Conventional PCR, PNA probe, HPLC detection	[150]

Note: HPLC = high-performance liquid chromatography; AGE-PCR = *Abiotrophia* genus-specific PCR.

for the detection of fish, crustaceans, milk, and dairy products are mentioned in the food authenticity section. Recently, conventional and real-time PCR methods for the detection of six important food allergens (soybean, sesame, mustard, peanut, hazelnut, and almond) have been developed [133].

Outlook

The last decade has witnessed tremendous developments in the area of DNA-based food analysis. The expected instrumental progress, further simplifications of the methodologies, and the obvious potential for automation will further increase the attractiveness of these approaches. Considering the high degree of information provided by DNA analysis, applications will definitely spread, and analyses of foods on the molecular biological level will be indispensable cornerstones of modern food diagnostics.

References

[1] Saiki, R. K., Scharf, S., Faloona, F., Mullis, K. B., Horn, G. T., Erlich, H. A., Arnheim, N. (1985) Enzymatic amplification of beta-globin genomic sequences and restriction site analysis for diagnosis of sickle cell anemia. Science, 230(4732), 1350–54.

[2] Mullis, K. B., Faloona, F. A. (1987) Specific synthesis of DNA in vitro via a polymerase-catalyzed chain reaction. Methods Enzymol., 155, 335–50.

[3] ILSI Europe (2001) Novel Food Task Force in collaboration with the European Commission's Joint Research Centre (JRC) and ILSI International Food Biotechnology Committee. Method development in relation to regulatory requirements for detection of GMOs in the food chain. ILSI Europe report series. ILSI Europe, Brussels.

[4] European Commission (1997) Regulation (EC) No 258/97 of the European Parliament and of the Council of 27 January 1997 concerning novel foods and novel food ingredients. Official Journal of the European Communities 14.02.1997, No L 043, pp. 1–7.

[5] European Commission (1998) Council Regulation (EC) No 1139/98 of 26 May 1998 concerning the compulsory indication of the labelling of certain foodstuffs produced from genetically modified organisms of particulars other than those provided for in Directive 79/112/EEC. Official Journal of the European Communities 03.06.1998, No L 159, pp. 4–7.

[6] European Commission (2000) Commission Regulation (EC) No 49/2000 of 10 January 2000 amending Council Regulation (EC) No 1139/98 concerning the compulsory indication of the labelling of certain foodstuffs produced from genetically modified organisms of particulars other than those provided for in Directive 79/112/EEC. Official Journal of the European Communities 11.01.2000, No L 6, pp. 13–14.

[7] European Commission (2000) Commission Regulation (EC) No 50/2000 of 10 January 2000 on the labelling of foodstuffs and food ingredients containing additives and flavourings that have been genetically modified or have been produced from genetically modified organisms. Official Journal of the European Communities 11.01.2000, No L 6, pp. 15–17.

[8] European Union (2003) Regulation (EC) No 1829/2003 of the European Parliament and of the Council of 22 September 2003 on genetically modified food and feed. Official Journal of the European Union, L 268/1.

[9] European Union (2003) Regulation (EC) No 1830/2003 of the European Parliament and of the Council of 22 September 2003 concerning the traceability and labelling of genetically modified organisms and the traceability of food and feed products produced from genetically modified organisms and amending Directive 2001/18/EC. Official Journal of the European Union, L 268/24.

[10] European Union (2004) Commission Regulation (EC) No 65/2004 of 14 January 2004 establishing a system for the development and assignment of unique identifiers for genetically modified organisms. Official Journal of the European Union, L10/5.

[11] Baumforth, K. R. N., Nelson, P. N., Digby, J. E., O'Neil, J. D., Murray, P. G. (1999) Demystified . . . The polymerase chain reaction. Mol. Pathol. 52(1), 1–10.

[12] Innis, M. A., Gelfand, D. H., Sninsky, J. J. (1990) PCR Protocols—A Guide to Methods and Applications. Academic Press, San Diego.

[13] Miraglia, M., Berdal, K. G., Brera, C., Corbisier, P., Holst-Jensen, A., Kok, E. J., Marvin, H. J. P., Schimmel, H., Rentsch, J., van Rie, J. P. P. F., Zagon, J. (2004) Detection and traceability of genetically modified organisms in the food production chain. Food Chem. Toxicol., 42, 1157–80.

[14] Anklam, E., Gadani, F., Heinze, P., Pijnenburg, H., Van Den Eede, G. (2002) Analytical methods for detection and determination of genetically modified organisms in agricultural crops and plant-derived food products. Eur. Food Res. Technol., 214(1), 3–26.

[15] ENGL (2003) European Network of GMO Laboratories. Institute for Health and Consumer Protection. Joint Research Centre, European Commission.

[16] Zimmermann, A., Luethy, J., Pauli, U. (1998) Quantitative and qualitative evaluation of nine different extraction methods for nucleic acids on soya bean food samples. Z. Lebensm. Unters. Forsch. A, 207(2), 81–90.

[17] Anklam, E., Gadani, F., Heinze, P., Pijnenburg, H., Van Den Eede, G. (2002) Analytical methods for detection and determination of genetically modified organisms in agricultural crops and plant-derived food products. Eur. Food Res. Technol., 214(1), 3–26.

[18] Holden, M. J., Blasic, J. R., Jr., Bussjaeger, L., Kao, C., Shokere, L. A., Kendall, D. C., Freese, L., Jenkins, G. R. (2003) Evaluation of extraction methodologies for corn kernel (*Zea mays*) DNA for detection of trace amounts of biotechnology-derived DNA. J. Agric. Food Chem., 51(9), 2468–74.

[19] Peano, C., Samson, M. C., Palmieri, L., Gulli, M., Marmiroli, N. (2004) Qualitative and quantitative evaluation of the genomic DNA extracted from GMO and non-GMO foodstuffs with four different extraction methods. J. Agric. Food Chem., 52(23), 6962–68.

[20] Gadani, F., Bindler, G., Pijnenburg, H., Rossi, L., Zuber, J. (2000) Current PCR methods for the detection, identification and quantification of genetically modified organisms (GMOs): a brief review. Contrib. Tob. Res., 19(2), 85–96.

[21] Wilson, I. G. (1997) Inhibition and facilitation of nucleic acid amplification. Appl. Environ. Microbiol., 63(10), 3741–51.

[22] Englund, S., Ballagi-Pordani, A., Bölske, G., Johansson, K.-E. (1999) Improved reliability of diagnostic PCR with an internal positive control molecule. Proceedings of the 6th International Colloqium on Paratuberculosis, Manning, E. J. B., Collins, M. T. (eds.). Section 4, Diagnostic Applications and Approaches.

[23] Holst-Jensen, A., Ronning, S. B., Lovseth, A., Berdal, K. G. (2003) PCR technology for screening and quantification of genetically modified organisms (GMOs). Anal. Bioanal. Chem., 375, 985–93.

[24] Wolf, C., Scherzinger, M., Wurz, A., Pauli, U., Hübner, P., Lüthy, J. (2000) Detection of cauliflower mosaic virus by the polymerase chain reaction: testing of food components for false-positive 35S-promoter screening results. Eur. Food Res. Technol., 210(5), 367–72.

[25] Ronning, S. B., Vaitilingom, M., Berdal, K. G., Holst-Jensen, A. (2003) Event specific real-time quantitative PCR for genetically modified Bt11 maize (*Zea mays*). Eur. Food Res. Technol., 216(4), 347–54.

[26] Zimmermann, A., Lüthy, J., Pauli, U. (2000) Event specific transgene detection in Bt11 corn by quantitative PCR at the integration site. Lebensm. Wiss. Und Technol., 33(3), 210–16.

[27] Taverniers, I., Windels, P., Vaitilingom, M., Milcamps, A., Van Bockstaele, E., Van den Eede, G., De Loose, M. (2005) Event-specific plasmid standards and real-time PCR methods for transgenic Bt11, Bt176, and GA21 maize and transgenic GT73 canola. J. Agric. Food Chem., 53(8), 3041–52.

[28] Collonnier, C., Schattner, A., Berthier, G., Boyer, F., Coue-Philippe, G., Diolez, A., Duplan, M. N., Fernandez, S., Kebdani, N., Kobilinsky, A., Romaniuk, M., de Beuckeleer, M., de Loose, M., Windels, P., Bertheau, Y. (2005) Characterization and event specific-detection by quantitative real-time PCR of T25 maize insert. J. AOAC Int., 88(2), 536–46.

[29] Hernandez, M., Rodriguez-Lazaro, D., Zhang, D., Esteve, T., Pla, M., Prat, S. (2005) Interlaboratory transfer of a PCR multiplex method for simultaneous detection of four genetically modified maize lines: Bt11, MON810, T25, and GA21. J. Agric. Food Chem., 53(9), 3333–37.

[30] Yang, L., Xu, S., Pan, A., Yin, C., Zhang, K., Wang, Z., Zhou, Z., Zhang, D. (2005) Event specific qualitative and quantitative polymerase chain reaction detection of genetically modified MON863 maize based on the 5′-transgene integration sequence. J. Agric. Food Chem., 53(24), 9312–18.

[31] Holck, A., Vaitilingom, M., Didierjean, L., Rudi, K. (2002) 5′-Nuclease PCR for quantitative event-specific detection of the genetically modified Mon810 MaisGuard maize. Eur. Food Res. Technol., 214(5), 449–53.

[32] Hernandez, M., Pla, M., Esteve, T., Prat, S., Puigdomenech, P., Ferrando, A. (2003) A specific real-time quantitative PCR detection system for event MON810 in maize YieldGard based on the 3′-transgene integration sequence. Transgenic Res., 12(2), 179–89.

[33] Margarit, E., Reggiardo, M. I., Vallejos, R. H., Permingeat, H. R. (2006) Detection of BT transgenic maize in foodstuffs. Food Res. Int., 39(2), 250–55.

[34] Huang, H. Y., Pan, T. M. (2004) Detection of genetically modified maize MON810 and NK603 by multiplex and real-time polymerase chain reaction methods. J. Agric. Food Chem., 52(11), 3264–68.

[35] Nielsen, C. R., Berdal, K. G., Holst-Jensen, A. (2004) Characterization of the 5′ integration site and development of an event-specific real-time PCR assay for NK603 maize from a low starting copy number. Eur. Food Res. Technol., 219(4), 421–27.

[36] Windels, P., Taverniers, I., Depicker, A., Van Bockstaele, E., De Loose, M. (2001) Characterisation of the Roundup Ready soybean insert. Eur. Food Res. Technol., 213(2), 107–12.

[37] Berdal, K. G., Holst-Jensen, A. (2001) Roundup Ready soybean event-specific real-time quantitative PCR assay and estimation of the practical detection and quantification limits in GMO analyses. Eur. Food Res. Technol., 213(6), 432–38.

[38] Terry, C. F., Harris, N. (2001) Event-specific detection of Roundup Ready soya using two different real time PCR detection chemistries. Eur. Food Res. Technol., 213(6), 425–31.

[39] Huang, C.-C., Pan, T.-M. (2005) Event-specific real-time detection and quantification of genetically modified Roundup Ready soybean. J. Agric. Food Chem., 53(10), 3833–39.

[40] Xu, X., Li, Y., Zhao, H., Wen, S. Y., Wang, S. Q., Huang, J., Huang, K. L., Luo, Y. B. (2005) Rapid and reliable detection and identification of GM events using multiplex PCR coupled with oligonucleotide microarray. J. Agric. Food Chem., 53(10), 3789–94.

[41] Jonas, D. A., Elmadfa, I., Engel, K.-H., Heller, K. J., Kozianowski, G., König, A., Müller, D., Narbonne, J. F., Wackernagel, W., Kleiner, J. (2001) Safety considerations of DNA in food. Ann. Nutr. Metab., 45, 235–54.

[42] Bauer, T., Weller, P., Hammes, W. P., Hertel, C. (2003) The effect of processing parameters on DNA degradation in food. Eur. Food Res. Technol., 217(4), 338–43.
[43] Hupfer, C., Hotzel, H., Sachse, K., Engel, K.-H. (1998) Detection of the genetic modification in heat-treated products of Bt maize by polymerase chain reaction. Z. Lebensm. Unters. Forsch. A, 206(3), 203–7.
[44] Straub, J. A., Hertel, C., Hammes, W. P. (1999) Limits of a PCR-based detection method for genetically modified soya beans in wheat bread production. Z. Lebensm. Unters. Forsch. A, 208, 77–82.
[45] Zimmermann, K., Mannhalter, J. W. (1996) Technical aspects of quantitative competitive PCR. Biotechniques, 21(2), 268–72, 274–79.
[46] Raeymaekers, L. (1993) Quantitative PCR: theoretical considerations with practical implications. Anal. Biochem., 214(2), 582–85.
[47] Hübner, P., Studer, E., Luthy, J. (1999) Quantitative competitive PCR for the detection of genetically modified organisms in food. Food Control, 10(6), 353–58.
[48] Studer, E., Rhyner, C., Luethy, J., Huebner, P. (1998) Quantitative competitive PCR for the detection of genetically modified soybean and maize. Z. Lebensm. Unters. Forsch. A, 207(3), 207–13.
[49] Hardegger, M., Brodmann, P., Herrmann, A. (1999) Quantitative detection of the 35S promoter and the NOS terminator using quantitative competitive PCR. Eur. Food Res. Technol., 209(2), 83–87.
[50] Wurz, A., Bluth, A., Zelz, P., Pfeifer, C., Willmund, R. (1999) Quantitative analysis of genetically modified organisms (GMO) in processed food by PCR based methods. Food Control, 10(6), 385–89.
[51] Hupfer, C., Hotzel, H., Sachse, K., Moreano, F., Engel, K.-H. (2000) PCR-based quantification of genetically modified Bt maize: single-competitive versus dual-competitive approach. Eur. Food Res. Technol., 212(1), 95–99.
[52] Wolf, C., Lüthy, J. (2001) Quantitative competitive (QC) PCR for quantification of porcine DNA. Meat Sci., 57, 161–68.
[53] Dahinden, I., Von Buren, M., Lüthy, J. (2001) A quantitative competitive PCR system to detect contamination of wheat, barley or rye in gluten-free food for coeliac patients. Eur. Food Res. Technol., 212(2), 228–33.
[54] Wilhelm, J., Pingoud, A. (2003) Real-time polymerase chain reaction. Chem. Bio. Chem., 4, 1120–28.
[55] Schmittgen, T. D. (2001) Real-time quantitative PCR. Methods, 25, 383–85
[56] Arya, M., Shergill, I. S., Williamson, M., Gommersall, L., Arya, N., Patel, H. R. H. (2005) Basic principles of real-time quantitative PCR. Expert Rev. Mol. Diagn., 5(2), 209–19.
[57] Ginzinger, D. G. (2002) Gene quantification using quantitative real-time PCR: an emerging technology hits the mainstream. Exp. Hematol., 30, 503–12.
[58] Taverniers, I., Van Bockstaele, E., De Loose, M. (2004) Cloned plasmid DNA fragments as calibrators for controlling GMOs: different real-time duplex quantitative PCR methods. Anal. Bioanal. Chem., 378, 1198–1207.
[59] Pardigol, A., Guillet, S., Pöpping, B. (2003) A simple procedure for quantification of genetically modified organisms using hybrid amplicon standards. Eur. Food Res. Technol., 216(5), 412–20.
[60] Moreano, F., Pecoraro, S., Bunge, M., Busch, U. (2005) Development of synthetic DNA-standards for the quantitative screening of different genetically modified rapeseed lines via real-time PCR. Proceedings of the Euro. Food Chem. XIII Conference, Macromolecules and Their Degradation Products in Food—Physiological, Analytical and Technological Aspects, Hamburg, Germany, 21–23 September 2005, 1, pp. 154–58.
[61] Trappmann, S., Schimmel, H., Kramer, G. N., Van den Eede, G., Pauwels, J. (2002) Production of certified reference materials for the detection of genetically modified organisms. J. AOAC Int., 85(3), 775–79.
[62] Vaitilingom, M., Pijnenburg, H., Gendre, F., Brignon, P. (1999) Real-time PCR detection of genetically modified Maximizer maize and Roundup Ready soybean in some representative foods. J. Agric. Food Chem., 47(12), 5261–66.
[63] Höhne, M., Santisi, C. R., Meyer, R. (2002) Real-time multiplex PCR: an accurate method for the detection and quantification of 35S-CaMV promoter in genetically modified maize containing food. Eur. Food Res. Technol., 215(1), 59–64.
[64] Taverniers, I., Windels, P., Van Bockstaele, E., De Loose, M. (2001) Use of cloned DNA fragments for event-specific quantification of genetically modified organisms in pure and mixed food products. Eur. Food Res. Technol., 213(6), 417–27.
[65] Pietsch, K., Waiblinger, H.-U. (2000) Quantification of genetically modified soybeans in food with the LightCycler system. In Rapid Cycle Real-Time PCR, Springer, Berlin, pp. 385–89.

[66] Waiblinger, H. U., Gutmann, M., Hadrich, J., Pietsch, K. (2001) Validation of real-time PCR for the quantification of genetically modified soybeans. Deutsche Lebensmittel-Rundschau, 97(4), 121–25.
[67] Broll, H., Zagon, J., Butschke, A., Grohmann, L. (2002) GVO Analytik: Validierung und Ringversuche. GMO Analytik heute. Symposium organized by Sci. Diagnostics GmbH and GeneScan Europe AG. Frankfurt am Main, Germany, 23 January 2002.
[68] Lipp, M., Shillito, R., Giroux, R., Spiegelhalter, F., Charlton, S., Pinero, D., Song, P. (2005) Polymerase chain reaction technology as analytical tool in agricultural biotechnology. J. AOAC Int., 88(1), 136–55.
[69] Trifa, Y., Zhang, D. (2004) DNA content in embryo and endosperm of maize kernel (*Zea mays* L.): impact on GMO quantification. J. Agric. Food Chem., 52, 1044–48.
[70] European Commission's Fifth Frameword Program, Quality of Life and Management of Living Resources. (1999) Project title: "Reliable, standardised, specific, quantitative detection of genetically modified food." Final report of workpackage 2: contract no. QLK1-CT-1999-01301.
[71] Kuribara, H., Shindo, Y., Matsuoka, T., Takubo, K., Futo, S., Aoki, N., Hirao, T., Akiyama, H., Goda, Y., Toyoda, M., Hino, A. (2002) Novel reference molecules for quantitation of genetically modified maize and soybean. J. AOAC Int., 85(5), 1077–89.
[72] Weighardt, F., Barbati, C., Paoletti, C., Querci, M., Kay, S., De Beuckeleer, M., Van den Eede, G. (2004) Real-time polymerase chain reaction-based approach for quantification of the pat gene in the T25 *Zea mays* event. J. AOAC Int., 87(6), 1342–55.
[73] Holden, M. J., Blasic, J. R., Jr., Bussjaeger, L., Kao, C., Shokere, L. A., Kendall, D. C., Freese, L., Jenkins, G. R. (2003) Evaluation of extraction methodologies for corn kernel (*Zea mays*) DNA for detection of trace amounts of biotechnology-derived DNA. J. Agric. Food Chem., 51(9), 2468–74.
[74] Prokish, J., Zeleny, R., Trapmann, S., Le Guern, L., Schimmel, H., Kramer, G. N., Pauwels, J. (2001) Estimation of the minimum uncertainty of DNA concentration in a genetically modified maize sample candidate certified reference material. Fresenius J. Anal. Chem., 370, 935–39.
[75] Moreano, F., Busch, U., Engel, K.-H. (2005) Distortion of genetically modified organism quantification in processed foods: influence of particle size compositions and heat-induced DNA degradation. J. Agric. Food Chem., 53(26), 9971–79.
[76] ISO/DIS (2003) Foodstuffs—methods of analysis for the detection of genetically modified organisms and derived products—quantitative nucleic acid based methods. European Committee for Standardization (21570).
[77] Moreano, F. (2005) Development of techniques for the quantification of DNA from genetically modified organisms in processed foods. Dissertation, Technical University of Munich, Germany.
[78] Anklam, E., Heinze, P., Kay, S., Van den Eede, G., Popping, B. (2002) Validation studies and proficiency testing. J. AOAC Int., 85(3), 809–15.
[79] Forte, V. T., Di Pinto, A., Martino, C., Tantillo, G. M., Grasso, G., Schena, F. P. (2005). A general multiplex-PCR assay for the general detection of genetically modified soya and maize. Food Control, 16, 535–39.
[80] Germini, A., Zanetti, A., Salati, C., Rossi, S., Forré, C., Schmid, S., Marchelli, R. (2004) Development of a seven-target multiplex PCR for the simultaneous detection of transgenic soybean and maize in feeds and foods. J. Agric. Food Chem., 52(11), 3275–80.
[81] James, D., Schmidt, A.-M., Wall, E., Green, M., Masri, S. (2003) Reliable detection and identification of genetically modified maize, soybean, and canola by multiplex PCR analysis. J. Agric. Food Chem., 51, 5829–34.
[82] Matsuoka, T., Kuribara, H., Akiyama, H., Miura, H., Goda, Y., Kusakabe, Y., Isshiki, K., Toyoda, M., Hino, A. (2001) A multiplex PCR method of detecting recombinant DNAs from five lines of genetically modified maize. J. Hyg. Soc. Japan, 42(1), 24–32.
[83] Matsuoka, T., Kuribara, H., Takubo, K., Akiyama, H., Miura, H., Goda, Y., Kusakabe, Y., Isshiki, K., Toyoda, M., Hino, A. (2002) Detection of recombinant DNA segments introduced to genetically modified maize (*Zea mays*). J. Agric. Food Chem., 50(7), 2100–2109.
[84] Permingeat, H. R., Reggiardo, M. I., Vallejos, R. H. (2002) Detection and quantification of transgenes in grains by multiplex and real-time PCR. J. Agric. Food Chem., 50(16), 4431–36.
[85] Onishi, M., Matsuoka, T., Kodama, T., Kashiwaba, K., Futo, S., Akiyama, H., Maitani, T., Furui, S., Oguchi, T., Hino, A. (2005) Development of a multiplex polymerase chain reaction method for simultaneous detection of eight events of genetically modified maize. J. Agric. Food Chem., 53(25), 9713–21.

[86] Germini, A., Mezzelani, A., Lesignoli, F., Corradini, R., Marchelli, R., Bordoni, R., Consolandi, C., De Bellis, G. (2004) Detection of genetically modified soybean using peptide nucleic acids (PNAs) and microarray technology. J. Agric. Food Chem., 52(14), 4535–40.

[87] Germini, A., Rossi, S., Zanetti, A., Corradini, R., Fogher, C., Marchelli, R. (2005) Development of a peptide nucleic acid array platform for the detection of genetically modified organisms in food. J. Agric. Food Chem., 53(10), 3958–62.

[88] Jensen, K. K., Orum, H., Nielsen, P. E., Norden, B. (1997) Kinetics for hybridization of peptide nucleic acids (PNA) with DNA and RNA studied with the BIAcore technique. Biochemistry, 36(16), 5072–77.

[89] Gerry, N. P., Witowski, N. E., Day, J., Hammer, R. P., Barany, G., Barany, F. (1999) Universal DNA microarray method for multiplex detection of low abundance point mutations. J. Mol. Biol., 292(2), 251–62.

[90] Bordoni, R., Mezzelani, A., Consolandi, C., Frosini, A., Rizzi, E., Castiglioni, B., Salati, C., Marmiroli, N., Marchelli, R., Bernardi, L. R., Battaglia, C., De Bellis, G. (2004) Detection and quantitation of genetically modified maize (Bt-176 transgenic maize) by applying ligation detection reaction and universal array technology. J. Agric. Food Chem., 52(5), 1049–54.

[91] Bordoni, R., Germini, A., Mezzelani, A., Marchelli, R., De Bellis, G. (2005) A microarray platform for parallel detection of five transgenic events in foods: a combined polymerase chain reaction-ligation detection reaction-universal array method. J. Agric. Food Chem. 53(4), 912–18.

[92] Peano, C., Bordoni, R., Gulli, M., Mezzelani, A., Samson, M. C., De Bellis, G., Marmiroli, N. (2005) Multiplex polymerase chain reaction and ligation detection reaction/universal array technology for the traceability of genetically modified organisms in foods. Anal. Biochem., 346(1), 90–100.

[93] Rudi, K., Rud, I., Holck, A. (2003) A novel multiplex quantitative DNA array based PCR (MQDA-PCR) for quantification of transgenic maize in food and feed. Nucleic Acids Res., 31(11), e62.

[94] Schouten, J. P., McElgunn, C. J., Waaijer, R., Zwijnenburg, D., Diepvens, F., Pals, G. (2002) Relative quantification of 40 nucleic acid sequences by multiplex ligation-dependent probe amplification. Nucl. Acids Res., 30(12), e57.

[95] Moreano, F. et al. (2005) Ligation-dependent probe amplification for the simultaneous event-specific detection and relative quantification of DNA from two genetically modified organisms. Eur. Food Res. Technol., DOI 10.1007/s00217-005-0169-9.

[96] Garcia-Canas, V., Gonzalez, R., Cifuentes, A. (2004) Sensitive and simultaneous analysis of five transgenic maizes using multiplex polymerase chain reaction, capillary gel electrophoresis, and laser-induced fluorescence. Electrophoresis, 25(14), 2219–26.

[97] Butler, J. M., Riutberg, C. M., Vallone, P. M. (2001) Capillary electrophoresis as a tool for optimization of multiplex PCR reactions. Fresen. J. Anal. Chem., 369(3–4), 200–205.

[98] European Commission. 2000. Directive 2000/13/EC of the European Parliament and of the Council of 20 March 2000 on the approximation of the laws of the member states relating to the labelling, presentation and advertising of foodstuffs. Official Journal of the European Communities 06.05.2000, L 109, pp. 29–42.

[99] Cheftel, J. C. (2005) Food and nutrition labelling in the European Union. Food Chem., 93, 531–50.

[100] Wolf, C., Burgener, M., Hübner, P., Lüthy, J. (2000) PCR-RFLP analysis of mitochondrial DNA: differentiation of fish species. Lebensm. Wiss. Technol. 33(2), 144–50.

[101] Garcia-Canas, V., Gonzalez, R., Cifuentes, A. (2004) The combined use of molecular techniques and capillary electrophoresis in food analysis. Trends Anal. Chem., 23(9), 637–43.

[102] Garcia-Beneytez, E., Moreno-Arribas, M. V., Borrego, J., Polo, M. C., Ibanez, J. (2002) Application of a DNA analysis method for the cultivar identification of grape musts and experimental and commercial wines of *Vitis vinifera* L. using microsatellite markers. J. Agric. Food Chem., 50, 6090–96

[103] Pascoal, A., Prado, M., Castro, J., Cepeda, A., Barros-Velazquez, J. (2004) Survey of authenticity of meat species in food products subjected to different technological processes, by means of PCR-RFLP analysis. Eur. Food Res. Technol., 218, 306–12.

[104] Sun, Y. L., Lin, C. S. (2003) Establishment and application of a fluorescent polymerase chain reaction-restriction fragment length polymorphism (PCR-RFLP) method for identifying porcine, caprine, and bovine meats. J. Agric. Food Chem., 51(7), 1771–76.

[105] Krcmar, P., Rencova, E. (2003) Identification of species-specific DNA in feedstuffs. J. Agric. Food Chem., 51(26), 7655–58.

[106] Carrera, E., Garcia, T., Cespedes, A., Gonzalez, I., Fernandez, A., Hernandez, P. E., Martin, R. (1999) Salmon and trout analysis by PCR-RFLP for identity authentication. J. Food Sci., 64(3), 410–13.

[107] Hold, G. L., Russell, V. J., Pryde, S. E., Rehbein, H., Quinteiro, J., Vidal, R., Rey-Mendez, M., Sotelo, C. G., Perez-Martin, R. I., Santos, A. T., Rosa, C. (2001) Development of a DNA-based method aimed at identifying the fish species present in food products. J. Agric. Food Chem., 49(3), 1175–79.
[108] Klossa-Kilia, E., Papasotiropoulos, V., Kilias, G., Alahiotis, S. (2002) Authentication of Messolongi (Greece) fish roe using PCR-RFLP analysis of 16s rRNA mtDNA segment. Food Control, 13(3), 169–73.
[109] Ram, J. L., Ram, M. L., Baidoun, F. F. (1996) Authentication of canned tuna and bonito by sequence and restriction site analysis of polymerase chain reaction products of mitochondrial DNA. J. Agric. Food Chem., 44(8), 2460–67.
[110] Laube, I., Spiegelberg, A., Butschke, A., Zagon, J., Schauzu, M., Kroh, L., Broll, H. (2003) Methods for the detection of beef and pork in foods using real-time polymerase chain reaction. Int. J. Food Sci. Tech., 38(2), 111–18.
[111] Hird, H., Goodier, R., Schneede, K., Boltz, C., Chisholm, J., Lloyd, J., Popping, B. (2004) Truncation of oligonucleotide primers confers specificity on real-time polymerase chain reaction assays for food. Food Addit. Contam., 21(11), 1035–40.
[112] Peter, C., Brünen-Nieweler, C., Cammann, K., Börchers, T. (2004) Differentiation of animal species in food by oligonucleotide microarray hybridization. Eur. Food Res. Technol., 219(3), 286–93.
[113] Lockley, A. K., Bardsley, R. G. (2000) DNA-based methods for food authentication. Trends Food Sci. Tech., 11, 67–7.
[114] Sefc, K. M., Lopes, M. S., Mendonca, D., Dos Santos, M. R., Da Camara Machado, M. L., Da Camara Machado, A. (2000) Identification of microsatellite loci in olive (*Olea europaea*) and their characterization in Italian and Iberian olive trees. Mol. Ecol., 9(8), 1171–73.
[115] Popping, B. (2002) The application of biotechnological methods in authenticity testing. J. Biotechnol. 98(1), 107–12.
[116] Woolfe, M., Primrose, S. (2004) Food forensics: using DNA technology to combat misdescription and fraud. Trends Biotechnol., 22(5), 222–26.
[117] Siret, R., Boursiquot, J. M., Merle, M. H., Cabanis, J. C., This, P. (2000) Toward the authentication of varietal wines by the analysis of grape (*Vitis vinifera* L.) residual DNA in must and wine using microsatellite markers. J. Agric. Food Chem., 48(10), 5035–40.
[118] Corbett, G., Lee, D., Donini, P., Cooke, R. J. (2001) Identification of potato varieties by DNA profiling. Acta Hortic., 546, 387–90.
[119] Knight, A. (2000) Development and validation of a PCR-based heteroduplex assay for the quantitative detection of mandarin juice in processed orange juices. Agro Food Ind. Hi-Tech, 11(2), 7–8.
[120] Bryan, G. J., Dixon, A., Gale, M. D., Wiseman, G. (1998) A PCR-based method for the detection of hexaploid bread wheat adulteration of durum wheat and pasta. J. Cereal Sci., 28(2), 135–45.
[121] European Commission (2005) Commission Regulation (EC) No 2073/2005 of 15 November 2005 on microbiological criteria for foodstuffs. Official Journal of the European Communities 15.11.2005, L 338, pp. 1–26.
[122] Lampel, K. A., Jagow, J. A., Trucksess, M., Hill, W. E. (1990) Polymerase chain reaction for detection of invasive *Shigella flexneri* in food. Appl. Environ. Microbiol., 56(6), 1536–40.
[123] Amtliche Sammlung von Untersuchungsverfahren nach §35 LMBG. (2005) Bundesamt für Verbraucherschutz und Lebensmittelsicherheit. Berlin, Germany.
[124] Kong, R. Y., Lee, S. K., Law, T. W., Law, S. H., Wu, R. S. (2002) Rapid detection of six types of bacterial pathogens in marine waters by multiplex PCR. Water Res., 36(11), 2802–12.
[125] Alarcon, B., Garcia-Canas, V., Cifuentes, A., Gonzalez, R., Aznar, R. (2004) Simultanous and sensitive detection of three foodborne pathogens by multiplex PCR, capillary gel electrophoresis, and laser-induced fluorescence. J. Agric. Food Chem., 52(23), 7180–86.
[126] Ellman, L. K., Chatchatee, P., Sicherer, S. H., Sampson, H. A. (2002) Food hypersensitivity in two groups of children and young adults with atopic dermatitis evaluated a decade apart. Pedriatr. Allergy Immunol., 13, 295–98.
[127] Poms, R. E., Klein, C. L., Anklam, E. (2004) Methods for allergen analysis in food: a review. Food Addit. Contam., 21(1), 1–31.
[128] Breiteneder, H., Ebner, C. (2000) Molecular and biochemical classification of plant-derived food allergens. J. Allergy Clin. Immunol., 106(1), 27–36.
[129] Codex Alimentarius Commission (2005) Food Labelling—Complete Texts. Joint FAO/WHO Food Standards Programme.

[130] European Commission (2003) Directive 2003/89/EC of the European Parliament and of the Council of 10 November 2003 amending Directive 2000/13/EC as regards indication of the ingredients present in foodstuffs. Official Journal of the European Communities 25.11.2003, L 308, pp. 15–18.

[131] Goodwin, P. R. (2004) Food allergen detection methods: a coordinated approach. J. AOAC Int., 87(6), 1383–90.

[132] Poms, R. E., Anklam, E. (2004) Polymerase chain reaction techniques for food allergen detection. J. AOAC Int., 87(6), 1391–97.

[133] Pancaldi, M., Paganellil, A., Righini, G., Carboni, E., Salvi, A., Rainieri, M., Villa, C., Benda, S. (2005) Molecular detection of plant-derived food allergens. Ingedienti Alimentari, 4(1), 21–27.

[134] Taverniers, I., Windels, P., Van Bockstaele, E., De Loose, M. (2001) Use of cloned DNA fragments for event-specific quantification of genetically modified organisms in pure and mixed food products. Eur. Food Res. Technol., 213(6), 417–24.

[135] Hernandez, M., Duplan, M.-N., Berthier, G., Vaietilingom, M., Hauser, W., Freyer, R., Pla, M., Bertheau, Y. (2004) Development and comparison of four real-time polymerase chain reaction systems for specific detection and quantification of *Zea mays* L. J. Agric. Food Chem., 52(15), 4632–37.

[136] Windels, P., Bertrand, S., Depicker, A., Moens, W., Van Bockstaele, E., De Loose, M. (2003) Qualitative and event-specific PCR real-time detection methods for StarLink maize. Eur. Food Res. Technol., 216(3), 259–63.

[137] Yang, L., Pan, A., Zhang, K., Yin, C., Qian, B., Chen, J., Huang, C., Zhang, D. (2005) Qualitative and quantitative PCR methods for event-specific detection of genetically modified cotton Mon1445 and Mon531. Transgenic Res., 14(6), 817–31.

[138] Dovicovicova, L., Olexova, L., Pangallo, D., Siekel, P., Kuchta, T. (2004) Polymerase chain reaction (PCR) for the detection of celery (*Apium graveolens*) in food. Eur. Food Res. Technol., 218(5), 493–95.

[139] Stephan, O., Weisz, N., Vieths, S., Weiser, T., Rabe, B., Vatterott, W. (2004) Protein quantification, sandwich ELISA, and real-time PCR used to monitor industrial cleaning procedures for contamination with peanut and celery allergens. J. AOAC Int., 87(6), 1448–57.

[140] Olexova, L., Dovicovicova, L., Svec, M., Siekel, P., Kuchta, T. (2005) Detection of gluten-containing cereals in flours and "gluten-free" bakery products by polymerase chain reaction. Food Control, 17(3), 234–37.

[141] Hernandez, M., Esteve, T., Pla, M. (2005) Real-time polymerase chain reaction based assays for quantitative detection of barley, rice, sunflower, and wheat. J. Agric. Food Chem., 53(18), 7003–9.

[142] Iida, M., Yamashiro, S., Yamakawa, H., Hayakawa, K., Kuribara, H., Kodama, T., Furui, S., Akiyama, H., Maitani, T., Hino, A. (2005) Development of taxon-specific sequences of common wheat for the detection of genetically modified wheat. J. Agric. Food Chem., 53(16), 6294–6300.

[143] Ronning, S. B., Berdal Knut, G., Boydler Andersen, C., Holst-Jensen, A. (2006) Novel reference gene, PKABA1, used in a duplex real-time polymerase chain reaction for detection and quantitation of wheat- and barley-derived DNA. J. Agric. Food Chem., 54(3), 682–87.

[144] Hird, H., Lloyd, J., Goodier, R., Brown, J., Reece, P. (2003) Detection of peanut using real-time polymerase chain reaction. Eur. Food Res. Technol., 217(3), 265–68.

[145] Rossi, S. et al. (2005) A PNA-array platform for the detection of hidden allergens in foodstuffs. Eur. Food Res. Technol., DOI 10.1007/s00217-005-0034-x.

[146] Sforza, S., Scaravelli, E., Corradini, R., Marchelli, R. (2005) Unconventional method based on circular dichroism to detect peanut DNA in food by means of a PNA probe and a cyanine dye. Chirality, 17(9), 515–21.

[147] Stephan, O., Vieths, S. (2004) Development of a real-time PCR and a sandwich ELISA for detection of potentially allergenic trace amounts of peanut (*Arachis hypogaea*) in processed foods. J. Agric. Food Chem., 52(12), 3754–60.

[148] Brezna, B. et al. (2005) A novel real-time polymerase chain reaction (PCR) method for the detection of walnuts in food. Eur. Food Res. Technol., DOI 10.1007/s00217-005-0214-8.

[149] Holzhauser, T., Wangorsch, A., Vieths, S. (2000) Polymerase chain reaction (PCR) for detection of potentially allergenic hazelnut residues in complex food matrixes. Eur. Food Res. Technol., 211(5), 360–65.

[150] Germini, A., Scaravelli, E., Lesignoli, F., Sforza, S., Corradini, R., Marchelli, R. (2005) Polymerase chain reaction coupled with peptide nucleic acid high-performance liquid chromatography for the sensitive detection of traces of potentially allergenic hazelnut in foodstuffs. Eur. Food Res. Technol. 220(5–6), 619–24.

10 Protein-Based Detection of GM Ingredients

A. Rotthier, M. Eeckhout, N. Gryson, K. Dewettinck, and K. Messens

Introduction

The use of genetically modified organisms (GMOs) in food products is regulated in many countries around the world. The regulations usually contain labeling rules if the GMO content is above a certain threshold. In order to comply with these regulations, analytical tools must be in place to monitor and verify the presence and the amount of GMOs in agricultural crops and in products derived thereof. These methods detect, identify, and quantify either the DNA introduced or the protein(s) expressed in the transgenic plant material.

Consequently, several laboratories have developed methods either based on DNA detection using the polymerase chain reaction (PCR) technique or based on protein detection using enzyme-linked immunosorbent assays (ELISAs). These methods, however, vary in their reliability, robustness and reproducibility, cost, complexity, and speed. Especially in Europe, the PCR technique is widely applied, whereas in the United States protein-based methods are preferred. Although PCR-based methods are known to be highly sensitive, the use of protein-based methods is in some cases the better choice. In this chapter, the advantages and limitations of protein detection for GMOs are discussed. Specific attention is also focused on the use of these methods in the traceability of GMOs along the production chain.

GM Crops in Food

The first genetically modified product for food use that was approved for commercial sale in an industrialized country was the FlavrSavr™ tomato in May 1994 (James 2005). Since then, the amount of GM crops has increased significantly. Nowadays, more than 80 different genetically modified plants are commercially available for food purposes worldwide.

The newly introduced traits are diverse and give a certain advantage, or combination of advantages, to the crop. The dominant trait used nowadays is herbicide tolerance by expression of a modified form of the plant enzyme 5-enolpyruvylshikimate-3-phosphate synthase (EPSPS). When the herbicide Roundup® is applied, the active component glyphosate will bind to the native form of EPSPS. This binding results in an inactivation of the enzyme, which is essential in the shikimate pathway. As a result the production of aromatic amino acids and other aromatic compounds is inhibited, and the plant dies. Due to a slight modification of the EPSPS enzyme, glyphosate can no longer bind it. Consequently, the plant can survive the inhibiting application of Roundup®. The gene responsible for the production of this modified EPSPS enzyme in plants has been isolated from the CP4 strain of the common soil bacterium *Agrobacterium tumefaciens* and has been introduced so far in soy, maize, canola, cotton, alfalfa, wheat, and sugar beet (http://www.agbios.com).

The second-most-introduced trait is insect resistance due to *cry* proteins, which are toxins. The gene responsible for this insect resistance is derived from the bacterium *Bacillus thuringiensis*. The toxic proteins act by selectively binding to protein receptors on the epithelial cells of the larval midgut of susceptible insect species. Following binding, ion channels are formed, leading to the loss of cellular adenosine triphosphate (ATP) and insect death. Their specificity of action can be directly attributed to the presence of specific protein receptors in the target insects: *cry1ab* is lethal only when eaten by the larvae of lepidopteran insects such as moths and butterflies, whereas the *cry3a* protein is insecticidal only when eaten by the larvae of coleopteran insects such as the Colorado potato beetle and yellow mealworm (http://www.agbios.com).

Other novel proteins include the PAT (phosphinothricin N-acetyltransferase) enzyme, which converts L-phosphinothricin (PPT), the active component in gluphosinate ammonium, into an inactive form. The gene responsible for production of PAT has been isolated from a common soil actinomycete *Streptomyces viridochromogenes* (http://www.agbios.com).

An overview of all newly introduced characteristics in crops is given in table 10.1.

Protein-Based Detection Methods

Introduction

The majority of protein detection methods is based on immunoassays, which rely on the interaction between an antibody and its antigen. An antibody is a protein used by the immune system to identify and neutralize foreign objects like bacteria and viruses. Each antibody recognizes a specific antigen, which elicits the production of the antibody. The antibody has a Y shape, and each half of the forked end consists of a variable domain that can bind very strongly with its antigen (fig. 10.1). Because of this strong interaction, immunoassays are highly specific, and generally samples only need a simple preparation before analysis. In the case of GM detection, the protein encoded by the newly introduced gene acts as the antigen.

The most important factor for the success of immunological detection is the availability of high-affinity antibodies directed toward the protein to be detected. The antibodies can be polyclonal (raised in animals) or monoclonal (produced by a cell culture). Polyclonal antibodies are heterogeneous, meaning that they are made up of different kinds of antibodies, all specific for another protein. Monoclonal antibodies are homogeneous, thus specific for one antigen. The choice for monoclonal and polyclonal antibodies depends on the amount and specificity of the antibodies required and on limitations of cost.

The success of immunoassays also depends on the active expression of the newly introduced gene in the tissue to be analyzed and on the characteristics of the targeted proteins. These must fulfill certain requirements related to size, hydrophobicity, and tertiary structure.

A third criterion in the design and development of immunoassays is the format of the assay. In general, two formats exist: competitive assays and sandwich assays. In the competitive assay, the unknown analyte and a known amount of a marker compete

Table 10.1. Traits introduced in genetically modified crops.

Trait	Genetic Element	Crop
Amino acid composition	Dihydrodipicolinate synthase	Maize
Fatty acid composition	Delta(12) fatty acid dehydrogenase	Soybean
	Fatty acid desaturase	Argentine canola
	Thioesterase	Argentine canola
Fertility restoration	Barnase ribonuclease inhibitor	Argentine canola
Herbicide tolerance	5-enolpuruvylshikimate-3-phosphate synthase	Soybean, cotton, Argentine canola, Polish canola, sugar beet, creeping bentgrass, alfalfa, wheat, maize
	Acetolactate synthase	Carnation, cotton, flax, linseed
	Glyphosate oxidoreductase	Sugar beet, Argentine canola, Polish canola, maize
	Nitrilase	Argentine canola, cotton, tobacco
	Phosphinothricin N-acetyltransferase	Sugar beet, Argentine canola, Polish canola, chicory, soybean, cotton, rice, maize
Insect resistance	*cry1Ab* delta-endotoxin (*Btk* HD-1)	Maize
	cry1Ac delta-endotoxin	Cotton, tomato, maize
	cry1F delta-endotoxin	Cotton, maize
	cry2Ab delta-endotoxin	Cotton
	cry34Ab1 delta-endotoxin	Maize
	cry35Av1 delta-endotoxin	Maize
	cry3A delta-endotoxin	Potato
	cry3Bb1 delta-endotoxin	Maize
	cry9c delta-endotoxin	Maize
	Protease inhibitor	Maize
	VIP3A vegetative insecticidal protein	Cotton
Lepidopteran resistance	*cry1F* delta-endotoxin	Cotton, maize
Male sterility	Barnase ribonuclease	Argentine canola, chicory, maize
	DNA adenine methylase	Maize
Mutations	Acetolactate synthase	Argentine canola, sunflower, lentil, rice, wheat, maize
	Acetyl-CoA-carboxylase	Maize
Nicotine reduction	Nicotinate-nucleotide pyrophosphorylase (carboxylating)	Tobacco
Ripening delay	1-amino cyclopropane-1-carboxylic acid synthase	Carnation, tomato
	Aminocyclopropane cyclase synthase	Carnation, tomato
	Polygalacturonase	Tomato
	S-adenosylmethionine hydrolase	Melon, tomato
Virus resistance	Helicase	Potato
	Replicase (RNA-dependent RNA polymerase)	Potato
	Viral coat protein	Papaya, squash, potato

Source: http://www.agbios.com.

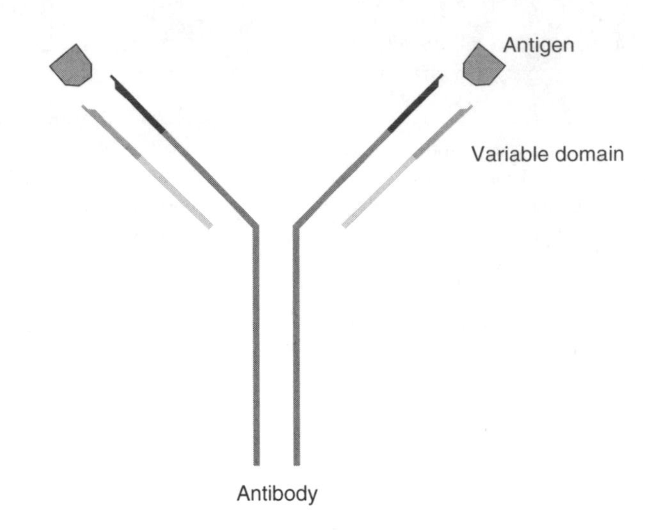

Fig. 10.1. Antibody structure.

for the antibody-binding sites. The amount of unknown analyte present in the sample determines the amount of marker that binds to the antibody. Sandwich assays—also called *reagent excess assays* or *two-site assays*—are characterized by a "sandwich" structure formed between the analyte and antibodies. An immobilized first antibody ("capture antibody") will bind the analyte present in a sample. These complexes will capture a second labeled antibody ("detector antibody"). Almost always, a sandwich assay is preferable to a competitive assay (Brett et al. 1999).

Methods

Western Blot

The western blot is a highly specific method that provides qualitative results suitable for determining whether a sample contains the target protein below or above a predetermined threshold level (Lipton et al. 2000). It is particularly useful for the analysis of insoluble proteins (Brett et al. 1999). Since this method is labor-intensive, it is preferred for research purposes rather than for routine analysis.

Figure 10.2 gives a schematic representation of western blot analysis. The samples being assayed are first subjected to an extraction procedure, followed by the solubilization of the extracted proteins in detergents and reducing agents. The proteins are then separated by sodium dodecyl sulfate (SDS)–polyacrylamide gel electrophoresis (SDS-PAGE). The denaturing conditions during the electrophoresis eliminate any problems of solubilization, aggregation, and coprecipitation of the target protein with adventitious proteins. The proteins are transferred and immobilized on a membrane, usually nitrocellulose, where they retain the same pattern of separation they had on the gel. The membrane is then immersed in a solution containing an antibody that specifically recognizes the target protein. Finally, the bound antibody is stained with Ponceau, silver nitrate or Coomassie, or a secondary immunological reagent, which catalyzes a color reaction (Sambrook and Russel 2000). The intensity of the color developed on the membrane is proportional to the amount of protein detected by the antibody and thus the amount of GM protein in the sample.

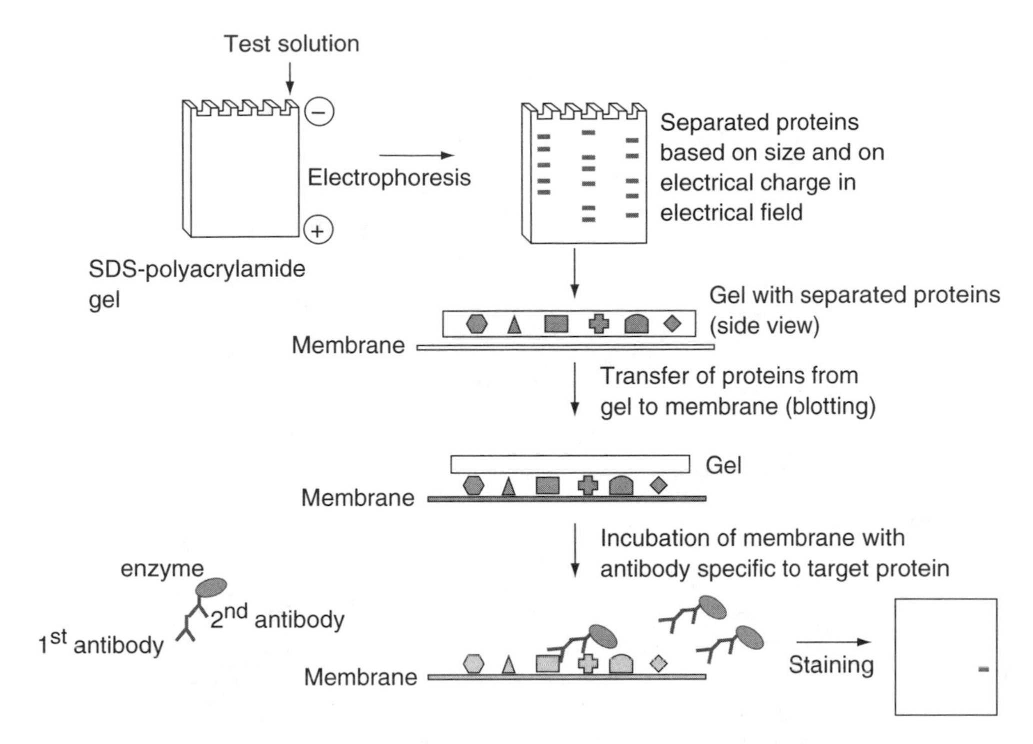

Fig. 10.2. Western blot.

ELISA

The enzyme-linked immunosorbent assay is the most common type of immunoassay. It covers any enzyme immunoassay involving an enzyme-labeled immunoreactant (antigen or antibody) and an immunosorbent (antigen or antibody bound to a solid support). So several variants of the ELISA method exist, with the sandwich assay being the most widely used and most flexible type of ELISA. Tests to detect GM proteins in food, based on commercially available ELISA kits, are mostly sandwich assays, although some are competitive ELISAs.

For the sandwich ELISA, antibodies specific to the target protein are bound to the surface of, typically, a microtiter well plate, as shown in figure 10.3. When the solution containing the test material is added, the antibody will work as a capture molecule for the target protein. Following an incubation period, a washing step removes all unbound components. Then a second specific antibody, chemically bound to an enzyme that catalyzes a color reaction, is added. If the target protein is present, the second labeled antibody binds to it, and any unbound labeled antibody is washed away. Enzyme substrate is then added to yield a colored product. The intensity of the signal produced is proportional to the amount of target protein present (Lipton et al. 2000).

In a competitive ELISA, the wells of the plate are coated with the target protein (fig. 10.4). A solution containing a limited number of the first antibody together with the test sample is added. A competition for the first antibody will then occur between the target protein in the sample and the target protein coated on the wells. The antibodies that are not bound to the antigens bound on the well will be washed away. A second

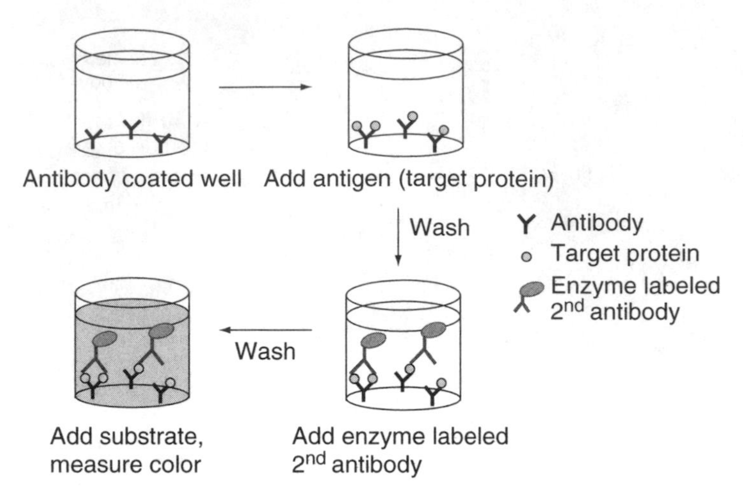

Fig. 10.3. Sandwich ELISA.

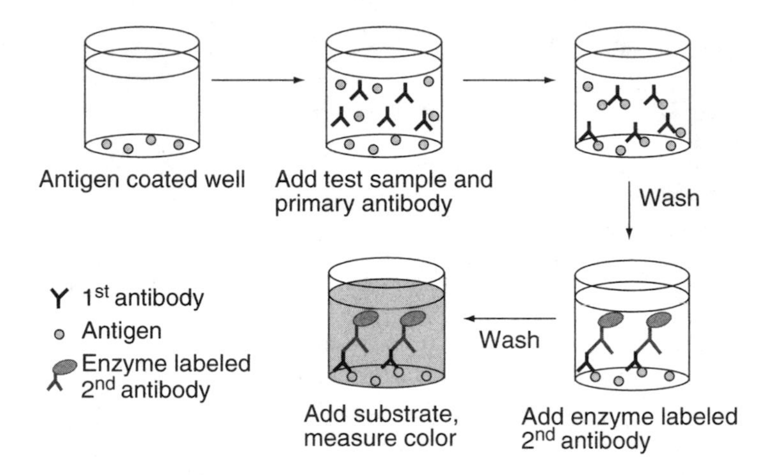

Fig. 10.4. Competitive ELISA.

antibody-enzyme complex is then added that specifically binds the antigen-antibody complex. After a second washing step, an enzyme substrate is added, resulting in color production. For the competitive assay, the intensity of the color is inversely proportional to the concentration of the GM protein in the sample (Yeung 2006).

The intensity of the color (optical density, OD) is measured with a spectrophotometer at the appropriate wavelength. By comparing the optical density of the samples to the value of the negative control (known to be free of the target analyte), it can be established whether the sample contains GM proteins or not.

Some commercial ELISA plate kits are supplied with calibrators (known concentrations of the target analyte in solution) and a negative control. These standards are run concurrently with each sample set and allow a standard curve to be set up. By using the spectrophotometer for all samples and all standards at the same time, a quantitative interpretation can be performed. The user can then calculate the protein concentration in the sample from the standard curve. ELISA test kits provide quantitative results in hours, with

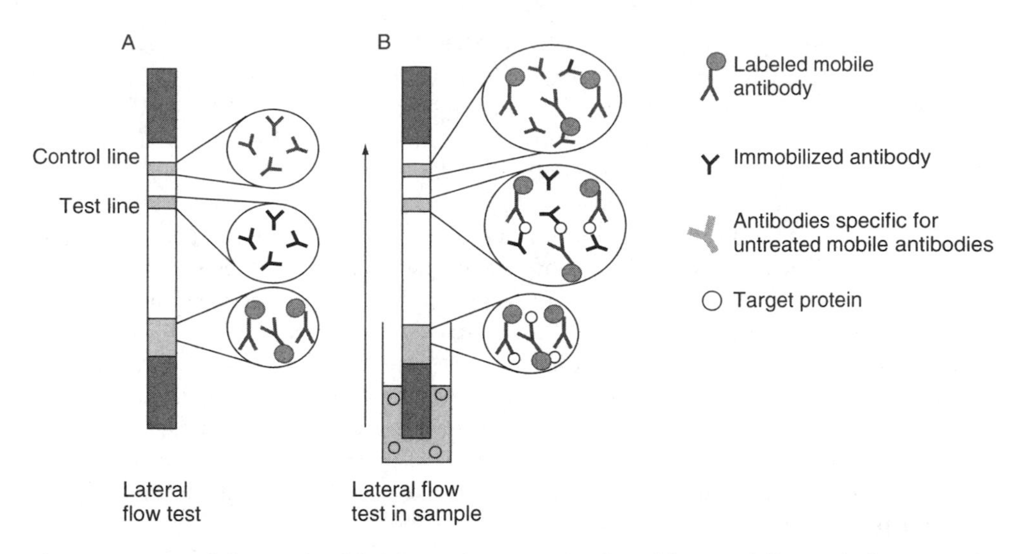

Fig. 10.5. Lateral flow strip. **(A)** Schematic representation. **(B)** Lateral flow strip in a sample containing the target protein.

detection limits sometimes less than 0.1 percent GMO. A semiquantitative interpretation can be made by comparing the color of the sample against the standards without the use of a spectrophotometer and determining the concentration range of the sample.

Lateral Flow Strips

The lateral flow strip is a variation on ELISA, using strips rather than microtiter well plates. The lateral flow test strip is placed in an Eppendorf vial containing the test solution. The sample migrates up the strip by capillary action. As it moves up the strip, the sample passes through a zone that contains mobile antibodies, usually labeled with colloidal gold, as shown in figure 10.5. This labeled antibody binds to the target protein, if present in the sample. The antibody-protein complex then continues to move up the strip through a porous membrane that contains two capture zones. The first capture zone of immobilized antibodies binds the antibody-protein complex, and the gold becomes visible as a red line. The second capture zone acts as a control zone and contains antibodies specific for untreated antibodies coupled to the color reagent. If there is no target GM protein present, only one single line will appear at the control zone. A result is called positive when both the control line and the line indicating presence of target protein change color.

Other Methods

Another format of immunoassays uses magnetic beads as a solid surface. The principle is the same as that of the ELISA format, with the magnetic particles being coated with capture antibodies and the reaction being performed in a test tube. The target protein, bound to the magnetic particles, can be separated using a magnet. The advantages include superior kinetics, as the particles are free to move in the reaction solution, and increased precision, due to uniformity of the particles (Ahmed 2002).

A new evolution in protein-based assays is high-throughput technologies, which have gained a great deal of attention over the last several years. The high throughput is obtained through the use of miniature and highly sensitive microfluidic devices (Burns et al. 1996). Although very small volumes are used, the technique remains reliable compared to the traditional immunoassays (Bernard et al. 2001). The device consists of a flat surface, usually glass or silicon, onto which narrow bands of antigen are deposited. Perpendicular to these stripes, a second set of lines is engraved. When diluted proteins flow through this second set of small channels, induced by capillary forces, they bind with their specific antigen, and a mosaic pattern of tiny squares occurs. This can then be analyzed using fluorescence microscopy (Bernard et al. 2001). The advantages of this technique are numerous. One sample can be screened for several proteins in one test, and several assays can be performed simultaneously. The amount of sample needed is reduced to a nanoliter range, thereby reducing the time to perform the analysis. Additionally this technique is highly suitable for automation. On the other hand, the equipment needed is expensive, and skilled personnel is required for the analysis of results. Therefore, this technique is currently not used for routine analysis of food samples.

Advantages of Protein-Based Detection Methods

Genetically modified organisms in food can be detected on the basis of the altered genome, thus by DNA analysis, or on the basis of the novel protein. As mentioned before, the protein-based assays comprise a very large and diverse group of assays. Thanks to their performance characteristics inherent to the technology, they are commercially very successful. The strong interaction between the antibody and the antigen is translated into high-sensitivity assays, and antibody specificity minimizes sample preparation. Their cost, ease of use, and flexible test format have resulted in wide-scale use in highly diverse markets. The cost per test of an immunoassay compared to other analytical methods is low, and a number of test formats exist that require little or no training to execute.

One-step "strip" tests can be performed by untrained personnel and give yes/no type results in minutes. The time-to-result or turnaround time may be one of the most critical attributes of a test method in applications where a large lot of material needs to be screened before being pooled with other lots or proceeded to the next stage in a process. For on-site testing the lateral flow test is generally the most cost-effective solution (Stave 1999).

Immunoassays can yield quantitative results in about an hour, or they can be incorporated into fully automated instruments capable of running hundreds of samples an hour. Quantification of proteins is easier than quantification of DNA if one looks to the expression unit of GMO levels. Quantitative results from protein analyses are expressed on a weight/weight basis (molar concentrations), while those from DNA assays represent genome equivalents. The influence of gene copy numbers makes DNA quantification more complex and the results more uncertain (Anklam and Neumann 2002).

Limitations of Protein-Based Methods for Detection of GMOs in Food

Despite the advantages protein-based methods offer, many drawbacks need to be considered when it comes to their application for the detection of GMOs.

One of the difficulties of an immunoassay system is experienced at the outset. Generating a specific antibody against the antigen of interest can be a difficult and time-consuming process and requires many skills and a lot of experience. This is where the use of recombinant antibody technology could offer benefits. Using this technique will ease the selection of antibodies with rare properties and the manipulation of the characteristics of already existing antibodies (Brett et al. 1999). Once a specific (monoclonal or polyclonal) antibody with high affinity for its target protein is generated, careful standardization and testing for unexpected cross-reactivity must be performed. Nonetheless, once established, the flexibility and turn-around time for immunoassay systems are excellent (Bindler et al. 1998).

The ELISA may be around 100 times less sensitive than the PCR method (Meyer et al. 1996). This lower sensitivity is not an automatic disadvantage because a false-positive result from a minor contaminant is less likely to occur (Bindler et al. 1998). Since the binding between antibody and antigen is based on the native structure of the target, any conformational change in the epitope of the antigen renders the assay ineffective. Food processing, such as heating or exposure to strong acids or alkalis, can cause this conformational change, by which detection of the protein will no longer be possible. Other processing steps, however, such as grinding, do not have an influence on the structure of a protein, and thus do not affect the effectivity of the assay, unless the grinding process leads to the heating of the product. To overcome this problem, some antibodies have been developed to recognize a protein in its unfolded shape and can therefore be used for the detection of GMOs in highly processed foods (Griffiths et al. 2002).

Nevertheless, the sensitivity of proteins to denaturation through processing limits the application of immunoassays to grains, raw materials, and fresh and unprocessed food products (Anklam et al. 2002). Furthermore, the accuracy and precision of the assay can be affected by substances present in food matrices such as surfactants, phenolic compounds, fatty acids, endogenous phosphatases, or enzymes that may inhibit the specific antigen-antibody interaction (Bindler et al. 1998; Brett et al. 1999; Spiegelhalter et al. 2001; Anklam et al. 2002).

Another disadvantage of the immunoassays is due to their reliance on the expression of the newly introduced gene. After all, genetic modification does not always result in the production of a new protein. This is, for example, the case for the FlavrSavr™ genetically modified tomato, where RNA is produced to suppress the production of the protein polygalacturonase in order to delay the ripening process. In other cases, the expression level of the introduced DNA can be too low for detection. Moreover, the protein expression levels can vary from tissue to tissue. The expression can be higher in certain parts of the plant or in certain phases in the physiological development. As a result, the transgenic protein might not be present in the part of the plant that is used in food production. For example, the endotoxin *cry1A* of GM maize Bt176 is expressed in the green tissues and in the pollen of the plant. Since the novel protein is not expressed in the maize kernels, protein analysis of the kernels will not reveal the presence of the foreign protein (Popping and Broll 2001). Expression is furthermore influenced by external factors, such as weather, soil, and other cultivation conditions (Wilson et al. 2001). This complicates the quantification of the GMO by protein-based detection methods. Quantification of GMOs by the use of immunoassays is even more complex, since the assays generate absolute values, such as the total amount of a novel protein present. To comply with GMO-labeling legislation however, not the absolute quantity but the relative quantity of the GM trait is important—that is, the relative ratio of the GM trait to the conventional counterpart of the same ingredient (e.g., the percentage of GM soy out of total soy in the food product). The relative protein-based quantification is only possible if a species- or taxon-specific protein is measured simultaneously (Van Duijn et al. 2002) or if the sample exists of one single ingredient and an appropriate reference material for the standard dose response curve is available.

Since there are no structures common to all GM proteins or groups of proteins and a single antibody will only bind to one particular protein, the immunoassays are less suited for a general GM screening (Griffiths et al. 2002). Moreover, protein-based methods cannot distinguish between different GM varieties that express the same protein (Gachet et al. 1999; Lüthy 1999; Kok et al. 2002).

The Role of Protein-Based Detection Methods in GMO Traceability

In spite of their limitations, immunoassays play an important role in the traceability of genetically modified organisms in the food production chain. The goal of traceability systems is to guarantee the differentiation among foods with different attributes. Traceability systems represent instruments to provide such differentiation in a reliable and documented manner (Miraglia et al. 2004). The traceability of transgenic crops is based on a combination of paper-based trails through the entire production chain and analytical tests at the critical points of the food production scheme.

While the seed company typically performs seed-quality analysis in-house by PCR, the farmer will use the lateral flow test. After harvest, the products are transported from the farm to the grain elevator, where lateral flow devices are also used. From here, the commodity is shipped to the processors (e.g., ingredient producers). They typically have the incoming goods tested by ELISA or real-time PCR using a contract laboratory. From the ingredient producer, products are shipped to the food processor, who also has the incoming goods tested by a contract laboratory, using real-time PCR. The retailer displaying the finished products on the shelves also has the products tested to comply with the due diligence requirements.

In every stage, all products will be accompanied by appropriate documentation to certify their origin. There is a significant probability of having cross-contamination because the same elevators, ships, trucks, barges, and production lines are used for transgenic and nontransgenic material. The favored option is the product chain where transport and production lines for transgenic and nontransgenic materials are separated and a complete audit trail including analytical test points is available (Popping 2003).

Conclusion

Different protein-based formats to analyze products for the presence of genetically modified organisms are available. Due to the diversity of the immunoassays and their sensitivity, specificity, user-friendliness, and speed, they are quite successful. Nevertheless, some characteristics limit their wide application. The protein-based methods can not be used on processed and complex matrices. Highly processed products should be analyzed with the DNA-based polymerase chain reaction (PCR technique). However, immunoassays are very suitable for screening for GMOs at critical control points earlier in the food production chain, on raw and unprocessed materials, to prevent mixing of non-GMO with GMO ingredients. In this context they are widely used to conform with traceability legislation.

As more GMOs are developed and enter the food and feed chain, the development of high-throughput methods will become more and more important. Attention should also be focused on the development of tests capable of detecting several GMOs simultaneously, thereby minimizing expenses and time for analysis. Nowadays, lateral flow tests are already available that can detect up to three different GMOs at the same time. In the future, it is likely that tests will be developed that detect multiple novel proteins using a single lateral flow strip.

References

Ahmed, F.E. 2002. Detection of genetically modified organisms in food. *Trends in Biotechnology* 20(5):215–23.

Anklam, E., and Neumann, D.A. 2002. Method development in relation to regulatory requirements for detection of GMOs in the food chain. *Journal of AOAC International* 85(3):754–56.

Anklam, E., Gadani, F., Heinze, P., Pijnenburg, H., Van Den Eede, G. 2002. Analytical methods for detection and determination of genetically modified organisms in agricultural crops and plant-derived food products. *European Food Research and Technology* 214:3–26.

Bernard, A., Michel, B., Delamarche, E. 2001. Micromosaic immunoassays. *Analytical Chemistry* 73:8–12.

Bindler, G., Dorlhac de Borne, F., Dadani, F., Gregg, E., Guo, Z., Klus, H., Maunders, M. 1998. Report of the CORESTA Task Force, Genetically modified tobacco: detection methods. CORESTA, 1–114.

Brett, G.M., Chambers, S.J., Huang, L., Morgan, M.R.A. 1999. Design and development of immunoassays for detection of proteins. *Food Control* 10:401–6.

Burns, M.A., Mastrangelo, C.H., Sammarco, T.S., Man, F.P., Webster, J.R., Johnson, B.N., Foerster, B., Jones, D., Fields, Y., Kaiser, A.R., Burke, D. 1996. Microfabricated structures for integrated DNA analysis. *Proceedings of the National Academy of Sciences of the United States of America* 93(11):5556–61.

Gachet, E., Martin, G.G., Vigneau, F., Meyer, G. 1999. Detection of genetically modified organisms (GMOs) by PCR: a brief review of methodologies available. *Trends in Food Science and Technology* 9:380–88.

Griffiths, K.D., Partis, L.M., Croan, D.D., Wang, N.M., Emslie, K.R.D. 2002. Review of technologies for detecting genetically modified materials in commodities and food. Research and Development Section, Australian Government Analytical Laboratories, 1–118.

James, C. 2005. Preview: global status of commercialized biotech/GM crops. ISAAA Briefs 32, ISAAA, Ithaca, NY.

Kok, E.J., Aarts, H.J.M., Van Hoef, A.M.A., Kuiper, H.A. 2002. DNA methods: critical review of innovative approaches. *Journal of AOAC International* 85(3):797–800.

Lipton, C.R., Dautlick, J.X., Grothaus, G.D., Hunst, P.L., Magin, K.M., Mihaliak, C.A., Rubio, F.M., Stave, W. 2000. Guidelines for the validation and use of immunoassays for determination of introduced proteins in biotechnology enhanced crops and derived food ingredients. *Food and Agricultural Immunology* 12:153–64.

Lüthy, J. 1999. Detection strategies for food authenticity and genetically modified foods. *Food Control* 10:359–61.

Meyer, R., Chardonnens, F., Hübner, P., Lüthy, J. 1996. Polymerase chain reaction (PCR) in the quality and safety assurance of food of soya in processed meat products. *Zeitschrift für Lebensmittel Untersuchung und Forschung A* 203:339–44.

Miraglia, M. Berdal, K.G., Brera, C., Corbisier, P., Holst-Jensen, A., Kok, E.J., Marvin, H.J.P., Schimmel, H., Rentsch, J., van Rie, J.P.P.F., Zagon, J. 2004. Detection and traceability of genetically modified organisms in the food production chain. *Food and Chemical Toxicology* 42(7):1157–80.

Popping, B. 2003. Identifying genetically modified organisms (GMOs). In *Food Authenticity and Traceability*, edited by Lees, M., pp. 415–25. Cambridge: Woodhead Publishing in Food Science and Technology.

Popping, B., and Broll, H. 2001. Detection of genetically modified foods: past and future. *L'actualité chimique* 11:3–12.

Sambrook, J., and Russel, D. 2000. *Molecular Cloning: A Laboratory Manual*, 3rd ed. Cold Spring Harbor, NY: Cold Spring Harbor Laboratory Press.

Spiegelhalter, F., Lauter, F.R., Russel, J.M. 2001. Detection of genetically modified food products in a commercial laboratory. *Journal of Food Science* 66(5):634–40.

Stave, J.W. 1999. Detection of new or modified proteins in novel foods derived from GMO—future needs. *Food Control* 10:367–74.
Van Duijn, G., van Biert, R., Bleeker-Marcelis, H., van Boeijen, I., Adan, A.J., Jhakrie, S., Hessing, M. 2002. Detection of genetically modified organisms in food by protein- and DNA-based techniques: bridging the methods. *Journal of AOAC International* 85(3):787–91.
Wilson, R.F., Hou, C.T., Hildebrand, D.F. (eds.). 2001. *Dealing with Genetically Modified Crops*. Champaign, IL: AOCS Press, p. 144
Yeung, J. 2006. Enzyme-linked immunosorbent assays (ELISAs) for detecting allergens in foods. In *Detecting Allergens in Food*, edited by Koppelman, S., and Hefle, S., pp. 109–24. Cambridge: Woodhead Publishing in Food Science and Technology.

Web Sites

http://www.agbios.com
http://www.isaaa.org

11 Immunodiagnostic Technology and Its Applications

Didier Levieux

Introduction

Immunoassays are a very powerful tool for the analyst faced with the increasing complexity of food ingredients and processed food. In comparison to other analytical methods such as high-performance liquid chromatography (HPLC), electrophoresis, gas chromatography (GC), or mass spectrometry, immunoassays can provide highly sensitive and specific analyses that are rapid, economical, and relatively simple to perform. Immunoassays can be defined as a binding reaction between an antibody and a target antigen (analyte to be detected or quantified). The qualities of immunoassays rely first on the antibody quality in terms of specificity and affinity and, second, on the format used to visualize the antigen-antibody reaction. A wide variety of immunoassay formats are discussed in this chapter: their general principles are first described, and then a critical appraisal of their applications in food analysis is given.

Principles of Immunoassays

Immunoassays utilize the exquisite specificity and sensitivity of the antigen-antibody reaction. An antigen is a molecule that is recognized by the immune system and particularly by antibodies. Proteins, polysaccharides, pesticides, antibiotics, toxins, and hormones are all antigens. An immunogen is an antigen that can elicit an immune response, and particularly antibody synthesis, upon injection. Proteins with molecular weights higher than 5–10 kDa are immunogens, while small peptides, pesticides, antibiotics, or hormones are not. These low-molecular-weight substances are called *haptens* and must be chemically coupled to larger carrier molecules such as bovine serum albumin or keyhole limpet hemocyanin in order to elicit specific antibody formation.

Antibodies of the IgG class are Y-shaped large proteins (160 kDa) bearing two combining sites (*paratopes*) that specifically bind to a limited surface (*epitope*) of the antigen (fig. 11.1). Epitopes are mostly located on the hydrophilic part of the antigen, and their size is 10–15 amino acid residues that can be contiguous on the protein chain (*linear epitope*) or spatially close while discontinuous on the protein chain (*conformational epitopes*). Most of the antibodies raised in animals against native proteins are directed against conformational epitopes, which are lost upon chemical or thermal denaturation of the proteins.

The antigen-antibody (Ag-Ab) reaction obeys the Law of Mass Action: Ag + Ab $\rightarrow$ AgAb. The higher the antibody affinity, the greater the amount of Ag-Ab complex formed. Therefore, the antibody affinity affects the detection limit, the sensitivity, and the incubation time of the assay. Moreover, it must be remembered that simultaneous binding of

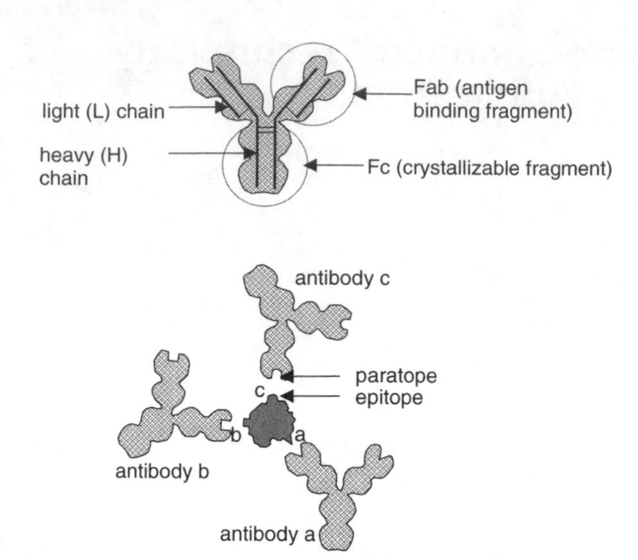

Fig. 11.1. Schematic representation (*top*) of an antibody with its two heavy and light chains linked by disulfide bonds and (*bottom*) of a protein antigen (MW around 50 kDa) bearing three epitopes (or antigenic sites), a, b, and c, surrounded by three specific antibodies (MW 160 kDa) bearing the corresponding paratopes (or antibody sites).

the two paratopes of the same antibody molecule on the antigen increases the affinity by a factor of 100 to 1000. This may occur with polymeric molecules, proteins oligomerized by mild heating or chemical crosslinking, and proteins coated at high concentration onto solid surfaces.

Antibodies are generally raised by monthly repeated intradermal or subcutaneous injections of immunogen at 20–50 μg/kg. Adjuvants are used to increase antibody production. The most popular adjuvants are the ones developed by Freund (1947), which are emulsions of the antigen-water solution in mineral oil. Complete Freund's adjuvant contains killed *Mycobacterium tuberculosis* and is mainly used for primary immunization. Freund's incomplete adjuvant, which is without bacteria, is employed for subsequent boosting. Depending on the antibody quantity required, polyclonal antisera are produced from rabbits, chickens, goats, sheep, or horses. To induce a good immune response, one should avoid the use of antigens or carriers too closely related to the host's constituents.

Serum of immunized animals contains a mixture of antibodies synthesized by the many lymphocytes that have been activated by the different epitopes of the immunogen and have divided as clones. These antibodies are thus called *polyclonal antibodies*. Antibodies obtained from a single lymphocyte clone are called *monoclonal antibodies*. Such antibodies can be exquisitely specific since they react with only one of the antigen epitopes. Monoclonal antibodies are mostly produced by fusing with polyethylene glycol (PEG), the splenic antibody-secreting cells of an immunized mouse or rat, with a long-lived cancerous lymphocyte called *myeloma*. The obtained suspension of hybrid cells, or "hybridomas," is then highly diluted and plated into 96-well culture plates in order to statistically obtain one cell per well. The culture supernatant of hybridomas growing as a single clone per well is tested by ELISA techniques (see later) for specificity and affinity of the secreted antibodies. The selected clones are amplified by culture in wells and vials of increasing volume. Antibody production at the 10–100 μg/ml range is obtained by culture of hybridomas in flasks or tanks, and at the 5–10 mg/ml range in ascites after intraperitoneal injection of mice or rats.

Alternatively, antibody engineering has allowed for the construction of large antibodies libraries that can be used for the in vitro selection of many different molecules. In recent years, phage display technology has also become increasingly powerful and has transformed the way antibodies are produced for a given target. To produce a recombinant antibody phage display library, RNA is extracted from the spleen of naïve (Moghaddam et al. 2001) or immunized mice and reverse transcripted to produce cDNA. The cDNA genes coding for the region of the H and L chain involved in the antibody site (VH and VL) are amplified by polymerase chain reaction (PCR), purified, and assembled using a 20-amino-acid linker (Krebber et al. 1997). The products are purified, digested by a restriction enzyme, and ligated in a vector. This vector is transformed into *Escherichia coli* by electroporation and results in an antibody phage display library used to produce antigen-specific fragments of antibodies.

Purification of the antibodies is generally required for their labeling or coating onto solid phases. IgG fraction of antisera contains the specific antibodies raised by immunization mixed with low levels of natural antibodies directed to a variety of antigens previously encountered by the host. IgG fraction is usually obtained by ammonium sulfate fractionation, ion exchange chromatography, or affinity chromatography on immobilized protein A. In order to obtain more specific reactives, the specific antibodies could be purified by affinity chromatography on the relevant antigen immobilized on chromatographic beads such as cyanogen-bromide-activated Sepharose 4B (Amersham). However, desorption of antibodies requires stringent conditions such as glycine–HCl 0.1 M pH 2.3–2.5, which are deleterious for antibodies. Moreover very high affinity antibodies cannot be generally eluted with these buffers because they are the most active in immunodiagnostic tests.

Antigens or antibodies are generally labeled when very low detection limits are necessary. The first labels used were radioisotopes ^{131}I or ^{125}I for proteins and polypeptide hormones, and ^{3}H or ^{14}C for steroids and drugs. These labels are now progressively replaced by enzyme labels such as peroxidase, alkaline phosphatase, and β-galactosidase. Peroxidase, for instance, is one of the most sensitive and least expensive enzymes and has many properties required for sensitive enzyme immunoassays. However, the majority of the chromogens previously used in the peroxidase determination were mutagenic or carcinogenic. This is not the case for tetramethylbenzidin (TMB), which is now commonly used. Peroxidase labeling by periodate oxidation (Tijssen and Kurstak 1984) is widely used and reported to provide conjugates of useful quality. The enzyme-labeled reagent is then purified by gel filtration to remove unlabeled or too highly labeled analyte. Alternative labels include fluorescent, chemiluminescent, and bioluminescent probes; latex particles for use in plate agglutination tests; and colloidal gold for lateral flow immunoassay or electronic microscopy.

Immunoassays are frequently amplified by use of a secondary labeled antibody directed against the antigen-specific antibody. This secondary antibody is obtained by immunization of nonphylogenetically related animals with IgG purified from animals of the species in which the specific antibodies have been raised. If the primary specific antibodies have been obtained from rabbits, sheep or goats are used for producing the secondary antibody. For mouse monoclonal antibodies, secondary antibodies are mostly produced in rabbits.

A variety of methods have been developed for visualizing the primary Ag-Ab reaction. Precipitation of large cross-linked Ag-Ab complexes can be directly visible to the naked eye, without the use of labeled reagents for amplification. In liquid media, the quantity of Ag-Ab complexes is quantified by turbidimetry using a colorimeter or, more sensitively, by using a nephelometer, which measures the light diffracted by the Ag-Ab complexes. Using permanent stirring and a buffer containing polyethylene glycol (PEG), proteins can be specifically quantified down to the μg/ml level using rate nephelometers in about 1 min. Precipitation of Ag-Ab complexes requires the availability of a minimum of two to three epitopes on the antigen molecule, which are cross-linked by antibodies directed against these different epitopes.

Precipitation of Ag-Ab complexes in agar gel is more frequently used than precipitation in liquid media. In the double-diffusion technique of Ouchterlöny, the antigen and the antibody are allowed to diffuse in the agar gel from two separate wells. A straight line of precipitate appears between the two wells after one night's diffusion. The density of the precipitate allows a qualitative estimation of the antigen or antibody concentration.

Mancini et al. (1965) demonstrated that quantitative results are obtained by allowing antigen solutions to diffuse from the wells of an antibody-containing agar gel. The surface of the ring-shaped precipitates is directly proportional to the antigen concentration (fig. 11.2). This single radial immunodiffusion technique (SRID) has gained wide acceptance for the routine quantification of proteins down to the μg/ml level in food such as meat or dairy products (Levieux et al. 1995, 2002, 2005; Levieux and Ollier 1999) or for the control of the animal origin of raw materials in the pharmaceutical industry (Levieux and Levieux 2001).

Advances in SRID analysis have been made by shortening the total analysis time to 4–15 h (fig. 11.3) by diffusion at 37°C and direct measurement of the precipitate from an image obtained with a numeric camera or a scanner. The washing, drying, and dying steps that required 1 to 2 days are thus unnecessary. Moreover, to fulfill quality assurance requirements, cheap software has been recently marketed (IDBiotech) for the measurement and recording of the ring diameters displayed on the computer screen from the image of the

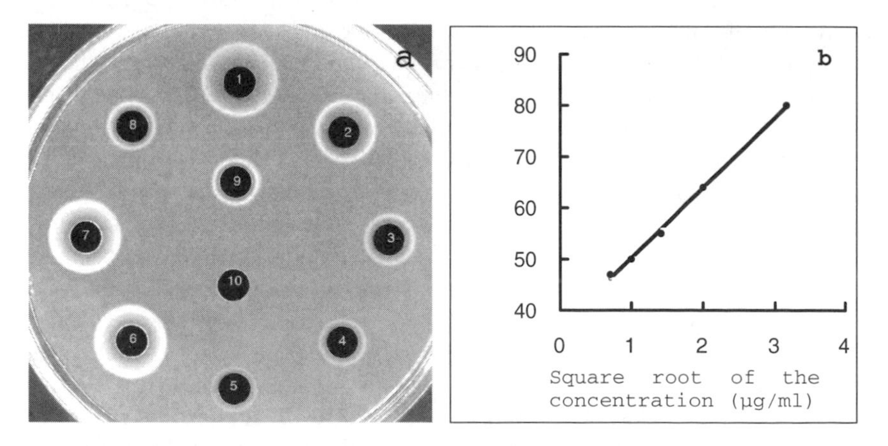

Fig. 11.2. SRID quantification of bovine IgG for the detection of cow's milk in goat's milk. **(a)** 1–5: IgG standards equivalent to 10, 4, 2, and 0.5 percent cow's milk in goat's milk; 6–9: adulterated goat's milk samples; 10: pure goat's milk. **(b)** Standard curve obtained with diameters 1–5 in **a**. (14:12SRID plate was kindly provided by IDBiotech.)

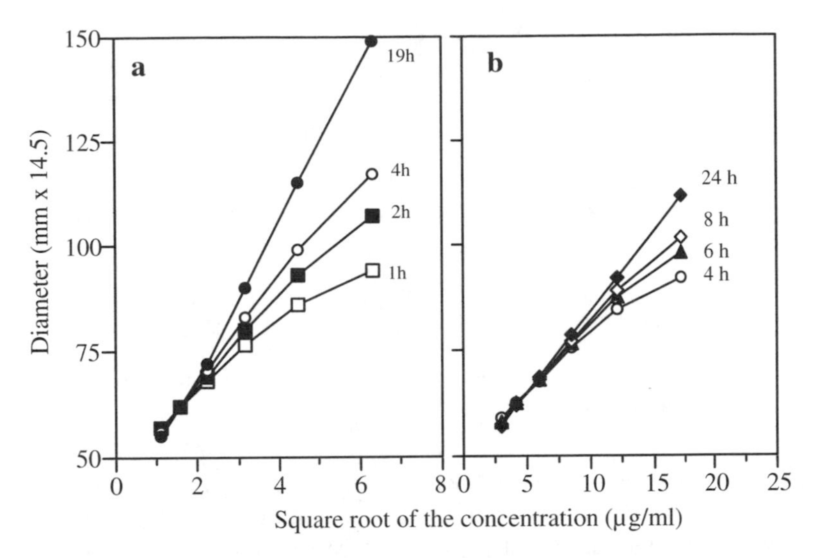

Fig. 11.3. Effect of the diffusion time on the ring diameters obtained for the α-lactalbumin **(a)** and IgG **(b)** quantitation on SRID plates (D. Levieux, unpublished results).

plate. The software automatically calculates the protein concentration in the samples from the regression line obtained for the standards. The technique has been fully automatized for high-sample throughput by using a laboratory robot for dispensing standards and samples in the agar plates and image analysis for automatically measuring the ring diameters from a scan of the SRID plate.

In the rocket-immunoelectrophoresis technique, an electric field is applied to the agar gel containing the antibodies. The negatively charged proteins migrate from the wells toward the anode and are precipitated by antibodies. The length of the rocket-like precipitates is directly proportional to the protein concentration. This technique has the same sensitivity as SRID but is more antibody consuming and must be used for very large and slow-diffusing antigens such as lipoproteins.

If Ag or Ab is labeled by particles such as colored latex, a visible agglutination can be observed within a few seconds or minutes after mixing of the reactants. Rapid card tests are routinely used in the medical field, and applications have been made to food analysis.

For the quantification of haptens and proteins down to the ng/ml level, the antigen or the antibody must be labeled with sensitive tags. The most popular method in use today is the enzyme-linked immunosorbent assay (ELISA), which refers to the fact that one of the binding elements is enzyme labeled and another is attached to a solid support (test tubes, microtitration plates, nitrocellulose or nylon membranes, paper disks, magnetic particles, etc.) for an easy separation of unbound from bound reactants by washing between the addition of reactants. The 96-well polystyrene microplate is particularly well suited for high throughput due to the wide variety of equipment that allows automation of reagents dispensing, washing, reading, and interpretation of results.

Numerous ELISA formats have been devised for antigen quantification, depending on which of the reactants is immobilized or labeled and on the use or not of a secondary labeled antibody (fig. 11.4). Immobilized antibodies can be used to quantify haptens and proteins by direct competition between these antigens and a labeled antigen. For

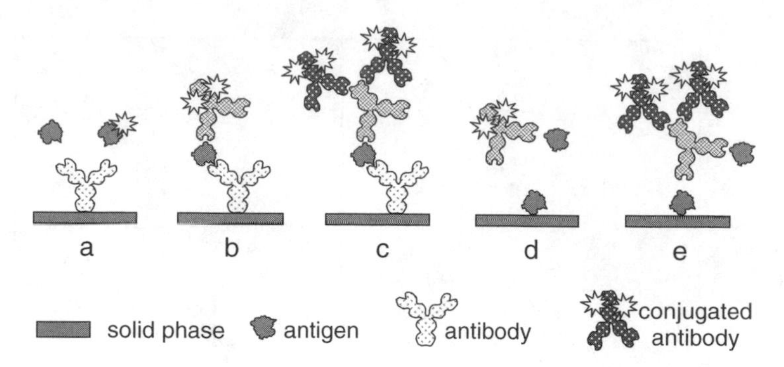

Fig. 11.4. Main formats of ELISA techniques: **(a)** antigen competitive, **(b)** sandwich direct, **(c)** sandwich indirect, **(d)** antibody competitive direct, **(e)** antibody competitive indirect.

multiepitopic proteins, the sandwich direct assay combines an immobilized capture antibody and a second enzyme-labeled antibody both raised to the target protein. The highest specificity and sensitivity are usually obtained when a monoclonal antibody is used as a capture antibody and a polyclonal labeled antibody is used as a tracer antibody. To perform a sandwich ELISA, samples are added to the plastic wells coated with the specific antibody; after a short incubation period, wells are washed, and the target protein remains bound to the solid phase. The second enzyme-linked antibody is added, which also binds to the target protein forming the third layer of the "sandwich." Wells are again washed, and the substrate solution specific to the enzyme is added. The color change of the substrate is directly proportional to the amount of enzyme and thus to the amount of captured protein.

In the sandwich indirect ELISA format the capture and second primary antibodies must have been raised in two phylogenetically distant hosts in order that the labeled secondary antibody does not cross-react with the capture antibody. This difficulty can be overcome by labeling the noncoated primary antibody with biotin and using enzyme-labeled streptavidin as a tracer of high affinity for biotin ($10^{15} M^{-1}$). If the antigen can be easily coated onto the microplate, competitive direct and competitive indirect ELISAs are performed. The indirect format provides lower detection limits and avoids purification and labeling of IgG from each primary antibody. The detection limit can also be lowered by preincubation of antigen with antibody and then adding the mixture to the antigen-coated plate.

Orbital shakers with controlled temperature, such as Eppendorf Thermomixer Comfort, allow a 50 percent reduction of the incubation time while keeping the same detection limit. The use of magnetic beads as a solid phase greatly increases the coated surface and thus allows performance of tests with detection limits after 10 min incubation steps equivalent to those limits obtained with 60 min incubation steps in the classical microplate ELISAs.

Rapid on-site ELISAs have been developed for a qualitative estimation of the presence or absence of the target analyte in a limited number of samples. In the dip stick format, the binding element is immobilized to the end of a plastic stick or a rigidified nitrocellulose or nylon membrane. The stick or the membrane is successively transferred from the sample to the conjugate and into the substrate. The rinsing steps between transfers can be done with rinse solutions provided with the kit or under tap water. More recently, the lateral flow immunoassay that has become popular for home pregnancy testing has been

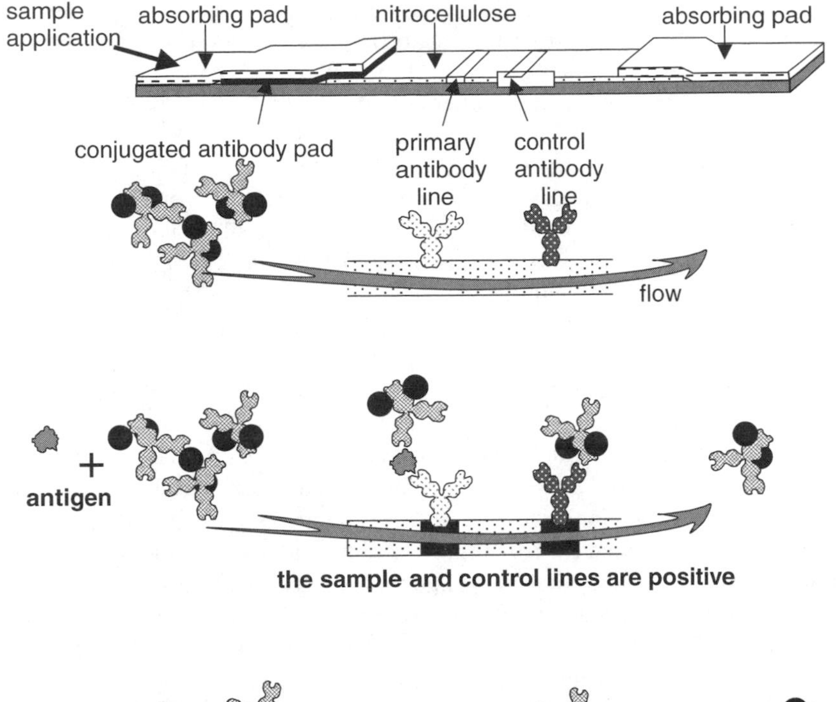

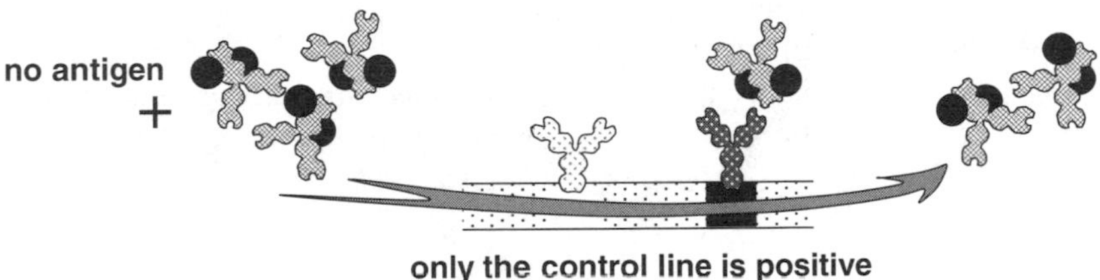

Fig. 11.5. Lateral flow immunoassay in the configuration of a sandwich-type ELISA. The sample applied on the first absorbing pad migrates through the conjugated antibody pad where it dissolves the dried colloidal-gold-labeled antibody. If present in the sample, the target antigen reacts with the conjugated antibody during diffusion. The Ag-Ab complexes migrate through the nitrocellulose membrane where they are captured by the immobilized primary antibody. The excess conjugated antibody is then captured by the control secondary antibody (antibody directed against the conjugated antibody). Thus two red lines are clearly visible with the naked eye if the target antigen is present in the sample. In the absence of antigen, only the control line is positive.

extended to food analysis. In this format the assay is reduced to the single step of sample addition since the necessary reagents are dried onto different areas of the solid phase. As the sample moves through the solid phase, it solubilizes and carries the appropriate reagents (fig. 11.5). In order to eliminate the need of the substrate step, the tags generally used are colloidal gold or blue-dyed latex particles.

Advances in Ag-Ab reaction monitoring have been obtained with the development of devices (biosensors) that respond to the presence of a specific analyte by producing an electrical signal proportional to the analyte concentration. They are usually constructed from three components: receptor, transducer, and electronic for amplification and display.

While many papers have been published on the use of different biosensor technologies, commercial development has been limited to those based on the surface plasmon resonance (SPR) analysis. Equipment such as that marketed by Biacore allows real-time quantification of the Ag-Ab reaction. It is particularly suited for the characterization of antibody specificity and affinity and for studies on the epitopic structure of antigens. Applications in food science have been published for the characterization of heated milk (Jeanson et al. 1999), for structural studies of proteins in emulsions (Vénien et al. 2000), and for the quantification of mycotoxins (Mullett et al. 1998; Daly et al. 2002) and pesticides (Alcocer et al. 2000). However, biosensors such as those manufactured by Biacore are expensive and have not come into routine use.

Microarray immunoassays for simultaneous detection of proteins, bacteria, or pesticides have been described and are under intensive investigation (reviewed by Marquette and Blum 2005). However, the methods are not yet commercially available, and published results are mostly qualitative.

Immunoaffinity chromatography has been extensively used for the purification of antigens or antibodies. In the last years the technique has been introduced in commercial kits for immunoconcentration of toxins. The target antigen, present in the sample extract injected through the column, is captured by specific antibodies immobilized on chromatographic beads (generally cyanogen-bromide-activated Sepharose 4B). After rinsing the column, the antigen is eluted in a highly concentrated form by using an Ag-Ab dissociating buffer and can be characterized by HPLC. Similar methods have been developed for rapid immunoconcentration of bacteria before plating on growth media.

The type of format that best fits an application is determined by the detection limit required, the need to be or not be quantitative, the sample throughput and turnaround time requirements, the final cost of the assay, and who performs the assay. Typical examples are presented in the following sections on the applications of immunodiagnostics in food analysis.

Immunoassays for Monitoring the Protein Quality in Raw Materials and in Processed Food

In raw materials the concentration of proteins of interest in product quality is generally higher than the μg/g level. Thus the SRID technique is particularly suited for their quantification since, due to its great simplicity, it can be performed in plants devoid of laboratory equipment as well as in control laboratories. Numerous applications have been found in the milk industry. The level of payment to the producer for raw milk depends on criteria such as the cell count and lipid and protein contents. High protein levels are attained during early lactation mostly due to an increase of IgG, a whey protein highly concentrated in the colostrum of ruminants (Levieux and Ollier 1999; Levieux et al. 2002). However, the presence of high IgG levels in milk, as observed during winter for goat's milk, creates several problems, such as reduced heat stability, low yield of cheese production, weak curd formation, and poor curd characteristics. As a consequence, IgG concentration in goat's milk is today routinely controlled in France by SRID analysis. SRID is also a reference method for the quantification of IgG in bovine colostrum or derived products with high IgG contents (Li-Chan et al. 1994; Quigley et al. 1994, 1995; Tyler et al. 1999; Mainer et al. 2000) and for the quantification of other native whey proteins such as β-lactoglobulin (β-Lg), α-lactalbumin, serum albumin, and lactoferrin. Kits are now commercially available for each of these proteins (IDBiotech).

Heat denaturation of individual whey proteins has been easily quantified by SRID to study their heat sensitivity or to define a posteriori the time/temperature applied to the product (Levieux 1980b; Sanchez et al. 1992; Delplace et al. 1997; Mainer et al. 1997; Levieux et al. 2005). Oligomers formed at the beginning of the proteins' denaturation (Laligant et al. 1991, 1995) are not quantified in SRID since they cannot diffuse in the agar layer or they diffuse very slowly. In contrast, these oligomers are misinterpreted in gel permeation chromatography or electrophoresis and highly overestimated when ELISA techniques are used (Levieux 1980b; Heppell 1985; Rumbo et al. 1996). However, Jeanson et al. (1999) have proposed two ELISAs based on monoclonal antibodies specific to the native or the denatured form of α-lactalbumin for the classification of milk according to the heat treatment it has been submitted to. Combining the two ELISAs avoids the need to know the α-lactalbumin concentration of the original raw milk.

The control of residual native milk β-Lg in milk formulated with tryptic hydrolysates and used for allergic children is usually performed in plants with commercially available SRID plates when the required detection limit is 0.5–1 percent residual native protein. For lower detection limits ELISA techniques are used.

Changes in the tertiary structure of β-Lg upon lactosylation or emulsification can be studied with immunochemical techniques using monoclonal antibodies in ELISAs (Morgan et al. 1999) or with surface plasmon resonance analysis (Vénien et al. 2000).

Immunoassays for analyses of meat proteins follow the same general steps described for milk proteins. The quantification of selected soluble proteins in meat extracts and the study of their heat denaturation have been described using SRID analysis (Levieux et al. 1995). In contrast, fundamental studies on the epitopic changes in myoglobin upon heating have required ELISA techniques using monoclonal antibodies (Levieux and Levieux 1996a, 1996b).

The solubility of proteins extracted from cereals is generally low, and the SRID technique cannot be easily used for their quantification. Thus ELISA techniques are mostly used for the control of wheat quality and for the detection of residues in cereals (reviewed by Skerritt et al. 1991; Howes 1995; Laurière and Dénery 2005). A lateral-flow-based immunoassay has been marketed (WheatRite) for the on-site estimation of basic α-amylase and is most routinely used in the United States and Australia.

Immunoassays for the Detection of Food Adulteration

Food products are prone to accidental and deliberate abuse that may happen at any stage from the raw material to the final, processed product. Accidental contamination may occur particularly when different kinds of food are manufactured with the same equipment. Moreover, the producer/manufacturer can deliberately add material to the product, usually to reduce costs.

A model example of the evolution of the panel of analytical tools is the control of the adulteration of goat's and ewe's milk with cheaper cow's milk. This adulteration is one of the main problems that affect goat's and ewe's cheese manufacture, and for many years a lot of methods have been proposed for the control of the milk's authenticity (recently reviewed for the last decade by Borkova and Snaselova 2005). Methods based on the analysis of constituents such as caprylic acid, β-carotene, or volatile fractions have been replaced by physical analysis of the proteic fractions by chromatographic or electrophoretic analysis of caseins or whey proteins. However, such methods cannot be used for the routine control of milk in plants, and in our opinion, they are not specific enough for the analysis of highly proteolysed cheese: in view of the large variety of cheese fabrications, maturation lengths, or ratios and concentrations of proteolytic enzymes in milk, the presence of a faint peak in chromatography (whatever the technique used) or of a faint band in electrophoresis (polyacrylamide gel electrophoresis [PAGE] or isoelectric focusing) cannot be unequivocally attributed to a contaminating cow's protein or peptide. Recently, more specific methods have been based on characterization of bovine β-Lg-specific sequences in goat's or ewe's milk by matrix-assisted laser desorption/ionization mass spectrometry (Cozzolino et al. 2001) or high-performance liquid chromatography coupled with electrospray ionization mass spectrometry (Chen et al. 2004). However, the detection limit obtained was 5 percent cow's milk in goat's or ewe's milk, which is far from the 1 percent level required by regulatory authorities during the last 30 years.

In contrast to these methods, immunoassays can provide highly sensitive and specific analyses that are rapid, economical, and easily performed in dairy plants or control laboratories. The first antisera produced were raised against whole cow's milk or blood and then absorbed in goat's or ewe's milk to eliminate cross-reacting antibodies. These antisera were utilized in a plate precipitation test (Solberg and Hadland 1953), passive hemagglutination (Berger 1959), or double immunodiffusion (Durand et al. 1974) with a detection limit of 2–5 percent cow's milk in goat's or ewe's milk. More specifically, antisera were obtained by immunizing goats or sheep with purified cow's IgG. They allowed the development of an SRID test with 1 percent detection limit and of a rapid (1 h) hemagglutination test with 0.1 percent detection limit (Levieux 1977, 1980a). Antisera produced in rabbits against caseins or whey proteins and absorbed in homologous goat's or ewe's proteins were then used in more sophisticated agar gel precipitation techniques such as the rocket or the crossed immunoelectrophoresis techniques (Radford et al. 1981; Elbertzhagen and Wenzel 1982) with a 1 percent detection limit. With the development of ELISA techniques, antisera against whey proteins (Garcia et al. 1990), caseins (Rodriguez et al. 1991), or more specific peptides of caseins (Bitri et al. 1993) allowed detection limits of 0.1 to 1 percent cow's milk in goat's or ewe's milk. The highest specificity and sensitivity were obtained with a sandwich ELISA based on a monoclonal capture antibody directed against a bovine-specific epitope of β-lactoglobulin (Vénien et al. 1997) and a peroxidase-labeled polyclonal antiserum raised in rabbits against the bovine β-lactoglobulin. The detection limit was 1 part cow's milk per 100,000 parts goat's or ewe's milk (Levieux and Vénien 1993).

However, ELISA techniques are not really suitable for routine control in dairy plants since they are relatively time-consuming for a low number of samples. For routine control of milk delivery in plants, SRID is today the best choice since only a few minutes are required to put a few microliters of the milk or whey samples in the wells of commercially available agar plates (IDBiotech) and to read and record the results after one night's diffusion. For an immediate control at the tank arrival, the latex agglutination test is best suited since the detection of 1 percent cow's milk in goat's or ewe's can be obtained after mixing one drop of milk with one drop of reagent for 1 min or 0.5 percent with 2 min mixing (fig. 11.6).

Adulteration of expensive meat with cheaper meat is reported as being a common phenomenon in many countries. In addition to the legal requirements, certain religious

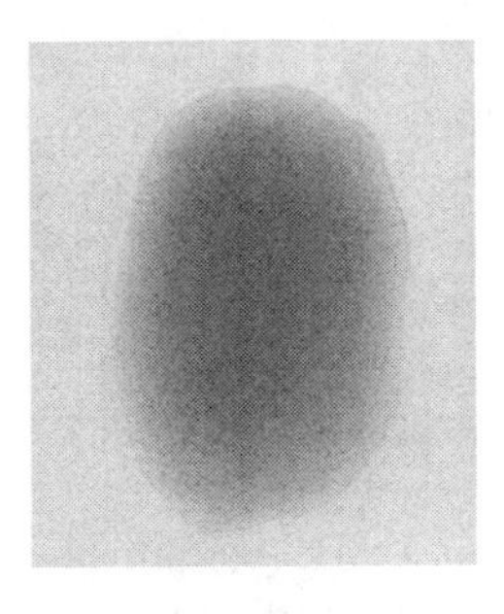

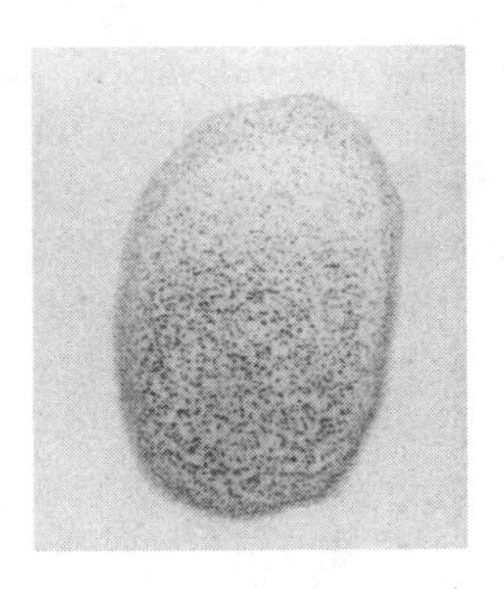

Fig. 11.6. One-minute detection of cow's milk in goat's milk by mixing one drop of milk and one drop of dyed latex microspheres coated with a monoclonal antibody bovine specific ("Cow milk Xpress" from IDBiotech).

groups observe prohibition from eating certain meats such as pork. As for milk adulteration control, several methods based on chromatographic (GC and HPLC), chemical, and electrophoretic techniques have been developed. These, however, lack rapidity and ability to handle large numbers of samples and, therefore, have not come into routine use (Smith 1991). Double immunodiffusion and SRID are frequently used with antisera against whole meat proteins; however, it has been reported that a number of commercial antibodies cross-react with other species, thus indicating the need to screen antibodies before their use (Pickering and Bazeley 1991).

Many different ELISA systems have been described for raw meat speciation. Some coat the solid phase with tissue extracts prepared from the sample to be analyzed (Kang'ethe et al. 1982; Whittaker et al. 1983). Other tests use antigen- or antibody-coated plates, such as commercially available kits manufactured by Cortecs that can be used for the identification of food proteins in raw and cooked products. However, none of them are really able to detect meat proteins in bone and meat meal obtained at a temperature >130°C for 30 min.

Detection of gluten in food is important because a significant number of individuals suffering from celiac disease cannot tolerate wheat, rye, and barley in their diets. ELISAs for laboratory use and lateral flow devices for home tests are commercially available from Diffchamb, Ingenasa, Medical Innovation, R-Biopharm, and Tepnel (reviewed by Laurière and Dénery 2005).

Immunoassays for the Detection of Pathogens and Their Toxins

Immunoassays have been developed for the detection of pathogenic bacteria such as *Salmonella*, *Listeria*, and *E. coli* O157. Most of the commercially available tests are ELISAs based on the microplate format (Biocontrol, Diffchamb, MAST, R-Biopharm, Tecra). Rapid tests based on latex agglutination (Oxoid), lateral flow immunoassays (Biocontrol, Diffchamb, Merck/VWR, Oxoid, SDI), or dip stick immunoassays (Tecra) have been recently marketed. Advances have been obtained by the use of antibody-coated magnetic particles that concentrate the bacteria on the bottom of tubes (Dynal Bead Retriever) or by using a patented recirculating device (Matrix) prior to the inoculation on agar. In addition, automation has been developed such as the VIDAS apparatus (BioMerieux), which uses a sensitive enzyme-linked fluorescent assay (ELFA). The technique uses a patented device with antibody-coated tips and strips of successive wells containing all the necessary reactives (wash buffer, conjugated antibody, substrate). The antigen-antibody and enzymatic reactions occur directly in the tips after successive aspiration of the sample and reactives.

Improvements in the specificity of immunochemical detection of *L. monocytogenes* have been obtained by immunization of rabbits with selected sequences of the P60 protein (Buber et al. 1994) or by use of conjugates obtained from the Fab of the antibodies instead of the full molecule in the VIDAS system.

Staphylococcal enterotoxins are today routinely and sensitively detected using ELISAs (Diffchamb, R-Biopharm, Tecra, or the VIDAS system from BioMerieux). The lower detection limit (LOD) for the Diffchamb kits (Transia Plate) is ≤1 ng/ml whatever the food analyzed (milk, cheese, meat, eggs, fish, pastries). The LOD can be decreased to ≤0.1 ng/ml and the specificity increased by immunoconcentration of the toxins on affinity chromatography columns prior to the ELISA (Lapeyre et al. 2001). Such columns are now commercially available (Diffchamb) for the staphylococcal enterotoxins A, B, C, D, E, and F.

Immunoassays for the Detection of Mycotoxins, Pesticides, and Drug Residues

Due to the widespread occurrence of mycotoxin-producing fungi in cereals, major efforts have been made to develop rapid and sensitive methods for their detection. Numerous ELISAs have been developed, and some of them are commercially available for aflatoxins, ochratoxins, deoxynivalenol, fumonisins, and zearalenones as antigen- or antibody-coated microplates (Beacon, Diagnostix, Diffchamb, Neogen, R-Biopharm, Strategic Diagnostics, Tepnel), lateral flow immunoassays (Envirologix, Neogen, R-Biopharm), or immunoaffinity columns (R-Biopharm, Romer Labs Diagnostic, Vicam). An elegant and sophisticated approach has been developed by Daly et al. (2002) for the quantification of aflatoxin B1. The authors generated single-chain fragment variable (scFv) antibodies from an established phage display system that incorporated a range of different plasmids for efficient scFv expression. The scFv antibodies were used for the development of surface plasmon resonance (SPR)-based inhibition immunoassays. SPR-based assays have also been developed for other mycotoxins, such as fumonisin B1 (Mullett et al. 1998).

The analysis of pesticides and drug residues in food by immunoassay has been developed within the past 10 years to the point that it will play a key role in any chemical food safety program at either the industrial or governmental level (Bushway and Fan 1995). The formats used for commercial immunoassay kits consist of polystyrene tubes, microtiter plates, or magnetic particles most often coated with the antigen or the antibody. Tests have been developed for the detection of the major insecticide groups (pyrethroids, carbamates, organophosphates, and cyclodienes) in fruits and vegetables, fruit juices, meat, fish, milk, honey, and grains and their products (reviewed by Bushway and Fan 1995). There is extensive interest in herbicides such as atrazine and alachlor because of their wide use and the fact that they have been shown to pollute water. Atrazine antibodies demonstrate cross-reactivity within the triazine class (simazine and propazine), making these antibodies very good for triazine screening. The lowest detection limit of atrazine by enzyme immunoassay has been reported at 2 ppb for solid foods and 0.5 ppb for liquid foods (Bushway et al. 1989).

The highest interest in food residues has been in the fungicides because of their frequent use at high levels. Immunoassays have been developed for the major fungicides that belong to the following classes: carboximides (procymidone, iprodione); benzimidazoles (benomyl/carbendazim, thiabendazole); imidazole and triazoles (triadimefon); carboxamides (metalaxyl); perhaloalkylmercapto-imides (captan); and aromatic carboxylic derivatives (chlorothalonil). Types of food analyzed range from alcohol products to fruits and vegetables and their processed products (reviewed by Bushway and Fan 1995).

Immunoassays for the detection of antihelminthics have been published, such as the detection of fenbendazole in bovine liver and milk by an antigen competitive ELISA based on a monoclonal antibody (Brandon et al. 2002).

Fast and inexpensive on-site tests are required to routinely detect antibiotics in dairy and/or meat products. Immunoassays are just such tests. Numerous techniques have been developed (reviewed by Bushway and Fan 1995), and several commercial kits with a microplate or card test format are on the market for sulfamethazine, chloramphenicol, streptomycin, β-lactams (amoxicillin, ampicillin, cephapirin, cloxacillin, oxacillin, penicillin), cephalosporins (cephalexin), and tetracyclines (chlortetracycline, oxytetracycline, tetracycline).

A striking demonstration of the suitability of enzyme immunoassays for high-sample-throughput detection of residues has been done by Ram et al. (1991). The authors developed a technique in which they were able to screen as many as 2400 swine plasmas for sulfamethazine in 8 h.

Immunoassays for the Detection of Allergens

Most allergens in food are proteins, and their quantification can be performed using the methods we have described for the control of protein quality and for the detection of food adulteration. These methods have been reviewed by Poms et al. (2004). However, one must remember that two antibodies responsible for anaphylactic reactions (mostly the IgE isotypes on the mastocyte membrane) must react together with the same antigen molecule in order to provoke the mastocyte degranulation with the release of histamine and other harmful mediators. Thus the detection of an epitope specific to a given protein potentially allergenic cannot be directly interpreted as an allergic risk. As an example, reconstituted milk for allergic children uses trypsic hydrolysates of β-lactoglobulin. In competitive ELISAs, some of these peptides are detected with polyclonal or monoclonal antibodies raised against native β-lactoglobulin, while being nonallergenic. Thus, the sandwich-type ELISA is preferred to the competitive format, while the latter can be used to detect the unwanted presence of an antigen indicative of cross-contamination.

Immunoassays for the Detection of Genetically Modified Organisms (GMOs)

The most frequently used method for the detection of GMOs in plants and food products is the amplification of GMO-specific DNA by polymerase chain reaction (PCR), followed by the identification of the amplification products by agarose gel electrophoresis, restriction fragment length analysis, Southern blot hybridization, or DNA sequencing. While highly sensitive, these PCR-based methods require expensive equipment and experienced investigators. Furthermore, PCR-based methods are time-consuming and difficult to quantify. Alternatively, immunoassays can be used for the detection and quantification of GMOs based on the determination of the protein product of the newly introduced gene. A number of transgenic insect-resistant maize lines, commonly referred to as *Bt-maize*, have been approved and commercialized for food and feed production in numerous countries but are not authorized in others. These maize lines express the gene for *Bacillus thuringiensis* toxins, whose toxicity is restricted to insects. ELISAs, using polyclonal or monoclonal antibodies, have been developed for the determination of these toxins in soil, commercial *B. thuringiensis* formulations, and transgenic cotton using polyclonal or monoclonal antibodies (reviewed by Walschus et al. 2002).

Conclusion

The applications of immunodiagnostic technology to food analysis have greatly expanded in the last decade, providing the analyst with powerful analytical tools. A wide variety of immunoassay formats and reagent configurations have been devised and are now commercially available. Thus, the major problem facing the analyst today is to choose the right format from single qualitative on-site tests to quantitative high-throughput laboratory tests. In addition, the analyst must always consider the possible interference of food components in the antigen-antibody reaction, particularly in ELISAs. That is why results from immunoassays for complex or processed foods should be supported by data from complementary techniques.

References

Alcocer M.M.J., Dillon P.P., Manning B.M., Doyen C., Lee H.A., Daly S.J., O'Kennedy R., Morgan M.R. 2000. Use of phosphonic acid as a generic hapten in the production of broad specificity anti-organophosphate pesticide antibody. *Journal of Agricultural Food Chemistry* 48:2228–33.

Berger V.E. 1959. Das haemagglutinationsverfahren als nachweismethode verfälschter frauenmilch. *Annals of Paediatrics* 193:365–70.

Bitri L., Rolland M.P., Besancon P. 1993. Immunological detection of bovine caseinomacropeptide in ovine and caprine dairy products. *Milchwissenschaft* 48:367–71.

Borkova M., Snaselova J. 2005. Possibilities of different animal milk detection in milk and dairy products—a review. *Czech Journal of Food Science* 23:41–50.

Brandon D.L., Bates A.H., Binder R.G., Montague W.C. 2002. Monoclonal antibody to fenbendazole: utility in residue studies. *Food and Agricultural Immunology* 14:275–83.

Buber A., Schubert P., Köhler S., Franck R., Goebel W. 1994. Synthetic peptides derived from the *Listeria monocytogenes* P60 protein as antigens for the generation of polyclonal antibodies specific for secreted cell-free *L. monocytogenes* P60 protein. *Applied Environmental Microbiology* 60:3120–27.

Bushway R.J., Fan T.S. 1995. Detection of pesticide and drug residues in food by immunoassay. *Food Technology* 52:108–15.

Bushway R.J., Perkins B., Savage S.A., Lekousi S.L., Ferguson B.S. 1989. Determination of atrazine in food by enzyme immunoassay. *Bulletin of Environmental Contamination Toxicology* 42:899–904.

Chen R.K., Chang L.W., Chung Y.Y., Lee M.H., Ling Y.C. 2004. Quantification of cow milk adulteration in goat milk using high-performance liquid chromatography with electrospray ionization mass spectrometry. *Rapid Communication in Mass Spectrometry* 18:1167–71.

Cozzolino R., Passalacqua S., Salemi S., Malvagna P., Spina E., Garozzo D. 2001. Identification of adulteration in milk by matrix-assisted laser desorption/ionization time-of-flight mass spectrometry. *Journal of Mass Spectrometry* 36:1031–37.

Daly S.J., Dillon P.P., Manning B.M., Dunne L., Killard A., O'Kennedy R. 2002. Production and characterization of murine single chain Fv antibody phage display library system. *Food and Agricultural Immunology* 14:255–74.

Delplace F., Leuliet J.C., Levieux D. 1997. A reaction engineering approach to the analysis of fouling by whey proteins of a six channels-per-pass plate heat exchanger. *Journal of Food Engineering* 34:91–108.

Durand M., Meusnier M., Delahaye J., Prunet P. 1974. Détection de l'addition frauduleuse de lait de vache dans les laits de chèvre et de brebis par la méthode d'immunodiffusion en gélose. *Bulletin de l'Académie Vétérinaire* 47:247–58.

Elbertzhagen H., Wenzel E. 1982. Detection of bovine milk in sheep's milk cheese by means of immunoelectrophoresis. *Zeitchrift für Lebensmittel-Untersuchung und Forschung* 175:15–16.

Freund J. 1947. Some aspects of active immunization. *Annual Review of Microbiology* 1:1–12.

Garcia T.B., Martin B., Rodriguez E., Morales P., Hernandez P.E., Sanz B. 1990. Detection of bovine milk in ovine milk by an indirect enzyme-linked immunosorbent assay. *Journal of Dairy Science* 73:1489–93.

Heppell L.M. 1985. "Determination of milk protein denaturation by an enzyme-linked immunosorbent assay." In *Immunoassays in Food Analysis*, edited by Morris B.A. and Clifford M.N., London: Elsevier Applied Science, pp. 115–23.

Howes N.K. 1995. "Antibody probes in cereal breeding for quality and disease resistance." In *New Diagnostics in Crop Science*, edited by Skerrit J.H. and Appels R., Wallingford/Oxon: CAB International, pp. 87–89.

Jeanson S., Dupont D., Grattard N., Rolet-Répécaud O. 1999. Characterization of the heat treatment undergone by milk using two inhibition ELISAs for quantification of native and heat denatured α-lactalbumin. *Journal of Agricultural and Food Chemistry* 47:2249–54.

Kang'ethe E.K., Jones S.J., Patterson R.L.S. 1982. Identification of the species origin of fresh meat using an enzyme linked immunosorbent assay procedure. *Meat Science* 7:229–40.

Krebber A., Bornhauser S., Burmester J., Honegger A., Willuda J., Bosshard H.R., Pluckthun A. 1997. Reliable cloning of functional antibody variable domains from hybridomas and spleen cell repertoires employing a reengineered phage display system. *Journal of Immunological Methods* 20:35–55.

Laligant A., Dumay E., Casas S., Valencia C., Cuq J.L., Cheftel J.C. 1991. Surface hydrophobicity and aggregation of β-lactoglobulin heated near neutral pH. *Journal of Agricultural and Food Chemistry* 39: 2147–55.

Laligant A., Marti J., Cheftel J.C., Dumay E., Cuq J.L. 1995. Detection of conformational modifications of heated β-lactoglobulin by immunochemical methods. *Journal of Agricultural and Food Chemistry* 43: 2896–2903.

Lapeyre C., Maire T., Messio S., Dragacci S. 2001. Enzyme immunoassay of staphylococcal enterotoxins in dairy products with cleanup and concentration by immunoaffinity column. *JAOAC International* 84:1587–92.

Laurière M., Dénery S. 2005. "Céréales et dérivés." In *Méthodes d'analyse immunochimique pour le contrôle de la qualité dans les IAA*, edited by Arbault P. and Daussant J., Paris: TecDoc, Lavoisier, pp. 293–328.

Levieux D. 1977. Une nouvelle technique de détection de l'adultération des laits de chèvre et de brebis. *Les Dossiers de l'Élevage* 2:37–46.

———. 1980a. The development of a rapid and sensitive method based on hemagglutination inhibition for the measurement of cow milk in goat milk. *Annales de Recherche Vétérinaire* 11:151–56.

———. 1980b. Heat denaturation of whey proteins: comparative studies with physical and immunological methods. *Annales de Recherche Vétérinaire* 11:89–97.

Levieux D., Levieux A. 1996a. Immunochemical quantification of myoglobin heat denaturation: comparative studies with monoclonal and polyclonal antibodies. *Food and Agricultural Immunology* 8:111–20.

———. 1996b. Localized conformational changes occurring in myoglobin upon heating as revealed by use of monoclonal antibodies. *Journal of Food Biochemistry* 20:295–309.

———. 2001. Immunochemical control of the species origin of intestinal mucosa used for heparin purification. *Journal of Immunoassay* 22:127–45.

Levieux D., Ollier A. 1999. Bovine Immunoglobulin G, β-lactoglobulin, α-lactalbumin and serum albumin in colostrum and milk during the early post partum period. *Journal of Dairy Research* 66:421–30.

Levieux D., Vénien A. 1993. Rapid, sensitive two-site ELISA for detection of cow's milk in goats' or ewes' milk using monoclonal antibodies. *Journal of Dairy Research* 61:91–99.

Levieux D., Levieux A., Vénien A. 1995. Immunochemical quantitation of heat denaturation of selected sarcoplasmic soluble proteins from bovine meat. *Journal of Food Science* 60:678–84.

Levieux A., Rivera V., Levieux D. 2001. A sensitive ELISA for the detection of bovine crude heparin in porcine heparin. *Journal of Immunoassay and Immunochemistry* 22:323–36.

Levieux D., Morgan F., Geneix N., Masle I., Bouvier F. 2002. Caprine immunoglobulin G, β-lactoglobulin, α-lactalbumin and serum-albumin in colostrum and milk during the early post partum period. *Journal of Dairy Research* 69:391–99.

Levieux D., Levieux A., El hatmi H., Rigaudière J.P. 2005. Immunochemical quantification of camel (*Camelus dromedarius*) whey proteins. *Journal of Dairy Research* 71:1–9.

Li-Chan E., Kummer A., Losso J.N., Nakai S. 1994. Survey of immunoglobulin G content and antibody specificity in cows' milk from British Columbia. *Food and Agricultural Immunolology* 6:443–51.

Li-Chan E., Kummer A., Losso J.N., Kitts D.D., Nakai S. 1995. Stability of bovine immunoglobulins to thermal treatment and processing. *Food Research International* 28:9–16.

Mainer G., Sanchez L., Ena J.M., Calvo M. 1997. Kinetic and thermodynamic parameters for heat denaturation of bovine milk IgG, IgA and IgM. *Journal of Food Science* 62:1034–38.

Mainer G., Perez M.D., Sanchez L., Puyol P., Millan M.A., Ena J.M., Dominguez E., Calvo M. 2000. Concentration of bovine immunoglobulins throughout lactation and effect of sample preparation on their determination. *Milchwissenschaft* 55:613–17.

Mancini G., Carbonara A.O., Heremans J.M. 1965. Immunological quantitation of antigens by single radial immunodiffusion. *Immunochemistry* 2:235–54.

Marquette C., Blum L. 2005. "Immunocateurs et immunopuces." In *Méthodes d'analyse immunochimique pour le contrôle de la qualité dans les IAA*, edited by Arbault P. and Daussant J., Paris: Editions TecDoc, Lavoisier, pp. 377–96.

Moghaddam A., Lobersli I., Gebhardt K., Braunagel M., Marvik O.J. 2001. Selection and characterisation of recombinant single-chain antibodies to the hapten aflatoxin B1 from naïve recombinant antibody libraries. *Journal of Immunological Methods* 254:169–81.

Morgan F., Vénien A., Bouhallab S., Molle D., Léonil J., Peltre G., Levieux D. 1999. Modification of bovine β-lactoglobulin by glycation in a powdered state or in an aqueous solution: immunochemical characterization. *Journal of Agricultural Food Chemistry* 47:4543–48.

Mullett W., Lai E.P., Yeung J.M. 1998. Immunoassays of fumonisins by a surface plasmon resonance biosensor. *Analytical Biochemistry* 258:161–67.

Pickering K., Bazeley J. 1991. "Immunoassays—their role in food analysis." In *Food Safety and Quality Assurance: Application of Immunoassays Systems*, edited by Morgan M.R.A, Smith C.J., and Williams P.A., London: Elsevier Aplied Science, pp. 467–72.

Poms R.E., Klein C.L., Anklam E. 2004. Methods for allergen analysis in food: a review. *Food Additive Contaminants* 21:1–31.

Quigley J.D., Martin K.R., Dowlen L.B., Wallis L.B., Kamar J. 1994. Immunoglobulin concentration, specific gravity, and nitrogen fractions of colostrum from Jersey cattle. *Journal of Dairy Science* 77:264–69.

Quigley J.D., Martin K.R., Dowlen L.B. 1995. Concentrations of trypsin inhibitor and immunoglobulins in colostrum of Jersey cows. *Journal of Dairy Science* 78:1573–77.

Radford D.V., Tchan Y.T., McPhillips J. 1981. Detection of cows' milk in goats' milk by immunoelectrophoresis. *Australian Journal of Dairy Technology* 36:144–46.

Ram B.P., Prithipal S., Lorelei M., Brock T., Sharkov N., Allison D. 1991. High-volume enzyme immunoassay test system for sulfamethazine in swine. *Journal of the Official Association of Analytical Chemistry* 74:43–45.

Rodriguez E., Martin R., Garcia T., Azcona J.I., Sanz B., Hernandez P.E. 1991. Indirect ELISA for detection of goats' milk in ewes' milk and cheese. *International Journal of Food Science and Technology* 26:457–65.

Rumbo M., Chirdo F.G., Fossati C.A., Anon M.C. 1996. Analysis of structural properties and immunochemical reactivity of heat-treated ovalbumin. *Journal of Agricultural Food Chemistry* 44:3793–98.

Sanchez L., Peiro J.M., Castillo H., Perez M.D., Ena J.M., Calvo M. 1992. Kinetic parameters for denaturation of bovine milk lactoferrin. *Journal of Food Science* 57:873–79.

Skerritt J.H., Andrews J.L., Hill A.S. 1991. "Elisa: a quality control tool for cereal food production and processing." In *Food Safety and Quality Assurance: Application of Immunoassays Systems*, edited by Morgan M.R.A, Smith C.J., and Williams P.A., London: Elsevier Applied Science, pp. 369–76.

Smith G.C. 1991. "Applications of immunoassay to the detection of food adulteration: an overview." In *Food Safety and Quality Assurance: Application of Immunoassays Systems*, edited by Morgan M.R.A, Smith C.J., and Williams P.A., London: Elsevier Applied Science, pp. 13–32.

Solberg P., Hadland G. 1953. Serological detection of cow's milk added to milk from goat. XII International Dairy Congress, La Haye, pp. 1287–90.

Tijssen P., Kurstak E. 1984. Highly efficient and simple methods for the preparation of peroxidase and active peroxidase-antibody conjugates for enzyme immunoassays. *Analytical Biochemistry* 136:451–57.

Tyler J.W., Stevens B.J., Hostetler D.E., Holle J.M., Denbigh J.L. 1999. Colostral immunoglobulin concentration in Holstein and Guernsey cows. *American Journal of Veterinary Research* 60:1136–39.

Vénien A., Levieux D., Astier C., Briand L., Chobert J.M., Haertlé T. 1997. Production and epitopic characterization of monoclonal antibodies against bovine β-lactoglobulin. *Journal of Dairy Science* 80:1977–87.

Vénien A., Levieux D., Dufour E. 2000. Oil/alkanethiol layers for the study of emulsified protein conformation by surface plasmon resonance using monoclonal antibodies. *Journal of Colloid Interface Science* 223:215–22.

Walschus U., Witt S., Wittmann C. 2002. Development of monoclonal antibodies against Cry1Ab protein from *Bacillus thuringiensis* and their application in an ELISA for detection of transgenic Bt-maize. *Food and Agricultural Immunology* 14:231–40.

Whittaker R.G., Spencer T.L., Copland J.W. 1983. An enzyme-linked immunosorbent assay for species identification of raw meat. *Journal of Science of Food and Agriculture* 34:1143–48.

12 Rapid Liquid Chromatographic Techniques for Detection of Key (Bio)Chemical Markers

M.-Concepción Aristoy, Milagro Reig, and Fidel Toldrá

Introduction

Modern agricultural practices and novel food processes are including new additives and processing aids and other changes in processing to improve productivity and thus increase competitiveness and profitability. These changes may lead to substantial changes in the nutritional value and sensory properties of foods. In addition, the presence of chemical contaminants and residues is getting increased attention due to environmental contamination and/or the use of illegal veterinary drugs or antibiotics for animal growth promotion.

Most countries have established official regulations to control the levels of chemical additives, residues, or contaminants in foods. Analytical methodologies have been developed for the qualitative and quantitative determination of these chemicals in foods. But most of these methodologies are expensive, tedious, and time-consuming. Due to the large amount of samples to be analyzed, high-performance liquid chromatography (HPLC) offers the possibility of relatively rapid analysis for a good number of chemicals. In spite of the swift development of other rapid immunological techniques, like enzyme-linked immunosorbent assays (ELISAs) or biosensors, HPLC is maintaining a number of applications due to its versatility, simplicity, and economy of use. Furthermore, these applications are being expanded through the continued development of better column packages and equipment (through automation, robustness, sensitive detectors, etc.). For instance, increased attention is being given to HPLC coupled to mass spectrometry for official confirmatory purposes. In addition, liquid chromatography can be used by the food industry as a routine technique for controlling processes, raw materials, and products. In fact, day by day a large number of new and rapid specific applications are being published in scientific and technical literature.

HPLC constitutes a technique that has been widely used for the analysis of foods. It has been used for a large number of applications, including the analysis of nutrients, chemical and biochemical contaminants, markers for processing control, detection of adulterations, and control of raw materials and products. The challenge is to improve the throughput for better competition with other techniques that have been appearing lately. The fundamentals of this technique and a summarized description of its applications are described in this chapter. As sample preparation is the most tedious and time-consuming step in food analysis, this chapter is mainly focused on those methodologies with less sample manipulation before HPLC analysis.

The Fundamentals of Liquid Chromatography

HPLC constitutes a useful technique for separation and detection at room temperature, and this is the reason why this technique is especially recommended for the analysis of thermally labile, nonvolatile compounds.

Analysis by HPLC requires some sample preparation, more or less complicated, before its injection. The procedure for analyte extraction and cleanup depends on the sample, matrix complexity, column efficiency, and detector selectivity. It usually includes extraction, centrifugation, isolation, derivatization, and concentration steps, prior to filtration and injection in the chromatograph.

A typical procedure for sample preparation and analysis is shown in figure 12.1 and includes a liquid extraction by homogenization, deproteinization, centrifugation, cleanup by solid phase extraction (SPE) or concentration by evaporation, filtration, chromatographic separation, and detection. Depending on the mode of separation to be used, an optional derivatization step, before or after separation, may be required. For example, amino acids need derivatization to enhance their detection. The method of choice for separating amino acids in their native form should be cation exchange with postcolumn derivatization, while if choosing partition (reverse phase) HPLC, the derivatization should be precolumn for both, enhancing hydrophobicity and suppressing the charge.

Based on the separation mechanisms, HPLC technique can be classified in four different types: adsorption, ion exchange, size exclusion, and partition HPLC.

Adsorption HPLC

In adsorption HPLC, separation of solutes is based on the differential electrostatic attraction to the stationary phase molecules. Its use has been relegated for sampling and cleanup more than for real HPLC analysis. For example, flavonoids from plant materials can be cleaned, fractionated, and enriched on alumina.

Ion Exchange HPLC

In ion exchange HPLC, separation of solutes is based on their charge. Depending on the stationary phase charge, ion exchange can be an anion or cation exchange. Thus, a positively charged stationary phase will separate cations in cation exchange HPLC, and a negatively charged stationary phase will separate anions in anion exchange HPLC. It is the method of choice for charged compounds analysis and especially for inorganic cations or anions analysis.

Size-Exclusion HPLC

In size-exclusion HPLC, separation of solutes is based on their size. This technique is mainly used for sample cleanup and fractionation, but it is also the method of choice in quality control of polymers where the molecular mass distribution is the quality marker.

Partition HPLC

In partition HPLC, separation of solutes is based on their different solubility between two solvents, the stationary and the liquid phase. Depending on the polarity of the phases,

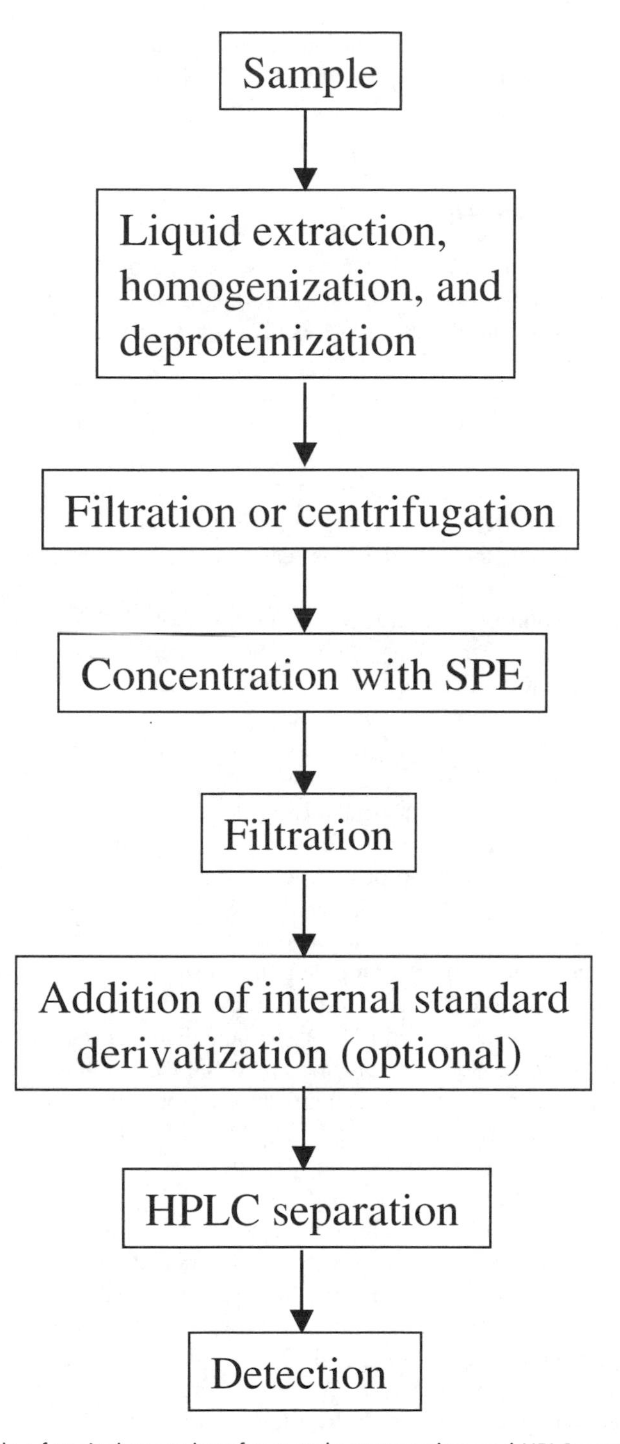

Fig. 12.1. Example of typical procedure for sample preparation and HPLC separation.

partition HPLC can be normal phase (stationary phase more polar than mobile phase) or reverse phase (stationary phase less polar than mobile phase). Reverse phase (RP)-HPLC is the chromatography most frequently used in food analysis, because of its efficiency and versatility.

Advances in Modern HPLC

Liquid chromatography offers a wide variety of separation modes and stationary-mobile phase combinations for optimizing the analysis. Two solutions have been proposed in order to improve quality control based on HPLC analysis throughput. These solutions are based on columns technology and equipment design.

Since the beginning of HPLC, the development of columns has focused on the enhancement of sensitivity, efficiency, and selectivity. The first proposal to improve sensitivity was using narrow-bore or microbore columns, which consisted of using <2.1 mm internal diameter columns, instead of standard-bore, around 4.6 mm internal diameter, columns. Benefits are higher sensitivity and savings in solvents because the flow rate is lower (<0.5 ml/min) than in the case of the classical analytical columns (around 1 ml/min). Requirements from HPLC equipment when using these columns are higher because the low flow rate used requires more precise and reproducible HPLC pumps, and the void volume of the whole system, capillary connections, and cell detector should be smaller to save the peak resolutions. Indeed, new designs for the detector cell, reducing its volume without losing sensitivity, have been developed.

Enhancements in efficiency were focused on the use of smaller-size stationary phase support particles. The limitations in the performance of the packed column are well recognized based on the pressure limit of a solvent delivery system, leading to the current compromise at about 3–5 μm between the column efficiency and the pressure drop. Thus, the current normally used totally porous particle for analytical purposes has a 5 μm size, though a 3 μm particle size is also available. The benefits of reducing particle size, in the chromatography known as fast-liquid chromatography, are partially neutralized because a pressure drop obliges reduction of column length and so efficiency remains unchanged.

Recently, a known liquid chromatography products company developed a series of chromatographic columns packed with 1.6 μm particle diameter that are well suited to the so-called ultraperformance liquid chromatographic (UPLC) system. These columns are designed to support very high pressure. Furthermore, the high-speed data reaching the detector led to the enhancement of its specifity through "high-speed" optical and mass detectors, capable of processing such rapid data by maintaining the resolution requirements of UPLC.

The recent development of monolithic columns may revolutionize food analysis. These columns contain highly porous rods of silica with a bimodal pore structure: (1) a macroporous structure, with pores that average 2 μm diameter and together form a dense network through which the mobile phase can rapidly flow at low pressure (up to 9 ml/min), dramatically reducing separation time, and (2) a mesoporous structure, which forms a fine porous structure (130 Å) and creates a very large surface area on which absorption of the target compounds can occur. The final result is a lack of intraparticular void volume, which improves mass transfer and separation efficiency, allowing for fast, high-quality separations. The possibility of working at such high flow rates without loss of efficiency and avoiding problems from pressure dropping makes run times and column reequilibration times much shorter, resulting in increased lab throughput. These columns may also be packed with a continuous rigid polymeric rod with a porous structure made of polystyrene divinylbenzene copolymer. This stationary phase possesses a stability superior (1–13) to that made on silica. By reducing column dimensions (5 cm long capillary column with 200 μ i.d.), it is possible to work with low flow rates (2.5 μl/min) to properly connect to the mass spectrometer detector.

The combination of both kinds of columns, monolithic and microbore, in a well-designed HPLC system in which void volume has been reduced to a minimum is ideally suited for on-line coupling to any mass spectrometer as detector.

Analysis of Biochemical Markers: Applications for Nutritional Quality

Amino Acids

Amino acids constitute important biochemical compounds in cells and organs for their growth, maintenance, and metabolic activity. Their requirements will vary depending on the stage of life (Aristoy and Toldrá 2004). The analysis of free amino acids in foods is very important because some amino acids, known as essential amino acids, can not be synthesized in the organism and must be supplied in the diet. Furthermore, the analysis of a specific essential amino acid that can be limiting is also very important in basic foods designed for populations with specific dietary needs like the elderly or infants. This is important in infant formulas where accurate compositional data for all the amino acids must be ensured. Some foods, for specific nutritional purposes, can be fortified with specific amino acids, and they must be analyzed to control the correct amount for each amino acid. In other cases, the control must ensure the absence of a specific amino acid (i.e., phenylalanine for phenylketonuria) in diets designed for specific populations with particular metabolic defects. HPLC amino acid analysis has been thoroughly studied, and many methods have been described (Fabiani et al. 2002; Herbert et al. 2000; Vicente et al. 2001; Cohen 2000; Cohen and De Antonis 1994; González-Castro et al. 1997; Stancher and Calabrese 1997; Diaz et al. 1996; Krause et al. 1995; Liu et al. 1995; Bartok et al. 1994; Calull et al. 1991b; Lemieux et al. 1990). A review of these methods in food analysis has been recently published (Aristoy and Toldrá 2004). Figure 12.2 shows a chromatogram of the phenylthiocarbamyl derivatives of free amino acids from dry-cured pork loin.

Carbohydrate and Carboxylic Acids

Knowledge of carbohydrate composition of foods is important because carbohydrates are the principal constituents of many foods, such as fruits, cereals, honey, and beverages, being responsible for their flavor and quality. Furthermore, these compounds may serve as indexes of maturity (aging), authenticity, and storage conditions. Sugars and organic acids are also routinely analyzed for their use as additives. This is the case of polyols in baked goods (Yang 1999) and confectionary products (Corradini et al. 1998).

A review of the analysis of low-molecular-weight carbohydrates (di- and trisaccharides, deoxy sugars, uronic and aldonic acids, alditols, polyols, amino sugars, and anhydrides) in food and beverages by high-performance liquid chromatography, capillary electrophoresis, and gas chromatography has been recently published (Martinez-Montero et al. 2004).

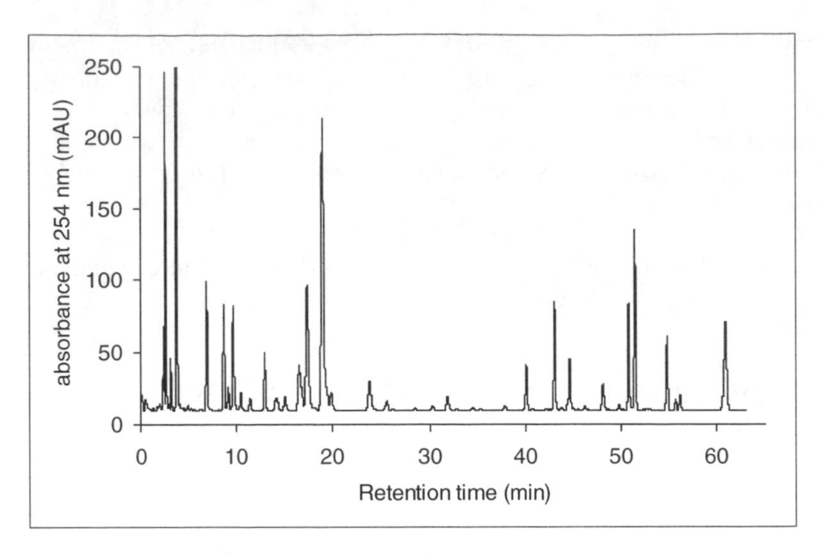

Fig. 12.2. RP-HPLC chromatogram of the phenylthiocarbamyl derivatives of free amino acids from dry-cured pork loin.

Some scientific papers using HPLC for the analysis of carbohydrates and organic acids in foods are compiled in table 12.1. In relation to the methodology used, this table shows the detection method only, because the possibility of achieving a specific HPLC analysis will depend, in many cases, on the ability of a specific detector. In fact, it is easier to purchase a chromatographic column than a detector to perform the required analysis.

Vitamins

Vitamins are biologically active compounds with relevant beneficial effects on health. Rapid methods for the analysis of fat-soluble and water-soluble vitamins in foods have been developed. Vitamins are found at very low concentrations in foods, and often are accompanied by an excess of compounds with similar characteristics or chemical properties. The use of specific detection like UV-visible diode array detection (spectral information), fluorescence detection, or electrochemical detection is thus necessary. Some vitamins, especially fat-soluble vitamins, are labile compounds, sensitive to high temperatures, oxygen, and light, which must be taken into account when analyzing them. The analysis of fat-soluble vitamins requires complex sample preparation procedures including alkaline hydrolysis, enzymatic hydrolysis, alcoholysis, direct solvent extraction, solid phase extraction (Heudi et al. 2004), or supercritical fluid extraction (Perretti et al. 2003a, 2003b) of the total lipid content. Indeed, supercritical fluid extraction (SFE) methods are adequate for fat-soluble vitamins because the extraction occurs at a lower temperature than in traditional methods, and without oxygen and light (Perretti et al. 2003a, 2003b). The use of organic solvents like methanol, ethanol, and hexane is often required for the extraction or for the analysis, and normal phase columns are sometimes preferred.

Fat-soluble vitamins include trans-retinol, retinol acetate, and retinol palmitate (vitamin A); gamma-tocopherol, delta-tocopherol, and tocopherol acetate (vitamin E); and vitamin D (table 12.2).

Table 12.1. Applications of HPLC to the analysis of carbohydrates and organic acids in foods.

Sample	Matrix	Detection Details	References
Carbohydrates	Cereals	ELSD[a]	Young 2002
Carbohydrates	Foods	ELSD	Wei and Ding 2002
Mono- and disaccharides	Cereal products	PAD[b]	Menezes et al. 2004
Sugars and alditols	Beverages and milk	PAD	Corradini et al. 2001
Monosaccharides	Fruits	UV (190 nm) or RI	Karkacier et al. 2003
Carboxylic acids and sugars	Fruit juices	UV (214 nm)	Chinnici et al. 2005
Carboxylic acids	Fruits/jam	UV (214 nm)	Silva et al. 2002
Carbohydrates, alcohols, and carboxylic acids	Wine	FT-IR[c]	Edelmann et al. 2003
Carbohydrates, alcohols, and carboxylic acids	Must and wine	UV (214 nm)	Castellari et al. 2000
Carboxylic acids, sugars, glycerol and ethanol	Must and wine	RI	Calull et al. 1991a
Carboxylic acids	Beer	UV or RI	Floridi et al. 2002
Carboxylic acids	Beer	UV or RI	Mancini et al. 2000
Carboxylic acids	Fruits and jam	UV (214 nm)	Silva et al. 2002
Alditols and sugars	Fruits and vegetables	PAD	Cataldi et al. 1998
Alditols and sugars	Beverages and confectionary products	PAD	Bruce 2003
Alditols	Confectionary products	PAD	Corradini et al. 1998
Alditols	Beverages and foods	Nitrobenzoylation and UV	Nojiri et al. 1999

[a]ELSD: evaporative light-scattering detector.
[b]PAD: pulsed amperometric detector.
[c]FT-IR: Fourier transform infrared.

On the other hand, water-soluble vitamins normally require more simple sample preparation that includes an extraction with water after homogenization of the food sample and in some cases a deproteinization or solid phase extraction. The most simple methodology described is the analysis of nine water-soluble vitamins (thiamine, B1; riboflavin, B2; calcium pantothenate, B5; pyridoxine, B6; biotin, B8; folic acid, B9; cyanocobalamin, B12; nicotinamide, PP; and ascorbic acid, C) from polyvitaminated premixes used for the fortification of infant nutrition, which have been simultaneously analyzed by RP-HPLC and UV detection (Heudi and Fontannaz 2005) in 17 min. Vitamins B1, B2, B6, and PP from Parma ham were analyzed by RP-HPLC and UV–fluorimetric detectors in a series (Consiglieri and Amendola 2003). Nicotinic acid and nicotinamide (niacin, PP) have also been determined in cereals by ion-paired RP-HPLC and UV detection (LaCroix et al. 2005).

Much more complex is the analysis of pantothenic acid in foods (cereals, eggs, meat, viscera, milk, legumes, and vegetables), which is cumbersome and requires previous hydrolysis with pepsin, pantetheinase, and alkaline phosphatase to release bound forms of pantothenic acid. Extracts are then purified by anion and cation exchange

Table 12.2. Applications of HPLC to the analysis of fat-soluble vitamins in foods.

Sample	Matrix	Separation/Detection	References
α- and β-carotene	Carrot juice	Reverse phase/DAD[a]	Marx et al. 2000
Vitamin E	Fruit juice and vegetable extract	Postcolumn photochemical reaction—DAD and FLD[b]	Terada and Tamura 2004
Vitamin E	Infant formulas	Normal phase/FLD	Rodrigo et al. 2002
Vitamin E	Meat	Normal phase/FLD	Hewavitharana et al. 2004
Vitamin E	Cereals	Normal phase/FLD	Ryynanen et al. 2004
Vitamin E	Edible oils	Normal phase/DAD and FLD	Gama et al. 2000
Vitamins A and E	Infant formulas	Reverse phase/DAD (292–235 nm)	Rodas-Mendoza et al. 2003
Vitamins A and E	Dairy products	Reverse phase/UV	Paixao and Campos 2003
Vitamins A and E	Beverages	Normal phase/FLD	Hoeller et al. 2003
Vitamins A, D, and E	Infant formulas	Normal phase/MS[c]	Heudi et al. 2004
Vitamin D	Baby food	Reverse phase/DAD (295 nm)	Konings 1994
Vitamins A, D, and E and β-carotene	Milk, pork, liver, infant formula	Photometric detection	Perretti et al. 2003b

[a]DAD: diode-array detector.
[b]FLD: fluorescence detector.
[c]MS: mass spectrometry.

chromatography, and the sample analyzed by RP-HPLC after postcolumn derivatization with OPA-3-mercaptopropionic acid and fluorimetric detection (Pakin et al. 2004).

The analysis of folic acid and folates is another complex assay due to the sample preparation, and it has been achieved in different ways. Zhang et al. (2004) propose the conversion of all forms of folate found in vegetables to p-aminobenzoic acid (PABA) by means of hydrogen peroxide followed by acid hydrolysis. After a purification step on an affinity column, PABA is then analyzed by HPLC and postcolumn derivatization with fluorescamine. Other examples of sample preparation based on solid phase extraction to analyze folic acid and folates in foods are presented by Nilsson et al. (2004) and Rodríguez-Bernaldo de Quirós et al. (2004), who use RP-HPLC for the separation and UV detection at 290 nm for folic acid and fluorescence (290 and 356 nm) for folates. However, Catharino et al. (2003) propose a very simple sample preparation method for the analysis of folic acid in enriched milk, followed by a reverse phase HPLC separation and UV detection (290 nm).

Mass spectrometry, after RP-HPLC, has been applied to the analysis of water-soluble vitamins B1, B2, B6 (pyridoxine, pyridoxal, and pyridoxamine), PP (nicotinamide and nicotinic acid), pantothenic acid, and folic acid in typical Italian pasta.

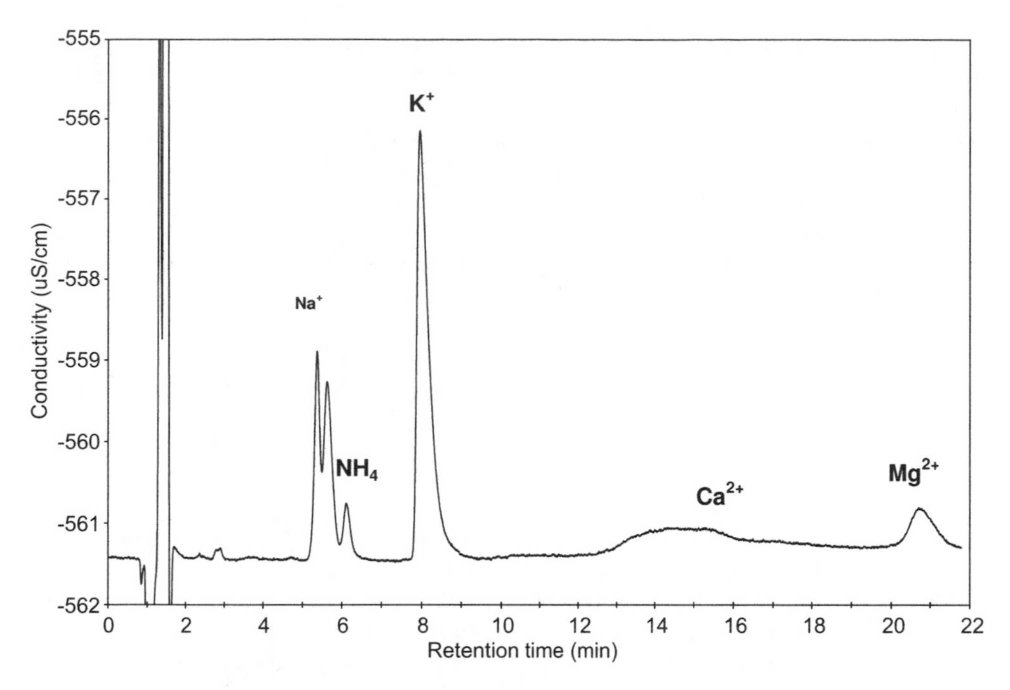

Fig. 12.3. Ion exchange HPLC chromatogram of mono- and divalent cations from a pork meat sample.

Minerals and Trace Elements

The determination of minerals and trace elements can be achieved with atomic absorption or atomic emission spectroscopy. However, the analysis of cations can be easily accomplished by ion exchange chromatography coupled to a conductivity detector. Column technology has been improved in recent years (Klampfl 2004). An example of a chromatogram with the separation of several cations in a single run is shown in figure 12.3. Several mono- and divalent cations have been analyzed in beverages (Munaf et al. 1999), fruit juices (Trifiro et al. 1996), and spinach (Vera et al. 1995).

Antioxidants

Control of raw materials for food processing includes the analysis of other nutritionally important compounds that support health benefits. There is a group of naturally present compounds in foods that do not possess nutritional characteristics on their own but are responsible for wellness and have beneficial effects on health as antioxidants, antihypertensives, antitumorals, promoters of immunodefense, and so on. Polyphenols are the most important bioactive compounds present in vegetal foods. They support health benefits as antioxidants (Burns et al. 2000), preventing oxidative damage diseases (coronary heart disease, cancer, etc.). Among them, flavonoids constitute a large group of plant polyphenols, which are present in plant tissues as sugar conjugates or as aglycones. When analyzed, glycosides are normally hydrolyzed (Nuutila et al. 2002), and the resulting aglycones are identified and quantified. Some methods for sample preparation and analysis of phenolic acids in foods are available in the scientific literature (Umphress et al. 2005; Hubert et al. 2005; Martínez-Ortega et al. 2004; Li et al. 2004; Chinnici et al. 2004; Belajova and Suhaj 2004; Alonso-García et al. 2004; Robbins 2003; Gyorik et al. 2003; Castellari et al. 2002; Nenadis and Tsimidou 2002; Merken and Beecher 2000).

Saponins and gingenosides are other important bioactive compounds, mainly present in soya and ginseng, with hypolipidemic (Rodrigues et al. 2005; Popovich et al. 2005) and protective against colon cancer (MacDonald et al. 2005) activity. Some HPLC methods for their analysis have been reported (Hubert et al. 2005).

Analysis of Biochemical Markers: Applications for Food Quality

Biochemical Compounds

Amino Acids

Free amino acids constitute important taste-active compounds that contribute characteristic tastes to foods (Nishimura and Kato 1988; Kato et al. 1989; Kemp and Birch 1992). On the other hand, some amino acids, like sulphur-containing amino acids, can react during food processing to aroma compounds (Flores et al. 1998; Klein et al. 2001), and sulfur-containing amino acids may reduce the adverse effects of certain food ingredients (Friedman 1994).

There are some negative effects of food processing that can be monitored through the HPLC analysis of amino acids (Anantharaman and Finot 1993). For instance, alkali treatment and/or heating can result in the destruction of essential amino acids and can induce racemization of amino acids (Friedman 1992, 1999) to their D-optical isomers, making them unavailable for protein synthesis and causing them to lose digestibility by cross-linking with the lysinoalanine or histidinoalanine formation (Henle et al. 1993; Wilkinson and Hewavitharana 1997). Some other effects are due to specific amino acids. Thus, the chromatographic determination of amino acids can be used to monitor the correct thermal processing of foods, like milk (Henle et al. 1993; Moret et al. 1997; Ferrer et al. 1999), contamination (Faist et al. 2000), or inadequate aroma development (Pfeiffer and Orben 2000).

Nucleotides and Nucleosides

The analysis of ATP-related compounds offers good possibilities for controlling post-mortem time in muscle foods. Nucleotides and nucleosides are generally formed from ATP breakdown in foods. Nucleotides and nucleosides are extracted and filtered and then injected into reverse phase HPLC coupled to a UV or diode array detector. The analysis is relatively easy and fast—the only limitation is the rapid breakdown of ATP when not properly managed under cold conditions. ATP disappears very fast in postmortem muscle, its disappearance being correlated with the development of rigor mortis (Batlle et al. 2000). Thus, the prediction of low-quality meats can be based on the analysis of ATP-related compounds within 6 h postmortem. In addition, the results of analysis of ATP metabolites, mainly inosine and hypoxanthine, is well correlated with aging time (Batlle et al. 2001). Some ratios have shown good utility for quality purposes. For instance, the 5′-inosine monophosphate (IMP)/ATP ratio may be used to detect exudative pork meats at 2 h postmortem (Batlle et al. 2000). Something similar was also observed with the R-value that consists of the ratio of inosine-related compounds to ATP-related compounds (Honikel and Fischer 1977). Finally, the Ko ratio, which is the ratio between inosine + hypoxanthine to the total ATP-related compounds + inosine + hypoxanthine, is a good indicator of fish freshness.

Additives

Nitrites and nitrates constitute two important additives used for the preservation of muscle foods. Also, their analysis is required in certain waters and vegetables. Anion chromatography coupled to an electrochemical detector allows good detection of both nitrates and nitrites in 15–30 min. Sample concentrations may range between 0.01 and 10 nM (Stratford 1999). The type of column depends on the composition of the sample. So standard anion exchange columns are valid for low-ionic-strength samples, while special high-capacity columns should be used for high-ionic-strength samples—for instance, those containing high buffer concentration (Kissner and Koppenol 2005). The analysis of chloride, which is the most common inorganic anion present in foods, is accomplished by ion exchange chromatography coupled to a conductivity detector. The use of hydroxide selective columns has been reported to give higher retention for commonly weakly retained solutes. Under these conditions, more than 30 anions, including chloride, nitrate, nitrite, and phosphate, have been efficiently separated. A higher sensitivity can be achieved when using suppressed conductivity detection. The suppressor, which is located between the column and the detector, reduces the background conductivity of the eluent and can also enhance the detectability of the analyte (Klampfl 2004).

Markers for Processing Control

Process quality control is important for detecting bad practices in food processing or in the use of preservatives containing nonpermitted additives or an excess of permitted ones. Quality control also includes the detection of food contaminants such as polycyclic aromatic hydrocarbons (PAHs) resulting from materials in contact with foods (packages), pesticides, and antibiotics and other drugs. Glucose content has been reported as a potential quality marker of green coffee. This content, determined by HPLC, is linked with the increase of a woody/rubbery note in coffee quality (Bucheli et al. 1998). RP-HPLC has been successfully applied to the control and follow-up of proteolysis, the generation of peptides and free amino acids, by different lactic acid bacteria typical of fermented meat products (Fadda et al. 1999a, 1999b; Sanz et al. 1999a, 1999b). These chromatograms were useful for selecting the most appropriate fermentation conditions to obtain the desired proteolysis.

Analysis of Biochemical Markers: Applications for the Detection of Food Adulterations

The analysis of certain compounds can be used as an index for the detection of food adulterations or fraudulent food additions. HPLC constitutes a useful technique for the analysis of many of these compounds. For instance, 4-hydroxyproline is used as a marker for the addition of collagen, gelatin, or any other low-value meat extract to meat products. Another practice consisting of the addition of nonmeat proteins or hydrolyzates to meat products can be easily detected by analyzing 3-methylhistidine (Lawrie 1988; Toldrá and Reig 2004). In other cases, the addition of cheap amino acids to mask fraudulent fruit juice dilution (Anantharaman and Finot 1993) or the addition of common wheat flour in making durum pasta products (Stancher et al. 1995) can be detected by HPLC analysis of free amino acid profiles.

Analysis of Biochemical Markers: Applications for Food Safety

Biochemical Compounds

Some compounds have been found to exert toxic effects on consumers and thus must be detected. This is the case for asparagine, which has been recently found responsible for the formation of acrylamide, a probable carcinogen and neurotoxic agent. Acrylamide has been detected in potato chips, french fries, and other cooked foods (Stadler et al. 2004). But certain amino acids can also be markers for food irradiation, like *o*-tyrosine, which is generated when phenylalanine is irradiated with gamma rays, being the conversion yield proportional to the absorbed dose and temperature during irradiation (Offermanns et al. 1993; Krach et al. 1997; Miyahara et al. 2000a, 2000b).

Some small peptides, like anserine, carnosine, and balenine, which are naturally present in meat and fish and not present in vegetal origin proteins, have been analyzed by HPLC. The content of these dipeptides differs according to the animal species, and some ratios have been proposed as markers for the presence of mammalian proteins in feeds (Aristoy and Toldrá 2004) and, in some cases, even differentiate the species of origin (Aristoy et al. 2004). Figure 12.4 shows how the presence of the dipeptide carnosine can be easily detected in cat feed, in which the presence of animal origin proteins is permitted, although they must be absent in ruminant (bovine) feed.

Biogenic amines are naturally occurring components in some foods. Problems arise because some specific amines can present a health hazard through their interaction with some medicaments, can have toxic properties at high concentrations, or can cause allergenic problems. Biogenic amines are mainly synthesized through decarboxylation of free amino acids. This pathway is frequent in fermented products such as meat, fish, cheese, and wine, and it is possibly potentiated by bad hygienic conditions of raw materials and manufacturing practices. Consequently, the level of biogenic amines in a food product is often considered as a marker for quality index (Vinci and Antonelli 2002; Chytiri et al. 2004). Some reviews of biogenic amines and the factors influencing their accumulation in meat and meat products (Suzi and Gardini 2003; Ruiz-Capillas and Jiménez-Colmenero 2004) and cheese (Stratton et al. 1991) have been published.

Sample preparation for the analysis of amines often includes a liquid-liquid or solid phase extraction. As amines do not possess chromophores or fluorophores, a derivatization, pre- or postseparation, is almost always required to enhance sensibility and selectivity in the UV or fluorescence detection. Some applications for HPLC analysis in foods are compiled in table 12.3. Derivatization will not be necessary when using mass spectrometry (Calbiani et al. 2005; Sacani et al. 2005). Detection of underivatized amines with electrochemical detectors has also been reported (Yashin and Yashin 2004).

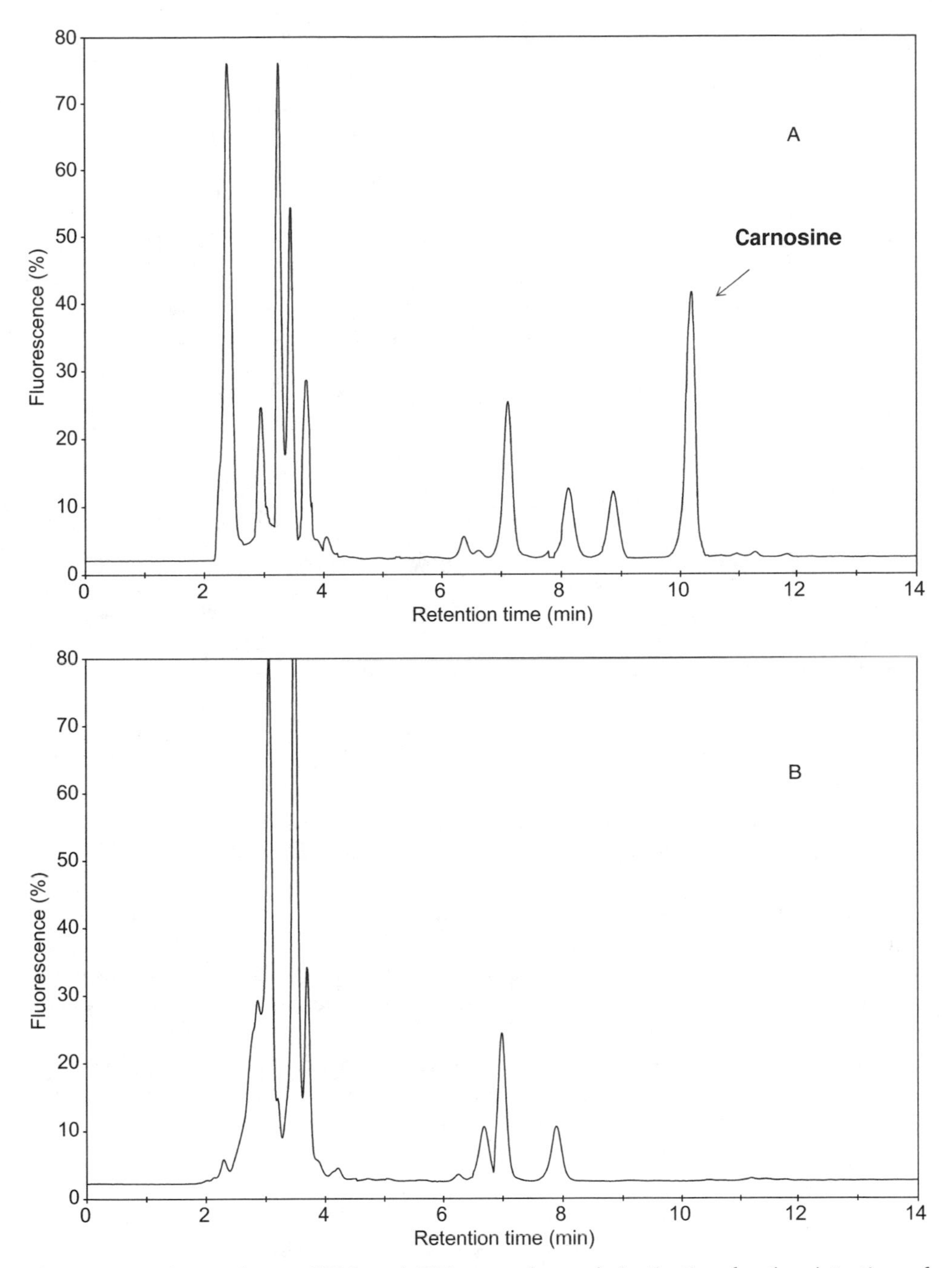

Fig. 12.4. Cation exchange HPLC and OPA postcolumn derivatization for the detection of mammalian dipeptides in feedstuffs for cats **(A)** and bovine **(B)**.

Table 12.3. Applications of HPLC to the analysis of amines in foods.

Matrix	Separation Technique	Detection	References
Wine/alcoholic beverages	Ion pair-reverse phase	OPA[a] postcolumn/DAD[b]	Vidal-Carou et al. 2003
Wine	Reverse phase	Dansyl precolumn/DAD	Dugo et al. 2006
Wine	Reverse phase	OPA precolumn/fluorescence	Marcobal et al. 2005
Wine	Reverse phase	AQC[c] precolumn/fluorescence	Busto et al. 1997
Meat	Cation exchange	MS[d]	Sacani et al. 2005
Sausages	Reverse phase	OPA postcolumn/fluorescence	Straub et al. 1993
Cheese	Reverse phase	FMOC[e] precolumn/fluorescence	Aygün et al. 1999
Cheese	Ion pair-reverse phase	OPA postcolumn/fluorescence	Vale and Glória 1997

[a]OPA: *o*-phthaldialdehyde.
[b]DAD: diode-array detector.
[c]AQC: 6-aminoquinolyl-n-hydroxysuccinimidyl carbamate.
[d]MS: mass spectrometry.
[e]FMOC: 9-fluorenilmethylchloroformate.

Veterinary Drug Residues in Foods of Animal Origin

Veterinary drugs having anabolic effect are used for therapeutic and prophylactic purposes, and they may be used for improved breeding efficiency, even though they are banned in the European Union and other countries. Some of these drugs can only be administered for therapeutic purposes and under the control of a responsible veterinarian (Van Peteghem et al. 2000). These substances improve the feed conversion efficiency and increase the lean to fat ratio, and due to this growth-promoting effect, they are illegally used (Toldrá and Reig, forthcoming).

The detection of veterinary drugs is difficult due to the large number of substances to be analyzed. Their presence in foods is controlled by official inspection and analytical services following European Community directive 96/23/EC on measures to monitor certain substances and residues in live animals and animal products. This directive was recently implemented by the commission decision 2002/657/EC, where detailed guidelines are provided for the analytical methods to be used in testing of official samples and specific common criteria are given for the interpretation of analytical results of official control laboratories for such samples.

Routinary inspection controls are based on rapid screening assays, and only those food samples suspected of containing veterinary drug residue are then confirmed through methods based on the use of gas or liquid chromatography coupled to mass spectrometry, or other sophisticated methodologies and analytical instrumentation, for accurate characterization and confirmation.

Liquid chromatography is getting extended use in the analysis of veterinary drug residues in foods (Van Peteghem et al. 2000). HPLC constitutes a useful technique for multiresidue screening due to the increasing demand for analysis and the need to speed up the analysis time. The analysis is based on a solid phase extraction cleanup followed by filtration and injection into a reverse phase HPLC with UV diode array detection. It has been applied for detection of veterinary drugs in several foods like eggs, milk, fish, and meat (Aerts et al. 1995; Horie et al. 1998), anabolic steroids in nutritional supplements (Gonzalo-Lumbreras and Izquierdo-Hornillos 2000; De Cock et al. 2001), and corticosteroids like dexamethasone in water and meat (Shearan et al. 1991; Mallinson et al. 1995; Stolker et al. 2000; Reig et al., forthcoming). HPLC allows the possibility to analyze a large number of residues for a given sample in a relatively short time. In addition, the availability of automation and computer-controlled facilities has contributed to the expansion of its use as a screening technique (Toldrá and Reig, forthcoming).

Antibiotic Residues in Foods of Animal Origin

Reverse phase HPLC in combination with a wide range of different detectors, like UV diode array or fluorescence, constitutes a useful technique for the screening of antibiotics. Antibiotics have been successfully detected by RP-HPLC with diode array in meat, kidney, and milk (Cooper et al. 1995; Furusawa 1999; Cinquina et al. 2003). Fluorescence detection after HPLC separation has also been used for the detection of penicillins in milk and liver, respectively (Bergwerff and Schloesser 2003), the detection of lasalocid in animal tissues and eggs (Matabudul et al. 2000), and the simultaneous determination of ten quinolone antibacterial residues in multispecies animal tissues (Verdon et al. 2005). Five types of antibiotics were determined by RP-HPLC with fluorescence and UV detection in meat, liver, kidney, and milk (Chonan et al. 2000). A chromatographic separation of carbadox, obtained in less than 15 min, is shown in figure 12.5. When the analytes are not detected by light absorption or fluorescence, they may require postcolumn chemical modifications to get a detectable derivative (Bergwerff and Schloesser 2003).

Other Residues

Multimycotoxin analysis of aflatoxins B1, B2, G1, and G2 and ochratoxin A has been performed by HPLC with fluorescence detection after adequate cleanup procedures by

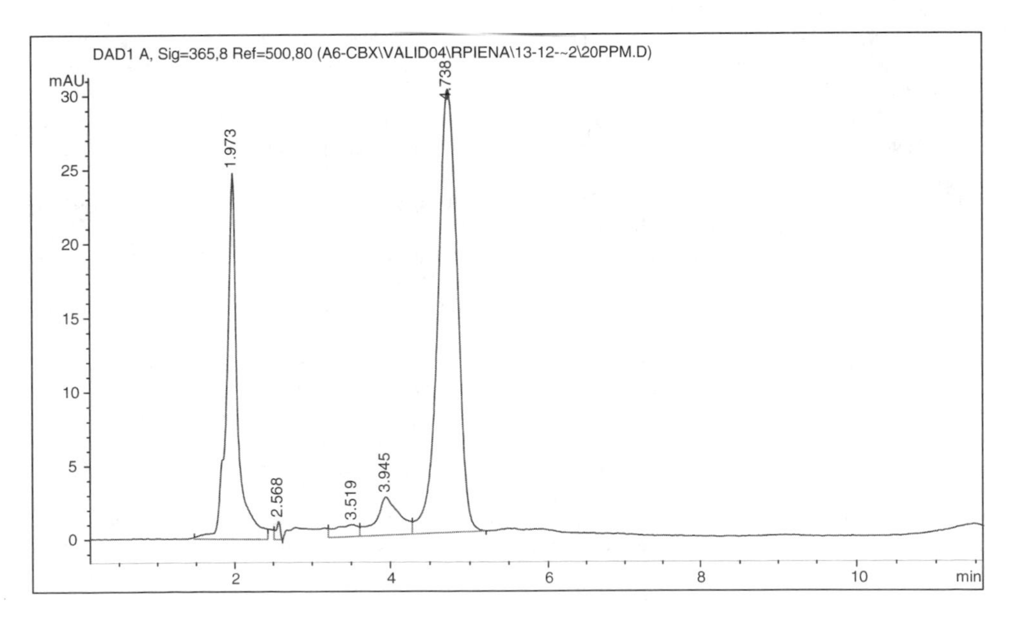

Fig. 12.5. RP-HPLC chromatogram of carbadox, eluting at 4.74 min, in animal tissues.

immunoaffinity column (García-Villanueva et al. 2004). Similar reverse phase HPLC methods for ochratoxin A have been reported (Chiavaro et al. 2002; Spotti et al. 2002). Fumonisins and protein-bound fumonisins have been analyzed in different foods like cornflakes, tortilla chips, and corn chips by HPLC after extraction and partial purification by solid phase extraction (Park et al. 2004). The determination of benzimidazole fungicides in fruits and vegetables has been normalized. This method involves a cleanup by liquid-liquid partition and analysis by HPLC (Prousalis et al. 2006).

Polycyclic aromatic hydrocarbons (PAHs) are considered cancer-causing agents. These compounds may be formed directly in food as a result of heating processes like charcoal grilling, roasting, or smoking (Simko 2002). The control of PAHs in smoke flavorings and smoked products is very important. Several chromatographic methodologies, including sample preparation by organic solvent extraction and cleanup through silica cartridges, have been proposed. So the use of RP-HPLC with fluorescence detection has been proposed for such control in coffee (García-Falcón et al. 2005), oils (Moret and Conte 2002), fishery products (Moret et al. 1999; Sobrado et al. 2004), and meat products (Simko 2002).

References

Aerts, M.M.L., Hogenboom, A.C., Brinkman, U.A.T. 1995. Analytical strategies for the screening of veterinary drugs and their residues in edible products. *Journal of Chromatography B—Biomedical Applications* 667: 1–40.

Alonso-García, A., Cancho-Grande, N., Simal-Gandara, J. 2004. Development of a rapid method based on solid-phase extraction and liquid chromatography and ultraviolet absorbance detection for the determination of polyphenols in alcohol-free beers. *Journal of Chromatography A* 1054: 175–80.

Anantharaman, A., Finot, P.A. 1993. Nutritional aspects of food proteins in relation to technology. *Food Reviews International* 9: 629–55.

Aristoy, M.C., Toldrá, F. 2004. Amino acids. In: *Handbook of Food Analysis*, 2nd edition, edited by L.M.L. Nollet, New York: Marcel-Dekker, pp. 83–123.

Aygün, O., Schneider, E., Scheuer, R., Usleber, E., Gareis, M., Märtlbauer, E. 1999. Comparison of ELISA and HPLC for the determination of histamine in cheese. *Journal of Agricultural and Food Chemistry* 47: 1961–64.

Bartok, T., Szalai, G., Lörincz, Z., Börcsök, G., Sagi, F. 1994. High-speed RP-HPLC/FL analysis of amino acids after automated two-step derivatization with o-phthaldialdehyde/3-mercaptopropionic acid and 9-fluorenylmethyl chloroformate. *Journal of Liquid Chromatography* 17(20): 4391–403.

Batlle, N., Aristoy, M.C., Toldrá, F. 2000. Early postmortem detection of exudative pork meat based on the nucleotide content. *Journal of Food Science* 65: 413–16.

———. 2001. ATP metabolites along aging of exudative and non-exudative pork meats. *Journal of Food Science* 66: 68–71.

Belajova, E., Suhaj, M. 2004. Determination of phenolic constituents in citrus juices: meted of high performance liquid chromatography. *Food Chemistry* 86: 339–43.

Bergwerff, A.A., Schloesser, J. 2003. Antibiotics and drugs: residue determination. In: *Encyclopedia of Food Sciences and Nutrition*, edited by B. Caballero, London: Elsevier, pp. 254–61.

Bruce, J. 2003. Analysis of sugars using ion chromatography. *International Sugar Journal* 105: 107–10.

Bucheli, P., Meyer, I., Pitet, A., Vuataz, G., Viani, R. 1998. Industrial storage of green Robusta coffee under tropical conditions and its impact on raw material quality and ochratoxin A content. *Journal of Agricultural and Food Chemistry* 46: 4507–11.

Burns, J., Gardner, P.T., O'Neill, J., Crawford, S., Morecroft, I., McPhail, D.B., Lister, C., Matthews, D., MacLean, M.R., Lean, M.E.J., Duthie, G.G., Crozier, A. 2000. Relationship among antioxidant activity, vasodilatation capacity, and phenolic content of red wines. *Journal of Agricultural and Food Chemistry* 48: 220–30.

Busto, O., Miracle, M., Guasch, J., Borrull, F. 1997. Solid phase extraction of biogenic amines from wine before chromatographic analysis of their AQC derivatives. *Journal of Liquid Chromatography and Related Technologies* 20: 743–55.

Calbiani, F., Careri, M., Elviri, L., Mangia, A., Pistara, L., Zagnoni, I. 2005. Rapid assay for analyzing biogenic amines in cheese: matrix solid-phase dispersion followed by liquid chromatography-electrospray-tandem mass spectrometry. *Journal of Agricultural and Food Chemistry* 53: 3779–83.

Calull, M., Borrull, F., Marce, R.M., Zamora, F. 1991a. HPLC analysis of fatty acids in wine. *American Journal of Enology and Viticulture* 42: 268–73.

Calull, M., Fábregas, J., Marcé, R.M., Borrull, F. 1991b. Determination of free amino acids by precolumn derivatization with phenylisothiocyanate: application to wine samples. *Chromatographia* 31: 272–76.

Castellari, M., Versari, A., Spinabelli, U., Galassi, S., Amati, A. 2000. An improved HPLC method for the analysis of organic acids, carbohydrates, and alcohols in grape musts and wines. *Journal of Liquid Chromatography and Related Technologies* 23: 2047–56.

Castellari, M., Sartini, E., Fabiani, A., Arfelli, G., Amati, A. 2002. Analysis of wine phenolics by high-performance liquid chromatography using monolithic type column. *Journal of Chromatography* 973: 221–27.

Cataldi, T.R.I., Margiotta, G., Zambonin, C.G. 1998. Determination of sugars and alditols in food samples by HPAEC with integrated pulsed amperometric detection using alkaline eluents containing barium or strontium ions. *Food Chemistry* 62: 109–15.

Catharino, R.R., Visentainer, Virgilio, J., Godoy, Teixeira, J. 2003. Evaluation of the experimental conditions of HPLC in the determination of folic acid in enriched milk. *Ciênc. Tecnol. Aliment.* 23(3): 389–95.

Chiavaro, E., Lepiani, A., Colla, F., Bettoni, P., Pari, E., Spotti, E. 2002. Ochratoxin A determination in ham by immunoaffinity clean-up and a quick fluorometric method. *Food Additives and Contaminants* 19(6): 575–81.

Chinnici, F., Gaiani, A., Natali, N., Riponi, C., Galassi, S. 2004. Improved HPLC determination of phenolic compounds in cv. Golden Delicious apples using monolithic column. *Journal of Agricultural and Food Chemistry* 52: 3–7.

Chinnici, F., Spinabelli, U., Riponi, C., Amati, A. 2005. Optimization of the determination of organic acids and sugars in fruit juices by ion-exclusion liquid chromatography. *Journal of Food Composition and Analysis* 18: 121–30.

Chonan, T., Nishimura, K., Hirama, Y. 2000. Rapid determination of residues of 5 kinds of veterinary drugs in livestock products by HPLC. *Journal of the Food Hygienic Society of Japan* 41: 326–29.

Chytiri, S., Paleologos, E., Savvaidis, I., Kontominas, M.G. 2004. Relation of biogenic amines with microbial and sensory changes of whole and filleted freshwater rainbow trout (*Onchorynchus mykiss*) stored on ice. *Journal of Food Protection* 67: 960–65.

Cinquina, A.L., Roberti, P., Gianetti, L., Longo, F., Draisci, R., Fagiolo, A., Brizioli, N.R. 2003. Determination of enrofloxacin and its metabolite ciprofloxacin in goat milk by high-performance liquid chromatography with diode-array detection: optimization and validation. *Journal of Chromatography A* 987: 221–26.

Cohen, S.A. 2000. Amino acid analysis using precolumn derivatization with 6-aminoquinolyl-N-hydroxysuccinimidyl carbamate. *Methods in Molecular Biology* 159: 39–47.

Cohen, S.A., De Antonis, K.M. 1994. Applications of amino acid derivatization with 6-aminoquinolyl-N-hydroxysuccinimidyl carbamate: analysis of feed grains, intravenous solutions and glycoproteins. *Journal of Chromatography A* 661: 25–34.

Comite Europeen de Normalisation. 2004. Non fatty foods—determination of benzimidazole fungicides carbendazim, thiabendazole and benomyl (as carbendazim)—part 3: HPLC method with liquid-liquid-partition clean up. *European Standard*: EN 14333-3, 14 pp.

Consiglieri, C., Amendola, F. 2003. Determinazione delle vitamine B1, B2, B6 e PP nel prosciutto di Parma mediante HPLC (High Performance Liquid Chromatography (HPLC) of hydro-soluble vitamins in Parma ham). *Industrie alimentari* 42(426): 602–4.

Cooper, A.D., Creaser, C.S., Farrington, W.H.H., Tarbin, J.A., Shearer, G. 1995. Development of multi-residue methodology for the HPLC determination of veterinary drugs in animal-tissues. *Food Additives and Contaminants* 12: 167–76.

Corradini, C., Cavazza, A., Canali, G., Nicoletti, I. 1998. Improved method for the analysis of alditols in confectionery products by capillary zone electrophoresis (CZE): comparison with high-performance anion-exchange chromatography with pulsed amperometric detection (HPAEC-PAD). *Italian Journal of Food Science* 10(3): 195–206.

Corradini, C., Canali, G., Galanti, R., Nicoletti, I. 2001. Determination of alditols and carbohydrates of food interest using a sulfonated monodisperse resin-based column, coupled with pulsed amperometric detection (PAD) and postcolumn pH adjustment. *Journal of Liquid Chromatography and Related Technologies* 24: 1073–88.

De Cock, K.J.S., Delbeke, F.T., Van Eenoo, P., Desmet, N., Roels, K., De Backer, P. 2001. Detection and determination of anabolic steroids in nutritional supplements. *Journal of Pharmaceutical and Biomedical Analysis* 25: 843–52.

Denli, Y., Ozkan, G. 1999. Determination of sorbic acid in wine using high performance liquid chromatography. *Gida* 24: 187–90.

Diaz, J., Lliberia, J.L., Comellas, L., Broto-Puig, F. 1996. Amino acid and amino sugar determination by derivatization with 6-aminoquinolyl-N-hydroxysuccinimidyl carbamate followed by high-performance liquid chromatography and fluorescence detection. *Journal Chromatography A* 719: 171–79.

Dugo, G., Vilasi, F., la Torre, G.L., Pellicano, T.M. 2006 Reverse phase HPLC/DAD determination of biogenic amines as dansyl derivatives in experimental red wines. *Food Chemistry* 95: 672–76.

Edelmann, A., Diewok, J., Rodríguez-Baena, J., Lendl, B. 2003. High-performance liquid chromatography with diamond ATR-FTIR detection for the determination of carbohydrates, alcohols and organic acids in red wine. *Analytical and Bioanalytical Chemistry* 376: 92–97.

Fabiani, A., Versari, A., Parpinello, G.P., Castellari, M., Galassi, S. 2002. High-performance liquid chromatographic analysis of free amino acids in fruit juices using derivatization with 9-fluorenylmethyl-chloroformate. *Journal of Chromatographic Science* 40: 14–18.

Fadda, S., Sanz, Y., Vignolo, G., Aristoy, M.-C., Oliver, G., Toldrá, F. 1999a. Hydrolysis of pork muscle sarcoplasmic proteins by *Lactobacillus curvatus* and *Lactobacillus sake. Applied and Environmental Microbiology* 65: 578–84.

———. 1999b. Characterization of muscle sarcoplasmic and myofibrillar protein hydrolysis caused by *Lactobacillus plantarum. Applied and Environmental Microbiology* 65: 3540–46.

Faist, V., Drusch, S., Kiesner, C., Elmadfa, I., Erbesdobler, H.F. 2000. Determination of lysinoalanine in foods containing milk protein by high-performance chromatography after derivatisation with dansyl chloride. *International Dairy Journal* 10: 339–46.

Ferrer, E., Alegría, A., Farré, R., Abellán, P., Romero, F. 1999. Indicators of damage of protein quality and nutritional value of milk. *Food Science and Technology International* 5: 447–61.

Flores, M., Spanier, A.M., Toldrá, F. 1998. Flavour analysis of dry-cured ham. In: *Flavor of Meat, Meat Products and Seafoods*, 2nd edition, edited by F. Shahidi, London: Blackie Academic and Professional, pp. 320–41.

Floridi, S., Perretti, G., Montanari, L., Fantozzi, P. 2002. Determination of organic acids in Italian beers by HPLC. *Industrie delle Bevande* 31: 546–49.

Friedman, M. 1992. Dietary impact of food processing. *Annual Reviews in Nutrition* 12: 119–37.

———. 1994. Improvement in the safety of foods by SH-containing amino acids and peptides: a review. *Journal of Agricultural and Food Chemistry* 42: 3–20.

———. 1999. Chemistry, nutrition, and microbiology of D-amino acids. *Journal of Agricultural and Food Chemistry* 47: 3457–79.

Furusawa, N. 1999. Rapid liquid chromatographic determination of oxytetracycline in milk. *Journal of Chromatography* 839: 247–51.

Gama, P., Casal, S., Oliveira, B., Ferreira, M.A. 2000. Development of an HPLC/diode-array/fluorimetric detector method for monitoring tocopherols and tocotrienols in edible oils. *Journal of Liquid Chromatography and Related Technologies* 23: 3011–22.

García-Falcón, M.S., Cancho-Grande, B., Simal-Gandara, J. 2005. Minimal clean-up and rapid determination of polycyclic aromatic hydrocarbons in instant coffee. *Food Chemistry* 90: 643–47.

García-Villanueva, R.J., Cordon, C., Gonález-Paramas, A.M., García-Rosales, M.E. 2004. Simultaneous immunoaffinity column cleanup and HPLC analysis of aflatoxins and ochratoxin A in Spanish bee pollen. *Journal of Agricultural Food Chemistry* 52: 7235–39.

González-Castro, M.J., López-Hernández, J., Simal-Lozano, J., Oruna-Concha, M.J. 1997. Determination of amino acids in green beans by derivatization with phenylisothiocianate and high-performance liquid chromatography with ultraviolet detection. *Journal of Chromatographic Science* 35: 181–85.

Gonzalo-Lumbreras, R., Izquierdo-Hornillos, R. 2000. High-performance liquid chromatographic optimization study for the separation of natural and synthetic anabolic steroids: application to urine and pharmaceutical samples. *Journal of Chromatography B: Biomedical Sciences and Applications* 742(1): 1–11.

Gyorik, M., Herpai, Z., Szecsenyi, I., Varga, L., Szigeti, J. 2003. Rapid and sensitive determination of phenol in honey by high-performance liquid chromatography with fluorescence detection. *Journal of Agricultural Food Chemistry* 51: 5222–25.

Henle, T., Walter, A.W., Klostermeyer, H. 1993. Detection and identification of the cross-linking amino acids N-tau- and N-pi-(2′-amino-2′-carboxy-ethyl)-L-histidine ("histidinoalanine," HAL) in heated milk products. *Z Lebensm Unters Forschung* 197: 114–17.

Herbert, P., Barros, P., Ratola, N., Alves, A. 2000. HPLC determination of amino acids in musts and port wine using OPA/FMOC derivatives. *Journal of Food Science* 65: 1130–33.

Heudi, O., Fontannaz, P. 2005. Determination of vitamin B_5 in human urine by high-performance liquid chromatography coupled with mass spectrometry. *Journal of Separation Science* 28(7): 669–72.

Heudi, O., Trisconi, M.J., Black, C.J. 2004. Simultaneous quantification of vitamins A, D3 and E in fortified infant formulae by liquid chromatography-mass spectrometry. *Journal of Chromatography A* 1022(1–2): 115–23.

Hewavitharana, A.K., Lanari, M.C., Becu, C. 2004. Simultaneous determination of vitamin E homologs in chicken meat by liquid chromatography with fluorescence detection. *Journal of Chromatography A* 1025(2): 313–17.

Hoeller, U., Wolter, D., Hofmann, P., Spitzer, V. 2003. Microwave-assisted rapid determination of vitamins A and E in beverages. *Journal of Agricultural and Food Chemistry* 51: 1539–42.

Honikel, K.O., Fischer, C. 1977. A rapid method for the detection of PSE and DFD porcine muscle. *Journal of Food Science* 42: 1633–36.

Horie, M., Yoshida, T., Saito, K., Nakazawa, H. 1998. Rapid screening method for residual veterinary drugs in meat and fish by HPLC. *Shokuhin Eiseigaku Zasshi* 39(6): 383–89.

Hubert, J., Berger, M., Dayde, J. 2005. Use of a simplified HPLC-UV analysis for soyasaponin B determination: study of saponin and isoflavone variability in soybean cultivars and soy-based health food products. *Journal of Agricultural and Food Chemistry* 53: 3923–30.

Jakob, E., Elmadfa, I. 2000. Rapid and simple HPLC analysis of vitamin K in food, tissues and blood. *Food Chemistry* 68: 219–21.

Karkacier, M., Erbas, M., Zulú, M.K., Aksu, M. 2003. Comparison of different extraction and detection methods for sugars using amino-bonded phase HPLC. *Journal of Chromatographic Science* 41: 331–33.

Kato, H., Rhue, M.R., Nishimura, T. 1989. Role of free amino acids and peptides in food taste. In: *Flavor Chemistry: Trends and Developments*, edited by R. Teranishi, R.G. Buttery, and F. Shahidi, ACS Symp. Series 358, Washington D.C.: ACS, pp. 158–74.

Kemp, S.E., Birch, G.G. 1992. An intensity/time study of the taste of amino acids. *Chemical Senses* 17: 151–68.

Kissner, R., Koppenol, W.H. 2005. Qualitative and quantitative determination of nitrite and nitrate with ion chromatography. *Methods in Enzymology* 396: 61–68.

Klampfl, C.W. 2004. Determination of cations and anions by chromatographic and electrophoretic techniques. In: *Handbook of Food Analysis*, 2nd edition, edited by L.M.L. Nollet, New York: Marcel-Dekker, pp. 1891–1918.

Klein, N., Maillard, M.B., Thierry, A., Lortal, S. 2001. Conversion of amino acids into aroma compounds by cell-free extracts of *Lactobacillus helveticus*. *Journal of Applied Microbiology* 91: 404–11.

Konings, E.J.M. 1994. Estimation of vitamin D in baby foods with liquid chromatography. *Nederlands melk en Zuiveltijdschrift* 48(1): 31–39.

Krach, C., Sontag, G., Solar, S., Getoff, N. 1997. HPLC with coulometric electrode array detection: determination of o- and m-tyrosine for identification of irradiated shrimps. *Z Lebensmmittel Unters Forschung* 204: 417–19.

Krause, I., Bockhardt, A., Neckermann, H., Henle, T., Klostermeyer, H. 1995. Simultaneous determination of amino acids and biogenic amines by reversed-phase high-performance liquid chromatography of the dabsyl derivatives. *Journal Chromatography A* 715: 67–79.

Kwanyuen, P., Burton, J.W. 2005. A simple and rapid procedure for phytate determination in soybeans and soy products. *Journal of the American Oil Chemists' Society* 82: 81–85.

LaCroix, D.E., Wolf, W.R., Kwansa, A.L. 2005. Rapid trichloroacetic acid extraction and liquid chromatography method for determination of nicotinamide in commercial cereals. *Cereal Chemistry* 82(3): 277–81.

Lau, A.-J., Woo, S.-O., and Koh, H.-L. 2003. Analysis of saponins in raw and steamed *Panax notoginseng* using high-performance liquid chromatography and diode array detection. *Journal of Chromatography A* 1011: 77–87.

Lawrie, R.A. 1988. 3-Methylhistidine as an index of meat content. *Food Science and Technology Today* 2: 208–10.

Lemieux, L., Puchades, R., Simard, R.E. 1990. Free amino acids in cheddar cheese: comparison of quantitation methods. *Journal of Food Science* 55: 1552–54.

Li, Y.H., Sun, Z.H., Zheng, P. 2004. Determination of vanillin, eugenol and isoeugenol by RP-HPLC. *Chromatographia* 60: 709–13.

Liu, H.J., Chang, B.Y., Yan, H.W., Yu, F.H., Lu, X.X. 1995. Determination of amino acids in food and feed by derivatization with 6-aminoquinolyl-N-hydroxysuccinimidyl carbamate and reversed-phase liquid chromatographic separation. *Journal AOAC International* 78: 736–44.

MacDonald, R.S., Guo, J., Copeland, J., Browning, J.D., Jr., Sleper, D., Rottinghaus, G.E., Berhow, M.A. 2005. Environmental influences on isoflavones and saponins in soybeans and their role in colon cancer. *Journal of Nutrition* 135: 1239–42.

Mallinson, E.T., Dreas, J.S., Wilson, R.T., Henry, A.C. 1995. Determination of dexamethasone in liver and muscle by liquid chromatography and gas chromatography/mass spectrometry. *Journal of Agricultural and Food Chemistry* 43: 140–45.

Mancini, F., Miniati, E., Montanari, L. 2000. Determination of organic acid anions in Italian beers by a new HPLC method. *Italian Journal of Food Science* 12: 443–50.

Marcobal, A., Polo, M.C., Martin-Alvarez, P.J., Moreno-Arribas, M.V. 2005. Biogenic amine content of red Spanish wines: comparison of a direct ELISA and an HPLC method for the determination of histamine in wines. *Food Research International* 38: 387–94.

Martínez-Montero, C., Rodríguez-Dodero, M.C., Guillén-Sánchez, D.A., Barroso, C.G. 2004. Analysis of low molecular weight carbohydrates in foods and beverages: a review. *Chromatographia* 59: 15–30.

Martínez-Ortega, M.V., García-Parrilla, M.C., Troncoso, A.M. 2004. Comparison of diffeent sample preparation treatments for the analysis of wine phenolic compounds in human plasma by reverse phase high-performance liquid chromatography. *Analytica Chimica Acta* 502: 49–55.

Marx, M., Schieber, A., Carle, R. 2000. Quantitative determination of carotene stereoisomers in carrot juices and vitamin supplemented (ATBC) drinks. *Food Chemistry* 70: 403–8.

Matabudul, D.K., Conway, B., Lumley, I.D. 2000. A rapid method for the determination of lasalocid in animal tissues and eggs by high performance liquid chromatography with fluorescence detection and confirmation by LC-MS-MS. *Analyst* 125: 2196–2200.

Menezes, E.W., de Melo, A.T., Lima, G.H., Lajolo, F.M. 2004. Measurement of carbohydrate components and their impact on energy value of foods. *Journal of Food Composition and Analysis* 17: 331–38.

Merken, H.M., Beecher, G.R. 2000. Measurement of food flavonoids by high-performance liquid chromatography: a review. *Journal of Agricultural and Food Chemistry* 48: 577–99.

Miyahara, M., Ito, H., Nagasawa, T., Kamimura, T., Saito, A., Kariya, M., Izumi, K., Kitamura, M., Toyoda, M., Saito, Y. 2000a. Determination of o-tyrosine production in aqueous solutions of phenylalanine irradiated with gamma ray, using high performance liquid chromatography with automated pre-column derivatization and LASER fluorometric detection. *Journal of Health Science* 46: 192–99.

———. 2000b. Detection of irradiation of meats by HPLC determination for o-tyrosine using novel laser fluorometric detection with automatic pre-column reaction. *Journal of Health Science* 46: 304–9.

Moret, S., Conte, L.S. 2002. A rapid method for polycyclic aromatic hydrocarbon determination in vegetable oils. *Journal of Separation Science* 25: 96–100.

Moret, S., Cherubin, S., Lercker, G. 1997. HPLC determination of lysinoalanine. *Latte* 22: 80–81.

Moret, S., Conte, L., Dean, D. 1999. Assessment of polycyclic aromatic hydrocarbon content of smoked fish by means of a fast HPLC/HPLC method. *Journal of Agricultural and Food Chemistry* 47: 1367–71.

Munaf, E., Zein, R., Takeuchi, T., Miwa, T. 1999. Indirect photometric detection of inorganic monovalent and divalent cations by microcolumn ion chromatography using 1,1′-dimethyl-4,4′-bipyridinium dichloride as visualization agent. *Analytica Chimica Acta* 379: 33–37.

Nenadis, N., Tsimidou, M. 2002. Determination of squalene in olive oil using fractional crystallization for sample preparation. *Journal of the American Oil Chemists' Society* 79(3): 257–59.

Nilsson, C., Johansson, M., Yazynina, E., Strålsjö, L., Jastrebova, J. 2004. Solid-phase extraction for HPLC analysis of dietary folates. *European Food Research and Technology* 219(2): 199–204.

Nishimura, T., Kato, H. 1988. Taste of free amino acids and peptides. *Food Reviews International* 4: 175–94.

Nojiri, S., Saito, K., Taguchi, N., Oishi, M., Maki, T. 1999. Liquid chromatographic determination of sugar alcohols in beverages and foods after nitrobenzoylation. *Journal of AOAC International* 82: 134–40.

Nuutila, A.M., Kammiovirta, K., Oskman-Caldentey, K.-M. 2002. Comparison of methods for the hydrolysis of flavonoids and phenolic acid from onion and spinach for HPLC analysis. *Food Chemistry* 76: 519–25.

Offermanns, N.C., McDougall, T.E., Guerrero, A.M. 1993. Validation of o-tyrosine as a marker for detection and dosimetry of irradiated chicken meat. *Journal of Food Protection* 56: 47–50.

Park, J.W., Scott, P.M., Lau, B.P.Y., Lewis, D.A. 2004. Analysis of heat processed corn foods for fumonisins and bound fumonisins. *Food Additives and Contaminants* 21: 1168–78.

Paixao, J.A., Campos, J.M. 2003. Determination of fat soluble vitamins by reversed-phase HPLC coupled with UV detection: a guide to the explanation of intrinsic variability. *Journal of Liquid Chromatography and Related Technologies*. 26(4): 641–63.

Pakin, C., Bergaentzle, M., Hubscher, V., Aoude-Werner, D., Hasselmann, C. 2004. Fluorimetric determination of pantothenic acid in foods by liquid chromatography with post-column derivatization. *Journal of Chromatography A* 1035(1): 87–95.

Patring, J.D.M., Jastrebova, J.A., Hjortmo, S.B., Andlid, T.A., Jagerstad, I.M. 2005. Development of a simplified method for the determination of folates in baker's yeast by HPLC with ultraviolet and fluorescence detection. *Journal of Agricultural and Food Chemistry* 53: 2406–11.

Perretti, G., Miniati, E., Montanari, L., Fantozzi, P. 2003a. Improving the value of rice by-products by SFE. *Journal of Supercritical Fluids* 26(1):63–71.

Perretti, G., Marconi, O., Montanari, L., Fantozzi, P. 2003b. Fat-soluble vitamin extraction by analytical supercritical carbon dioxide. *Journal of the American Oil Chemists' Society* 80(7): 629–33.

Pfeiffer, P., Orben, C. 2000. Pyroglutamic acid in wine and fruit juices: risk and avoidance of unwanted aroma influencing. *Deut Lebensm-Rundsch* 96: 4–8.

Popovich, D.G., Hu, C., Durance, T.D., Kitts, D.D. 2005. Retention of ginsenosides in dried ginseng root: comparison of drying methods. *Journal of Food Science* 70: S355–58.

Prousalis, K.P., Kaltsonoudis, C.K., Tsegenidis, T. 2006. A new sample clean-up procedure, based on ion-pairing on RP-SPE cartridges, for the determination of ionizable pesticides. *International Journal of Environmental Analytical Chemistry* 86: 33–43.

Reig, M., Mora, L., Navarro, J.L., Toldrá, F. Forthcoming. Development of a method for the screening and confirmatory detection of dexamethasone in feed and drinking water. *Food Chemistry*.

Riediker, S. 2002. Acrylamide from Maillard reaction products. *Nature* 419: 449.

Robbins, R.J. 2003. Phenolic acids in foods: an overview of analytical methodology. *Journal of Agricultural and Food Chemistry* 51(10): 2866–87.

Rodas-Mendoza, B., Morera-Pons, S., Castellote-Bargallo, A.I, López-Sabater, M.C. 2003. Rapid determination by reversed-phase high-performance liquid chromatography of vitamins A and E in infant formulas. *Journal of Chromatography A* 1018: 197–202.

Rodrigo, N., Alegría, A., Barberá, R., Farré, R. 2002. High-performance liquid chromatographic determination of tocopherols in infant formulas. *Journal of Chromatography A* 947: 97–102.

Rodrigues, H.G., Diniz, Y.S., Faine, L.A., Galhardi, C.M., Burneiko, R.C., Almeida, J.A., Ribas, B.O., Novelli, E.L.B. 2005. Antioxidant effect of saponin: potential action of a soybean flavonoid on glucose tolerance and risk factors for atherosclerosis. *International Journal of Food Sciences and Nutrition* 56: 79–85.

Rodríguez-Bernaldo de Quirós, A., Castro, R.C. de, Lopez-Hernandez, J., Lage-Yusty, M.A. 2004. Determination of folates in seaweeds by high-performance liquid chromatography. *Journal of Chromatography A* 1032(1–2): 135–39.

Ruiz-Capillas, C., Jiménez-Colmenero, F. 2004. Biogenic amines in meat and meat products. *Critical Reviews in Food Science and Nutrition* 44: 489–99.

Ryynanen, M., Lampi, A.M., Salo-Vaananen, P., Ollilainen, V., Piironen, V. 2004. A small-scale sample preparation method with HPLC analysis for determination of tocopherols and tocotrienols in cereals. *Journal of Food Composition and Analysis* 17: 749–65.

Sacani, G., Tanzi, E., Pastore, P., Cavalli, S., Rey, M. 2005. Determination of biogenic amines in fresh and processed meat by suppressed ion chromatography-mass spectrometry using a cation-exchange column. *Journal of Chromatography A* 1082: 43–50.

Salazar, M.T., Smith, T.K., Harris, A. 2000. High-performance liquid chromatographic method for determination of biogenic amines in feedstuffs, complete feeds, and animal tissues. *Journal of Agricultural and Food Chemistry* 48: 1708–12.

Sanz, Y., Fadda, S., Vignolo, G., Aristoy, M.-C., Oliver, G., Toldrá, F. 1999a. Hydrolysis of muscle myofibrillar proteins by *Lactobacillus curvatus* and *Lactobacillus sake*. *International Journal of Food Microbiology* 53: 115–25.

———. 1999b. Hydrolytic action of *Lactobacillus casei* CRL 705 on pork muscle sarcoplasmic and myofibrillar proteins. *Journal of Agricultural and Food Chemistry* 47: 3441–48.

Shearan, P., O'Keeffe, M., Smyth, M.R. 1991. Reversed-phase high-performance liquid chromatographic determination of dexamethasone in bovine tissues. *Analyst* 116(12): 1365–68.

Silva, B.M., Andrade, P.B., Mendes, G.C., Seabra, R.M., Ferreira, M.A. 2002. Study of the organic acids composition of quince (*Cydonia oblonga* Miller) fruit and jam. *Journal of Agricultural and Food Chemistry* 50: 2313–17.

Simko, P. 2002. Determination of polycyclic aromatic hydrocarbons in smoked meat products and smoke flavouring food additives. *Journal of Chromatography B* 770: 3–18.

Sobrado, C., Quintela, M.C., González, J.C., Vieites, J.M. 2004. Determination of heavy polycyclic aromatic hydrocarbons (PAH) in fishery products. *Journal of Aquatic Food Product Technology* 13: 93–102.

Spotti, E., Chiavaro, E., Bottazzi, R., Del Soldato, L. 2002. Monitoraggio di ocratossina A in carne suina fresca (Ochratoxin A monitoring in fresh pork meat). *Industria Conserve* 77(1): 3–13.

Stadler, R.H., Blank, I., Varga, N., Robert, F., Hau, J., Guy, P.A., Robert, M.C., Van Peteghem, C., Daeseleire, E. 2004. Residues of growth promoters. In: *Handbook of Food Analysis*, 2nd edition, edited by L.M.L. Nollet, New York: Marcel Dekker, pp. 1037–63.

Stancher, B., Calabrese, M. 1997. Anserine and free amino acid content in intensively reared *Onchorhyncus mykiss*. *International Journal of Food Science* 9: 215–22.

Stancher, B., Riccobon, P., Calabrese, M. 1995. Determination of the free amino acids content for the detection of common wheat flour additions in making durum pasta products. *Tecnica Molitoria* 46: 944–52.

Stolker, A.A.M., Stephany, R.W., van Ginkel, L.A. 2000. Identification of residues by LC-MS: the application of new EU guidelines. *Analusis* 28(10): 947–51.

Stratford, M.R.L. 1999. Measurement of nitrite and nitrate by high-performance ion chromatography. *Methods in Enzymology* 268: 130–35.

Stratton, J.E., Hutkins, R.W., Taylor, S.L. 1991. Biogenic amines in cheese and other fermented foods: a review. *Journal of Food Protection* 54: 460–70.

Straub, B., Schollenberger, M., Kicherer, M., Luckas, B., Hammes, W.P. 1993. Extraction and determination of biogenic amines in fermented sausages and other meat products using reversed-phase-HPLC. *Z. Lebensmittel Untersuchung und Forschung* 197: 230–32.

Suzi, G., Gardini, F. 2003. Biogenic amines in dry fermented sausages: a review. *International Journal of Food Microbiology* 88: 41–54.

Terada, H., Tamura, Y. 2004. Determination of dl-alpha-tocopherol acetate and dl-alpha-tocopherol in foods by HPLC using post-column photochemical reaction. *Shokuhin Eiseigaku Zasshi* 45(6): 289–94.

Toldrá, F., Reig, M. 2004. Analysis of meat-containing food. In: *Handbook of Food Analysis*, 2nd edition, L.M.L. Nollet, New York: Marcel Dekker, 1941–59.

———. Forthcoming. Methods for rapid detection of chemical and veterinary drug residues in animal foods. *Trends in Food Science and Technology*.

Trifiro, A., Saccani, G., Zanotti, A., Gherardi, S., Cavalli, S., Reschiotto, C. 1996. Determination of cations in fruit juices and purees by ion chromatography. *Journal of Chromatography A* 739: 175–81.

Umphress, S.T., Murphy, S.P., Franke, A.A., Custer, L.J., Blitz, C.L. 2005. Isoflavone content of foods with soy additives. *Journal of Food Composition and Analysis* 18: 530–50.

Vale, S.R., Glória, M.B.A. 1997. Determination of biogenic amines in cheese. *Journal of AOAC International* 80: 1006–12.

Van Peteghem, C., Daeseleire, E., and Heeremans, A. 2000. Residues of growth promoters. In: *Food Analysis by HPLC*, 2nd edition, edited by L.M.L. Nollet, New York: Marcel Dekker, pp. 965–85.

Vera, A., Murcia, M.A., García-Carmona, F. 1995. Ion levels of fresh and processed spinach using ion chromatography. *Journal of Food Quality* 18: 19–31.

Verdon, E., Couedor, P., Roudaut, B., Sandérs, P. 2005. Multiresidue method for simultaneous determination of ten quinolone antibacterial residues in multimatrix/multispecies animal tissues by liquid chromatography with fluorescence detection: single laboratory validation study. *Journal AOAC International* 88(4): 1179–92.

Vicente, M.S., Ibáñez, F.C., Barcina, Y., Barrón, L.J.R. 2001. Changes in the free amino acid content during ripening of Idiazabal cheese: influence of starter and rennet type. *Food Chemistry* 72: 309–17.

Vidal-Carou, M.C., Lahoz-Portolés, F., Bover-Cid, S., Mariné-Font, A. 2003. Ion-pair high-performance liquid chromatographic determination of biogenic amines and polyamines in wine and other alcoholic beverages. *Journal of Chromatography A* 998: 235–41.

Vinci, G., Antonelli, M.L. 2002. Biogenic amines: quality index of freshness in red and white meat. *Food Control* 13: 519–24.

Voss, K., Galensa, R. 2000. Determination of L- and D-amino acids in foodstuffs by coupling of high-performance liquid chromatography with enzyme reactors. *Amino Acids* 18: 339–52.

Wei, Y., Ding, M.Y. 2000. Analysis of carbohydrates in drinks by high-performance liquid chromatography with a dynamically modified amino column and evaporative light scattering detection. *Journal of Chromatography A* 904: 113–17.

———. 2002. Ethanolamine as modifier for analysis of carbohydrates in foods by HPLC and evaporative light scattering. *Journal of Liquid Chromatography and Related Technologies* 25: 1769–78.

Wilkinson, J., Hewavitharana, A.K. 1997. Lysino alanine determination in sodium caseinate using the LKB Alpha Plus Amino Acid Analyser. *Milchwissenschaft* 52: 423–27.

Yang, S.-W., Park, J.-B., Han, N.S., Ryu, Y.-W., and Seo, J.-H. 1999. Production of erythritol from glucose by an osmophilic mutant of *Candida magnoliae*. *Biotechnology Letters* 21(10): 887–90.

Yashin, Ya., Yashin, A. Ya. 2004. Analysis of food products and beverages using high-performance liquid chromatography and ion chromatography with electrochemical detectors. *Journal of Analytical Chemistry* 59: 1121–27.

Young, C.S. 2002. Evaporative light scattering detection methodology for carbohydrate analysis by HPLC. *Cereal Foods World* 47: 14–16.

Zhang, G.-F., Mortier, K.A., Storozhenko, S., Van De Steene, J., Van Der Straeten, D., Lambert, W.E. 2004. Free and total *para*-aminobenzoic acid analysis in plants with high-performance liquid chromatography/tandem mass spectrometry. *Rapid Communications in Mass Spectrometry* 19(8): 963–69.

13 Sampling Procedures with Special Focus on Automatization

K. K. Kleeberg, D. Dobberstein, N. Hinrichsen, A. Müller, P. Weber, and H. Steinhart

Introduction

Food is, almost without exception, a complex, nonhomogeneous mixture of a multitude of different chemical substances. Consequently, isolation from the matrix is a crucial step in the determination of the analytes of interest. Even with the implementation of advanced techniques of separation and identification, it is rarely possible to analyze food samples without sample pretreatment. In most cases sample extraction, removal of interferences, and preconcentration of analytes are necessary (Pawliszyn 2003). In recent years, the growing concern of the public for food safety and quality control necessitated the development of rapid and automated procedures in order to increase sample throughput and reduce operator manipulation and thereby sources of error. Modern analytical method development is currently focused on automation with the aim of gaining a number of undisputed advantages such as lower time consumption, increase of simplicity, lower probability of sample contamination, and higher repeatability (Mondello et al. 2005). In food analysis the classical sample preparation techniques are time-consuming and require large amounts of solvents that are expensive, generate considerable waste, and can contaminate the sample (Buldini et al. 2002). Traditional liquid-liquid extraction, in particular, is both labor- and time-consuming. Purge and trap techniques have been automated but are expensive. Solid phase extraction (SPE) has generally required either complicated valving or a robot for automation and is, consequently, quite expensive. In contrast, automation of solid phase microextraction (SPME) requires only slight modification of a normal gas chromatography (GC) autosampler (Arthur et al. 1992), but sometimes lacks sensitivity because it is a nonexhaustive equilibrium extraction.

This chapter gives an overview of modern sample preparation techniques with regard to subsequent analytical methods. In the majority of cases, sampling procedures cannot be discussed separately but have to be considered as a combination of analyte extraction and measurement. The most important extraction methods are presented on the basis of aroma and flavor analytics, since related compounds cover a wide range of different chemical classes and properties. Further sections focus on specific extraction and analytical methods for the main food ingredients: proteins, lipids, and carbohydrates. These methods are often characterized by conventional extraction procedures followed by specific analytical techniques. Coverage is not intended to be comprehensive but rather to outline those developments that are of particular relevance for food analysis. Extraction of trace compounds like vitamins, minerals, or contaminants is not explicitly mentioned in this chapter, as most sample preparation methods presented can also be used for these substances.

Analysis of Aromas and Flavors

Problems

The very high production rates of food products and the need for a standard product stable in their aroma demand efficient analytical methods for the characterization of such products (Pillonel et al. 2002). The development of methodologies for the determination of the chemical composition of aromas and flavors is a challenging task, for the following reasons:

1. The concentration of odor-active compounds can be extremely low. Typical perceptible concentrations can be less then 1 ng/liter. Therefore, analytical procedures must provide very high sensitivities (Augusto et al. 2003).
2. Most aromas and flavors are complex mixtures of odorous substances. For example, in different red wines more than 100 substances have been identified (Genovese et al. 2005), whereas in coffee as many as 800 compounds have been identified (Ranau et al. 2005).
3. Some odorants have limited chemical stability as a result of photolysis, oxidation, or other reactions (e.g., terpenes show degradation under daylight conditions when sampled on Tenax tubes (Schrader et al. 2001). Hence, the conditions during sample preparation should be as mild as possible to avoid oxidation or thermal degradation (Sides et al. 2000).

Historically, a number of techniques have been employed for the isolation of aroma compounds from food, including solvent extraction, distillation, and simultaneous distillation-extraction (SDE). In the literature, it is clearly demonstrated that the composition of aroma extracts depends on the isolation methods used (Augusto et al. 2003). The different sampling procedures offer a number of individual advantages but also suffer from specific limitations.

In the following sections, several extraction approaches such as liquid-liquid extraction (LLE), solid phase extraction (SPE), solid phase microextraction (SPME), and supercritical fluid extraction (SFE) as well as static and dynamic headspace techniques with special focus on automatization will be presented. The methods presented are classified on the basis of their extraction principle in figure 13.1 (Marsili 2002; Augusto et al. 2003; Pawliszyn 2003). Aroma and flavor extracts are typically analyzed by chromatographic techniques, especially GC coupled with mass selective detectors or a sniffing port for olfactory detection (Sides et al. 2000; Steinhart et al. 2000), but also high-performance liquid chromatography (HPLC) applications have been reported (Wilkes et al. 2000). The use of electronic noses as a tool for quality and safety control in the food industry is described in chapter 6.

Liquid-Liquid Extraction

Liquid-liquid extraction (LLE) followed by concentration under nitrogen flow or in a rotary evaporator was one of the earliest methods used for the extraction of aroma and flavor compounds from foods. However, the application of LLE is limited, due to the low concentrations of odorous compounds in foods and the problems with coextracted matrix components (Sides et al. 2000). Nevertheless, it remains useful as a simple method for direct extraction of odorous compounds from aqueous food samples like carbonated

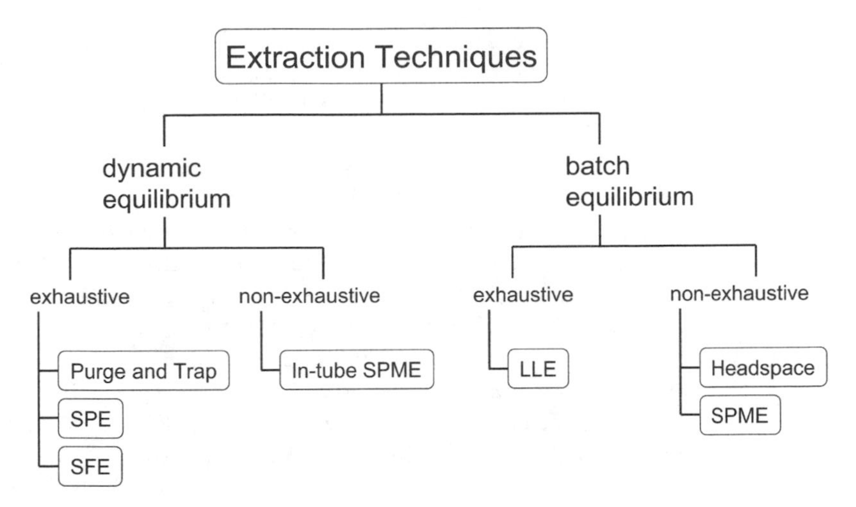

Fig. 13.1. Classification of commonly used methods for the extraction of aroma and flavor compounds from food.

beverages, caffeinated beverages, fruit juices, and wine. Fruits and vegetables can also be extracted by LLE after homogenization with water, treatment with pectinase to destroy the pectins, and filtration to remove particulates (Marsili 2002).

Batch and continuous extractors are commercially available. Solvents most commonly used today are diethylether, diethylether/pentane mixtures, hydrocarbons, Freons, and dichloromethane. Diethylether and dichloromethane are good general purpose solvents, whereas nonpolar solvents like Freons and hydrocarbons are suitable when the sample contains alcohol (Marsili 2002). Despite its simplicity, the modern tendency is to replace LLE by solvent-free techniques, in order to reduce environmental and health risks associated with the needed solvents (Augusto et al. 2003).

However, several applications of batch and continuous LLE in flavor and aroma analysis can be found in the literature. Genovese et al. (2005) described LLE procedures with Freons and dichloromethane for the extraction of aroma compounds from wine. Pihlsgard et al. (2000) used conventional batch LLE to analyze the behavior of volatile compounds during the manufacturing process of liquid beet sugar. Continuous LLE procedures were presented by Castro et al. (2004) for extraction of aroma compounds from sherry wine with 12 samples extracted in parallel using diethylether/pentane, and by Rocha et al. (2000) for the analysis of the aroma potential of Portuguese white grapes in order to optimize wine production. Most published applications of continuous LLE are limited to samples that are free-running liquids. Continuous LLE of viscous or heterogeneous samples such as fruit or vegetable pulps can lead to incomplete extraction because the solvent repeatedly follows the same channels through the sample. To overcome this problem Apps and Lim Ah Tock (2005) implemented a simple magnetic stirrer to the extractor to enhance flavor extraction of guava pulp.

Supercritical Fluid Extraction

Supercritical fluid extraction (SFE) has been used for many years for the extraction of volatile compounds (e.g., essential oils and aroma compounds from plant materials on an industrial scale) (Smith 2003). In recent years, the application of SFE on an analytical scale for sample preparation prior to chromatographic analysis has become more popular due to its advantages over traditional flavor extraction methods like steam distillation or solvent extraction (Huie 2002). The technique is similar to Soxhlet extraction, but the solvent used is a supercritical fluid, namely a substance above its critical temperature and pressure, which provides a particular combination of properties: supercritical fluids diffuse through solid samples like gases but dissolve analytes like liquids. As a consequence, the extraction rate is enhanced, and fewer degradation reactions occur (Buldini et al. 2002). Carbon dioxide is the fluid of choice for SFE of food components because of its low critical constants (31.1°C at 7.38 Mpa), low cost, nontoxic nature, and ease of use (Augusto et al. 2003). By adjusting the pressure and temperature of the extraction, properties of the solvent and thus the extraction efficiency can be varied. If a sample can be extracted at a predetermined density at which the solubility of the target analytes maximizes whereas that of potential interfering substances minimizes, class selective extractions can be achieved (Morales et al. 1998). In the dynamic mode, which is the most frequently used SFE technique, fresh supercritical fluid is continuously pumped through the sample, leading to more exhaustive extraction. The equipment for SFE is shown in figure 13.2. The food sample is introduced into an inert extraction cell in which fluid is pumped at a pressure above its critical point. The temperature of the cell is increased to overcome the critical value of the fluid. Liquid food must be adsorbed onto a porous and inert substrate. The water content of the sample must be strictly controlled because water is a strong cosolvent that can alter the SFE extraction strength and, in addition, can freeze as the fluid is evaporated, resulting in blocking of the flow restrictors (Buldini et al. 2002). After extraction, the extract is generally recovered in a cooled liquid solvent or by solid trapping and is analyzed off-line.

SFE has also been coupled on-line to a range of detection methods such as GC and supercritical fluid chromatography (SFC) (Huie 2002). A dynamic on-line coupling of

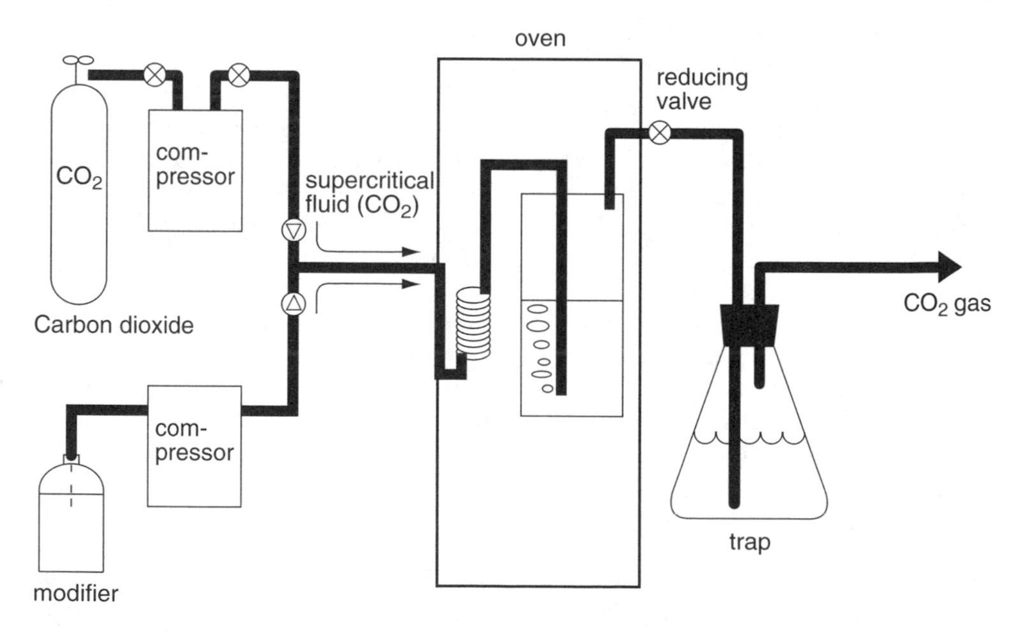

Fig. 13.2. SFE apparatus (Kellner et al. 2004, reprinted with permission).

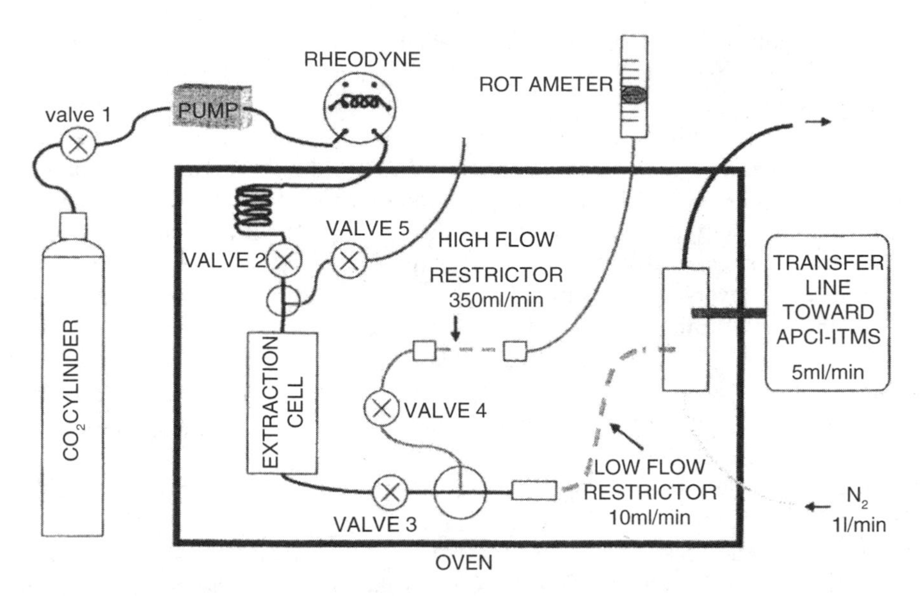

Fig. 13.3. Coupling of the SFE apparatus to gas APCI-ITMS (Jublot et al. 2004, reprinted with permission).

SFE with a gas phase atmospheric pressure chemical ionization (APCI) linked to an ion trap mass spectrometer (ITMS), presented in figure 13.3, has been successfully used for the analysis of volatiles from mint as a fast extraction-analysis device, which could find application as a rapid quality control method or for on-line monitoring of large-scale extractions. CO_2 is pressurized using a pump while the extraction cell is prewarmed and maintained at a constant temperature in an oven. Valve 2 allows CO_2 to enter the cell. Valve 3 controls the extract release. Two flow restrictors are implemented to allow static and dynamic extraction. In the case of static extraction valve 4 is closed, with only the low-flow restrictor opened. The very low flow of 0.2 percent of the cell volume per minute does not disturb the equilibrium in the cell, and the extraction can therefore be considered as quasi static. In the dynamic mode both the low-flow restrictor linked to the MS and the high-flow restrictor are used. A rotameter is fixed at the end of the high-flow restrictor to check the flow stability (Jublot et al. 2004).

Typical applications for SFE are the extraction of different essential oils from plant materials and spices (e.g., ginger, chamomile, pepper) (Diaz-Maroto et al. 2002; Huie 2002) and flavor volatiles from several foods (e.g., olive oil, beef) (Morales et al. 1998; Buldini et al. 2002).

Supercritical carbon dioxide has good solvent properties for nonpolar compounds but is a poor extractor for polar compounds. Thus, for the extraction of such analytes, addition of modifiers like methanol or use of other fluids such as supercritical water is advisable (Sides et al. 2000).

Solid Phase Extraction

Solid phase extraction (SPE) is a simple preparation technique based on the principles used in liquid chromatography. A common drawback of classical extraction methods in sample preparation for complex matrices is that additional cleanup steps are often necessary prior to chromatographic analysis. SPE has been applied both in the isolation and cleanup of aroma extracts. In typical applications, aroma compounds are extracted directly from liquid samples such as beverages and fruit pulps, or aroma compounds are first recovered by solvent extraction or distillation and then passed through a suitable SPE cartridge for cleanup (Sides et al. 2000; Augusto et al. 2003). The SPE procedure involves following four steps: First, prior to sample application, conditioning of the SPE cartridge is necessary. In the second step, the sample is passed through the sorbent bed, and the analytes are exhaustively extracted from the matrix into the solid sorbent. Then, the interfering substances are selectively desorbed from the cartridge by washing with a solution capable of eluting unwanted compounds, but leaving desired analytes retained on the sorbent. In the final step the wash solution is changed for one able to desorb the analytes of interest, which are eluted and collected for analysis (Buldini et al. 2002). The choice of the sorbent material is dependent on the food matrix, the analytes of interest, and their interferents. Alkyl-silica and polymer-based reversed phase sorbents are the most widely used SPE sorbent materials. However, other applications utilize different principles for SPE sample cleanup, including normal phase, gel filtration, affinity SPE, molecular-imprinted stationary phases, restricted access sorbents, ion exchange materials, and mixed-mode sorbents (Gilar et al. 2001). Several applications of SPE for the extraction and cleanup of aroma extracts can be found in the literature. For example, Genovese et al. (2005) used polystyrene-divinylbenzene cartridges to extract aroma compounds directly from red wines. Elution was achieved by pentane/dichloromethane (20 : 1) and dichloromethane. SPE has also been applied to characterize butter aroma. Adahchour et al. (1999) tested the suitability of SPE cartridges packed with C_{18}-, C_8-, NH_2-, and CN-bonded silica, as well as a polystyrene-divinylbenzene copolymer. The butter samples were melted, and the fat fraction was separated by centrifugation. The obtained aqueous fraction was passed through the cartridges and the aroma compounds eluted with methyl acetate. Best recoveries were obtained with the polystyrene-divinylbenzene copolymer cartridge. SPE is probably best suited for extraction of semivolatile aroma compounds (Sides et al. 2000).

Two approaches have been taken to automating SPE: robotic systems and column switching. Robotic systems automate extraction by using traditional cartridges or extraction plates. The operator supplies the robotic systems with a processing algorithm, appropriate solutions, and cartridges. The systems typically comprise an autosampler capable of injecting the resulting extracts into an analytical device (e.g., HPLC, capillary electrophoresis (CE), or (MS) (Gilar et al. 2001). The second approach for SPE automation uses the principle of column switching. It requires, in addition to standard HPLC, an additional mobile phase pump and one or more additional switching valves. The sample is loaded onto a modified SPE cartridge capable of high-pressure operation, which is located at the injection sample loop. Interfering substances are washed off the cartridge to a waste reservoir using mobile phase from the second pump. Subsequently, a valve is switched, and the remaining content of the SPE cartridge is eluted onto the HPLC column. During HPLC analysis the cartridge is switched back to the load position (Gilar et al. 2001). Figure 13.4 shows a schematic of on-line SPE coupled with an ion chromatograph (IC), following the principle described in the preceding (Buldini et al. 2002).

Further examples of SPE automation for extraction of aroma and flavor compounds as well as other food components can be found in the literature. Papagiannopoulos et al. (2002) described an on-line coupling of pressurized liquid extraction (PLE), SPE, and HPLC for automated analysis of proanthocyanidins in malt, which are related to flavor and

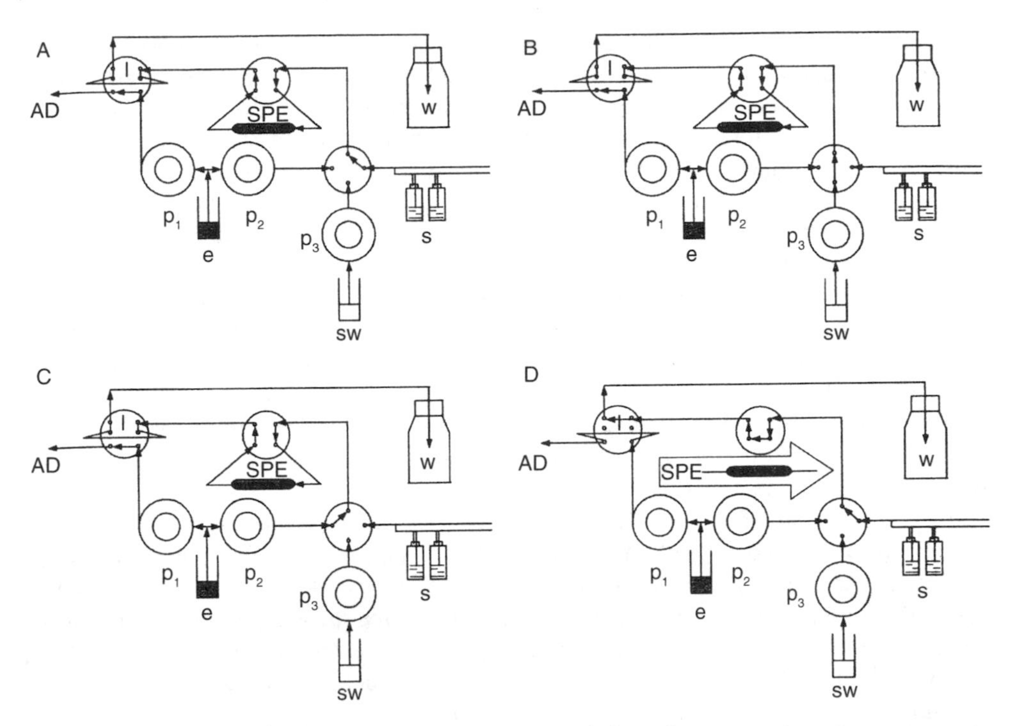

Fig. 13.4. Operative sequence of an SPE unit coupled on-line to an ion chromatograph. **(A)** Sample is transferred from the autosampler (s) to the SPE cartridge (SPE); the IC column is kept in a steady flow of eluent (e) from pump p_1; **(B)** interfering substances are selectively washed off the cartridge (SPE) to the waste reservoir (w), by means of the proper solution (sw) pressurized from pump p_3; the IC column remains in the same condition as in step **A**; **(C)** analytes are eluted from the cartridge (SPE) by means of the eluent (e) via pump p_2 and the loop (1) is filled; the column still remains in the same condition as in steps **A** and **B**; **(D)** the loop (1) is connected to the IC column and analytes are eluted by means of the eluent (e); at the same time a new cartridge (SPE) is inserted on-line for the cleanup of the next sample (modified from Buldini et al. 2002, reprinted with permission).

foam stability in beer. The presented method allows a completely automated analysis from the extraction of the solid malt sample to the chromatogram. Purification and concentration of the PLE extracts were achieved by using commercial polyamide SPE cartridges and elution with a mixture of NaH_2PO_4 buffer and acetonitrile, which also served as mobile phase in the HPLC analysis. An application of automated SPE for the determination of biogenic amines in wine was presented by Arce et al. (1998). They coupled automated SPE with capillary electrophoresis (CE) by using the column switching technique with C_{18} minicolumns. Chilla et al. (1996) used automated on-line SPE-HPLC for the detection of phenolic compounds in sherry wine. The robotic system was equipped with disposable polystyrene-divinylbenzene cartridges.

Static Headspace Analysis

Static headspace analysis is a simple sampling method dependent upon the formation of equilibrium conditions in a closed system. Henry's law can describe the equilibrium between a liquid or a solid sample and the gaseous phase above. After equilibrium is established, an aliquot of the gaseous headspace is sampled by means of a gas tight syringe or a sample loop and then transferred to a gas chromatograph for analysis (Miller and Stuart 1999). Numerous automated instruments for static headspace analysis are available. In most cases the syringe is replaced by a heated transfer line, and the headspace is sampled by pressurizing the equilibrated headspace with carrier gas to a pressure above atmospheric pressure and then allowing the pressurized headspace to vent to atmosphere via a sample loop and thereby filling the loop.

This technique allows for a more inert sampling and a rapid sample transfer to the GC (Snow and Slack 2002). Typical applications of static headspace extractions are the identification of flavor compounds in juices, alcoholic fruit beverages, frankfurters, and cheese (Chevance and Farmer 1999; Miller and Stuart 1999; Valero and Martinez-Castro 2001). Buecking and Steinhart (2002) developed a static headspace method for the investigation of the influence of different milk additives on the release of flavor impact compounds from coffee beverages.

Key odorants are often present in very low concentrations, making it difficult to place enough analyte mass on the column without first concentrating the sample. The full evaporation technique (FET) can address the complex matrix problem by using equilibration above the boiling point of the analytes of interest to force the analytes out of the matrix and into the headspace. These conditions can pose problems because the boiling points are much too high ($<170°C$) to safely use this technique on aqueous samples. Cryogenically trapping and concentrating the analytes in the GC inlet is another possibility, but this poses difficulties when large amounts of water are present, since ice buildup can interfere with this process (Miller and Stuart 1999).

Dynamic Headspace Analysis (Purge and Trap)

The purge and trap or dynamic headspace technique involves passing an inert gas through the sample and collecting the stripped volatile compounds in a trap. The constant removal of the equilibrium between the food and headspace leads to enhanced sensitivity. Different types of traps including loading and unloading techniques are available: cryogenic trapping, sorption on a sorbent bed, on-column vapor traps, and whole-column cryotrapping (Sides et al. 2000; Baltussen et al. 2002; Pillonel et al. 2002). Cold traps have the disadvantage that water is collected with the aroma compounds. Most commonly used sorbent traps use charcoal or porous polymers like Tenax as trapping material. Aroma compounds are thermally desorbed, which allows this technique to be used for a wide variety of volatile compounds; however, it can cause degradation reactions in sensitive molecules. For example, Baltussen et al. (1999) reported several artifact-forming reactions of volatile sulphur compounds, including hydrogen sulphide elimination and dimerization on Tenax and Carbotrap. Solvent extraction is a milder procedure, but very volatile compounds are lost or obscured. The effectiveness of the technique can be increased by closed loop stripping and binding of water with excess sodium sulfate (Sides et al. 2000). Extracts obtained with headspace techniques are generally cleaner than those obtained by solvent extraction or distillation methods (Sides et al. 2000). Numerous purge and trap applications have been published for the analysis of aroma and flavor compounds in food (e.g., beet sugar, soya oil, olive oil, milk, and cheese) (Marsili et al. 1994; Pinnel and Vandegans 1996; Angerosa et al. 1997; Marsili 1999; Valero et al. 2001).

Solid Phase Microextraction

Solid phase microextraction (SPME) is a simple, rapid, solvent-free, and easy to automate technique for the extraction of analytes from gaseous, liquid, and solid matrices that was developed by Pawliszyn and coworkers (Arthur and Pawliszyn 1990; Pawliszyn 1997). SPME combines sampling, extraction, preconcentration, and sample introduction into an analytical instrument in a single step (Chai and Pawliszyn 1995) and has therefore gained increasing popularity for flavor and aroma analysis in recent years. Two different implementations of SPME are frequently used: the fiber SPME and the in-tube SPME. A conventional fiber SPME device consists of a fiber holder and fiber assembly with a built-in coated fiber inside that looks like a modified syringe (Zhang et al. 1994; Lord and Pawliszyn 2000) (fig. 13.5).

The process of fiber SPME is presented in figure 13.6. The sample is placed in a vial sealed with a septum-type cap. When the SPME needle is pierced through the septum and the fiber is extended into the sample, the analytes partition between the sample matrix and

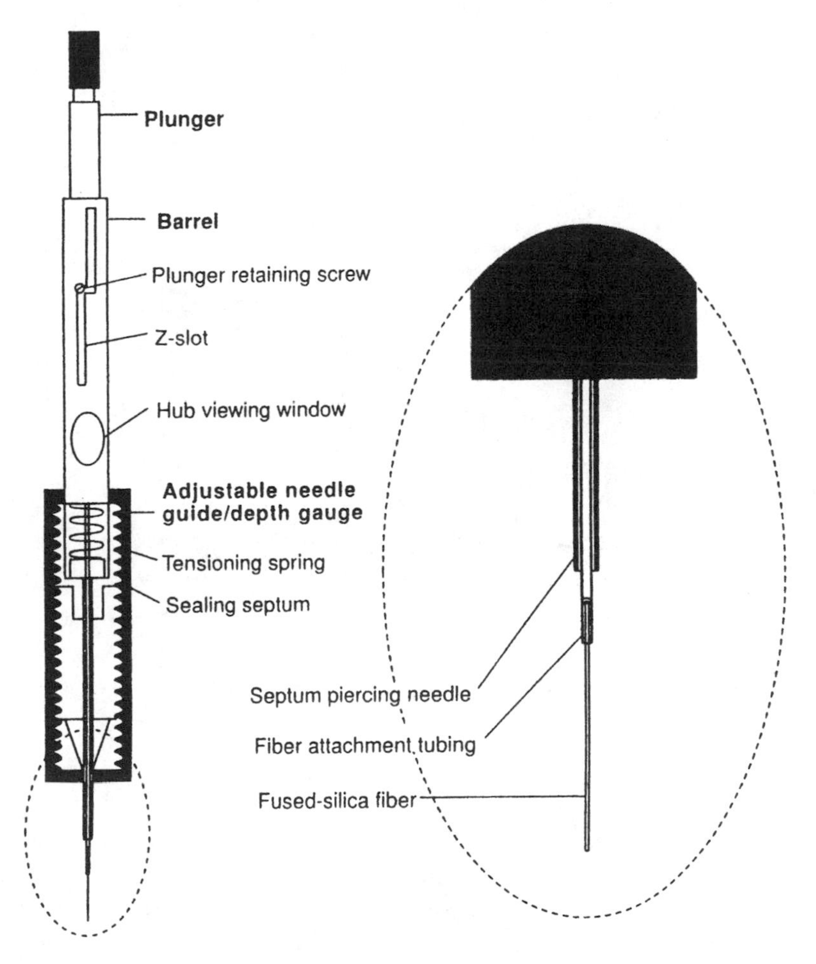

Fig. 13.5. Commercially available fiber SPME device (Zhang et al. 1994, reprinted with permission).

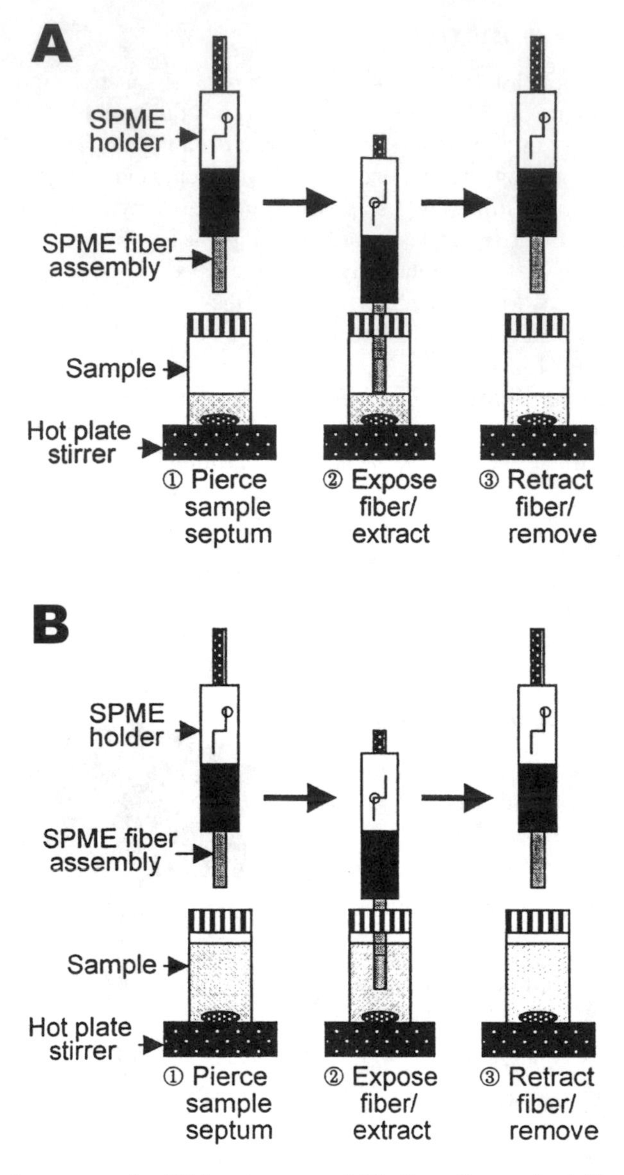

Fig. 13.6. Extraction process by **(A)** headspace and **(B)** direct (immersion) fiber SPME (Kataoka et al. 2000, reprinted with permission).

the extraction phase. If the extraction time is long enough, a concentration equilibrium is established between the sample matrix and the extraction phase. Although SPME has its maximum sensitivity at this equilibrium, a proportional relationship is obtained between the amount of analyte extracted by the SPME fiber and its initial concentration in the sample matrix before equilibrium is reached (Ai 1997a, 1997b). Thus, full equilibration is not necessary for quantification in SPME. Two types of fiber SPME can be used to extract the analytes from the food matrix: direct or immersion SPME or headspace (HS)-SPME.

In direct SPME the fiber is directly exposed in a gaseous or liquid sample, whereas in HS-SPME the fiber is exposed to the vapor phase above a liquid or solid sample. Agitation of the sample using small stirring bars leads to accelerated equilibration.

After a suitable extraction time the fiber is retracted into the needle, and the needle is removed from the septum and inserted directly into the analytical device. The desorption of the analytes is performed either thermally by heating the fiber in the GC injection port or by solvent extraction in the desorption chamber of an HPLC interface (Kataoka et al. 2000) (fig. 13.7). Fiber SPME-GC can easily be automated by modifying a normal GC autosampler. For automated SPME-HPLC special equipment is needed (Lord and Pawliszyn 2000).

The newer in-tube SPME consists of the extraction phase coated on the internal surface of a capillary tube and was developed by Eisert and Pawliszyn (1997) for coupling with HPLC or liquid chromatography (LC)-MS. The in-tube SPME technique is suitable for automation and can continuously perform extraction, desorption, and injection using a normal autosampler (Kataoka et al. 2000; Lord and Pawliszyn 2000). An overview of automated sample preparation using in-tube SPME is given by Kataoka (2002). Figure 13.8 shows a schematic diagram of in-tube SPME coupled with LC-MS. While the injection syringe repeatedly draws and ejects sample from the vial under computer control, the analytes partition from the sample matrix into the internally coated stationary phase of a capillary column until equilibrium is almost reached. Subsequently, the extracted analytes are directly desorbed from the capillary coating by the mobile phase or by a special desorption solvent after switching the six-port valve. The desorbed analytes are transported to the HPLC column for separation and then detected by the mass selective detector.

Several applications of automated in-tube SPME for the analysis of endocrine disruptors, pesticides, and aromatic compounds in environmental samples have been published (Gou et al. 2000; Kataoka 2002; Globig and Weickhardt 2005), whereas only a few applications for the analysis of food components (e.g., caffeine, polyphenols) can be found in the literature (Kataoka 2002).

Further developments of the in-tube SPME technique are wire-in-tube and fiber-in-tube techniques (Saito and Jinno 2003) (fig. 13.9). By inserting a stainless steel wire into the extraction capillary of the in-tube SPME, the internal volume of the capillary is significantly reduced while the surface area of the coating material remains unchanged, resulting in a more effective extraction. In the fiber-in-tube technique several hundred fine filaments of polymeric material that serve as the extraction phase are packed longitudinally into a short capillary, leading to enhanced preconcentration of analytes.

Two distinct types of fiber coatings for both fiber and in-tube SPME are commercially available, with the extraction principle based on either absorption or adsorption. A number of different fiber coatings offer a wide range of analyte solubilities and porosities. The most widely used examples of absorption or liquid fiber coatings are polydimethylsiloxane (PDMS) and polyacrylate (PA). PDMS is a rubbery liquid with a very high viscosity, whereas PA is a solid crystalline coating that turns into liquid at desorption temperatures. PDMS/divinylbenzene (DVB), Carbowax/DVB, Carboxen (CAR)/PDMS, and DVB/CAR/PDMS are mixed coatings in which the primary extracting phase is a porous solid extracting analytes via adsorption (Gorecki and Pawliszyn 1999).

In recent years, a wide variety of applications of SPME for analysis of food volatiles have been published. The vast majority of publications deal with HS-SPME, because of its

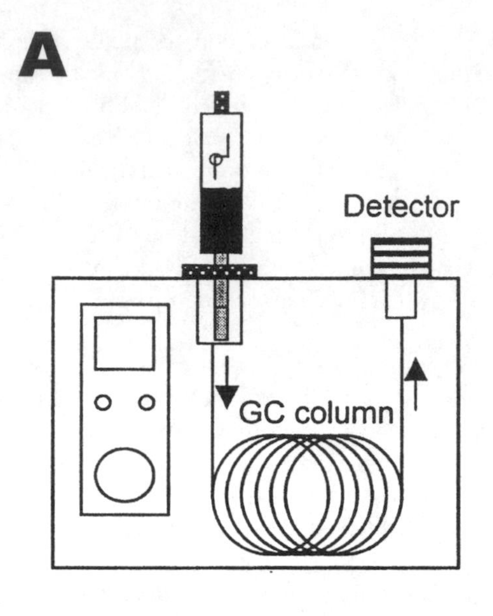

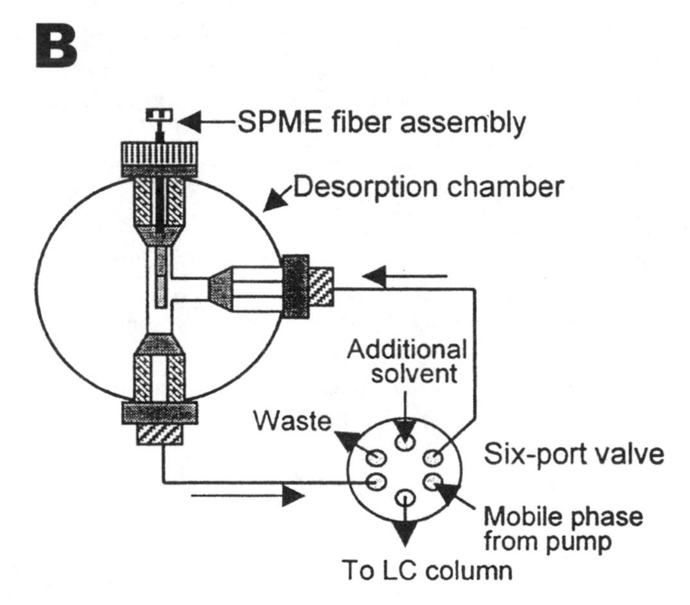

Fig. 13.7. SPME desorption systems for subsequent GC and HPLC analyses: **(A)** thermal desorption on GC injection port and **(B)** solvent desorption using HPLC interface (Kataoka et al. 2000, reprinted with permission).

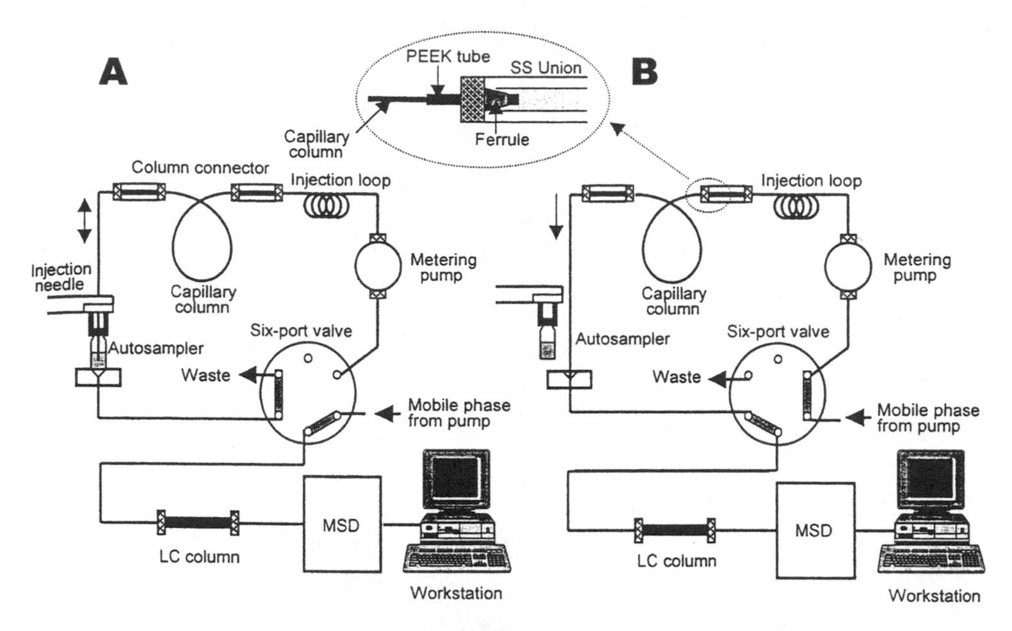

Fig. 13.8. Schematic diagram of an in-tube SPME-LC-MS system: **(A)** load position (extraction) and **(B)** injection position (desorption) (Kataoka et al. 2002, reprinted with permission).

ease of use. Examples of different SPME techniques for the extraction of aroma and flavor compounds from foods are presented in table 13.1. CAR/PDMS and DVB/CAR/PDMS fibers show the best extraction efficiencies for a wide range of analytes with different polarities and molecular weights, whereas PDMS fibers are more suitable for nonpolar analytes and PA fibers for polar analytes.

Figure 13.10 shows the GC-MS chromatograms of waste gas from fat and oil processing containing more than 70 odorous substances after preconcentration with five different commercially available SPME fibers. Highest extraction efficiencies (up to 200-fold) can be obtained by using the CAR/PDMS fiber (Kleeberg et al. 2005).

Although the CAR/PDMS fiber shows the highest sample capacity, it has some disadvantages. Displacement effects of analytes with a lower affinity to the coating have been observed (Cho et al. 2003; Frank et al. 2004; Kleeberg et al. 2005). Therefore, a valid quantification of odorants using CAR/PDMS fibers is only possible if the composition and total amount of compounds in the sample is known (Kleeberg et al. 2005).

This matrix effect can be overcome in liquid samples by standard addition, but this calibration method does not take into account the recovery differences when solid samples are extracted. In that case, multiple SPME can be used. The technique involves sampling the same vial by HS-SPME at equilibrium conditions. The total peak area of an exhaustive extraction of the analytes can be estimated using the peak areas in each individual extraction (Ezquerro et al. 2003). Another possibility was reported by Chen and Pawliszyn (2004): Calibration can be accomplished by exposing the SPME fiber, preloaded with a standard, to an agitated sample matrix, during which desorption of the standard and sorption of analytes occur simultaneously. When the standard is the isotopically labeled analogue of the analyte, the information from the desorption process can be used for estimation of the analyte concentration.

Table 13.1. SPME methods for the analysis of flavor compounds in food.

		SPME Conditions							
Analyte	Food Sample	Extraction Principle[a]	Fiber[b]	Time (min)	Temp. (°C)	Salt Addition	Desorption Temp. (°C)	Detection[c]	Reference
Vegetables, fruits, spices									
Flavor compounds	Fermented cucumbers	HS-SPME	CAR/PDMS	20.0	50	NaCl	275	GC-MS	Marsili and Miller 2000
Aroma compounds	Truffle	HS-SPME	DVB/CAR/PDMS	13.6	53	—	200	GC-MS	Diaz et al. 2003
Aroma volatiles	Cupuassu liquor	HS-SPME	CAR/PDMS	15.0	45	NaCl	280	GC-ITMS	de Oliveira et al. 2004
Eucalyptus volatiles	Eucalyptus	Automated HS-SPME	PDMS	30.0	30	—	250	GC-ITMS	Zini et al. 2002
Off notes	Mint oil	Automated HS-SPME	DVB/CAR/PDMS	0.1	RT	—	230	GC-MS	Coleman et al. 2002
Flavor compounds	Mustard paste	HS-SPME	DVB/PDMS	40.0	50	—	250	GC-MS	Cai et al. 2001
Aroma compounds	Black pepper	HS-SPME	DVB/CAR/PDMS	240.0	RT	—	250	GC-FID/MS/O	Jirovetz et al. 2002
Flavor compounds	Garlic	HS-SPME	DVB/CAR/PDMS	60.0	RT	—	250	GC-ITMS	Lee et al. 2003
Volatile compounds	Aromatic plants	HS-SPME	DVB/CAR/PDMS	60.0	RT	—	230	GC-MS	Bicchi et al. 2000
Juices and alcoholic beverages									
Bouquet	Wine	DI-SPME	PA	15.0	60	NaCl	300	GC-FID	de la Calle Garcia et al. 1996
Esters	Wine	Automated HS-SPME	PDMS	40.0	RT	NaCl	250	GC-FID	Rodriguez-Bencomo et al. 2002
Volatile compounds	Wine	Automated HS-SPME	Carbowax/DVB	10.0	RT	NaCl/ $(NH_4)_2SO_4$	220	GC-FID/MS	Cabredo-Pinillos et al. 2004
Volatile compounds	Wine	HS-SPME	DVB/CAR/PDMS	40.0	35	NaCl	270	GC-FID	Torrens et al. 2004
Sulphur compounds	Wine	Automated HS-SPME	CAR/PDMS	15.0	30	—	300	GC-PFPD	Fang and Qian 2005
Sulphur compounds	Beer	Automated HS-SPME	CAR/PDMS	32.0	45	—	250	GC-PFPD	Hill and Smith 2000
Flavor compounds	Fruit juices	HS-SPME	DVB/PDMS	30.0	40	—	220	GC-ITMS	Miller and Stuart 1999
Flavor compounds	Beverages	HS-SPME	PA	60.0	RT	NaCl	250	GC-MS	Ebeler et al. 2001
Aroma profile	Coffee beans	Automated HS-SPME	DVB/CAR/PDMS	40.0	60	—	260	GC-MS	Mondello et al. 2005

Dairy products									
Lipid oxidation products	Milk	HS-SPME	CAR/PDMS	15.0	45	—	250	GC-MS	Marsili 1999
Volatile fatty acids	Cheese	HS-SPME	PA	20.0	65	—	220	GC-MS	Pinho et al. 2002
Volatile compounds	Cheese	HS-SPME	DVB/CAR/PDMS	40.0	60	—	260	GC-MS	Verzera et al. 2004
Aroma compounds	Cheese	HS-SPME	CAR/PDMS	960.0	RT	—	250	GC-MS/O	Frank et al. 2004
Sulphur compounds	Cheddar cheese	Automated HS-SPME	CAR/PDMS	30.0	50	—	300	GC-PFPD	Burbank and Qian 2005
Fats and oils									
Volatile compounds	Sunflower oil	HS-SPME	PDMS	45.0	40	—	275	GC-ITMS	Keszler et al. 1998
Volatile compounds	Vegetable oils	HS-SPME	DVB/CAR/PDMS	30.0	RT	—	270	GC-MS/FID	Jelen et al. 2000
Volatile compounds	Soybean oil, corn oil	HS-SPME	PDMS	60.0	60	—	250	GC-MS	Steenson et al. 2002
Volatile compounds	Olive oil	HS-SPME	DVB/CAR/PDMS	30.0	40	—	260	GC-FID/MS	Vichi et al. 2003
Flavor compounds	Butter	HS-SPME	CAR/PDMS and Carbowax/DVB	20.0–60.0	40	NaCl	250	GC × GC-FID/ TOF-MS	Adahchour et al. 2005
Others									
Aroma compounds	Vinegar	HS-SPME	CAR/PDMS	60.0	70	NaCl	280	GC-FID/MS	Natera Marin et al. 2002
Furan derivatives	Vinegar	HS-SPME	DVB/CAR/PDMS	40.0	50	NaCl	280	GC-MS	Giordano et al. 2003
Volatile fatty acids	Cane and beet sugar	Automated HS-SPME	DVB/CAR/PDMS	15.0	70	—	270	GC-MS	Batista et al. 2002
Flavor compounds	Honey	HS-SPME	CAR/PDMS	30.0	RT	—	280	GC-MS	de la Fuente et al. 2005
Volatile compounds	Fish	Automated HS-SPME	CAR/PDMS	40.0	50	NaCl	250	GC-MS	Duflos et al. 2005
Furan	Heated foodstuffs	Automated HS-SPME	CAR/PDMS	20.0	50	NaCl	300	GC-MS	Goldmann et al. 2005

[a]HS, headspace; DI, direct immersion.
[b]CAR, Carboxen; PDMS, polydimethylsiloxane; DVB, divinylbenzene; PA, polyacrylate.
[c]ITMS, ion trap mass spectromer; FID, flame ionization detector; O, olfactometry; PFPD, pulsed flame photometric detector; TOF-MS, time-of-flight mass spectrometer.

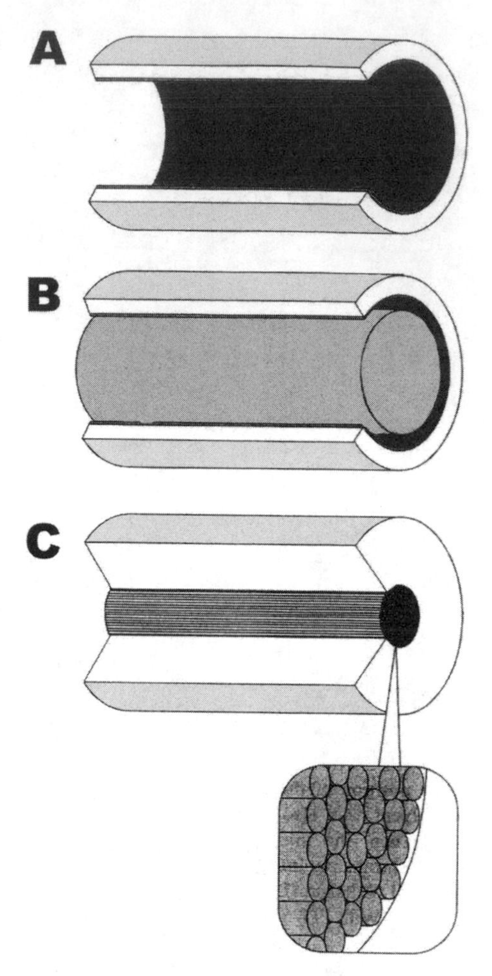

Fig. 13.9. Illustration of three types of extraction capillaries: **(A)** in-tube, **(B)** wire-in-tube, and **(C)** fiber-in-tube.

A further problem that occurs using CAR/PDMS fibers is the formation of artifacts. Lestremau et al. (2001, 2004) reported the formation of unsaturated derivatives from volatile amines and reaction of mercaptans to the corresponding amines during thermal desorption of CAR/PDMS fibers in the GC injection port. Haberhauer-Troyer et al. (2000) explained these artifact formations by reaction of the analytes with metallic particles contaminating the fiber surface in the hot GC injector. Furthermore, CAR/PDMS fibers have been shown to rearrange some monoterpenes during extraction due to adsorbed moisture catalyzing this reaction (Zabaras and Wyllie 2002).

Stir-Bar-Sorptive Extraction

SPME fibers have a small volume of bound extraction phase and therefore have a relatively low sample capacity and sensitivity. For aqueous samples a new technique—referred to as *stir-bar-sorptive extraction* (*SBSE*)—was recently described, which uses a magnetic stirring bar coated with a bonded adsorbent layer (e.g., PDMS) (Baltussen et al. 2002).

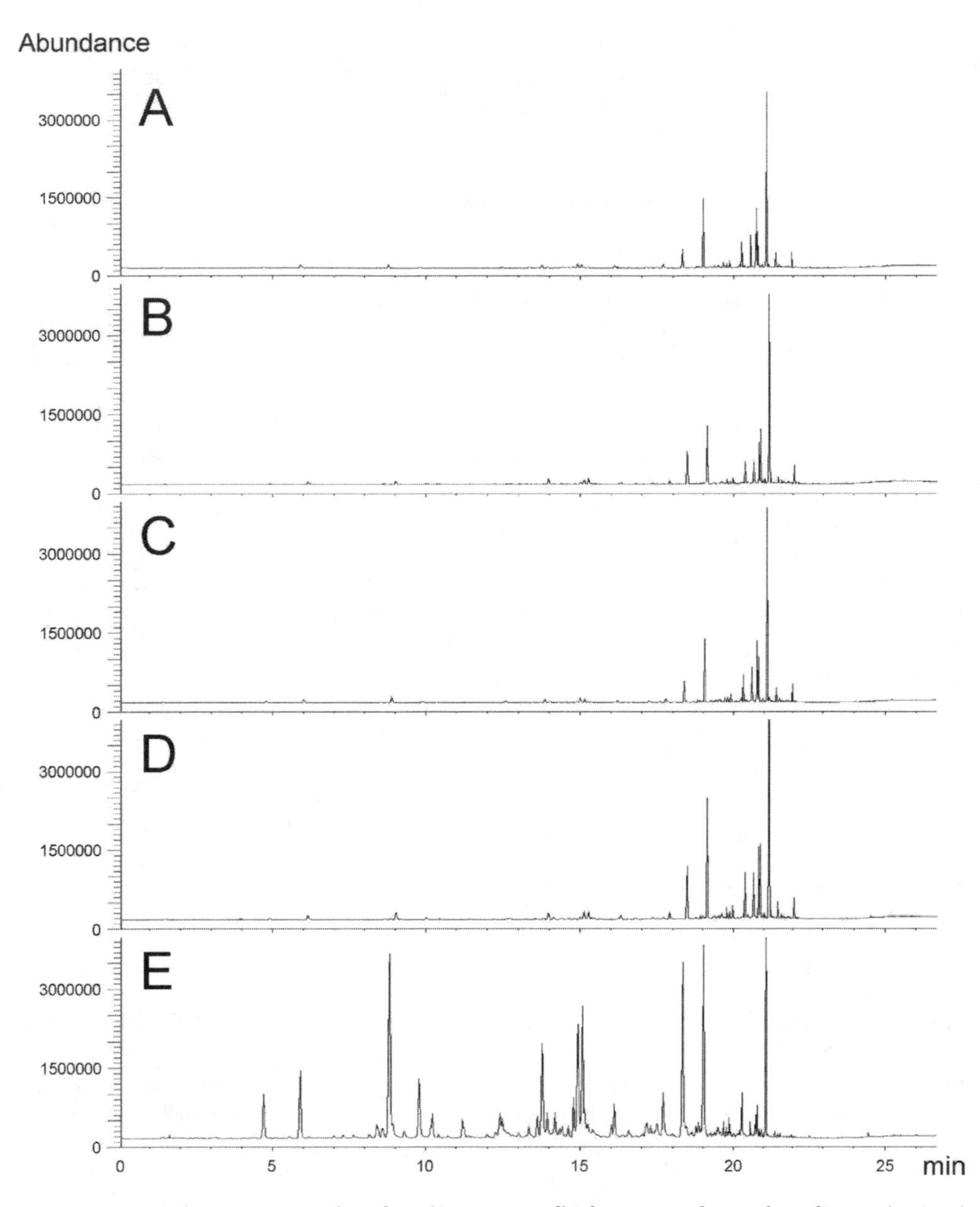

Fig. 13.10. MS-chromatograms (TICs [total ion currents]) of waste gas from a fat refinery obtained using various fibers: **(A)** PA, **(B)** PDMS, **(C)** Carbowax/DVB, **(D)** DVB/PDMS, **(E)** CAR/PDMS.

SBSE has a higher extraction capacity compared to SPME due to the higher surface area of the stirring bar and the much larger adsorbent layer (Smith 2003). The stirring bar is simply rotated in the sample, removed, and extracted thermally for subsequent GC analysis or placed into a solvent for HPLC. Although this technique is still relatively new, several interesting applications have already been reported in the literature. Alves et al. (2005) developed an SBSE method for the analysis of ultratrace volatile flavor compounds in Madeira wine followed by GC-MS analysis. Enhanced extraction of volatiles from malt whiskey has been achieved by SBSE in comparison with SPME (Demyttenaere et al. 2003). Stale carbonyl compounds in beer have been determined by means of SBSE with in situ derivatization followed by GC-MS (Ochiai et al. 2003).

Extraction and Purification of Proteins

Problems

The extraction of proteins plays an important role for several applications in food science and quality control. Common tasks are the detection or determination of proteins (e.g., enzymes or allergens) to appreciate the risk of allergic reactions and to detect and quantify food ingredients or contaminations from other foodstuffs, bacteria, or viruses using methods like polymerase chain reaction (PCR), HPLC-MS, electrophoresis, or immunological methods such as enzyme-linked immunosorbent assay (ELISA), radioallergosorbent test (RAST), or immunoblotting.

The determination of proteins requires a thorough purification to eliminate disturbing effects from the food matrix. For many tasks, like the characterization of proteins for research projects, isolated proteins are needed. But purification is not always an easy job. In many cases only a small amount of protein is present in the food source, or different compounds such as fats, oil, or phenols make protein extraction difficult (Pastello and Trambaioli 2001).

Another problem for extraction purposes is the insolubility of some proteins (e.g., scleroproteins such as collagen, keratin, or elastin). To transfer these proteins into solution, heavy treatment is necessary (e.g., with strong bases or acids, high temperature, or organic solvents). Since those treatments lead to denaturation or fragmentation of the protein, further studies with those extracts are often not expedient.

For soluble proteins, various gentle extraction and purification methods were developed in the past, which are briefly described in this chapter.

Extraction of Soluble Proteins

A simple but effective method is the extraction of the foodstuff in buffer solutions like PBS or TRIS buffer. This method plays a major role in the detection and characterization of proteins using immunological methods or electrophoresis and has been confirmed by countless studies for several foodstuffs like fish (Bernhisel-Broadbent et al. 1992), eggs (Langeland 1982; Bernhisel-Broadbent et al. 1994), nuts (Yeung and Collins 1996; Holzhauser and Vieths 1999), or other plant materials such as lupine (Holden et al. 2005), olives (Arilla et al. 2002) or peaches (Duffort et al. 2002). For extraction, the homogenized foodstuff is stirred with buffer solution for at least 1 h, and the supernatant is isolated by centrifugation. Protein extracts can be stored at –20°C after freeze-drying.

A subsequent purification step of the buffer extracts may be dialysis (Duffort et al. 2002). With this method, small molecules such as sugars or phenols that can have an effect on the determinations are washed out, while larger molecules can not diffuse through the porous membrane of the dialysis chamber into the outer solvent. Special phenolic compounds, which are frequently present in plant materials, can cause problems since they are often colored and can react with proteins. Another widely used method to inhibit problems provoked by phenolic compounds is buffer extraction with the ingredients polyvinylpolypyrrolidone (PVPP), ethylenediaminetetraacetic acid (EDTA), or diethyldithiocarbamate (DIECA) (Bjorksten et al. 1980).

There are of course further methods for the purification of aqueous protein extracts, which are briefly summarized here. One method is the precipitation of proteins with a high salt concentration (Sun et al. 1987). A high ion concentration in the solution leads to a decreasing solvatization of proteins caused by the competitive solvatization of the ions. Another method is the precipitation through addition of organic solvents to the aqueous protein extract (Bleumink 1966). Both methods are easy to handle and result in protein precipitates that are completely free of sugars, lipids, and phenolic compounds. Otherwise the recovery rates are not unusually insufficient (Vieths et al. 1994).

For samples with a high fat content, the extraction of the homogeneous material with organic solvents like hexane is recommended and has been successfully applied in several studies (Burks et al. 1991; Pastorello et al. 1998). After defatting, normal extraction with the previously described methods can be performed.

One drawback of the buffer extraction is that only water-soluble proteins such as albumins or globulins can be extracted. These include allergenic proteins, for which the buffer extraction is used for measuring allergic potentials of foods, among other applications.

Fractionation of Proteins

Fractionation of proteins on the basis of their solubility can be performed, for instance, with cereals. Albumins and globulins are soluble in neutral solutions and are easy to extract with aqueous salt solutions, as described before (Sandiford et al. 1994). Other protein fractions are glutelins, which can be extracted with diluted acid (Shewry et al. 1995), and prolamines, which can be extracted with an alcohol-water mixture (Fido et al. 2004). All these fractions are well-known as the Osborne fractions (fig. 13.11). However, prolamines are only present in plants from the grasses family and may not be a frequent objective in protein extraction tasks. Nevertheless, cereals play an important role in human nutrition.

Fractionation can also be performed by the differences in protein solubility at different pH values (Sun et al. 1987). For example, the major milk protein groups casein and whey can be separated by lowering the pH value below 4.65. The casein precipitate can be resolved after separation from the acid-soluble whey under neutral conditions (Bleumink 1966).

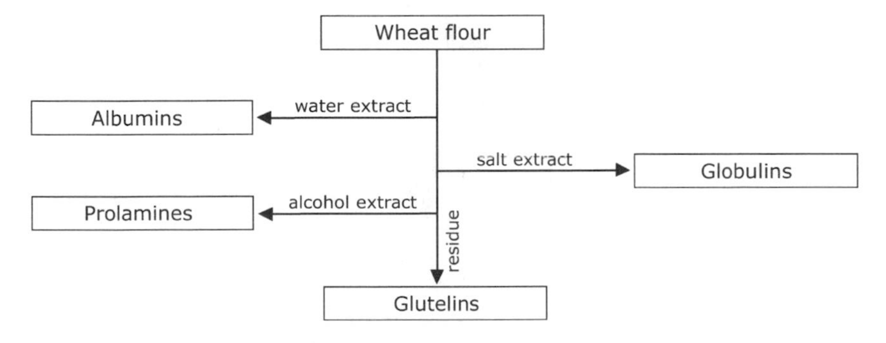

Fig. 13.11. Classical scheme for wheat protein fractionation according to Osborne.

Purification of Single Proteins from Aqueous Extracts

The previously described methods are suitable for the extraction of water-soluble proteins from different foodstuffs. Since these methods are not very specific, further purification steps are needed for isolation of single proteins such as enzymes or allergens. Commonly used methods for the isolation of single proteins are subsequently described.

Electrophoresis

A widely applied method for purification and characterization of single proteins is electrophoresis. Monodimensional assays consist of a polyacrylamide gel electrophoresis

Fig. 13.12. Partition of pea proteins by electrophoresis.

(PAGE) in which the single proteins of an extract are isolated by their mass/load ratio (fig. 13.12). A specific application is the sodium dodecyl sulfate (SDS)-PAGE for isolation of proteins only by their molecular weight. Isolated proteins can be eluted from the gel using special solvents or electro-elution techniques for preparative tasks (Vieths et al. 1995) or by using blotting techniques especially for immunochemical investigations (Koppelmann et al. 1999; Besler et al. 2002). But in many cases the resolution of this monodimensional electrophoresis is insufficient, particularly for complex protein mixtures. In these cases, bidimensional techniques give improvement. Bidimensional electrophoresis combines separation based on molecular mass with that based on isoelectric points for the different proteins. Gels for that kind of electrophoresis consist of PAGE with immobilized pH gradients (Goerg et al. 1988) and afford a high resolution for complex protein mixtures.

Ion Exchange

With regard to the charge of proteins, ion exchange chromatography was established several years ago and has since been improved to a powerful tool for protein purification. Ion exchangers consisting of anionic or cationic groups are able to bind converse-loaded proteins. The protein charge depends on the pH value. On this basis, a suitable chromatography can be set up for each protein depending on the pH value or ion concentration of the solvent and the charge of the ion exchanger (fig. 13.13). For example, anion exchange chromatography was performed for the isolation of various allergens, such as peanut allergens, using a salt gradient (Burks et al. 1992). However, the purity of the ion-exchanged extracts is not always sufficient. Therefore, combined two-step methods using an ion exchanger and gel filtration were introduced to improve the purity and desalt the ion-exchanged extracts for further investigations (Urisu et al. 1991; Ebbehoej et al. 1995).

Gel Filtration

As described before, gel filtration can also be used for protein purification. In fact, gel filtration is one of the most known and used separation techniques for protein mixtures. The proteins are very gently isolated in relation to their molecular mass. Since the resolution of gel filtration is not very high and therefore only well-distinguished molecular mass

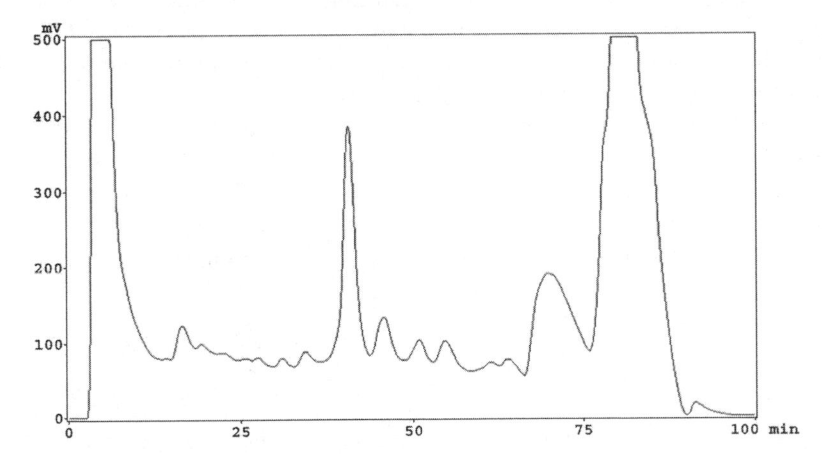

Fig. 13.13. Partition of tomato proteins from a TRIS buffer extract by anion exchange chromatography.

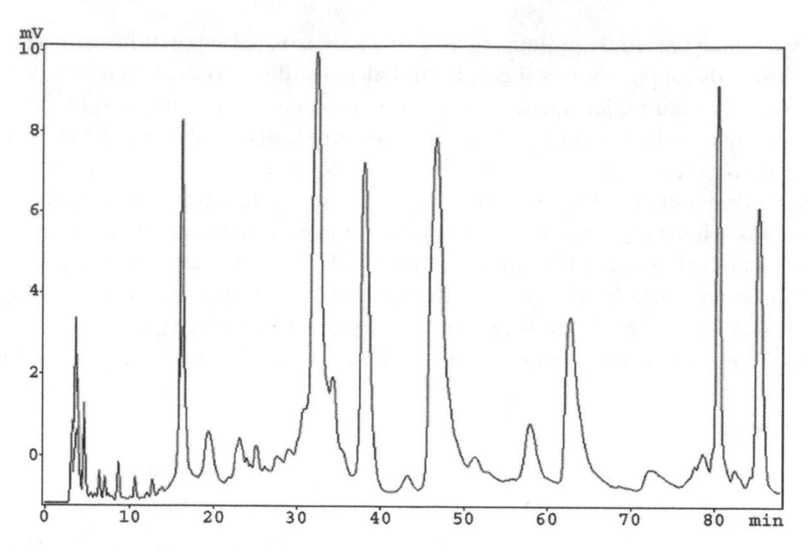

Fig. 13.14. Partition of milk proteins from a PBS buffer extract by reverse phase chromatography.

proteins can be isolated, gel filtration is commonly used as the last purification step after previous isolation steps (e.g., ion exchange or reverse phase chromatography) (Pastello and Trambaioli 2001). However, in some cases it may be possible to isolate proteins only by gel filtration (Sun et al. 1987; Menendez-Arias et al. 1988).

Reverse Phase Chromatography

Reverse phase chromatography is also an often-used tool for protein purification. Hydrophobic sections of proteins can interact with reverse phases, prevalently improved by the addition of ion-pair reagents (Hearn 1991). Accordingly, the isolation of proteins using reverse phase chromatography is based on the differences between protein polarities (fig. 13.14). Therefore, organic solvent gradients are often used to improve the resolution of this method. But it must be noted that many matrix compounds, particularly fats, can also interact with the reverse phase and can consequently lead to inadequate results. Therefore, thorough purification steps (as described before) are needed prior to performing reverse phase chromatography. Thus, reverse phase chromatography is often performed as a second-step purification—for example, in combination with ion exchange or gel chromatography (Fahlbusch et al. 2003). But just as for gel chromatography, there are also known applications that use reverse phase chromatography as a single purification step after extensive processing of the raw protein extract (Inschlag et al. 1998; Sanchez-Monge et al. 1999).

Affinity Chromatography

Almost the best purification method for proteins is affinity chromatography. The great advantages of this method are an absolute specificity to the target protein and the possibility to concentrate small amounts of a protein. Affinity chromatography is based on specific biochemical interactions between the protein and a ligand, well-known as key-lock interactions. Ligands are, for example, receptor proteins, enzyme substrates, activators, or inhibitors and are linked to a solid or gel phase such as activated sepharose. This means that a detailed knowledge of the target protein is required to select the specific ligand for sufficient purification (Palosuo et al. 1999).

A special practice of affinity chromatography is immunoaffinity chromatography. In this case, the ligand is an antibody that interacts highly specifically with the antigenic protein or vice versa. Modern biotechnology is able to generate antibodies to a broad range of proteins or other target molecules, which is why affinity chromatography is commonly used in purification of proteins (Holzhauser and Vieths 1999; Palosuo et al. 1999).

One drawback of affinity chromatography is the high amount of ligands needed to coat the solid or gel phase (Pastello and Trambaioli 2001). This can be a problem especially for immunoaffinity chromatography since specific antibodies are very expensive.

Immunoprecipitation

A similar but more economical method based on immunochemical interactions is the immunoprecipitation method. The main difference from affinity chromatography is the lack of a solid or gel phase. Antibodies are incubated together with the appropriate target protein to form the antibody-antigen complex. Afterward, the antibody-antigen complex can be isolated by centrifugation, and the antigenic protein can be discharged by changing the pH value (Johansen and Svensson 2002).

Reversed Micelles

A more recent method for purification of protein extracts is liquid-liquid extraction with reversed micelles. The system consists of two immiscible phases, that is, an aqueous salt buffer and an organic phase, forming an emulsion by the addition of a surfactant (Göklen and Hatton 1985). Proteins from the aqueous phase can interact with the surfactant by hydrophilic forces and are consequently transferred into the organic phase. The extent of interactions is protein specific and depends mainly on the pH value, ionic strength, and ion type of the aqueous phase and on the concentration of the surfactant (Pires et al. 1996). Increasing the concentration of the surfactant also increases the protein molecular size that will be extracted into the organic phase. By increasing the ionic strength (i.e., increasing the buffer concentration) interactions between proteins and surfactants or between surfactant molecules themselves will decrease (Pires et al. 1996). After the proteins are extracted into the organic phase, the reversed step is performed by extracting the proteins back into a second aqueous phase. The recovery of the proteins from the surfactant-protein complex can be provoked by changing the pH value, increasing the salt concentration (Göklen and Hatton 1987), or simply the addition of polar alcohols (Aires-Barros and Cabral 1991).

The reversed micelles method has been performed for many proteins with very different results regarding the recovery rate (Krei and Hustedt 1992; Rahaman et al. 1988). Furthermore, there are still some unknown factors related to the existing interactions. Besides the mentioned surfactant-protein interactions, there are other forces like hydrophobic interactions (Aires-Barros and Cabral 1991) or ion-pair mechanisms (Hatton 1989) to be considered, making the establishment of this method somewhat incomplete.

Extraction and Purification of Lipids

Problems

Lipids, a very important class of natural compounds, occur as membrane compounds, energy sources, and as signaling molecules. The group of lipids consists of different lipid classes (neutral lipids, free fatty acids, phospholipids, ceramides, sphingolipids, gangliosides) covering a wide range in polarity. Lipids normally do not appear in their free form, but embedded in a matrix. Thus an extraction step is necessary before further analysis. In fact this step frequently generates mistakes that lead to a false analytical conclusion. Therefore a well-chosen method of extraction for the lipid that needs to be determined is required, as differences in polarity require multiple extraction procedures to obtain clean extracts for further analysis.

Extraction of Lipids

The components obtained in the lipid extract depend on the extraction methods used, especially on the solvent. Nonpolar solvents (hexane, ether, or supercritical carbon dioxide) can be adopted for the extraction of simple neutral lipids—for example, esters of fatty acids and acylglycerols. More complex and more polar lipids (phospholipids, lipoproteins, glycolipids, free fatty acids, etc.) require more polar solvents such as methanol or acetonitrile. Generally, solid phase extraction methods are advisable for complex polar and nonpolar lipid components (Ruiz et al. 2004). This technique has been automated and coupled not only for the analysis of lipids but also for other nonpolar compounds such as dioxins (Focant et al. 2004), phenols (Rosenberg et al. 1999), or pesticides (Planas et al. 2004) in foods or biological matrices. There are possibilities to couple automated SPE to HPLC and GC.

However, most biological matrices cannot be analyzed or extracted without a previous digestion. There are different "classical" digestion methods that have been used for many years. The choice of technique depends on the surrounding matrix. The Weibull-Stoldt method is used for fat extraction from meat, fish, or oilseed. After the surrounding proteinogenic material has been decomposed by hot hydrochloric acid, the fat is extracted in a Soxhlet apparatus with either ether or hexane. For dairy products an alkaline decomposition is normally used. Although these methods have proved to provide exact results regarding the total fat content (Eckoldt et al. 1999; Kolar et al. 1993), the extreme conditions (high temperature, very high or very low pH value) often lead to unintentional modifications of the molecular structure of the analytes. Therefore, if further analysis of certain lipid compounds is required, an extraction method has to be used that does not alter the structure of the analytes.

In 1959 Bligh and Dyer introduced a method for total lipid isolation from fish muscle using chloroform and methanol as solvents (Bligh and Dyer 1959), which became very popular and has often been modified and improved. The disadvantages of this method are the toxicity and cost of the solvents. The toxicity can be avoided by using other, preferentially nonhalogenated, solvents (Hara and Radin 1978; Jensen et al. 2003).

However, the cost factor remains with these applications, but it can be eliminated by using supercritical fluid extraction (SFE) (Neff et al. 2002). The marginal alteration of lipid compounds during extraction, the ease of solvent removal from the extract, and the lack of toxicity are other advantages of the SFE (Johnson and Barnett 2003). Furthermore, the SFE equipment can be coupled with chromatographic systems, which provides the opportunity to automate nearly the complete lipid analysis. Although the length of the "classical" lipid extraction has been successfully shortened, for example, by using an ultrasonic- (Luque-Garcia and Luque De Castro 2004) or microwave-assisted (Garcia-Ayuso et al. 1999; Priego-Capote et al. 2004; Priego-Capote and Luque De Castro 2005) Soxhlet extraction, SFE remains one of the fastest methods. Modern extraction methods provide reliable results within 30–50 min. However, SFE requires expensive equipment, and the savings achieved by avoiding solvents are only profitable after a multitude of extractions.

Finally it should be emphasized that no single extraction method is applicable to all matrices or all analytes. The right choice of application depends on various factors and is an important element of a successful analysis.

Fractionation of Lipids

Different analysis and extraction methods have to be chosen. Extraction and quantitation of total lipids is mainly the first step in analysis. For metabolic studies detailed information on the amount of different lipid classes requires tedious extraction separation steps. A schematic presentation for a (nearly) complete lipid analysis is presented in figure 13.15 (Dasgupta and Hogan 2001; Hoffmann et al. 2005).

Fast Methods

Some methods of food analysis do not need sophisticated sampling procedures, as the analysis can be carried out directly in the matrix.

Infrared (IR) and Fourier transform infrared (FT-IR) techniques are important tools for the resolution of the configuration and structure analysis of lipids. Functional groups, such as double bonds (conjugated as well as isolated), hydroxyl, epoxy, and ester functions produce unique absorption bands in the midinfrared spectral region (wavenumbers 600–4000 cm^{-1}). The solvent commonly used is carbon disulfide.

FT-IR is frequently used to measure the total *trans* fatty acids in oils (Ma et al. 1999). The determination is based on the measurement of the 966 cm^{-1} out-of-plane deformation vibration. This band is characteristic for isolated *trans* fatty acid bonds. Regrettably, it overlaps with other features in the IR spectrum, leading to an inconsistent background. Actually the band turns into a shoulder in the spectrum if the sample measured contains amounts of *trans* fatty acids under 2 percent. The resulting low accuracy can be improved by using attenuated total reflection (ATR) cells (Mossoba et al. 2004). The method requires neither weighing nor quantitative dilution in any solvents and delivers a radical improvement of the sensitivity at low *trans* fatty acid concentrations (Milosevic et al. 2004). The McGill IR Group also developed FT-IR applications for the measurements of free fatty acids (Al-Alawi et al. 2004), iodine value (Man et al. 1999), and saponification number (Van De Voort et al. 1992). With the application of an FT-IR method for the measurement of the peroxide value (Ma et al. 2000), an accurate and fast tool for monitoring the deterioration of fats was obtained.

Another interesting application area of FT-IR spectroscopy is the determination of the content of solids in fats. This parameter influences the characteristics of margarines, shortenings, and other fat blends. It is normally obtained using dilatometry or nuclear magnetic resonance (NMR). Both techniques involve measurements at a series of set temperatures

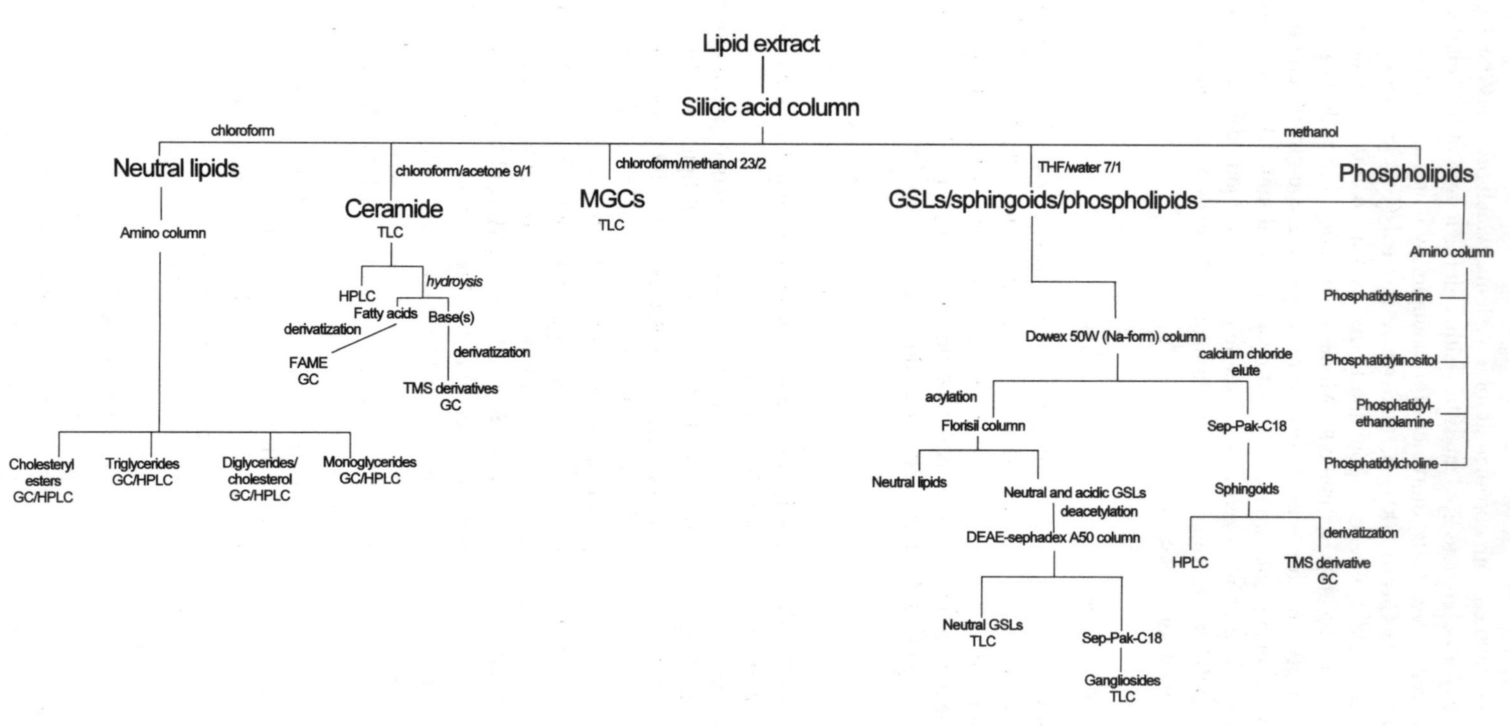

Fig. 13.15. Schematic presentation of lipid class fractionation by solid phase extraction (modified from Dasgupta and Hogan 2001).

and accordingly are relatively time-consuming. In contrast, Van de Voort et al. (1992) presented an FT-IR method that requires only a single measurement of the cleaned/purified and melted sample. The method delivered a reproducibility and an accuracy comparable to the conventional methods. Thus, the FT-IR method could be used as a substitute for either the dilatometric or the NMR method to determine the content of solids. In addition it has the advantage of the shortened time of analysis by the elimination of the tempering steps required for the traditional methods.

Related to FT-IR is near-infrared (NIR) technology. It is frequently used in food analysis and provides fast results with only small sample amounts. NIR applications normally require little or even no sample preparation. The spectra are recorded in the area between 700 and 2500 nm, where combination and overtone bands of carbon-hydrogen, oxygen-hydrogen, and nitrogen-hydrogen are displayed. The technique is applied for the determination of moisture, fat, and protein content in various matrices—for instance, oilseeds, grains, meat, or flour. NIR techniques are primarily employed in food production, where raw materials have to be checked within seconds before and during the production process to prevent faulty products and to save costs resulting from any out-of-specification batch (Kurowski et al. 1998).

The spectral region examined in this technique is complex; thus sophisticated calibrations are necessary. Although most NIR instruments have associated software to assist with development of calibrations, adjustments have to be performed prior to every measurement.

There are a large number of NIR applications available for food analysis. Moisture, protein, and fat content can be measured simultaneously in different matrices (Cozzolino et al. 1996; H.-Z. Zhang 2000). Christy et al. (2004) developed a rapid method using NIR technology for the detection and quantification of adulterations in olive oil. Although it was possible to detect an adulteration with a certainty of nearly 100 percent, it was not possible to identify the adulterant exclusively with NIR. Other analytical techniques, for instance GC of fatty acid methyl esters (FAME), are necessary.

Nevertheless, various works show that the estimation of the fatty acid composition from fats and oils can also be performed by NIR (Sato 2002; Sato et al. 2003; Baye and Becker 2004; Gonzalez-Martin et al. 2004). The method provides a simple, rapid, and nondestructive means of estimating the fatty acid composition. However, not all fatty acids can be measured with the accuracy yielded with other analytical methods, like GC of FAME.

NMR spectroscopy is a frequently used method in the fat and oil industry for monitoring and quality control applications. It is employed for the in-complex and rapid measurement of the fat concentration in food and oilseeds, as no extraction steps are necessary in many cases. Guthausen et al. (2004) described a portable NMR analyzer to measure the fat content in a packaged product without destruction of the material. The processed NMR signal was comparable to the fat content obtained by reference methods.

^{1}H-NMR is used for the characterization of different lipids. For example, the technique enables food controllers to distinguish milk samples from different species of animals (Brescia et al. 2004). Aerts and Jacobs (2004) detected the yields of mono- and diepoxidized products emerging from diunsaturated substrates during epoxidation reactions of fatty acid methyl esters.

In fundamental research, ^{1}H- and ^{13}C-NMR are used for structure determination and confirmation of identity. In ^{13}C spectra characteristic bands are generated by certain carbon atoms, for instance olefinic, allylic, or ω1-3 carbons. The technique requires large amounts of high-purity samples (1–50 mg). Sometimes the plotting of different spectra against each other is essential for the clear identification of the structure. Thus, 2D spectra are generated. This technique identifies the atoms that are adjacent and clarifies which atoms couple with each other. Both NMR methods (^{1}H and ^{13}C) are very useful for structure determination of individual compounds but provide only medium accuracy in analyzing complex mixtures.

In contrast to this, the ^{31}P-NMR can be used for the exact determination of different phospholipids in parallel (Helmerich and Koehler 2003). As these only contain one phosphor atom per molecule, which generates characteristic signals for the substances, the intensity of the signals correlates with the analyte content in the measured sample. ^{31}P-NMR techniques have been used for the determination of phospholipids in different matrices, for instance human and boar spermatozoa (Lessig et al. 2004), milk (Murgia et al. 2003) or lecithins and flour (Helmerich and Koehler 2003). Cremonini et al. (2004) reported that not all employed solvents provide exact results. The so-called CUBO solvent (a ternary mixture of N,N-dimethylformamide (DMF), triethylamine, and guanidinium hydrochloride) adulterated the results obtained for the phospholipid content. In general ^{31}P-NMR displays a fast and precise method for the measurement of phospholipids in different matrices. Regrettably, high sample amounts are required in all NMR techniques. When only small amounts are available, normal phase high-performance liquid chromatography (NP-HPLC) with evaporative light-scattering detection (ELSD) is the more appropriate choice for the analysis of these compounds.

Extraction and Purification of Carbohydrates

Problems

Carbohydrates are one of the major components of foodstuffs along with lipids and proteins. The term *carbohydrate* summarizes a huge range of structures. The most common food carbohydrates are classified in mono-, oligo-, and polysaccharides. These groups of carbohydrates differ not only in the degree of polymerization but also in their function and physical and chemical properties. The complexity of carbohydrate analysis is caused by the diversity in structures and solubility. In addition, the analytical methodology is influenced by the desired information (e.g., detailed structural information or fast on-line measurement).

For extraction purposes, carbohydrates are divided into soluble and insoluble carbohydrates. Extraction of soluble carbohydrates, monosaccharides and oligosaccharides, is usually carried out with water or aqueous alcoholic solutions, whereas the extraction of polysaccharides is much more complex. Methods are time-consuming, often remaining part of the research. On the other hand, methodologies to analyze dietary fiber contents as a sum parameter are widely used in food analysis. These are based on the determination of indigestible food constituents, which contain structural polysaccharides, for example, cellulose and hemicelluloses, as well as lignin.

A short outline is given here summarizing recent methods in the analysis of mono- and oligosaccharides: colorimetric and chromatographic methods and biosensors. NIR methods have already been reviewed in former sections (e.g., to analyze lipid samples) and therefore will not be mentioned, although they are frequently used in carbohydrate analysis.

Colorimetric Methods

Colorimetric methods for carbohydrate analysis are well established. Simplicity, rapidity, and sensitivity are their main advantages, whereas a limited specificity is often a drawback. Among a multitude of existing methods to determine the total sugar content, the phenol-sulfuric acid method (Dubois et al. 1956) is widely used because of its high sensitivity and simplicity. This method is based on the formation of furfural derivatives during the reaction of saccharides and sulfuric acid, followed by condensation with phenol using strong acidic conditions.

The reaction can be used for the quantitative determination of all kinds of carbohydrates (e.g., ketoses, aldoses, uronic acids, and oligo- and polysaccharides) but also for glycoproteins, DNA, or RNA. The absorption maximum of the measured products depends on the type of saccharide (e.g., hexoses: 485 to 490 nm; pentoses or RNA: 480 nm).

A recently scaled-down and modified version of the original test using 96-well microplates requires only 1–150 nmol of sugars and can be completed in less than 15 min (Masuko et al. 2005), showing its practicability as a fast screening method. Unfortunately the application of the phenol-sulfuric acid method is often limited by its poor specificity.

Chromatographic Methods

As most foodstuffs contain a mixture of different sugars, separation methods have become necessary tools in carbohydrate analysis. Various gas chromatographic and high-performance chromatographic methods have been established in carbohydrate analysis.

Gas Chromatographic Methods

Since mono- and oligosaccharides have to be converted to volatile derivatives prior to GC analysis, these methods could hardly be automated. The most common derivatization methods for gas chromatographic determination of mono- and oligosaccharides are trimethylsilylation, acetylation, and trifluoroacetylation. Detection is usually carried out using flame ionization or mass spectrometry. In a first step the carbonyl group is often reduced, oximated, or converted into diethyl dithioacetals. The most common derivatization method for monosaccharides includes a reduction with sodium borohydride to the corresponding sugar alcohols, followed by acetylation with acetic anhydride, catalyzed by anhydrous sodium acetate (Albersheim et al. 1967), water-free pyridine (Blake and Richards 1970), or preferably 1-methylimidazole (Connors and Pandit 1978; Bittner et al. 1980). Acetylation using the former catalysts requires prior removal of borate, usually done by evaporation with methanol (Zill et al. 1953). By using 1-methylimidazole as catalyst, removal of borate becomes unnecessary (Blakeney et al. 1983). The alternatively used derivatization of aldoses into dithioacetals provides a single peak per compound. The reaction is carried out with ethanethiol under acidic conditions. Subsequently, hydroxyl groups of the dithioacetals are converted into trimethylsilyl ethers (Honda et al. 1979). Oximation of reducing sugars with, for example, hydroxylamine is mostly carried out in pyridine or methanol, followed by silylation of the remaining hydroxyl groups. This procedure provides two isomers, the syn- and antioxims. Oximation is of interest due to the possibility to determine ketoses and aldoses as well as oligosaccharides (Zürcher et al. 1975). Trifluoroacetylation, using trifluoroacetic anhydride as derivatization reagent, provides the most volatile sugar derivatives allowing determinations up to tetrasaccharides (König et al. 1981).

High-Performance Liquid Chromatography

High-performance liquid chromatography (HPLC), usually combined with refractive index (RI) detection or pulsed amperometric detection (PAD), is one of the most common separation techniques in carbohydrate analysis. To obtain optimal results for each particular purpose, different stationary phases have been developed, including unmodified silica, chemically modified silica, reversed phase, and pressure stable cation and anion exchange phases. Due to their polarity unmodified silica phases are limited in their application in carbohydrate analysis, whereas modified silica and ion exchange stationary phases are widely used.

Frequently applied modified silica phases are amine-bonded silica phases as well as diol and cyano phases. A major disadvantage of the successfully used amine-bonded silica phases is the limited column life mainly caused by the formation of Schiff bases between the stationary amino groups and carboxyl groups of the sample (Verhaar and Kuster 1981). Compared to amino-bonded stationary phases, cyano or diol phases are characterized by a higher stability.

In the majority of cases the separation of carbohydrates on the mentioned modified silica phases is performed with a mixture of acetonitrile and water as eluents. As the separation requires relatively high contents of acetonitrile (>70 percent), difficulties regarding solubility may arise. A separation with water is possible if reversed phase stationary phases are used. These are most suitable to less polar sugars (e.g., oligosaccharides).

A commonly used chromatographic technique to separate carbohydrates is the ion exchange chromatography. Numerous applications have been published using high-pressure stable cation and anion exchange stationary phases. Both types are either based on silica gel or on polymeric matrices. Regarding high-pressure cation exchange chromatography (HPCEC), polymer-based stationary phases are more common than silica gel matrices, leading to excellent separation results. Polymeric cation exchanger resins are usually based on styrene-divinylbenzene copolymers modified with sulfonate groups and with different degrees of cross-linking (Dorfner 1991). Higher cross-linked resins are designed for the separation of mono- and disaccharides, whereas lower cross-linked matrices are used for the analysis of oligosaccharides. The separation of carbohydrates with HPCEC is also highly influenced by the type of counter ion. For instance Ag^+ is the favored counter ion in the separation of oligosaccharides.

As in the case of HPCEC, in high-performance anion exchange chromatography (HPAEC) the preferred stationary phases are based on polymeric resins, mainly because of their higher stability under alkaline conditions. Particularly poly(styrene-divinylbenzene) copolymers combined with PAD are applied in carbohydrate analysis. For example prebiotics like fructooligosaccharides can be analyzed rapidly and accurately using HPAEC-PAD (Sangeetha et al. 2005).

A special application of HPAEC is the separation of sugars as borate complexes, using strongly basic anion exchangers and a borate buffer as the mobile phase. This methodology is based on the formation of anionic carbohydrate-borate complexes that interact specifically with the stationary phase (Bauer and Voelter 1976). One advantage of this application is the continuous regeneration of the anion exchanger due to the borate buffer. Nonetheless, because high buffer concentrations and gradients of temperature or borate concentration are used, a postcolumn derivatization is necessary to perform the detection (Sinner and Puls 1978).

Biosensors

Biochemical methods, in particular enzymatic methods, are important analytical tools in carbohydrate analysis. They are frequently used, and several official analytical methodologies are based on enzymatic assays. Although biosensors represent a relatively new detection method, their application in carbohydrate analysis is already well established. High specificity of the reactions and fast and easy operability are the predominant advantages of biosensors. Biosensors are generally defined as devices combining a biological recognition element (biochemical receptor) in direct spatial contact with an electrochemical transduction element (Thevenot et al. 1999). In carbohydrate analysis, enzymes are mostly used as biological recognition elements. In the majority of cases the quantitative measurement is carried out either by amperometric or by potentiometric detection. The first amperometric "enzyme electrode" in which glucose oxidase was entrapped at an oxygen electrode using a dialysis membrane was described by Clark and Lyons (1962). The decrease in oxygen concentration is proportional to glucose concentration. The first potentiometric enzyme electrode was described by Guilbault and Montalvo (1969).

Many advancements and improvements in biosensor applications have since been made. One objective in research was the improvement of the stability of biosensors, keeping in mind the maintenance and optimization of the activity of the enzymes. An important precondition for stability and optimal enzyme activity is the immobilization of the enzymes on a carrier. However, immobilized enzymes normally show a lower specific activity than native enzymes. Enzyme immobilization is classified in two main groups. Either the enzymes are adsorbed or covalently linked to water-insoluble matrices or they are entrapped in a matrix or by a membrane (Hartmeier 1986; Stein and Schwedt 1993).

Further research has been done in the development of biosensors for the determination of sugars other than glucose: for example, fructose (Hanke et al. 1996), galactose (Wen et al. 2005), disaccharides, such as maltose (Marconi et al. 2004), sucrose (Mohammadi et al. 2005) and lactose (Eshkenazi et al. 2000; Stoica et al. 2006), and polysaccharides (e.g., cellulose) (Eberhardt and Galensa 1998). This is achieved by a combination of different enzyme reactions.

Coupling of biosensors is an important issue in flow systems like flow injection (FI) or HPLC. FI analysis is an additional facility to enhance the performance of biosensors. In practical applications biosensors often have to be run under suboptimal conditions. Major advantages in the use of FI are the possible periodic calibration, conditioning, and therefore regeneration of the biosensor (Hansen 1996). As a biosensor is defined by direct spatial connection of a biochemical receptor and an electrochemical transducer as well as by being a self-contained device, the combination with HPLC and FI requires a broader definition of biosensors. More accurately, they should be described as an immobilized enzyme reactor (IMER) (Stadler and Nesselhut 1986) followed by electrochemical detection (ECD). Combining the advantages of HPLC (separation), enzymatic reaction (specificity), and electrochemical detection (sensitivity), HPLC/IMER is an important method for precise, sensitive, but also fast analysis of carbohydrates in foodstuffs. Its importance will presumably further increase (Galensa 2004). For example, Sprenger et al. (1999) described a method for simultaneous determination of cellobiose, maltose, and maltotriose in juices using HPLC/IMER. Oligosaccharides are separated by HPAEC; a postcolumn suppressor is used to adjust the required pH for the following enzymatic reaction. Six serially operated enzyme reactors are used for biochemical conversions, followed by electrochemical detection. To exclude glucose, naturally occurring in juices, it is oxidized in a first step using glucoseoxidase (GOD). Resulting H_2O_2 is transformed using two serial catalase reactors. Subsequently, cellubiose, maltose, and maltotriose are cleaved on-line using amyloglucosidase and cellulase followed by oxidation of the liberated glucose with GOD. Finally H_2O_2 is determined electrochemically.

References

Adahchour M., Vreuls R.J.J., van der Heijden A., Brinkmann U.A.T. 1999. Trace-level determination of polar flavour compounds in butter by solid-phase extraction and gas chromatography-mass spectrometry. *Journal of Chromatography A* 844:295–305.

Adahchour M., Wiewel J., Verdel R., Vreuls R.J.J., Brinkmann U.A.T. 2005. Improved determination of flavour compounds in butter by solid-phase (micro)extraction and comprehensive two-dimensional gas chromatography. *Journal of Chromatography A* 1086:99–106.

Aerts H.A.J., Jacobs P.A. 2004. Epoxide yield determination of oils and fatty acid methyl esters using ^{1}H NMR. *Journal of the American Oil Chemists' Society* 81(9):841–6.

Ai J. 1997a. Solid phase microextraction for quantitative analysis in nonequilibrium situations. *Analytical Chemistry* 69:1230–6.

———. 1997b. Headspace solid phase microextraction: dynamics and quantitative analysis before reaching a partition equilibrium. *Analytical Chemistry* 69:3260–6.

Aires-Barros M.R., Cabral J.M.S. 1991. Selective separation and purification of two lipases from *Chromobacterium viscosum* using AOT reversed micelles. *Biotechnology and Bioengineering* 38:1302–7.

Al-Alawi A., Van De Voort F.R., Sedman J. 2004. New FTIR method for the determination of FFA in oils. *Journal of the American Oil Chemists Society* 81:441–6.

Albersheim P., Nevins D.J., English P.D., Karr A. 1967. Analysis of sugars in plant cell-wall polysaccharides by gas-liquid chromatography. *Carbohydrate Research* 5:340–5.

Alves R.F., Nascimento A.M.D., Nogueira J.M.F. 2005. Characterisation of the aroma profile of Madeira wine by sorptive extraction techniques. *Analytica Chimica Acta* 546:11–21.

Angerosa F., Di Giacinto L., d'Alessandro N. 1997. Quantitation of some flavor components responsible for the "green" attributes in virgin olive oils. *Journal of High Resolution Chromatography* 20:507–10.

Apps P., Lim Ah Tock M. 2005. Enhanced flavour extraction in continuous liquid-liquid extractors. *Journal of Chromatography A* 1083:215–18.

Arce L., Rios A., Valcarcel M. 1998. Direct determination of biogenic amines in wine by integrating continuous flow clean-up and capillary electrophoresis with indirect UV detection. *Journal of Chromatography A* 803:249–60.

Arilla M.C., Erase E., Ibarrola I., Algorta J., Martínez A., Asturias J.A. 2002. Monoclonal antibody-based method for measuring olive pollen major allergen Ole e 1. *Annals of Allergy, Asthma and Immunology* 89: 83–9.

Arthur C.L., Pawliszyn J. 1990. Solid phase microextraction with thermal desorption using fused silica optical fibers. *Analytical Chemistry* 62:2145–8.

Arthur C.L., Killam L.M., Buchholz K.D., Pawliszyn J., Berg J.R. 1992. Automation and optimization of solid-phase microextraction. *Analytical Chemistry* 64:1960–6.

Augusto F., Leite e Lopes A., Alcaraz Zini C. 2003. Sampling and sample preparation for analysis of aromas and fragrances. *Trends in Analytical Chemistry* 22:160–9.

Baltussen E., David F., Sandra P., Cramers C. 1999. On the performance and inertness of different materials used for the enrichment of sulfur compounds from air and gaseous samples. *Journal of Chromatography A* 864:345–50.

Baltussen E., Cramers C.A., Sandra P.J.F. 2002. Sorptive sample preparation—a review. *Analytical and Bioanalytical Chemistry* 373:3–22.

Batista R.B., Grimm C.C., Godshall M. 2002. Semiquantitative determination of short-chain fatty acids in cane and beet sugars. *Journal of Chromatographic Science* 40:127–32.

Bauer H., Voelter W. 1976. Carbohydrates. 52. Ion exchange chromatography of carbohydrates. *Chromatographia* 9:433–9.

Baye T., Becker H.C. 2004. Analyzing seed weight, fatty acid composition, oil, and protein contents in *Vernonia galamensis* germplasm by near-infrared reflectance spectroscopy. *Journal of the American Oil Chemists Society* 81:641–5.

Bernhisel-Broadbent J., Scanlon S.M., Sampson H.A. 1992. Fish hypersensitivity. I. In vitro and oral challenge results in fish-allergic patients. *Journal of Allergy and Clinical Immunology* 89:730–7.

Bernhisel-Broadbent J., Dintzis H.M., Dintzis R.Z., Sampson H.A. 1994. Allergenicity and antigenicity of chicken egg ovomucoid (Gal d III) compared with ovalbumin (Gal d I) in children with egg allergy and in mice. *Journal of Allergy and Clinical Immunology* 93:1047–59.

Besler M., Kasel U., Wichmann G. 2002. Determination of hidden allergens in foods by immunoassays. *Internet Symposium on Food Allergens [online computer file]* 4:1–18.

Bicchi C., Drigo S., Rubiolo P. 2000. Influence of fibre coating in headspace solid-phase microextraction-gas chromatographic analysis of aromatic and medicinal plants. *Journal of Chromatography A* 892: 469–85.

Bittner A.S., Harris L.E., Campbell W.F. 1980. Rapid N-methylimidazole-catalyzed acetylation of plant cell wall sugars. *Journal of Agricultural and Food Chemistry* 28:1242–5.

Bjorksten F., Halmepuro L., Hannuksela M., Lahti A. 1980. Extraction and properties of apple allergens. *Allergy (Oxford, United Kingdom)* 35:671–7.

Blake J.D., Richards G. 1970. Critical re-examination of problems inherent in compositional analysis of hemicelluloses by gas-liquid chromatography. *Carbohydrate Research* 14:375–87.

Blakeney A.B., Harris P.J., Henry R.J., Stone B.A. 1983. A simple and rapid preparation of alditol acetates for monosaccharide analysis. *Carbohydrate Research* 113:291–9.

Bleumink E. 1966. Preparation of β-lactalbumin from milk whey by precipitation with acetone. *Netherlands Milk and Dairy Journal* 20:13–16.

Bligh E.G., Dyer W.J. 1959. A rapid method of total lipid extraction and purification. *Canadian Journal of Biochemistry and Physiology* 37:911–17.

Brescia M.A., Mazzilli V., Sgaramella A., Ghelli S., Fanizzi F.P., Sacco A. 2004. ^{1}H NMR characterization of milk lipids: a comparison between cow and buffalo milk. *Journal of the American Oil Chemists Society* 81:431–6.

Buecking M., Steinhart H. 2002. Headspace GC and sensory analysis characterization of the influence of different milk additives on the flavor release of coffee beverages. *Journal of Agricultural and Food Chemistry* 50:1529–34.

Buldini P.L., Ricci L., Sharma J.L. 2002. Recent applications of sample preparation techniques in food analysis. *Journal of Chromatography A* 975:47–70.

Burbank H.M., Qian M.C. 2005. Volatile sulfur compounds in cheddar cheese determined by headspace solid-phase microextraction and gas chromatograph-pulsed flame photometric detection. *Journal of Chromatography A* 1066:149–57.

Burks A.W., Williams L.W., Helm R.M., Connaughton C., Cockrell G., O'Brien T. 1991. Identification of a major peanut allergen, Ara h I, in patients with atopic dermatitis and positive peanut challenges. *Journal of Allergy and Clinical Immunology* 88:172–9.

Burks A.W., Williams L.W., Connaughton C., Cockrell G., O'Brien T.J., Helm R.M. 1992. Identification and characterization of a second major peanut allergen, Ara h II, with use of the sera of patients with atopic dermatitis and positive peanut challenge. *Journal of Allergy and Clinical Immunology* 90:962–9.

Cabredo-Pinillos S., Cedron-Fernandez T., Parra-Manzanares A., Saenz-Barrio C. 2004. Determination of volatile compounds in wine by automated solid-phase microextraction and gas chromatography. *Chromatographia* 59:733–8.

Cai J., Liu B., Su Q. 2001. Comparison of simultaneous distillation extraction and solid-phase microextraction for the determination of volatile flavor components. *Journal of Chromatography A* 930:1–7.

Castro R., Natera R., Benitez P., Barroso C.G. 2004. Comparative analysis of volatile compounds of "fino" sherry wine by rotatory and continuous liquid-liquid extraction and solid-phase microextraction in conjunction with gas chromatography-mass spectrometry. *Analytica Chimica Acta* 513:141–50.

Chai M., Pawliszyn J. 1995. Analysis of environmental air samples by solid-phase microextraction and gas chromatography/ion trap mass spectrometry. *Environmental Science and Technology* 29:693–701.

Chen Y., Pawliszyn J. 2004. Kinetics and the on-site application of standards in a solid-phase microextration fiber. *Analytical Chemistry* 76(19):5807–15.

Chevance F.F.V., Farmer L.J. 1999. Identification of major volatile odor compounds in frankfurters. *Journal of Agricultural and Food Chemistry* 47:5151–60.

Chilla C., Guillen D.A., Barroso C.G., Perez-Bustamante J.A. 1996. Automated on-line solid-phase extraction-high-performance liquid chromatography-diode array detection of phenolic compounds in sherry wine. *Journal of Chromatography A* 750:209–14.

Cho H.-J., Baek K., Lee H.-H., Lee S.-H., Yang J.-W. 2003. Competitive extraction of multi-component contaminants in water by Carboxen-polydimethylsiloxane fiber during solid-phase microextraction. *Journal of Chromatography A* 988:177–84.

Christy A.A., Kasemsumran S., Du Y., Ozaki Y. 2004. The detection and quantification of adulteration in olive oil by near-infrared spectroscopy and chemometrics. 20(6):935.

Clark L.C. Jr., Lyons C. 1962. Electrode systems for continuous monitoring in cardiovascular surgery. *Annals of the New York Academy of Sciences* 102:29–45.

Coleman W.M. III., Lawrence B.M., Cole S.K. 2002. Semiquantitative determination of off-notes in mint oils by solid-phase microextraction. *Journal of Chromatographic Science* 40:133–9.

Connors K.A., Pandit N.K. 1978. N-Methylimidazole as a catalyst for analytical acetylations of hydroxy compounds. *Analytical Chemistry* 50:1542–5.

Cozzolino D., Murray I., Paterson R., Scaife J.R. 1996. Visible and near-infrared reflectance spectroscopy for the determination of moisture, fat and protein in chicken breast and thigh muscle. *Journal of Near Infrared Spectroscopy* 4:213–23.

Cremonini M.A., Laghi L., Placucci G. 2004. Investigation of commercial lecithin by ^{31}P NMR in a ternary CUBO solvent. *Journal of the Science of Food and Agriculture* 84:786–90.

Dasgupta S., Hogan E.L. 2001. Chromatographic resolution and quantitative assay of CNS tissue sphingoids and sphingolipids. *Journal of Lipid Research* 42:301–8.

de la Calle Garcia D., Magnaghi S., Reichenbächer M., Danzer K. 1996. Systematic optimization of the analysis of wine bouquet components by solid-phase microextraction. *Journal of High Resolution Chromatography* 19:257–62.

de la Fuente E., Martinez-Castro I., Sanz J. 2005. Characterization of Spanish unifloral honeys by solid phase microextraction and gas chromatography-mass spectrometry. *Journal of Separation Science* 28:1093–100.

Demyttenaere J.C.R., Sanchez Martinez J.I., Verhe R., Sandra P., De Kimpe N. 2003. Analysis of volatiles of malt whisky by solid-phase microextraction and stir bar sorptive extraction. *Journal of Chromatography A* 985:221–32.

de Oliveira A.M., Pereira N.R., Marsaioli Jr. A., Augusto F. 2004. Studies on the aroma of cupuassu liquor by headspace solid-phase microextraction and gas chromatography. *Journal of Chromatography A* 1025:115–24.

Diaz P., Ibanez E., Senorans F.J., Reglero G. 2003. Truffle aroma characterization by headspace solid-phase microextraction. *Journal of Chromatography A* 1017:207–14.

Diaz-Maroto M.C., Perez-Coello M.S., Cabezudo M.D. 2002. Supercritical carbon dioxide extraction of volatiles from spices: comparison with simultaneous distillation-extraction. *Journal of Chromatography A* 947:23–9.

Dorfner K. 1991. *Ion Exchangers*. Walter de Gruyter, Berlin.

Dubois M., Gilles K.A., Hamilton J.K., Rebers P.A., Smith F. 1956. Colorimetric method for determination of sugars and related substances. *Analytical Chemistry* 28:350–6.

Duffort O.A., Polo F., Lombardero M., Diaz-Perales A., Sanchez-Monge R., Garcia-Casado G., Salcedo G., Barber D. 2002. Immunoassay to quantify the major peach allergen Pru p 3 in foodstuffs: differential allergen release and stability under physiological conditions. *Journal of Agricultural and Food Chemistry* 50:7738–41.

Duflos G., Moine F., Coin V.M., Malle P. 2005. Determination of volatile compounds in whiting (*Merlangius merlangus*) using headspace-solid-phase microextraction-gas chromatography-mass spectrometry. *Journal of Chromatographic Science* 43:304–12.

Ebbehoej K., Dahl A.M., Froekiaer H., Noergaard A., Poulsen L.K., Barkholt V. 1995. Purification of egg-white allergens. *Allergy (Oxford, United Kingdom)* 50:133–41.

Ebeler S.E., Sun G.M., Datta M., Stremple P., Vickers A.K. 2001. Solid-phase microextraction for the enantiomeric analysis of flavors in beverages. *Journal of AOAC International* 84:479–85.

Eberhardt A., Galensa R. 1998. Pflanzenfasernachweis in Wurstwaren. *Deutsche Lebensmittelrundschau* 94:75–6.

Eckoldt J., Buschmann R., Wauschkuhn C. 1999. Methods for determination of the total fat content in meat and sausage in comparison. *Lebensmittelchemie* 53:120.

Eisert R., Pawliszyn J. 1997. Automated in-tube solid-phase microextraction coupled to high-performance liquid chromatography. *Analytical Chemistry* 69:3140–7.

Eshkenazi I., Maltz E., Zion B., Rishpon J. 2000. A three-cascaded-enzymes biosensor to determine lactose concentration in raw milk. *Journal of Dairy Science* 83:1939–45.

Ezquerro O., Pons B., Tena M.T. 2003. Evaluation of multiple solid-phase microextraction as a technique to remove the matrix effect in packing analysis for determination of volatile organic compounds. *Journal of Chromatography A* 1020:189–97.

Fahlbusch B., Rudeschko O., Schlott B., Henzgen M., Schlenvoigt G., Schubert H., Kinne R.W. 2003. Further characterization of IgE-binding antigens from guinea pig hair as new members of the lipocalin family. *Allergy (Oxford, United Kingdom)* 58:629–34.

Fang Y., Qian M.C. 2005. Sensitive quantification of sulfur compounds in wine by headspace solid-phase microextraction technique. *Journal of Chromatography A* 1080:177–85.

Fido R.J., Mills E.N.C., Rigby N.M., Shewry P.R. 2004. Protein extraction from plant tissues. *Methods in Molecular Biology (Totowa, NJ)* 244:21–7.

Focant J.-F., Pirard C., De Pauw E. 2004. Automated sample preparation-fractionation for the measurement of dioxins and related compounds in biological matrices: a review. *Talanta* 63:1101–13.

Frank C.D., Owen M.C., Patterson J. 2004. Solid phase microextraction (SPME) combined with gas-chromatography and olfactometry-mass spectrometry for characterization of cheese aroma compounds. *Lebensmittel-Wissenschaft und -Technologie* 37:139–54.

Galensa R. 2004. Anwendungen der HPLC-Biosensorkopplung. In: *Schnellmethoden zur Beurteilung von Lebensmitteln und ihren Rohstoffen,* edited by Baltes W., Kroh L.W., Behr's Verlag, Hamburg, pp. 169–85.

Garcia-Ayuso L.E.V.J., Dobarganes M.C., Luque de Castro M.D. 1999. Accelerated extraction of the fat content in cheese using a focused microwave-assisted Soxhlet device. *Journal of Agricultural and Food Chemistry* 47:2308–15.

Genovese A., Dimaggio R., Lisanti M.T., Piombino P., Moio L. 2005. Aroma composition of red wines by different extraction methods and gas chromatography-SIM/mass spectrometry analysis. *Annales di Chimica* 95:383–94.

Gilar M., Bouvier E.S.P., Compton B.J. 2001. Advances in sample preparation in electromigration, chromatographic and mass spectrometric separation methods. *Journal of Chromatography A* 909:111–35.

Giordano L., Calabrese R., Davoli E., Rotilio D. 2003. Quantitative analysis of 2-furfural and 5-methylfurfural in different Italian vinegars by headspace solid-phase microextraction coupled to gas chromatography-mass spectrometry using isotope dilution. *Journal of Chromatography A* 1017:141–9.

Globig D., Weickhardt C. 2005. Fully automated in-tube solid-phase microextraction for liquid samples coupled to gas chromatography. *Analytical and Bioanalytical Chemistry* 381:656–9.

Goerg A., Postel W., Guenther S. 1988. The current state of two-dimensional electrophoresis with immobilized pH gradients. *Electrophoresis* 9:531–46.

Göklen K.E., Hatton T.A. 1985. Protein extraction using reverse micelles. *Biotechnology Progress* 1:69–74.

———. 1987. Liquid liquid extraction of low molecular proteins by selective solubilization in reversed micelles. *Separation Science and Technology* 22:831–41.

Goldmann T., Perisset A., Scanlan F., Stadler R.H. 2005. Rapid determination of furan in heated foodstuffs by isotope dilution solid phase micro-extraction-gas chromatography-mass spectrometry (SPME-GC-MS). *Analyst* 130:878–3.

Gonzalez-Martin I., Gonzalez-Perez C., Alvarez-Garcia N., Gonzalez-Cabrera J.M. 2004. On-line determination of fatty acid composition in intramuscular fat of Iberian pork loin by NIRs with a remote reflectance fibre optic probe. *Meat Science* 69:243–8.

Gorecki T., Yu X., Pawliszyn J. 1999. Theory of analyte extraction by selected porous polymer SPME fibres. *Analyst* 124:643–9.

Gou Y., Tragas C., Lord H., Pawliszyn J. 2000. On-line coupling of in-tube solid phase microextraction (SPME) to HPLC for analysis of carbamates in water samples: comparison of two commercially available autosamplers. *Journal of Microcolumn Separation* 12:125–34.

Guilbault G.G., Montalvo J.G., Jr. 1969. A urea-specific enzyme electrode. *Journal of the American Chemical Society* 91:2164–5.

Guthausen A., Guthausen G., Kamlowski A., Todt H., Burk W., Schmalbein D. 2004. Measurement of fat content of food with single-sided NMR. *Journal of the American Oil Chemists' Society* 81(8):727–31.

Haberhauer-Troyer C., Crnoja M., Rosenberg E., Grasserbauer M. 2000. Surface characterization of commercial fibers for solid-phase microextraction and related problems in their application. *Fresenius Journal of Analytical Chemistry* 366:329–31.

Hanke A., Eberhardt A., Bilitewski U., Galensa R., Künnecke W. 1996. Fructosebestimmung in Säften mittels Fructosedehydrogenase-Dickschichtelektroden. *Deutsche Lebensmittelrundschau* 92:35–9.

Hansen E.H. 1996. Principles and applications of flow injection analysis in biosensors. *Journal of Molecular Recognition* 9:316–25.

Hara A., Radin N.S. 1978. Lipid extraction of tissues with a low-toxicity solvent. *Analytical Biochemistry* 90:420–6.

Hartmeier W. 1986. *Immobilisierte Biokatalysatoren*. Springer, Berlin.

Hatton T.A. 1989. Reversed micellar extraction of proteins In: *Surfactant-Based Processes*, vol. 33, edited by Scamehorn J.F., Horwell J.H., Marcel-Dekker, New York, pp. 55–90.

Hearn M.T.W. 1991. *HPLC of Proteins, Peptides and Polynucleotides*. VCH, New York.

Helmerich G., Koehler P. 2003. Comparison of methods for the quantitative determination of phospholipids in lecithins and flour improvers. *Journal of Agricultural and Food Chemistry* 51:6645–51.

Hill P.G., Smith R.M. 2000. Determination of sulphur compounds in beer using headspace solid phase micro-extraction and gas chromatographic analysis with pulsed flame photometric detection. *Journal of Chromatography A* 872:203–13.

Hoffmann K., Blaudszun J., Brunken C., Hoepker W.W., Tauber R., Steinhart H. 2005. Lipid class distribution of fatty acids including conjugated linoleic acids in healthy and cancerous parts of human kidneys. *Lipids* 40:1057–62.

Holden L., Faeste C.K., Egaas E. 2005. Quantitative sandwich ELISA for the determination of lupine (*Lupinus* spp.) in foods. *Journal of Agricultural and Food Chemistry* 53:5866–71.

Holzhauser T., Vieths S. 1999. Indirect competitive ELISA for determination of traces of peanut (*Arachis hypogaea* L.) protein in complex food matrices. *Journal of Agricultural and Food Chemistry* 47:603–11.

Honda S., Yamauchi N., Kakehi K. 1979. Rapid gas chromatographic analysis of aldoses as their diethyl dithioacetal trimethylsilylates. *Journal of Chromatography* 169:287–93.

Huie C.W. 2002. A review of modern sample-preparation techniques for the extraction and analysis of medical plants. *Analytical and Bioanalytical Chemistry* 373:23–30.

Inman D.J., Hornby W.E. 1974. Preparation of some immobilized linked enzyme systems and their use in the automated determination of disaccharides. *Biochemical Journal* 137:25–32.

Inschlag C., Hoffmann-Sommergruber K., O'Riordain G., Ahorn H., Ebner C., Scheiner O., Breiteneder H. 1998. Biochemical characterization of Pru a 2, a 23-kD thaumatin-like protein representing a potential major allergen in cherry (*Prunus avium*). *International Archives of Allergy and Immunology* 116:22–8.

Jelen H.H., Obuchowska M., Zawirska-Wojtasiak R., Wasowicz E. 2000. Headspace solid-phase microextraction use for the characterization of volatile compounds in vegetable oils of different sensory quality. *Journal of Agricultural and Food Chemistry* 48:2360–7.

Jensen S., Häggberg L., Jörundsdóttir H., Odham G. 2003. A quantitative lipid extraction method for residue analysis of fish involving nonhalogenated solvents. *Journal of Agricultural and Food Chemistry* 51:5607–11.

Jirovetz L., Buchbauer G., Ngassoum M.B., Geissler M. 2002. Aroma compound analysis of *Piper nigrum* and *Piper guineense* essential oils from Cameroon using solid-phase microextraction-gas chromatography, solid-phase microextraction-gas chromatography-mass spectrometry and olfactometry. *Journal of Chromatography A* 976:265–75.

Johansen K., Svensson L. 2002. Immunoprecipitation. In: *Protein Protocols Handbook*, 2nd edition, edited by Walker J.M., Humana Press, Totowa, N.J., pp. 1097–1106.

Johnson R.B., Barnett H.J. 2003. Determination of fat content in fish feed by supercritical fluid extraction and subsequent lipid classification of extract by thin layer chromatography-flame ionization detection. *Aquaculture* 216:263–82.

Jublot L., Linforth R.S.T., Taylor A.J. 2004. Direct coupling of supercritical fluid extraction to a gas phase atmospheric pressure chemical ionisation source ion trap mass spectrometer for fast extraction and analysis of food components. *Journal of Chromatography A* 1056:27–33.

Kataoka H. 2002. Automated sample preparation using in-tube solid-phase microextraction and its application—a review. *Analytical and Bioanalytical Chemistry* 373:31–45.

Kataoka H., Lord H.L., Pawliszyn J. 2000. Applications of solid-phase microextraction in food analysis. *Journal of Chromatography A* 880:35–62.

Kellner R., Mermet J.M., Otto M., Valcarcel M., Widmer H.M. 2004. *Analytical Chemistry*. 2nd edition, Wiley-VCH, Weinheim.

Keszler A., Heberger K., Gude M. 1998. Identification of volatile compounds in sunflower oil by headspace SPME and ion-trap GC/MS. *Journal of High Resolution Chromatography* 21:368–70.

Kleeberg K.K., Liu Y., Jans M., Schlegelmilch M., Streese J., Stegmann R. 2005. Development of a simple and sensitive method for the characterisation of odorous waste gas emissions by means of solid-phase microextraction (SPME) and GC-MS/olfactometry. *Waste Management* 25:872–9.

Koenig W.A., Benecke I. 1983. Enantiomer separation of polyols and amines by enantioselective gas chromatography. *Journal of Chromatography* 269:19–21.

Koenig W.A., Benecke I., Sievers S. 1981. New results in the gas chromatographic separation of enantiomers of hydroxy acids and carbohydrates. *Journal of Chromatography* 217:71–9.

Koerner H.U., Gottschalk D., Wiegel J., Puls J. 1984. The degradation pattern of oligomers and polymers from lignocelluloses. *Analytica Chimica Acta* 163:55–66.

Kolar K., Faure U., Torelm I., Finglas P. 1993. An intercomparison of methods for the determination of total fat in a meat reference material. *Fresenius Journal of Analytical Chemistry* 347:393–5.

König W.A., Benecke I., Bretting H. 1981. Gas chromatographic separation of carbohydrate enantiomers on a new chiral stationary phase. *Angewandte Chemie, International Edition in English* 20(8):693–4.

Koppelman S.J., Knulst A.C., Koers, W.J., Penninks A.H., Peppelman H., Vlooswijk R., Pigmans I., Van Duijn G., Hessing M. 1999. Comparison of different immunochemical methods for the detection and quantification of hazelnut proteins in food products. *Journal of Immunological Methods* 229:107–20.

Krei G.A., Hustedt H. 1992. Extraction of enzymes by reverse micelles. *Chemical Engineering Science* 47:99–111.

Kurowski C., Timm D., Grummisch U., Meyhack U., Grunewald H. 1998. The benefits of near infrared analysis for food product quality. *Journal of Near Infrared Spectroscopy* 6:A343–8.

Langeland T.A. 1982. Clinical and immunological study of allergy to hen's egg white. III. Allergens in hen's egg white studied by crossed radio-immunoelectrophoresis (CRIE). *Allergy* 37:521–30.

Lee S.-N., Kim N.-S., Lee D.-S. 2003. Comparative study of extraction techniques for determination of garlic flavor components by gas chromatography-mass spectrometry. *Analytical and Bioanalytical Chemistry* 377:749–56.

Lessig J., Gey C., Suss R., Schiller J., Glander H.-J., Arnhold J. 2004. Analysis of the lipid composition of human and boar spermatozoa by MALDI-TOF mass spectrometry, thin layer chromatography and ^{31}P NMR spectroscopy. *Comparative Biochemistry and Physiology*, Part B: *Biochemistry and Molecular Biology* 137B:265–77.

Lestremau F., Desauziers V., Fanlo J.L. 2001. Formation of artifacts during air analysis of volatile amines by solid-phase micro extraction. *Analyst* 126:1969–73.

Lestremau F., Andersson F.A.T., Desauziers V. 2004. Investigation of artifact formation during analysis of volatile sulphur compounds using solid phase microextraction (SPME). *Chromatographia* 59: 607–13.

Lord H., Pawliszyn J. 2000. Evolution of solid-phase microextraction technology. *Journal of Chromatography A* 885:153–93.

Luque-Garcia J.L., Luque de Castro M.D. 2004. Ultrasound-assisted Soxhlet extraction: an expeditive approach for solid sample treatment; application to the extraction of total fat from oleaginous seeds. *Journal of Chromatography A* 1034:237–42.

Ma K., Van de Voort F.R., Sedman J., Ismail A.A. 1999. Trans fatty acid determination in fats and margarine by Fourier transform infrared spectroscopy using a disposable infrared card. *Journal of the American Oil Chemists Society* 76:1399–404.

Ma K., Van de Voort F.R., Ismail A.A., Zhuo H., Cheng B. 2000. Monitoring peroxide value in fatliquor manufacture by fourier transform infrared spectroscopy. *Journal of the American Oil Chemists Society* 77:681–5.

Man Y.B.C., Setiowaty G., Van de Voort F.R. 1999. Determination of iodine value of palm oil by Fourier transform infrared spectroscopy. *Journal of the American Oil Chemists Society* 76:693–9.

Marconi E., Messia M.C., Palleschi G., Cubadda R. 2004. A maltose biosensor for determining gelatinized starch in processed cereal foods. *Cereal Chemistry* 81:6–9.

Marsili R.T. 1999. Comparison of solid-phase microextraction and dynamic headspace methods for the gas chromatographic-mass spectrometric analysis of light-induced lipid oxidation products in milk. *Journal of Chromatographic Science* 37:17–23.

———. 2002. *Flavor, Fragrance, and Odor Analysis.* Marcel Dekker, New York.

Marsili R.T., Miller N. 2000. Determination of major aroma impact compounds in fermented cucumbers by solid-phase microextraction-gas chromatography-mass spectrometry-olfactometry detection. *Journal of Chromatographic Science* 38:307–14.

Marsili R.T., Miller N., Kilmer G.J., Simmons R.E. 1994. Identification and quantitation of the primary chemicals responsible for the characteristic malodor of beet sugar by purge and trap GC-MS-OD techniques. *Journal of Chromatographic Science* 32:165–71.

Masuko T., Minami A., Iwasaki N., Majima T., Nishimura S.-I., Lee Y.C. 2005. Carbohydrate analysis by a phenol-sulfuric acid method in microplate format. *Analytical Biochemistry* 339:69–72.

Mello L.D., Kubota L.T. 2002. Review of the use of biosensors as analytical tools in the food and drink industries. *Food Chemistry* 77:237–56.

Menendez-Arias L., Moneo I., Dominguez J., Rodriguez R. 1988. Primary structure of the major allergen of yellow mustard (*Sinapis alba* L.) seed, Sin a I. *European Journal of Biochemistry* 177:159–66.

Miller M.E., Stuart J.D. 1999. Comparison of gas-samples and SPME-sampled static headspace for the determination of volatile flavor components. *Analytical Chemistry* 71:23–7.

Milosevic M., Milosevic V., Kramer J.K.G., Azizian H., Mossoba M.M. 2004. Determining low level of trans fatty acids in foods using an improved ATR-FTIR procedure. *Lipid Technology* 16:252–5.

Mohammadi H., Amine A., Cosnier S., Mousty C. 2005. Mercury-enzyme inhibition assays with an amperometric sucrose biosensor based on a trienzymatic-clay matrix. *Analytica Chimica Acta* 543:143–9.

Mondello L., Costa R., Tranchida P.Q., Dugo P., Presti M.L., Fazio A., Dugo G. 2005. Reliable characterisation of coffee bean aroma profiles by automated headspace solid phase microextraction-gas chromatography-mass spectrometry with the support of a dual-filter mass spectra library. *Journal of Separation Science* 28:1101–9.

Morales M.T., Berry A.J., McIntyre P.S., Aparicio R. 1998. Tentative analysis of virgin oil aroma by supercritical fluid extraction-high resolution gas chromatography-mass spectrometry. *Journal of Chromatography A* 819:267–75.

Mossoba M.M., Yurawecz M.P., Delmonte P., Kramer J.K.G. 2004. Overview of infrared methodologies for trans fat determination. *Journal of AOAC International* 87:540–4.

Murgia S., Mele S., Monduzzi M. 2003. Quantitative characterization of phospholipids in milk fat via ^{31}P NMR using a monophasic solvent mixture. *Lipids* 28:585–91.

Natera Marin R., Castro Mejias R., de Valme Garcia Moreno M., Garcia Rowe F., Garcia Barroso C. 2002. Headspace solid-phase microextraction analysis of aroma compounds in vinegar: validation study. *Journal of Chromatography A* 967:261–7.

Neff W.E., Eller F., Warner K. 2002. Composition of oils extracted from potato chips by supercritical fluid extraction. *European Journal of Lipid Science and Technology* 104:785–91.

Ochiai N., Sasamoto K., Daishima S., Heiden A.C., Hoffmann A. 2003. Determination of stale-flavor carbonyl compounds in beer by stir bar sorptive extraction with in-situ derivatization and thermal desorption-gas chromatography-mass spectrometry. *Journal of Chromatography A* 986:101–10.

Palosuo K., Alenius H., Varjonen E., Koivuluhta M., Mikkola J., Keskinen H., Kalkkinen N., Reunala T. 1999. A novel wheat gliadin as a cause of exercise-induced anaphylaxis. *Journal of Allergy and Clinical Immunology* 103:912–17.

Papagiannopoulos M., Zimmermann B., Mellenthin A., Krappe M., Maio G., Galensa R. 2002. Online coupling of pressurized liquid extraction, solid-phase extraction and high-performance liquid chromatography for automated analysis of proanthocyanidins in malt. *Journal of Chromatography A* 958:9–16.

Pastello E.A., Trambaioli C. 2001. Isolation of food allergens. *Journal of Chromatography B* 756:71–84.

Pastorello E.A., Farioli L., Pravettoni V., Ispano M., Conti A., Ansaloni R., Rotondo F., Incorvaia C., Bengtsson A., Rivolta F., Trambaioli C., Previdi M., Ortolani C. 1998. Sensitization to the major allergen of Brazil nut is correlated with the clinical expression of allergy. *Journal of Allergy and Clinical Immunology* 102:1021–7.

Pawliszyn J. 1997. *Solid Phase Microextraction—Theory and Practice*. Wiley-VCH, New York.

———. 2003. Sample preparation: quo vadis? *Analytical Chemistry* 75:2543–58.

Pihlsgard P., Larsson M., Leufven A., Lingnert H. 2000. Volatile compounds in the production of liquid beet sugar. *Journal of Agricultural and Food Chemistry* 48:4844–50.

Pillonel L., Bosset J.O., Tabacchi R. 2002. Rapid preconcentration and enrichment techniques for the analysis of food volatile: a review. *Lebensmittel-Wissenschaft und -Technologie* 35:1–14.

Pinho O., Ferreira I.M.P.L.V.O., Ferreira M.A. 2002. Solid-phase microextraction in combination with GC/MS for quantification of the major volatile free fatty acids in ewe cheese. *Analytical Chemistry* 74:5199–204.

Pinnel V., Vandegans J. 1996. GC-MS headspace analysis of the volatile components of soya oil without heating the sample. *Journal of High Resolution Chromatography* 19:263–6.

Pires M.J., Aires-Barros M.R., Cabral J.M.S. 1996. Liquid-liquid extraction of proteins with reversed micelles. *Biotechnology Progress* 12:290–301.

Planas C., Saulo J., Rivera J., Caixach J. 2004. Analysis of halogenated and priority pesticides at different concentration levels: automated SPE extraction followed by isotope dilution-GC/MS. *Organohalogen Compounds* 66:125–30.

Priego-Capote F., Luque de Castro M.D. 2005. Focused microwave-assisted Soxhlet extraction: a convincing alternative for total fat isolation from bakery products. *Talanta* 65:98–103.

Priego-Capote F., Ruiz-Jimenez J., Garcia-Olmo J., Luque de Castro M.D. 2004. Fast method for the determination of total fat and trans fatty-acids content in bakery products based on microwave-assisted Soxhlet extraction and medium infrared spectroscopy detection. *Analytica Chimica Acta* 517:13–20.

Rahaman R.S., Chee J.Y., Cabral J.M.S., Hatton T.A. 1988. Recovery of an extracellular alkaline protease from whole fermentation broth using reverse micelles. *Biotechnology Progress* 4:217–24.

Rajakyla E. 1986. Use of reversed-phase chromatography in carbohydrate analysis. *Journal of Chromatography* 353:1–12.

Ranau R., Kleeberg K.K., Schlegelmilch M., Streese J., Stegmann R., Steinhart H. 2005. Analytical determination of the suitability of different processes for the treatment of odorous waste gas. *Waste Management* 25:908–16.

Rocha S., Coutinho P., Barros A., Coimbra M.A., Delgadillo I., Cardoso A.D. 2000. Aroma potential of two Bairrada white grape varieties: Maria Gomes and Bical. *Journal of Agricultural and Food Chemistry* 48:4802–7.

Rodriguez-Bencomo J.J., Conde J.E., Rodriguez-Delgado M.A., Garcia-Montelongo F., Perez-Trujillo J.P. 2002. Determination of esters in dry and sweet white wines by headspace solid-phase microextraction and gas chromatography. *Journal of Chromatography A* 963:213–23.

Rosenberg E., Wissiack R., Grasserbauer M. 1999. Possibilities and problems in determining phenols by on-line SPE-HPLC-MS. *Biologische Abwasserreinigung* 11:73–91.

Ruiz J., Antequera T., Andres A.I., Petron M.J., Muriel E. 2004. Improvement of a solid phase extraction method for analysis of lipid fractions in muscle foods. *Analytica Chimica Acta* 520:201–5.

Saito Y., Jinno K. 2003. Miniaturized sample preparation combined with liquid phase separations. *Journal of Chromatography A* 1000:53–67.

Sanchez-Monge R., Lombardero M., Garcia-Selles F.J., Barber D., Salcedo G. 1999. Lipid-transfer proteins are relevant allergens in fruit allergy. *Journal of Allergy and Clinical Immunology* 103:514–19.

Sandiford C.P., Tee R.D., Taylor A.J. 1994. The role of cereal and fungal amylases in cereal flour hypersensitivity. *Clinical and Experimental Allergy: Journal of the British Society for Allergy and Clinical Immunology* 24:549–57.

Sangeetha P.T., Ramesh M.N., Prapulla S.G. 2005. Recent trends in the microbial production, analysis and application of fructooligosaccharides. *Trends in Food Science and Technology* 16:442–57.

Sato T. 2002. New estimation method for fatty acid composition in oil using near infrared spectroscopy. *Bioscience, Biotechnology, and Biochemistry* 66:2543–8.

Sato T., Maw A.A., Katsuta M. 2003. NIR reflectance spectroscopic analysis of the FA composition in sesame (*Sesamum indicum* L.) seeds. *Journal of the American Oil Chemists Society* 80:1157–61.

Scheller F.W., Schubert F. 1989. *Biosensoren*. Birkhäuser Verlag, Basel.

Schrader W., Geiger J., Klockow D., Korte E.-H. 2001. Degradation of α-pinene on Tenax during sample storage: effects of daylight radiation and temperature. *Environmental Science and Technology* 35:2717–20.

Shewry P.R., Napier J.A., Tatham A.S. 1995. Seed storage proteins: structures and biosynthesis. *Plant Cell* 7:945–56.

Sides A., Robards K., Helliwell S. 2000. Developments in extraction techniques and their application to analysis of volatiles in food. *Trends in Analytical Chemistry* 19:322–9.

Sinner M., Puls J. 1978. Non-corrosive dye reagent for detection of reducing sugars in borate complex ion-exchange chromatography. *Journal of Chromatography* 156:197–204.

Smith R.M. 2003. Before the injection—modern methods of sample preparation for separation techniques. *Journal of Chromatography A* 1000:3–27.

Snow N.H., Slack G.C. 2002. Head-space analysis in modern gas chromatography. *Trends in Analytical Chemistry* 21:608–17.

Sprenger C., Galensa R., Jensen D. 1999. Simultanbestimmung von Cellobiose, Maltose und Maltotriose in Fruchtsäften mittels HPLC-Biosensorkopplung. *Deutsche Lebensmittelrundschau* 95:499–504.

Stadler H., Nesselhut T. 1986. Simple and rapid measurement of acetylcholine and choline by HPLC and enzymatic-electrochemical detection. *Neurochemistry International* 9:127–9.

Steenson D.F., Lee J.H., Min D.B. 2002. Solid phase microextraction of volatile soybean oil and corn oil compounds. *Journal of Food Science* 67:71–6.

Stein K., Schwedt G. 1993. Comparison of immobilization methods for the development of an acetylcholinesterase biosensor. *Analytica Chimica Acta* 272:73–81.

Steinhart H., Stephan A., Bücking M. 2000. Advances in flavor research. *Journal of High Resolution Chromatography* 23:489–96.

Stoica L., Ludwig R., Haltrich D., Gorton L. 2006. Third-generation biosensor for lactose based on newly discovered cellobiose dehydrogenase. *Analytical Chemistry* 78:393–8.

Sun S.S.M., Leung F.W., Tomic J.C. 1987. Brazil nut (*Bertholletia excelsa* HBK) proteins: fractionation, composition, and identification of a sulfur-rich protein. *Journal of Agricultural and Food Chemistry* 35:232–5.

Sweeley C.C., Bentley R., Makita M., Wells W.W. 1963. Gas-liquid chromatography of trimethylsilyl derivatives of sugars and related substances. *Journal of the American Chemical Society* 85:2497–507.

Thevenot D.R., Toth K., Durst R.A., Wilson G.S. 1999. Electrochemical biosensors: recommended definitions and classification. *Pure Applied Chemistry* 71:2333–48.

Torrens J., Riu-Aumatell M., Lopez-Tamames E., Buxaderas S. 2004. Volatile compounds of red and white wines by headspace-solid-phase microextraction using different fibers. *Journal of Chromatographic Science* 42:310–16.

Urisu A., Yamada K., Masuda S., Komada H., Wada E., Kondo Y., Horiba F., Tsuruta M., Yasaki T., Yamada M. 1991. 16-Kilodalton rice protein is one of the major allergens in rice grain extract and responsible for cross-allergenicity between cereal grains in the *Poaceae* family. *International Archives of Allergy and Applied Immunology* 96:244–52.

Valero E., Sanz J., Martinez-Castro I. 2001. Direct thermal desorption in the analysis of cheese volatiles by gas chromatography and gas chromatography-mass spectrometry: comparison with simultaneous distillation-extraction and dynamic headspace. *Journal of Chromatographic Science* 39:222–8.

Van de Voort F.R., Sedman J., Emo G., Ismail A.A. 1992. Rapid and direct iodine value and saponification number determination of fats and oils by attenuated total reflectance/Fourier transform infrared spectroscopy. *Journal of the American Oil Chemists Society* 69:1118–23.

Verhaar L.A.T., Kuster B.F.M. 1981. Liquid chromatography of sugars on silica-based stationary phases. *Journal of Chromatography* 220:313–28.

Verzera A., Ziino M., Condurso C., Romeo V. 2004. Solid-phase microextraction and gas chromatography-mass spectrometry for rapid characterisation of semi-hard cheeses. *Analytical and Bioanalytical Chemistry* 380:930–6.

Vichi S., Castellote A.I., Pizzale L., Conte L.S., Buxaderas S., Lopez-Tamames E. 2003. Analysis of virgin olive oil volatile compounds by headspace solid-phase microextraction coupled to gas chromatography with mass spectrometric and flame ionization detection. *Journal of Chromatography A* 983:19–33.

Vieths S., Schoening B., Petersen A. 1994. Characterization of the 18-kDa apple allergen by two-dimensional immunoblotting and microsequencing. *International Archives of Allergy and Immunology* 104:399–404.

Vieths S., Janek K., Aulepp H., Petersen A. 1995. Isolation and characterization of the 18-kDa major apple allergen and comparison with the major birch pollen allergen (Bet v I). *Allergy* 50:421–30.

Wagner G., Guilbault G.G. 1994. *Biosensors for Food Analysis*, Marcel Dekker Verlag, New York.

Wen G., Zhang Y., Zhou Y., Shuang S., Dong C., Choi M.M.F. 2005. Biosensors for determination of galactose with galactose oxidase immobilized on eggshell membrane. *Analytical Letters* 38:1519–29.

Wilkes J.G., Conte E.D., Kim Y., Holcomb M., Sutherland J.B., Miller D.W. 2000. Sample preparation for the analysis of flavors and off-flavors in foods. *Journal of Chromatography A* 880:3–33.

Yeung J.M., Collins P.G. 1996. Enzyme immunoassay for determination of peanut proteins in food products. *Journal of AOAC International* 79:1411–16.

Zabaras D., Wyllie S.G. 2002. Rearrangement of p-menthane terpenes by Carboxen during HS-SPME. *Journal of Separation Science* 25:685–90.

Zhang H.-Z., Wie Z., Rutman M., Lee, T.-C. 2000. Simultaneous determination of moisture, protein and fat in fish meal using near-infrared spectroscopy. *Food Science and Technology Research* 6:19–23.

Zhang Z., Yang M.J., Pawliszyn J. 1994. Solid-phase microextraction. *Analytical Chemistry* 66:844A–53A.

Zill L.P., Khym J.X., Cheniae G.M. 1953. The separation of the borate complexes of sugars and related compounds by ion-exchange chromatography. *Journal of the American Chemical Society* 75:1339–42.

Zini C.A., Lord H., Christensen E., de Assis T.F., Caramao E.B., Pawliszyn J. 2002. Automation of solid-phase microextraction-gas chromatography-mass spectrometry extraction of eucalyptus volatiles. *Journal of Chromatographic Science* 40:140–7.

Zürcher K., Hadorn H., Strack C. 1975. Simplified method of preparing sugar oxime silyl derivatives for gas chromatographic analysis. *Deutsche Lebensmittelrundschau* 71:393–9.

14 Data Processing

Riccardo Leardi

Introduction

In this chapter the fundamentals of chemometrics will be presented by means of a quick overview of the most relevant techniques for data display, classification, modeling, and calibration. The goal of the chapter is to make people aware of the great superiority of multivariate analysis over the commonly used univariate approach. Mathematical and algorithmical details will not be presented, since the chapter is mainly focused on the general problems to which chemometrics can be successfully applied in the field of food chemistry.

As a matter of fact, many of the readers of this book may not be familiar with chemometrics, and a significant percentage of them may have never even heard of this "new" science (quite strange that it is still considered a new science, when the Chemometrics Society was founded 30 years ago and the most basic algorithms date back to the beginning of the 20th century). Furthermore, some of them could be quite put off by anything involving mathematical computations higher than a square root or statistical tests more complex than a *t*-test.

Therefore, the goal of this chapter is simply that of being read and understood by the majority of the readers of this book. This goal will be completely achieved if some of them, after having read it, could say: "Chemometrics is easy and powerful indeed, and from now on I will always think in a multivariate way."

Of course, to accomplish this goal in the limited space of a chapter, the attractive sides of chemometrics must be highlighted. Therefore, the intuitive aspects of each technique will be shown, without giving too much relevance to the algorithms.

First of all, what is chemometrics? According to the definition of the Chemometrics Society, it is "the chemical discipline that uses mathematical and statistical methods to design or select optimal procedures and experiments, and to provide maximum chemical information by analyzing chemical data."

One of the major mistakes people make about chemometrics is thinking that to use it one has to be a very good mathematician and to know the mathematical details of the algorithms being used. From the definition itself, it is clear instead that a chemometrician is a *chemist* who can *use* mathematical and statistical methods.

If we want to draw a parallel with everyday life, how many of us really know in detail how a TV set, a telephone, a car, or a washing machine works? But everybody watches TV programs, makes phone calls, drives a car, and starts a washing machine. Of course, what is important is that people know what each instrument is made for and that nobody tries to watch inside a telephone, or to drive a TV set, or to speak inside a washing machine, or to do the laundry in a car.

	var. 1	var. 2	var. 3	var. 4	var. 5	var. 6	var. 7	...	var. v
obj. 1									
obj. 2									
obj. 3									
obj. 4									
obj. 5									
obj. 6									
......									
obj. n									

Fig. 14.1. The structure of a chemometrical data set.

Though chemometrics makes available a very wide range of techniques, some of them being very difficult to fully understand and use correctly, the great majority of the real problems can be solved by applying one of the basic techniques. Understanding these, at least from an intuitive point of view, is relatively easy and does not require high-level mathematical skills.

Data Collection

Chemometrics works on data matrices. This means that on each sample a certain number of variables have been measured (in the "chemometrical jargon" we say that each object is described by v variables). Although some techniques can work with a limited number of missing values, a chemometrical data set must be thought of as a spreadsheet in which all the cells are full.

Sometimes, instead, if data are gathered without having any specific project, it happens that the result is a "sparse" matrix containing some blank cells. In such cases, if the percentage of missing data is quite high, the whole data set is not suitable for a multivariate analysis; as a consequence, the variables and/or the objects with the lowest number of data must be removed, and therefore a huge amount of experimental effort can be lost.

All the chemometrical software allows the import of data from ASCII files or from spreadsheets. It is therefore suggested to organize the data in matrix form from the start, as shown in figure 14.1, in such a way that the import can be performed in a single step.

If, on the contrary, the data are spread in several files or sheets (e.g., one file for each sample or for each variable), then the import procedure would be much longer and more cumbersome.

Data Display

The human mind can digest much more information when looking at plots rather than numbers. This is easily demonstrated by looking first at the sequence of numbers reported in table 14.1, and then the plot in figure 14.2.

It is very clear that, even with a very simple data set like this one (just ten samples, and only one variable), the information obtained by looking at the plot is superior and much more easily available than the information one can get by analyzing the raw numbers. From the plot, it becomes evident that the samples are clustered into two groups of the same size,

Table 14.1. Ten samples described by one variable.

Sample	1	2	3	4	5	6	7	8	9	10
Value	25.3	22.1	25.5	25.6	19.4	25.7	20.2	21.3	25.9	21.8

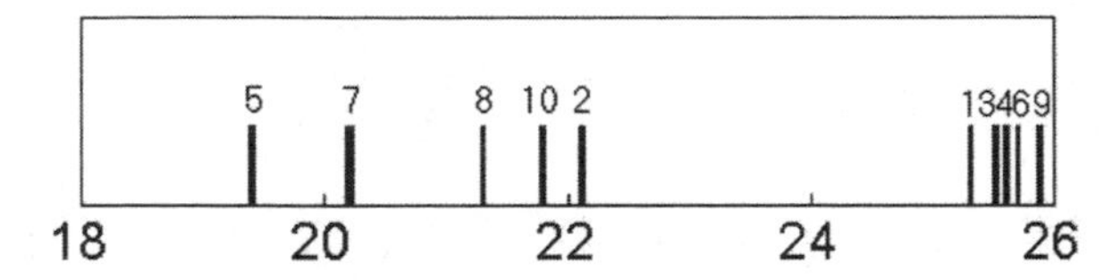

Fig. 14.2. Scatter plot of the data in table 14.1.

Table 14.2. Twenty samples described by two variables.

Sample	Variable 1	Variable 2
1	21.2	32.5
2	16.2	21.0
3	13.1	21.7
4	11.6	21.3
5	20.8	29.9
6	10.4	20.6
7	19.5	26.8
8	9.8	25.2
9	15.2	31.2
10	12.0	26.0
11	17.6	28.5
12	24.0	30.0
13	17.8	33.1
14	15.0	24.0
15	11.0	24.2
16	24.8	25.3
17	12.8	23.3
18	26.5	30.6
19	22.9	27.5
20	9.7	22.8

the one at higher values being much tighter than the one at low values. Much more time and effort are required when we want to get the same information from the table.

Let us now take into account a more complex data set—the one reported in table 14.2, where each object is described by two variables. The same data are plotted in figure 14.3.

This bivariate data set, beyond showing once more that a plot is much more easily handled by the human brain than a data table, demonstrates that when dealing with more than one variable the analysis of just one variable at a time can lead to wrong results. In this data set we have 20 samples, supposed to belong to the same population. When looking at the plot, we realize that we are in a situation very similar to what we found

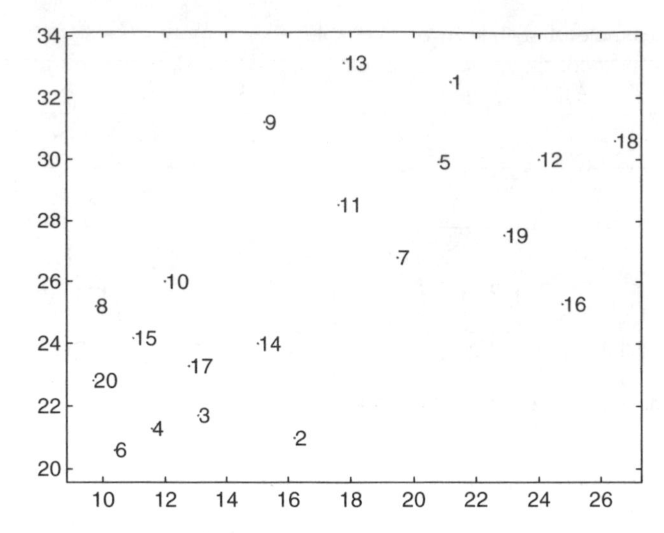

Fig. 14.3. Scatter plot of the data in table 14.2.

with the univariate data set. The samples are split into two clusters of the same size, with the objects of the first one more tightly grouped than the objects of the second one. This conclusion cannot be reached when looking at one variable at a time, since neither of the two variables is able to discriminate between the two groups.

If we had a data set with three variables, it would still be possible to visualize the whole information by a three-dimensional scatter plot, in which the coordinates of each object are the values of the variables. But what to do if there are more than three variables? What we need therefore is a technique permitting the visualization by simple bi- or tridimensional scatter plots of the majority of the information contained in a highly dimensional data set. This technique is principal component analysis (PCA), one of the simplest and most used methods of multivariate analysis. PCA is very important especially in the preliminary steps of an elaboration, when one wants to perform an exploratory analysis in order to have an overview of the data.

It is quite common to have to deal with large data tables with, for instance, a series of samples described by a number (v) of chemico-physical parameters. Examples of such data sets can be samples of olive oils from different origins described by their content in fatty acids and sterols, or samples of wines described by Fourier transform infrared (FT-IR) spectra. It is easy to realize how, especially in spectral data sets, v can be very high (>1000). In such cases it would be impossible to obtain valuable information without the help of multivariate techniques.

From a geometrical point of view, we can consider a v-dimensional space, in which each dimension is associated with one of the variables. In this space each sample (object) has coordinates corresponding to the values of the variables describing it.

Since it is impossible to visualize all the information at once, one should be content with the analysis of several bi- or tridimensional plots, each of them showing a different part of the global information.

It is also evident that not all possible combinations of two or three variables will give the same quality of information. For instance, if some variables are very highly correlated, then the information brought by each of them would be almost the same. If two variables are perfectly correlated, then one of them can be discarded, losing no information at all. In this way, the dimensionality of our space will be reduced from v to $v - 1$. If two variables are very highly correlated, then the elimination of one of them would produce only a slight loss of information, while the dimensionality of the space would be reduced to $v - 1$. So one can deduce that the information contained in the "lost" vth dimension was well below the average of the information contained in the other dimensions.

It is quite apparent now that not all the dimensions have the same importance, and that, owing to the correlations among the variables, the "real" dimensionality of our data matrix is somehow lower than v. Therefore, it would be very valuable to have a technique capable of concentrating in a few variables, and therefore in a few dimensions, the bulk of our information. This is exactly what is performed by PCA: it reduces the dimensionality of the data and extracts the most relevant part of the information, placing into the last dimensions the nonstructured information (i.e., the noise). According to these two characteristics, the information contained in very complex data matrices can be visualized in just one or a few plots.

From the mathematical point of view, the goal of PCA is to obtain, from v variables ($X_1, X_2, \ldots, X_v$), v linear combinations having two important features: to be uncorrelated and to be ordered according to the explained variance (i.e., to the information they contain). The lack of correlation among the linear combinations is very important, since it means that each of them describes different "aspects" of the original data. As a consequence, the examination of a limited number of linear combinations (generally the first two or three) allows us to obtain a good representation of the studied data set.

From a geometrical point of view, what is performed by PCA corresponds with looking for the direction which, in the v-dimensional space of the original variables, brings the greatest possible amount of information (i.e., explains the greatest variance). Once the first direction is identified, the second one is looked for: it will be the direction explaining the greatest part of the residual variance, under the constraint of being orthogonal to the first one. This process goes on until the vth direction has been found.

These new directions can be considered as the axes of a new orthogonal system, obtained after a simple rotation of the original axes. While in the original system each direction (i.e., each variable) brings with it, at least in theory, $1/v$ of total information, in the new system the information is concentrated in the first directions, and decreases progressively so that in the last ones no information can be found except noise.

The global dimensionality of the system is always that of the original data (v), but since the last dimensions explain only a very small part of the information, they can be neglected, and one can take into account only the first dimensions (the "significant components"). The projection of the objects in this space of reduced dimensionality retains almost all the information that can now also be analyzed in a visual way, by bi- or tridimensional plots. These new directions, linear combinations of the original ones, are the principal components (PCs) or eigenvectors.

With a mathematical notation, we can write:

$$\mathrm{var}(Z_1) > \mathrm{var}(Z_2) > \ldots > \mathrm{var}(Z_v)$$

where $\mathrm{var}(Z_i)$ is the variance explained by component i. Furthermore, since a simple rotation has been performed, the total variance is the same in the two systems of axes:

$$\sum \mathrm{var}(X_i) = \sum \mathrm{var}(Z_i).$$

The first PC is formed by the linear combination

$$Z_1 = a_{11}X_1 + a_{12}X_2 + \ldots + a_{1v}X_v$$

explaining the greatest variance, under the condition

$$\sum a_{1i}^2 = 1$$

This last condition notwithstanding, the variance of Z_1 could be made greater simply by increasing one of the values of *a*.

The second PC

$$Z_2 = a_{21}X_1 + a_{22}X_2 + \ldots + a_{2v}X_v$$

is the one having var(Z_2) as large as possible, under the conditions that

$$\sum a_{2i}^2 = 1$$

and that

$$\sum a_{1i}a_{2i} = 0$$

(this last condition ensures the orthogonality of components one and two).

The lower-order components are computed in the same way, always under the two conditions previously reported.

From a mathematical point of view, PCA is solved by finding the eigenvalues of the variance-covariance matrix; they correspond to the variance explained by the corresponding principal component. Since the sum of the eigenvalues is equal to the sum of the diagonal elements (trace) of the variance-covariance matrix, and since the trace of the variance-covariance matrix corresponds to the total variance, one has the confirmation that the variance explained by the principal components is the same as explained by the original data.

It is now interesting to locate each object in this new reference space. The coordinate on the first PC is computed simply by substituting into equation $Z_1 = a_{11}X_1 + a_{12}X_2 + \ldots + a_{1v}X_v$ the term X_i with the values of the corresponding original variables. The coordinates on the other principal components are then computed in the same way.

These coordinates are named *scores,* while the constants a_{ij} are named *loadings.* By taking into account the loadings of the variables on the different principal components, it is very easy to understand the importance of each single variable in constituting each PC. A high absolute value means that the variable under examination plays an important role for the component, while a low absolute value means that it has a very limited importance.

If a loading has a positive sign, it means that the objects with a high value of the corresponding variable have high positive scores on that component. If the sign is negative, then the objects with high values of that variable will have high negative scores. As already mentioned, after a PCA the information is mainly concentrated on the first components. As a consequence, a plot of the scores of the objects on the first components allows the direct visualization of the global information in a very efficient way. It is now very easy to detect similarity between objects (similar objects have a very similar position in the space) or the presence of outliers (they are very far from all other objects) or the existence of clusters. Taking into account at the same time scores and loadings, it is also possible to interpret very easily the differences among objects or groups of objects, since it is immediately understandable which are the variables giving the greatest contribution to the phenomenon under study.

Mathematically speaking, we can say that the original data matrix $\mathbf{X}_{o,v}$ (having as many rows as objects and as many columns as variables) has been decomposed into a matrix of scores $\mathbf{S}_{o,c}$ (having as many rows as objects and as many columns as retained components, with c usually $< v$) and a matrix of loadings $\mathbf{L}_{c,v}$ (having as many rows as retained components and as many columns as variables). If, as usual, $c < v$, a matrix of the residuals $\mathbf{E}_{o,v}$, having the same size as the original data set, contains the differences between the original data and the data reconstructed by the PCA model (the smaller the values of this matrix, the higher the variance explained by the model).

We can therefore write the following relationship:

$$\mathbf{X}_{o,v} = \mathbf{S}_{o,c} * \mathbf{L}_{c,v} + \mathbf{E}_{o,v}.$$

Now, let us see the application of PCA to a real data set (MacNamara 2005). Twelve variables have been measured by gas chromatography on 43 samples of Irish whiskeys, of two different types. Nineteen samples were from type A, while 24 samples were from type B, with the samples ordered according to the production time. The data are reported in table 14.3.

Since a trained assessor can easily discriminate a whiskey of type A from a whiskey of type B, it is interesting to know whether this discrimination is possible also on a chemical basis, just taking into account the variables obtained by a routine analysis. When looking separately at each of the twelve variables, it can be seen that none of them completely separates the two types. Therefore, when thinking on a univariate basis, one could say that it is not possible to discriminate between the two types of whiskey. As a consequence, one could look for different (and possible more expensive to determine) variables.

After a PCA (fig. 14.4), it is instead evident that the information present in the twelve variables is sufficient to clearly discriminate the two whiskeys. Once more, it has to be pointed out that taking into account all the variables at the same time gives much more information than just looking at one variable at a time.

Now, let us go one step back and try to understand how this result has been obtained. First, since the variables have different magnitudes and variances, a normalization has to be performed in such a way that each variable will have the same importance. Autoscaling is the most frequently used normalization, which is done by subtracting from each variable its mean value and then dividing the result by its standard deviation. After that, each normalized variable will have mean = 0 and variance = 1. Table 14.4 shows the data after autoscaling.

The results of the PCA are such that PC1 explains 38.4 percent of the total variance and PC2 26.4 percent. This means that the PC1-PC2 plots shown in figure 14.4 explain 64.8 percent of total variance.

Table 14.3. Chemical composition of 43 whiskey samples.

Sample	Type	1) Acetaldehyde	2) Ethyl Acetate	3) Acetal	4) Propanol	5) Isobutanol	6) Isoamyl Acetal	7) Butanol-1	8) 2-Me-1-butanol	9) 3-Me-1-butanol	10) Ethyl Caproate	11) Ethyl Caprylate	12) Ethyl Caprate
1	A	80	408	37	583	466	24	15	388	988	3	13	45
2	A	76	327	40	507	483	25	18	396	1033	3	12	46
3	A	79	296	43	467	397	20	17	323	859	4	13	44
4	A	74	415	28	569	407	24	15	352	921	4	13	46
5	A	69	381	29	510	367	21	14	329	870	4	13	46
6	A	66	340	35	428	387	26	13	339	910	4	14	50
7	A	82	373	17	401	337	23	11	297	813	4	13	42
8	A	78	385	34	459	371	19	12	313	843	3	12	41
9	A	67	374	34	458	385	22	12	326	868	3	13	47
10	A	50	331	32	422	345	17	12	307	835	3	12	42
11	A	66	342	30	423	341	17	13	305	846	3	13	43
12	A	54	321	28	408	354	20	13	310	874	4	13	41
13	A	68	344	33	429	333	16	12	300	824	3	11	38
14	A	69	358	37	446	347	17	13	311	855	3	11	37
15	A	78	346	40	411	320	16	12	287	796	3	11	36
16	A	77	387	51	427	345	22	12	290	805	3	10	32
17	A	104	322	72	432	353	18	13	303	823	3	10	35
18	A	84	333	55	421	340	17	13	292	787	3	10	31
19	A	82	382	47	457	328	18	10	278	765	3	10	31
20	B	65	403	18	496	529	19	19	365	1014	3	11	35
21	B	58	352	18	434	457	17	17	312	907	3	8	26

22	B	71	394	25	555	560	18	20	391	1083	3	11	33
23	B	69	369	25	497	500	16	18	349	1005	3	10	29
24	B	83	344	28	489	479	15	17	352	957	3	10	29
25	B	93	344	31	500	481	15	18	352	990	3	10	29
26	B	65	453	18	503	529	21	17	390	1017	3	10	31
27	B	62	405	17	500	488	18	17	357	965	3	9	27
28	B	58	435	16	501	548	21	17	415	1056	3	10	31
29	B	63	459	17	544	575	21	19	426	1100	3	10	28
30	B	99	462	26	490	500	22	16	403	1057	3	10	30
31	B	81	357	21	402	396	16	14	310	814	2	7	17
32	B	80	380	23	497	483	18	17	395	1041	3	10	28
33	B	76	425	22	486	475	22	17	379	1007	4	10	25
34	B	79	446	24	446	418	18	14	319	803	3	9	25
35	B	78	461	24	478	458	19	16	352	908	3	9	23
36	B	108	477	29	493	430	16	14	329	811	3	11	28
37	B	111	481	28	494	429	16	15	330	833	3	9	22
38	B	82	408	22	473	431	18	12	317	774	3	10	27
39	B	73	428	20	493	445	18	13	327	804	3	8	20
40	B	102	469	25	490	457	20	11	327	776	3	10	27
41	B	90	463	22	491	452	20	12	324	774	3	9	21
42	B	50	410	14	440	419	19	12	300	704	3	11	28
43	B	61	425	17	445	432	20	12	318	758	3	10	23

Table 14.4. Autoscaled data.

Sample	Type	1) Acetaldehyde	2) Ethyl Acetate	3) Acetal	4) Propanol	5) Isobutanol	6) Isoamyl Acetal	7) Butanol-1	8) 2-Me-1-butanol	9) 3-Me-1-butanol	10) Ethyl Caproate	11) Ethyl Caprylate	12) Ethyl Caprate
1	A	0.288	0.340	0.672	2.507	0.553	1.743	0.187	1.348	0.936	−0.338	1.441	1.418
2	A	0.013	−1.285	0.927	0.791	0.796	2.105	1.335	1.559	1.366	−0.338	0.821	1.536
3	A	0.219	−1.907	1.183	−0.112	−0.435	0.295	0.952	−0.365	−0.297	2.084	1.441	1.301
4	A	−0.125	0.481	−0.095	2.190	−0.291	1.743	0.187	0.399	0.296	2.084	1.441	1.536
5	A	−0.469	−0.202	−0.010	0.858	−0.864	0.657	−0.196	−0.207	−0.192	2.084	1.441	1.536
6	A	−0.675	−1.024	0.501	−0.993	−0.578	2.467	−0.579	0.056	0.190	2.084	2.060	2.005
7	A	0.426	−0.362	−1.032	−1.602	−1.293	1.381	−1.344	−1.051	−0.737	2.084	1.441	1.066
8	A	0.150	−0.121	0.416	−0.293	−0.806	−0.067	−0.961	−0.629	−0.450	−0.338	0.821	0.949
9	A	−0.606	−0.342	0.416	−0.316	−0.606	1.019	−0.961	−0.286	−0.211	−0.338	1.441	1.653
10	A	−1.776	−1.205	0.246	−1.128	−1.178	−0.791	−0.961	−0.787	−0.526	−0.338	0.821	1.066
11	A	−0.675	−0.984	0.075	−1.106	−1.236	−0.791	−0.579	−0.840	−0.421	−0.338	1.441	1.184
12	A	−1.501	−1.406	−0.095	−1.444	−1.050	0.295	−0.579	−0.708	−0.154	2.084	1.441	0.949
13	A	−0.537	−0.944	0.331	−0.970	−1.350	−1.153	−0.961	−0.972	−0.631	−0.338	0.202	0.597
14	A	−0.469	−0.663	0.672	−0.586	−1.150	−0.791	−0.579	−0.682	−0.335	−0.338	0.202	0.480
15	A	0.150	−0.904	0.927	−1.377	−1.536	−1.153	−0.961	−1.314	−0.899	−0.338	0.202	0.363
16	A	0.082	−0.081	1.864	−1.015	−1.178	1.019	−0.961	−1.235	−0.813	−0.338	−0.418	−0.106
17	A	1.939	−1.386	3.654	−0.903	−1.064	−0.429	−0.579	−0.893	−0.641	−0.338	−0.418	0.245
18	A	0.563	−1.165	2.205	−1.151	−1.250	−0.791	−0.579	−1.183	−0.985	−0.338	−0.418	−0.224
19	A	0.426	−0.182	1.524	−0.338	−1.422	−0.429	−1.727	−1.552	−1.195	−0.338	−0.418	−0.224
20	B	−0.744	0.240	−0.947	0.542	1.454	−0.067	1.718	0.742	1.184	−0.338	0.202	0.245
21	B	−1.225	−0.784	−0.947	−0.857	0.424	−0.791	0.952	−0.655	0.162	−0.338	−1.657	−0.810

22	B	−0.331	0.059	−0.351	1.874	1.897	−0.429	2.100	1.427	1.844	−0.338	0.202	0.011
23	B	−0.469	−0.442	−0.351	0.565	1.039	−1.153	1.335	0.320	1.098	−0.338	−0.418	−0.458
24	B	0.494	−0.944	−0.095	0.384	0.739	−1.515	0.952	0.399	0.640	−0.338	−0.418	−0.458
25	B	1.182	−0.944	0.160	0.633	0.767	−1.515	1.335	0.399	0.955	−0.338	−0.418	−0.458
26	B	−0.744	1.243	−0.947	0.700	1.454	0.657	0.952	1.401	1.213	−0.338	−0.418	−0.224
27	B	−0.950	0.280	−1.032	0.633	0.867	−0.429	0.952	0.531	0.716	−0.338	−1.037	−0.693
28	B	−1.225	0.882	−1.117	0.655	1.726	0.657	0.952	2.060	1.586	−0.338	−0.418	−0.224
29	B	−0.881	1.364	−1.032	1.626	2.112	0.657	1.718	2.350	2.006	−0.338	−0.418	−0.575
30	B	1.595	1.424	−0.265	0.407	1.039	1.019	0.570	1.743	1.595	−0.338	−0.418	−0.341
31	B	0.357	−0.683	−0.691	−1.580	−0.449	−1.153	−0.196	−0.708	−0.727	−2.759	−2.276	−1.866
32	B	0.288	−0.222	−0.521	0.565	0.796	−0.429	0.952	1.533	1.442	−0.338	−0.418	−0.575
33	B	0.013	0.681	−0.606	0.317	0.681	1.019	0.952	1.111	1.117	2.084	−0.418	−0.927
34	B	0.219	1.103	−0.436	−0.586	−0.134	−0.429	−0.196	−0.471	−0.832	−0.338	−1.037	−0.927
35	B	0.150	1.404	−0.436	0.136	0.438	−0.067	0.570	0.399	0.171	−0.338	−1.037	−1.162
36	B	2.214	1.725	−0.010	0.475	0.038	−1.153	−0.196	−0.207	−0.756	−0.338	0.202	−0.575
37	B	2.420	1.805	−0.095	0.497	0.023	−1.153	0.187	−0.181	−0.545	−0.338	−1.037	−1.279
38	B	0.426	0.340	−0.606	0.023	0.052	−0.429	−0.961	−0.524	−1.109	−0.338	−0.418	−0.693
39	B	−0.194	0.742	−0.777	0.475	0.252	−0.429	−0.579	−0.260	−0.823	−0.338	−1.657	−1.514
40	B	1.801	1.564	−0.351	0.407	0.424	0.295	−1.344	−0.260	−1.090	−0.338	−0.418	−0.693
41	B	0.976	1.444	−0.606	0.429	0.352	0.295	−0.961	−0.339	−1.109	−0.338	−1.037	−1.396
42	B	−1.776	0.380	−1.288	−0.722	−0.120	−0.067	−0.961	−0.972	−1.778	−0.338	0.202	−0.575
43	B	−1.019	0.681	−1.032	−0.609	0.066	0.295	−0.961	−0.497	−1.262	−0.338	−0.418	−1.162

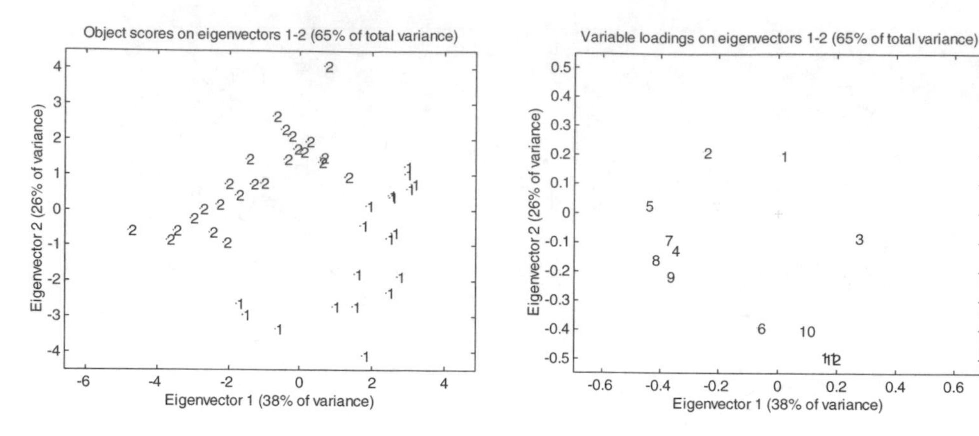

Fig. 14.4. PCA of the data in table 14.3. On the left, the score plot of the objects (coded according to the whiskey type), on the right the loading plot of the variables (coded according to the order in table 14.3).

Table 14.5 shows the loadings of the variables on PC1 and PC2. From it, the loading plot in figure 14.4 is obtained.

From the score plot in figure 14.4 it can be seen that the plane PC1-PC2 perfectly separates the two categories. By looking at the loading plot and at table 14.5 it is possible to know which are the variables mainly contributing to each of the PCs. Variables 4, 5, 7, 8, and 9 (propanol, isobutanol, butanol-1, 2-Me-1-butanol, and 3-Me-1-butanol, i.e., the alcohols) have the loadings with the highest absolute value on PC1, all of them being negative. This means that the alcohols are higher in those samples having the highest negative scores on PC1. Variables 6, 10, 11, and 12 (isoamyl acetal, ethyl caproate, ethyl caprylate, and ethyl caprate, i.e., the esters) have the loadings with the highest absolute value on PC2, all of them being negative. This means that the esters are higher in those samples having the highest negative scores on PC2. Therefore, it can be said that the esters are the main reason for the separation between the two types, while the alcohols are the main reason for the variability inside each type. The fact that all the alcohols have very similar loadings means that they are very much correlated, as is the case for the esters. This is a further demonstration of the superiority of multivariate analysis over univariate analysis. Indeed, it will be possible to adulterate a product in such a way that all the variables, singularly taken, fall inside their individual range of acceptance; much more difficult (not to say impossible) will be to have an adulterated product in which also the correlations among the variables will be preserved. Therefore, adulterated products that will be unnoticed by the "classical" univariate analysis will be easily detected by a multivariate analysis (see the section "Modeling").

Table 14.6 reports the scores of the objects on PC1 and PC2. As previously shown, the scores of an object are computed by multiplying the loadings of each variable by the value of the variable. As an example, let us compute the score of sample 1 on PC1 (since the autoscaled data have been used, these are the values that must be taken into account):

$$0.288 * 0.006 + 0.340 * (-0.253) + 0.672 * 0.261 + 2.507 * (-0.363) + 0.553 * (-0.452) + 1.743 * (-0.067) + 0.187 * (-0.385) + 1.348 * (-0.429) + 0.936 * (-0.378) + (-0.338) * 0.071 + 1.441 * 0.146 + 1.418 * 0.159 = -1.778.$$

Table 14.5. Loadings of the variables on PC1 and PC2.

	1) Acetaldehyde	2) Ethyl Acetate	3) Acetal	4) Propanol	5) Isobutanol	6) Isoamyl Acetal	7) Butanol-1	8) 2-Me-1-butanol	9) 3-Me-1-butanol	10) Ethyl Caproate	11) Ethyl Caprylate	12) Ethyl Caprate
PC1	0.006	−0.253	0.261	−0.363	−0.452	−0.067	−0.385	−0.429	−0.378	0.071	0.146	0.159
PC2	0.196	0.206	−0.086	−0.129	0.023	−0.395	−0.096	−0.162	−0.221	−0.404	−0.493	−0.498

Table 14.6. Scores of the objects on PC1 and PC2.

Object	Category	Score on PC1	Score on PC2
1	A	−1.778	−2.654
2	A	−1.581	−2.974
3	A	1.477	−2.730
4	A	−0.679	−3.359
5	A	0.921	−2.744
6	A	1.737	−4.096
7	A	2.675	−1.889
8	A	1.673	−0.432
9	A	1.534	−1.811
10	A	2.526	−0.650
11	A	2.395	−0.793
12	A	2.395	−2.350
13	A	2.488	0.350
14	A	1.850	0.113
15	A	3.080	0.722
16	A	2.446	0.409
17	A	2.955	0.603
18	A	2.891	1.031
19	A	2.903	1.231
20	B	−2.546	−0.658
21	B	−0.426	1.448
22	B	−3.730	−0.862
23	B	−1.805	0.400
24	B	−1.093	0.748
25	B	−1.392	0.723
26	B	−3.069	−0.258
27	B	−2.090	0.724
28	B	−3.556	−0.589
29	B	−4.814	−0.585
30	B	−2.815	0.020
31	B	0.677	4.098
32	B	−2.361	0.140
33	B	−2.148	−0.925
34	B	0.180	1.939
35	B	−1.527	1.444
36	B	−0.173	1.729
37	B	−0.747	2.663
38	B	0.575	1.484
39	B	−0.510	2.283
40	B	0.012	1.648
41	B	−0.314	2.114
42	B	1.251	0.927
43	B	0.514	1.368

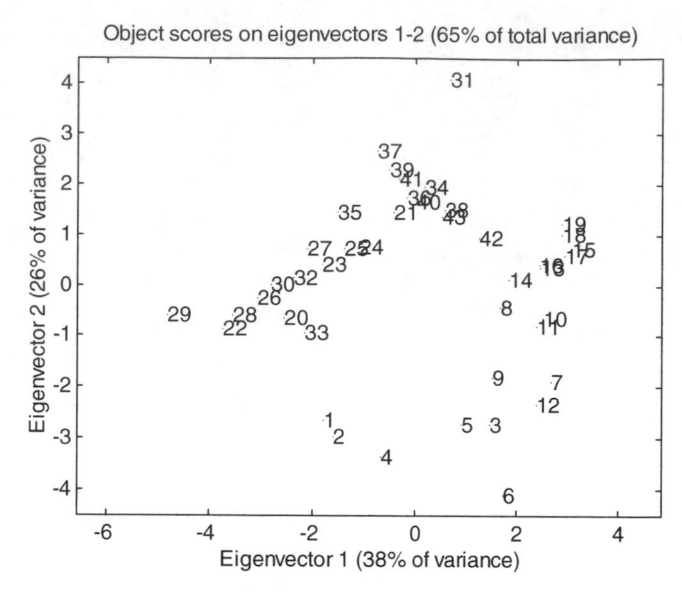

Fig. 14.5. Score plot of the data in table 14.3. (The samples are coded according to the order in table 14.3.)

So we have demonstrated that the two types of whiskeys are really also different from the chemical point of view.

Now, let us look at figure 14.5. In it, the samples are coded according to table 14.3 (i.e., following the production order). It can be seen that for both types there is a trend from the left-hand side of the plot (negative values of PC1) to the right-hand side of the plot (positive values of PC1), with this effect being much clearer for type 1. As has been previously said, PC1 is mainly related to the alcohols. Therefore, it can be concluded that throughout the production period there has been a progressive decrease of the alcohol content. While the previous finding was the answer to a question that was explicitly formulated by the producer ("are the two types of whiskey different?"), this result came out totally unexpected. This shows very well what is mentioned in a paper by Bro et al. (2002):

> Usually, data analysis is performed as a confirmatory exercise, where a postulated hypothesis is claimed, data generated accordingly and the data analysed in order either to verify or reject this hypothesis.
>
> No new knowledge is obtained in confirmatory analysis except the possible verification of a prior postulated hypothesis. Using exploratory analysis the data are gathered in order to represent as broadly and as well as possible the problem under investigation.
>
> The data are analysed and through the, often visual, inspection of the results, hypotheses are suggested on the basis of the empirical data. Consequently, exploratory data analysis is an extraordinary tool in displaying thus far unknown information from established and potential monitoring methods.

Process Monitoring and Quality Control

When running a process, it is very important to know whether it is under control (i.e., inside its natural variability) or out of control (i.e., in a condition that is not typical and therefore can lead to an accident).

Analogously, when producing a product, it is very important to know whether each single piece is inside specifications (i.e., close to the "ideal" product, inside its natural variability) or out of specifications (i.e., significantly different from the "standard" product and therefore in a condition possibly leading to a complaint by the final client).

PCA is the basis for a multivariate process monitoring and a multivariate quality control, much more effective than the usually applied univariate approaches (Kourti and MacGregor 1995).

After having collected a relevant number of observations describing the "normal operating" process (or the "inside specification" products), encompassing all the sources of normal variability, it will be possible to build a PCA model defining the limits inside which the process (or the product) should stay.

Any new set of measurements (a vector $\mathbf{x}_{1,v}$) describing the process in a given moment (or a new product) will be projected onto the previously defined model by using the following equation: $\mathbf{s}_{1,c} = \mathbf{x}_{1,v} * \mathbf{L}_{c,v}'$. From the computed scores, it can be estimated how far from the barycenter of the model (i.e., from the "ideal" process [or product]) it is.

Its residuals can also be easily computed: $\mathbf{e}_{1,v} = \mathbf{x}_{1,v} - \mathbf{s}_{1,c} * \mathbf{L}_{c,v}$ ($\mathbf{e}_{1,v}$ is the vector of the residuals, and each of its v elements corresponds to the difference between the measured and reconstructed value of each variable). From them, it can be understood how well the sample is reconstructed by the PCA model—that is, how far from the model space (a plane, in case c = 2) it lies.

Statistical tests make possible the automatic detection of an outlier in both cases (they are defined as T^2 outliers in the first case and Q outliers in the second case). With these simple tests it is possible to detect a fault in a process or to reject a bad product by checking just two plots, instead of as many plots as variables, as in the case of the Sheward charts commonly used when the univariate approach is applied. Furthermore, the multivariate approach is much more robust, since it leads to a lower number of false-negatives and false-positives, and much more sensitive, since it allows the detection of faults at an earlier stage.

Finally, the contribution plots will easily outline which variables are responsible for the sample being an outlier.

Three-Way PCA

It can happen that the structure of a data set is such that a standard two-way table (objects versus variables) is not enough to describe it. Let us suppose that the same analyses have been performed at different sampling sites on different days. A third way needs to be added to adequately represent the data set, which can be imagined as a parallelepiped of size $I \times J \times K$, where I is the number of sampling sites (objects), J is the number of variables, and K is the number of sampling times (conditions) (Geladi 1989; Smilde 1992).

To apply standard PCA, these three-way data arrays $\underline{\mathbf{X}}$ have to be matricized to obtain a two-way data table. This can be done in different ways, according to what one is interested in focusing on.

If we are interested in studying each "sampling," a matrix $\mathbf{X'}_b$ is obtained having $I \times K$ rows and J columns. This approach is very straightforward in terms of computation, but since $I \times K$ is usually a rather large number, the interpretation of the resulting score plot can give some problems.

To focus on the sampling sites, the data array $\underline{\mathbf{X}}$ can be matricized to $\mathbf{X'}_a$ (I rows, $J \times K$ columns). The interpretability of the score plot is usually very high, but since $J \times K$ is usually a rather large number, the interpretation of the loading plot is very difficult.

The same considerations can be made when focusing on the sampling times: in this case, X'_c is obtained (K rows, $I \times J$ columns).

Three-way PCA allows a much easier interpretation of the information contained in the data set, since it directly takes into account its three-way structure. If the Tucker3 model is applied, the final result is given by three sets of loadings together with a core array describing the relationship among them. If the number of components is the same for each way, the core array is a cube. Each of the three sets of loadings can be displayed and interpreted in the same way as a score plot of standard PCA.

In the case of a cubic core array a series of orthogonal rotations can be performed on the three spaces of the objects, variables, and conditions, looking for the common orientation for which the core array is as much body-diagonal as possible.

If this condition is sufficiently achieved, then the rotated sets of loadings can also be interpreted jointly by overlapping them.

An example of the application of three-way PCA is a data set from the field of sensory evaluation. In it, eight types of noodles, each corresponding to a different formulation, were produced in four independent replicates, with each replicate tested by the panel in two independent sessions. Each of the twelve panelists gave a score to eight descriptors (1: yellow color, 2: translucency, 3: shininess, 4: surface smoothness, 5: firmness, 6: chewiness, 7: surface stickiness, 8: elasticity). The data set can therefore be seen as a $64 \times 8 \times 12$ data set (Tang et al. 1999).

By taking into account the loading plots of the objects (fig. 14.6), it can be seen that the regions occupied by the eight samples of each noodle (4 replicates × 2 sessions) never overlap. This means that the global variability (production + sensory evaluation) of each noodle is always smaller than the differences among the noodles. The fact that the region spanned by each noodle is approximately the same (with the exception of noodle 2) indicates that the global variability can be considered as independent of the type of noodle.

It can also be seen that the variability between sessions is smaller than the variability among replicates, this meaning that the "instrumental error" of the judges is smaller than the variability of the production.

On the first axis, noodles 7 and 8 have the lowest loading, followed by noodle 6 and then by the remaining five types, all with very similar loadings. This ranking ($7 = 8 > 6$) corresponds to the content of glyceryl monostearate (GMS, 2.8 percent, 2.8 percent, and 1.4 percent, respectively, with the other noodles having no GMS). It can be concluded that the loadings of each noodle group on the first axis are directly related to the GMS content.

On the second axis, the five formulations having no GMS are discriminated, with noodle 3 having the highest loading and noodle 1 having by far the lowest loading. Noodle 3 is made only with durum wheat four (DWF), while noodle 1 is the only one containing wheat starch (WS). On the same axis, noodles 5, 4, and 2 have decreasing loadings, and this corresponds to their amount of wheat gluten (WG, 6 percent, 3 percent, and 0 percent, respectively).

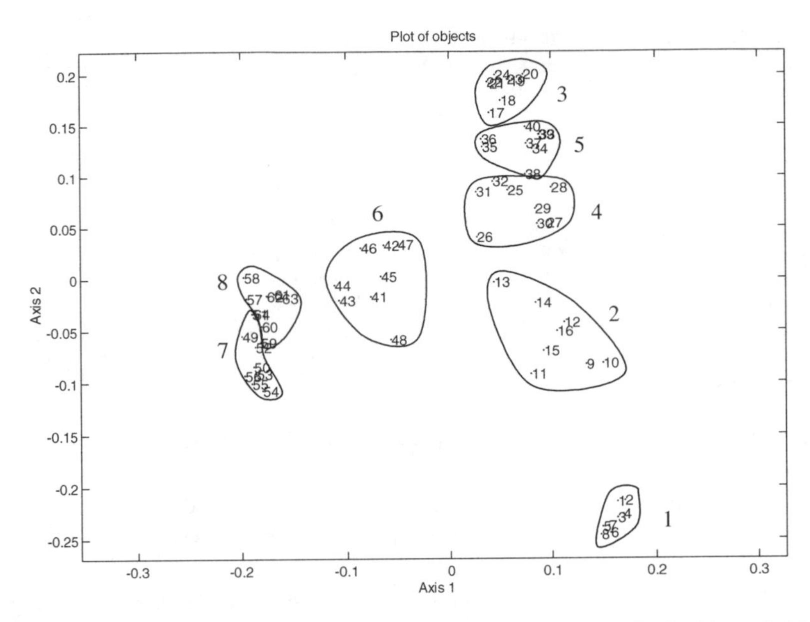

Fig. 14.6. Scatter plot of the loadings of the objects. Objects 1–8: noodle 1; objects 9–16: noodle 2; . . . ; objects 57–64: noodle 8.

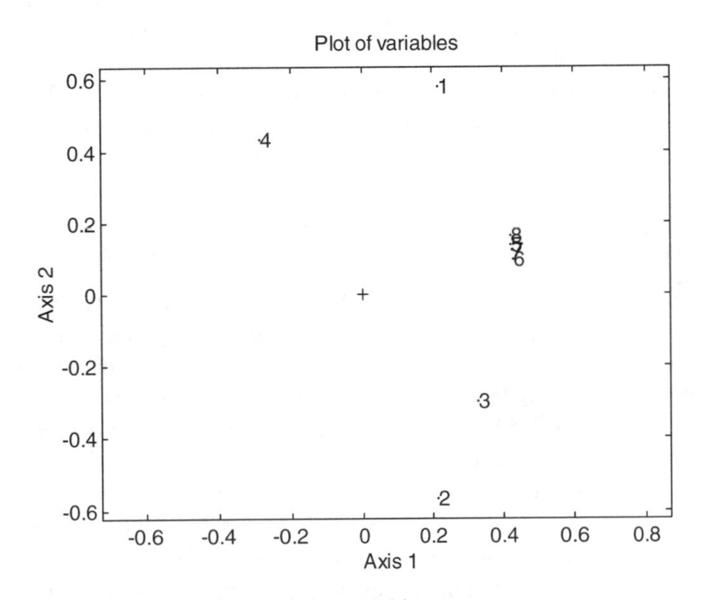

Fig. 14.7. Scatter plot of the loadings of the variables.

Figure 14.7 shows the scatter plot of the loadings of the variables. Variables 5–8 (the texture-related descriptors) have the highest values on the first axis. This means that the first axis is mainly related to the texture of the product.

Variables 1 and 4 (color and smoothness, both positive attributes) have positive loadings on axis 2, in contrast with variables 2 and 3 (translucence and shininess, both

Table 14.7. Data sets on which three-way PCA can be applied.

Field of Application	Objects	Variables	Conditions
Environmental analysis	Air or water samples	Chemico-physical analyses	Time
Environmental analysis	Water samples (different locations)	Chemico-physical analyses	Depth
Panel tests	Food products (oils, wines)	Attributes	Assessors
Food chemistry	Foods (cheeses, spirits, etc.)	Chemical composition	Aging
Food chemistry	Foods (oils, wines, etc.)	Chemical composition	Crops
Sport medicine	Athletes	Blood analyses	Time after effort
Process monitoring	Batches	Chemical analyses	Time

negative attributes). Therefore, the second axis is mainly related to the appearance attributes of the noodles.

It must also be noticed that variables 5–8 (the texture-related descriptors) have very similar loadings on both axes, and therefore are very highly correlated.

As a result, it can be concluded that axis 1 is related to the amount of GMS and to the texture of the product; it can be seen that the addition of GMS leads to a worse product.

Axis 2 is related to the aspect; it can be seen that noodle 3, made with DWF, is the product with the best appearance (the most yellow and the smoothest), while noodle 1, made with a large amount of WS, has the worst appearance (the most translucent and the most shiny). The addition of WG also improves the appearance, since it results in an increase of the yellow color and the smoothness.

By taking into account both axes, it is easy to detect noodle 3 as the best one.

Table 14.7 shows some types of data sets on which three-way PCA can be successfully applied.

Classification

In the third section of this chapter, "Data Display," we verified that the two types of whiskey are indeed well separated in the multivariate space of the variables. Therefore, we can say that we have two really different classes. Let us suppose we now get some unknown samples and we want to know what their class is. After having performed the chemical analyses, we can add these data to the previous data set, run a PCA, and see where the new samples are placed. This will be fine if the new samples fall inside one of the clouds of points corresponding to a category, but what if they fall in a somehow intermediate position? How can we say with "reasonable certainty" that the new samples are from type A or type B? We know that PCA is a very powerful technique for data display, but we realize that we need something different if we want to classify new samples. What we want is a technique producing some "decision rules" discriminating among the possible categories.

While PCA is an "unsupervised" technique, the classification methods are "supervised" techniques. In these techniques the category of each of the objects on which the model is built must be specified in advance.

The most commonly used classification techniques are linear discriminant analysis (LDA) and quadratic discriminant analysis (QDA). They define a set of delimiters (according to the number of categories under study) in such a way that the multivariate space of the objects is divided in as many subspaces as the number of categories, and that each point of the space belongs to one and only one subspace. Rather than describing in detail the algorithms behind these techniques, special attention will be given to the critical points of a classification.

As previously stated, the classification techniques use objects belonging to the different categories to define boundaries delimiting regions of the space. The final goal is to apply these classification rules to new objects for their classification into one of the existing categories. The performance of the techniques can be expressed as classification ability and prediction ability. The difference between "classification" and "prediction," though quite subtle at first glance, is actually very important, and its underestimation can lead to very bitter deceptions.

Classification ability is the capability of assigning to the correct category the same objects used to build the classification rules, while prediction ability is the capability of assigning to the correct category objects that have not been used to build the classification rules. Since the final goal is the classification of new samples, it is clear that predictive ability is by far the most important figure.

The results of a classification method can be expressed in several ways. The most synthetic one is the percentage of correct classifications (or predictions). Note that, in the following, only the term *classification* will be used, but it has to be understood as "classification or prediction." This can be obtained as the number of correct classifications (independently of the category) divided by the total number of objects, or as the average of the performance of the model over all the categories. The two results are very similar when the sizes of all the categories are very similar, but they can be very different if the sizes are quite different. Let us consider the case shown in table 14.8.

The very poor performance of category 3, by far the smallest one, almost does not affect the classification rate computed on the global number of classifications, while it produces a much lower result if the classification rate is computed as the average of the three categories.

A more complete and detailed overview of the performance of the method can be obtained by using the classification matrix, which also allows knowing the categories to which the wrongly classified objects are assigned (in many cases the cost of an error can be quite different according to the category the sample is assigned to). In it, each row corresponds to the true category, and each column to the category to which the sample has been assigned. Continuing with the previous example, a possible classification matrix is the one shown in table 14.9.

From it, it can be seen that the 112 objects of category 1 were classified in the following way: 105 correctly to category 1, none to category 2, and 7 to category 3. In the

Table 14.8. Example of the performance of a classification technique.

Category Number	Objects	Correct Classifications	Percentage of Correct Classifications
1	112	105	93.8
2	87	86	98.9
3	21	10	47.6
Total	220	201	91.4/80.1

Table 14.9. Example of a classification matrix.

Category	1	2	3
1	105	0	7
2	1	86	0
3	11	0	10

same way, it can be deduced that all the objects of category 3 which were not correctly classified have been assigned to category 1. Therefore, it is easy to conclude that category 2 is well-defined and that the classification of its objects gives no problems at all, while categories 1 and 3 are quite overlapping. As a consequence, to have a perfect classification more effort must be put into better separating categories 1 and 3. All this information cannot be obtained from just the percentage of correct classifications.

If overfitting occurs, then the prediction ability will be much worse than the classification ability. To avoid it, it is very important that the sample size is adequate for the problem and for the technique. A general rule is that the number of objects should be more than five times (at least, no less than three times) the number of parameters to be estimated. LDA works on a pooled variance-covariance matrix: this means that the total number of objects should be at least five times the number of variables. QDA computes a variance-covariance matrix for each category, which makes it a more powerful method than LDA, but this also means that each category should have a number of objects at least five times higher than the number of variables. This is a good example of how the more complex, and therefore "better," methods sometimes cannot be used in a safe way because their requirements do not correspond to the characteristics of the data set.

Modeling

In classification, the space is divided into as many subspaces as categories, and each point belongs to one and only one category. This means that the samples that will be predicted by such methods must belong to one of the categories used to build the models; if not, they will anyway be assigned to one of them. To make this concept clearer, let us suppose the use of a classification technique to discriminate between water and wine. Of course, this discrimination is very easy. Each sample of water will be correctly assigned to the category "water," and each sample of wine will be correctly assigned to the category "wine." But what happens with a sample of orange squash? It will be assigned either to the category "water" (if variables such as alcohol are taken into account) or to the category "wine" (if variables such as color are considered). The classification techniques are therefore not able to define a new sample as being "something different" from all the categories of the training set. This is instead the main feature of the modeling techniques.

Though several techniques are used for modeling purpose, UNEQ (one of the modeling versions of QDA) and SIMCA (soft independent model of class analogy) are the most used. While in classification every point of the space belongs to one and only one category, with these techniques the models (one for each category) can overlap and leave some regions of the space unassigned. This means that every point of the space can belong to one category (the sample has been recognized as a sample of that class), to more than one category (the sample has such characteristics that it could be a sample of more than one class), or to none of the categories (the sample has been considered as being different from all the classes).

Of course, the "ideal" performance of such a method would not only be to correctly classify all the samples in their category (as in the case of a classification technique) but also be such that the models of each category could be able to accept all the samples of that category and to reject all the samples of the other categories. The results of a modeling technique are expressed the same way as in classification, plus two very important parameters: specificity and sensitivity. For category *c*, its specificity (how much the model rejects the objects of different categories) is the percentage of the objects of categories different from *c* rejected by the model, while its sensitivity (how much the model accepts the objects of the same category) is the percentage of the objects of category *c* accepted by the model.

While the classification techniques need at least two categories, the modeling techniques can be applied when only one category is present. In this case the technique detects if the new sample can be considered as a typical sample of that category or not. This can be very useful in the case of Protected Denomination of Origin products, to verify whether a sample, declared as having been produced in a well-defined region, has indeed the characteristics typical of the samples produced in that region.

The application of a multivariate analysis will greatly reduce the possibility of frauds. While an "expert" can adulterate a product in such a way that all the variables, independently considered, still stay in the accepted range, it is almost impossible to adulterate a product in such a way that its multivariate "pattern" is still accepted by the model of the original product, unless the amount of the adulterant is so small that it becomes unprofitable from the economic point of view.

Calibration

Let us imagine we have a set of wine samples and that on each of them the FT-IR spectrum is measured, together with some variables such as alcohol content, pH, or total acidity. Of course, chemical analyses will require much more time than a simple spectral measurement. It would therefore be very useful to find a relationship between each of the chemical variables and the spectrum. This relationship, after having been established and validated, will be used to predict the content of the chemical variables. It is easy to understand how much time (and money) this will save, since in a few minutes it will be possible to have the same results as previously obtained by a whole set of chemical analyses.

Generally speaking, we can say that multivariate calibration finds relationships between one or more response variables y and a vector of predictor variables $\mathbf{x}$. As the previous example should have shown, the final goal of multivariate calibration is not just "to describe" the relationship between the $\mathbf{x}$ and the y variables in the set of samples on which the relationship has been computed, but to find a real, practical application for samples that in a following time will have the $\mathbf{x}$ variables measured.

The model is a linear polynomial ($y = b_0 + b_1x_1 + b_2x_2 + \ldots + b_Kx_K + f$), where b_0 is an offset, the b_k ($k = 1, \ldots, K$) are regression coefficients, and f is a residual. The "traditional" method of calculating $\mathbf{b}$, the vector of regression coefficients, is ordinary least squares (OLS). However, this method has two major limitations that make it inapplicable to many data sets:

1. It cannot handle more variables than objects.
2. It is sensitive to collinear variables.

It can be easily seen that both these limitations do not allow the application of OLS to spectral data sets, where the samples are described by a very high number of highly collinear variables. If one wants to apply OLS to such data anyway, the only way to do it is to reduce the number of variables and their collinearity through a suitable variable selection (see the section "Variable Selection").

When describing the PCA, it has been noticed that the components are orthogonal (i.e., uncorrelated) and that the dimensionality of the resulting space (i.e., the number of significant components) is much lower than the dimensionality of the original space. Therefore, it can be seen that both the aforementioned limitations have been overcome. As a consequence, it is possible to apply OLS to the scores originated by PCA. This technique is principal component regression (PCR).

It has to be considered that PCs are computed by taking into account only the **x** variables, without considering at all the y variable(s), and are ranked according to the explained variance of the "x space." This means that it can happen that the first PC has little or no relevance in explaining the response we are interested in. This can be easily understood by considering that, even when we have several responses, the PCs to which the responses have to be regressed will be the same.

Nowadays, the most favored regression technique is partial least squares regression (PLS, or PLSR). As happens with PCR, PLS is based on components (or "latent variables"). The PLS components are computed by taking into account both the **x** and the **y** variables, and therefore they are slightly rotated versions of the principal components. As a consequence, their ranking order corresponds to the importance in the modeling of the response. A further difference with OLS and PCR is that, while the former must work on each response variable separately, PLS can be applied to multiple responses at the same time.

Because both PCR and PLS are based on latent variables, a very critical point is the number of components to be retained. Though we know that information is "concentrated" in the first components and that the last components explain just noise, it is not always an easy task to detect the correct number of components (i.e., when information finishes and noise begins). Selecting a lower number of components would mean removing some useful information (underfitting), while selecting a higher number of components would mean incorporating some noise (overfitting).

Before applying the results of a calibration, it is very important to look for the presence of outliers. Three major types of outliers can be detected: outliers in the x space (samples for which the x variables are very different from those of the rest of the samples; they can be found by looking at a PCA of the x variables), outliers in the y space (samples with the y variable very different from those of the rest of the samples; they can be found by looking at a histogram of the y variable), and samples for which the calibration model is not valid (they can be found by looking at a histogram of the residuals).

The efficacy of a calibration can be summarized by two values, the percentage of variance explained by the model and the root mean square error in calibration (RMSEC). The former, being a "normalized" value, gives an initial idea about how much of the variance of the data set is "captured" by the model; the latter, being an absolute value to be interpreted in the same way as a standard deviation, gives information about the magnitude of the error.

As already described in the classification section and as pointed out at the beginning of this section, the goal of a calibration is essentially not to describe the relationship between the response and the x variables of the samples on which the calibration is computed (training, or calibration, set) but to apply it to future samples where only the cheaper x variables will be measured. In this case too, the model must be validated by using a set of samples different from those used to compute the model (validation, or test, set). The responses of the objects of the test set will be computed by applying the model obtained by the training set and then compared with their "true" response. From these values the percentage of variance explained in prediction and the root mean square error of prediction (RMSEP) can be computed. Provided that the objects forming the two sets have been selected flawlessly, these values give the real performance of the model on new samples.

Variable Selection

Usually, not all the variables of a data set bring useful and nonredundant information. Therefore, a variable (or feature) selection can be highly beneficial, since from it the following results are obtained:

1. Removal of noise and improvement of the performance
2. Reduction of the number of variables to be measured and simplification of the model

The removal of noisy variables should always be sought. Though some methods can give good results even with a moderate amount of noise disturbing the information, it is clear that their performance will increase when this noise is removed. So feature selection is now widely applied also for those techniques (PLS and PCR) that in the beginning were considered to be almost insensitive to noise.

While noise reduction is a common goal for any data set, the relevance of the reduction of the number of variables in the final model depends very much on the kind of data constituting the data set, and a very wide range of situations are possible. Let us consider the extreme conditions:

1. Each variable requires a separate analysis
2. All the variables are obtained by the same analysis (e.g., chromatographic and spectroscopic data)

In the first case, each variable not selected means a reduction in terms of costs and/or analysis time. The variable selection should therefore always be made on a cost/benefit basis, looking for the subset of variables leading to the best compromise between performance of the model and cost of the analyses. This means that, in the presence of groups of useful but highly correlated (and therefore redundant) variables, only one variable per group should be retained. With such data sets, it is also possible that a subset of variables giving a slightly worse result is preferred, if the reduction in performance is widely compensated by a reduction in costs or time.

In the second case, the number of retained variables has no effect on the analysis cost, while the presence of useful and correlated variables improves the stability of the model. Therefore, the goal of variable selection will be to improve the predictive ability of the model by removing the variables giving no information, without being worried by the number of retained variables.

Intermediate cases can happen, in which "blocks" of variables are present. As an example, take the case of olive oil samples, on each of which the following analyses have been run: a titration for acidity, the analysis of peroxides, a UV spectroscopy for ΔK, a GC for sterols, and another GC for fatty acids. In such a situation, what counts is not the final number of variables, but the number of analyses one can save.

The only possible way to be sure that "the best" set of variables has been picked up is the "all-models" techniques testing all the possible combinations. Since with k variables, the number of possible combinations is $2^k - 1$, it is easy to understand that this approach cannot be used unless the number of variables is really very low (e.g., with 30 variables more than 10^9 combinations should be tested).

The simplest (but least effective) way of performing a feature selection is to operate on a "univariate" basis, by retaining those variables having the greatest discriminating power (in the case of a classification) or the greatest correlation with the response (in the case of a calibration). By doing that each variable is taken into account by itself without considering how its information "integrates" with the information brought by the other (selected or unselected) variables. As a result, if several highly correlated variables are "good," they are all selected, without taking into account that, owing to their correlation, the information is highly redundant and therefore at least some of them can be removed without any decrease in the performance. On the other hand, those variables are not taken into account that, though not giving by themselves significant information, become very important when their information is integrated with that of other variables.

An improvement is brought by the "sequential" approaches. They select the best variable first, then the best pair formed by the first and second, and so on in a forward or backward progression. A more sophisticated approach applies a look back from the progression to reassess previous selections. The problem with these approaches is that only a very small part of the experimental domain is explored and that the number of models to be tested becomes very high in case of highly dimensional data sets, such as spectral data sets. For instance, with 1000 wavelengths, 1000 models are needed for the first cycle (selection or removal of the first variable), 999 for the second cycle, 998 for the third cycle, and so on.

More "multivariate" methods of variable selection, especially suited for PLS applied to spectral data, are currently available. Among them, we can cite interactive variable selection (Lindgren et al. 1994), uninformative variable elimination (Centner et al. 1996), iterative predictor weighting PLS (Forina et al. 1999), and interval PLS (Nørgaard et al. 2000).

Future Trends

In the future, multivariate analysis should be used more and more in everyday (scientific) life. Until recently, experimental work resulted in a very limited amount of data, the analysis of which was quite easy and straightforward. Nowadays, it is common to have instrumentation producing an almost continuous flow of data. One example is process monitoring performed by measuring the values of several process variables, at a rate of one measurement every few minutes (or even seconds). Another example is quality control of a final product of a continuous process, on which an FT-IR spectrum is taken every few minutes (or seconds).

In the section "Calibration" the case of wine FT-IR spectra was cited, from which the main characteristics of the product can be directly predicted. It is therefore clear that the main problem has shifted from obtaining a few data to the treatment of a huge amount of data. It is also clear that standard statistical treatment is not enough to extract all the information buried in them.

Many instruments already have some chemometric routines built into their software in such a way that their use is totally transparent to the final user (and sometimes the word *chemometrics* is not even mentioned, to avoid possible aversion). Of course, they are "closed" routines, and therefore the user cannot modify them. It is quite obvious that it would be much better if chemometric knowledge were much more widespread, in order that the user could better understand what kind of treatment the user's data have undergone and eventually modify the routines in order to make them more suitable to his or her requirements. As computers become faster and faster, it is nowadays possible to routinely apply some approaches requiring very high computing power. Two of them are genetic algorithms (GAs) and artificial neural networks (ANNs).

Genetic algorithms are a general optimization technique with good applicability in many fields, especially when the problem is so complex that it cannot be tackled with "standard" techniques. In chemometrics they have been applied especially in feature selection (Leardi 2000). GAs try to simulate the evolution of a species according to Darwinian theory. Each experimental condition (in this case, each model) is treated as an individual, whose "performance" (in the case of a feature selection for a calibration problem, it can be the explained variance) is treated as its "fitness." Through operators simulating the fights among individuals (the best ones have the greatest probability of mating and thus spreading their genome), the mating among individuals (with the consequent "birth" of "offspring" having a genome that is derived from both the parents), and the occurrence of mutations, the GAs result in a pattern of search that, by mixing "logical" and "random" features, allows a much more complete search of complex experimental domains.

Artificial neural networks try to mimic the behavior of the nervous system to solve practical computational problems. As in life, the structural unit of ANNs is the neuron. The input signals are passed to the neuron body, where they are weighted and summed. Then they are transformed, by passing through the transfer function into the output of the neuron. The propagation of the signal is determined by the connections between the neurons and by their associated weights. The appropriate setting of the weights is essential for the proper functioning of the network. Finding the proper weight setting is achieved in the training phase. The neurons are usually organized into three different layers: the input layer contains as many neurons as input variables, the hidden layer contains a variable number of neurons, and the output layer contains as many neurons as output variables. All units from one layer are connected to all units of the following layer. The network receives the input signals through the input layer. Information is passed to the hidden layer and finally to the output layer that produces the response.

These techniques are very powerful, but very often they are not applied in a correct way. In such cases, despite a very good performance on the training set (due to overfitting), they will show very poor results when applied to external data sets.

Conclusion: The Advantages and Disadvantages of Chemometrics

In one of his papers, Workman (2002) very efficiently depicts the advantages and disadvantages of multivariate thinking for scientists in industry. From the eight advantages of chemometrics he clearly outlines, special relevance should be given to the following ones:

1. Chemometrics provides speed in obtaining real-time information from data.
2. It allows high-quality information to be extracted from less-resolved data.
3. It promises to improve measurements.
4. It improves knowledge of existing processes.
5. It has very low capital requirements—it's cheap.

The last point, especially, should convince people to give chemometrics a try. No extra equipment is required: just an ordinary computer and some chemometrical knowledge (or a chemometrical consultancy). It is certain that in the very worst cases the same information as found from a classical analysis will be obtained in a much shorter time and with much more evidence. In the great majority of cases, instead, a simple PCA can provide much more information than was previously collected. So why are people so shy of applying chemometrics? In the same paper previously cited, Workman gives some very common reasons:

1. There is widespread ignorance about what chemometrics is and what it can realistically accomplish.
2. This science is considered too complex for the average technician and analyst.
3. Chemometrics requires a change in one's approach to problem solving from univariate to multivariate thinking.

So, while chemometrics leads to several real advantages, its "disadvantages" lie only in the general reluctance to use it and accepting the idea that the approach that has been followed over many years can turn out not to be the best one.

References

Bro, R, F van den Berg, A Thybo, CM Andersen, BM Jørgensen, H Andersen. 2002. Multivariate data analysis as a tool in advanced quality monitoring in the food production chain. *Trends in Food Science and Technology* 13:235–44.

Centner, V, DL Massart, OE de Noord, S de Jong, BM Vandeginste, C Sterna. 1996. Elimination of uninformative variables for multivariate calibration. *Analytical Chemistry* 68:3851–58.

Forina, M, C Casolino, C Pizarro Millán. 1999. Iterative predictor weighting (IPW) PLS: a technique for the elimination of useless predictors in regression problems. *Journal of Chemometrics* 13:165–84.

Geladi, P. 1989. Analysis of multi-way (multi-mode) data. *Chemometrics and Intelligent Laboratory Systems* 7:11–30.

Kourti, T, JF MacGregor. 1995. Process analysis, monitoring and diagnosis, using multivariate projection methods. *Chemometrics and Intelligent Laboratory Systems* 28:3–21.

Leardi, R. 2000. Application of genetic algorithm-PLS for feature selection in spectral data sets. *Journal of Chemometrics* 14:643–55.

Lindgren, F, P Geladi, S Rännar, S Wold. 1994. Interactive variable selection (IVS) for PLS. 1. Theory and algorithms. *Journal of Chemometrics* 8:349–63.

MacNamara, K. 2005. Personal communication.

Nørgaard, L, A Saudland, J Wagner, JP Nielsen, L Munck, SB Engelsen. 2000. Interval partial least-squares regression (iPLS): a comparative chemometric study with an example from near-infrared spectroscopy. *Applied Spectroscopy* 54:413–19.

Smilde, AK. 1992. Three-way analyses: problems and prospects. *Chemometrics and Intelligent Laboratory Systems* 15:143–57.

Tang, C, F Hsieh, H Haymann, HE Huff, 1999. Analyzing and correlating instrumental and sensory data: a multivariate study of physical properties of cooked white noodles. *Journal of Food Quality* 22:193–211.

Workman, J, Jr. 2002. The state of multivariate thinking for science in industry: 1980–2000. *Chemometrics and Intelligent Laboratory Systems* 60:13–23.

Further Reading

Books

Beebe, KR, RJ Pell, MB Seasholtz. 1998. *Chemometrics: A Practical Guide*. New York: Wiley and Sons.

Brereton, RG. 2003. *Chemometrics—Data Analysis for the Laboratory and Chemical Plant*. Chichester: Wiley, 2003.

Leardi, R (ed.). 2003. *Nature-Inspired Methods in Chemometrics: Genetic Algorithms and Artificial Neural Networks*, in Data Handling in Science and Technology Series, vol. 23. Amsterdam: Elsevier.

Manly, BFJ. 1986. *Multivariate Statistical Methods: A Primer*. London: Chapman and Hall.

Martens, H, T Naes. 1991. *Multivariate Calibration*. New York: Wiley and Sons.

Massart, DL, BGM Vandeginste, SN Deming, Y Michotte, L Kaufman. 1990. *Chemometrics: A Textbook*, in Data Handling in Science and Technology Series, vol. 2. Amsterdam: Elsevier.

Massart, DL, BGM Vandeginste, LMC Buydens, S de Jong, PJ Lewi, J Smeyers-Verbeke. 1997. *Handbook of Chemometrics and Qualimetrics: Part A*, in Data Handling in Science and Technology Series, vol. 20A. Amsterdam: Elsevier.

———. 1998. *Handbook of Chemometrics and Qualimetrics: Part B*, in Data Handling in Science and Technology Series, vol. 20B. Amsterdam: Elsevier.

Meloun, M, J Militky, M Forina. 1992. *Chemometrics for Analytical Chemistry. Vol. 1: PC-Aided Statistical Data Analysis*. Chichester: Ellis Horwood.

———. 1994. *Chemometrics for Analytical Chemistry. Vol. 2: PC-Aided Regression and Related Methods*. Hemel Hempstead: Ellis Horwood.

Sharaf, MA, DL Illman, BR Kowalski. 1986. *Chemometrics*, in Chemical Analysis: A Series of Monographs on Analytical Chemistry and Its Applications, vol. 82, edited by P.J. Elving and J.D. Winefordner. New York: Wiley and Sons, 1986.

Web Sites

http://ull.chemistry.uakron.edu/chemometrics/
http://www.chemometrics.se/
http://www.models.kvl.dk/
http://www.namics.nysaes.cornell.edu/
http://www.statsoft.com/textbook/stathome.html

15 Data Handling

Philippe Girard, Sofiane Lariani, and
Sébastien Populaire

Introduction

Data handling is all activities dealing with data, from their collection (how are experiments properly set up? how many samples/measurements are required?) to their transformation into useful information (which method to use?) to answer a specific question.

Data, More and More

The world is becoming more and more quantitative, and the food industry is certainly very much concerned with this trend toward using more numerous and more sophisticated measuring devices to dig into the infinitesimal, leading to dealing with more and more data. Indeed, as manufactured food is being more often consumed, food may have an unexpected impact on human beings from a health perspective and also from a well-being perspective. This automatically suggests that food manufacturers should put emphasis not only on better controlling their processes for consistent quality but also on better understanding what influence food (components) may have on human health (e.g., cardiovascular health, prebiotics, obesity) and/or well-being (e.g., skin health and reduction of risk of allergy). Whereas the first activity (quality control) has already been in place for a long time, research and confirmation of active food components (nutrition and omics research) are becoming more important in major food companies.

Microarray and Omics Technologies, at the Edge of Science

Microarray technology is part of a new and revolutionary class of biotechnologies for digging into the unknown world of DNA by understanding the expression of thousands of genes simultaneously. This powerful technology is based on the hybridization process or base pairing where the C always pairs with G and T always pairs with A.

The widest application of microarray is certainly to discover how genes are involved in cellular processes where nutritional additives are involved. Hence, using microarray, we attempt to learn which genes are turned on or off in treated (e.g., with nutritional additive intake) versus control tissues, cells, or animals.

Another application, which is the bioMérieux FoodExpert-ID® Array, powered by Affymetrix GeneChip® technology, allows scientists to identify at the gene level the presence or absence of different species of animals in any food product. We can imagine the impact of the latter on food diagnostics and the detection of food fraud. Genotyping and detection of SNPs (single-nucleotide polymorphisms due to single-nucleotide substitutions) are also applications of the genomics microarray (Brown and Bostein 1999).

Several microarray technologies currently exist, and the two most used approaches are the oligonucleotides array and the cDNA arrays. Both techniques are based on hybridization, but they differ in the length of the target (DNA sequence), and in how they are fixed on the array (refer to Schena 2000 for a comprehensive review of these two main technologies). It is clear that when dealing with DNA sequences, scientists will have to deal with a large amount of data.

The Necessity of Using Statistical Methods to Handle Data

In food industry, quality control, Omics, or, from a broader viewpoint, research in food industry requires data to be handled in a proper and relevant way so that they can be transformed straightforwardly. In essence, data contain information, but transforming them into knowledge requires their interpretation, and because of their inherent variability, recourse to statistics and statistical methods is required.

It's no secret that many people harbor a vague distrust toward statistics as commonly used. Indeed, statistics requires the ability to consider things from a probabilistic perspective, employing quantitative and technical concepts such as "confidence," "reliability," "significance," "variability," and "uncertainty," which may frighten some people. By the way, this is in contrast to the way nonmathematicians often cast problems. As a matter of fact, many nonmathematicians hold quantitative data in a sort of awe. They have been led to believe that numbers are, or at least should be, unquestionably correct (i.e., known or measured without uncertainty). This certainly takes its root in secondary school where there are clearly defined methods for finding the answer, and that answer is the only acceptable one. It comes, then, as a shock that different research studies can produce very different, often contradictory results. Statistical methods are then required as they better take into account variabilities by assessing them and integrating them in the decision-making process.

In order to provide useful and relevant data for a specific study, we should consider three major steps when handling data (see fig. 15.1). The first involves the data provider (i.e., the measuring device), which may supply biased data. We should ask ourselves, as a prerequisite to any study, whether the data collected are valid. The second step is to design the data collection so as to answer the following questions: How can I optimally combine factors to establish that one or two mainly influence variable(s) of interest? How many samples (measurements) should I collect to have a significant answer? The third step is to select the appropriate statistical method to describe the data and to specifically answer the question.

Are the Data Valid? Is the Analytical Method Fit for the Intended Purpose?

As most of activities rely on data provided by an analytical method, a qualitician or scientist must be sure that the analytical method has been thoroughly validated and does not exhibit a too high variability, which may reduce the significance of the results.

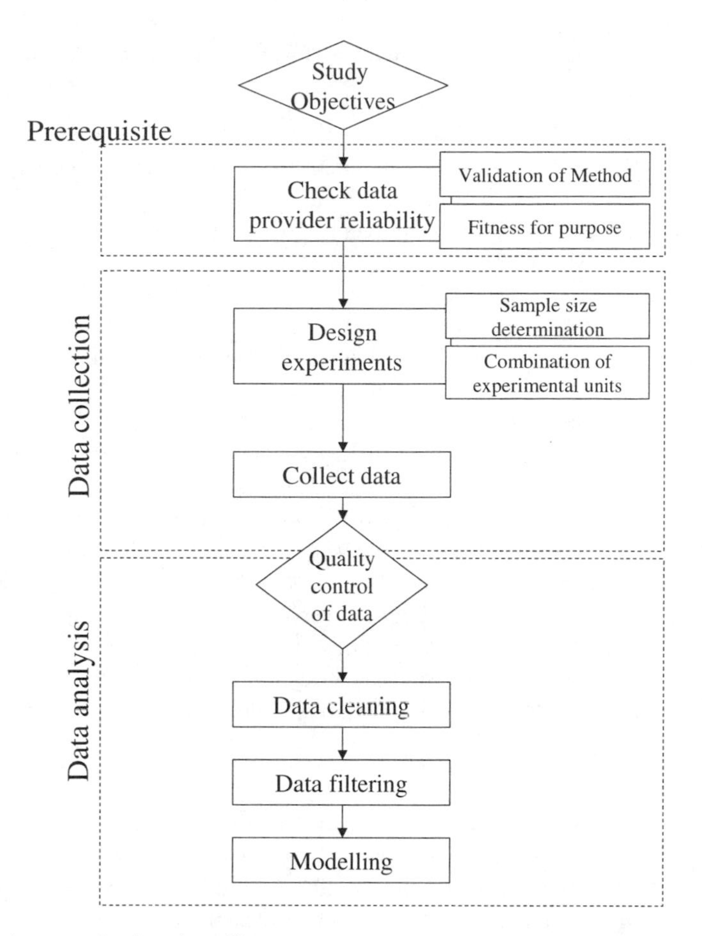

Fig. 15.1. Major steps in data handling.

Validation of Analytical Methods

Validation of analytical methods is a prerequisite when aiming to get data from analytical measurements. Indeed, the reliability of data is first based on the reliability of the analytical measurements. Method validation is the process of proving that a method is suitable for its intended purpose and will consequently provide reliable data. Method validation is necessary to verify that its performance characteristics are adequate for its use in one particular situation. It is also becoming more and more mandatory for legal authorities (for instance, to get the ISO/IEC 17025 accreditation, ISO 1999).

A method has to be validated.

- before using a published method, in order to check that the in-house performances of the method are not significantly different from the published performances;
- before a change in the method use conditions: new matrix, new sample types, new samples range, etc.; and
- following a period of nonuse.

For a given analytical method, the main performance characteristics to be evaluated are the following:

1. The method is applicable to the field of application and the working range.
2. The selectivity (also called *specificity*) represents its ability to distinguish between the analyte of interest and other substances in the matrix.
3. The linearity represents its ability to obtain test results proportional to the concentration of the analyte.
4. The detection and quantitation limits (also called *decision limit* and *detection capability*) are the lowest concentration that can be detected and quantified.
5. The trueness represents the closeness of agreement between the measured and the true value.
6. The precision represents the variability of the results obtained under stipulated conditions (repeatability [*r*]/reproducibility [*R*]). Repeatability measures the lowest observable variability when carrying out replicate determinations on the same sample (same analyst, same equipment, same laboratory, and short interval of time between replicates). Reproducibility measures the highest variability when carrying out replicate determinations on the same sample (different analysts, different equipment, different laboratories, long interval of time between replicates). Precision expressed in terms of standard deviation ($SD[r]$, $SD[R]$) estimates the variability between analytical results. Precision expressed in terms of limit (r, R) estimates the maximum tolerated spread between analytical results (see ISO 1994 for detailed calculations).
7. The ruggedness (or robustness) represents its capacity to remain unaffected by small variations.

Since 1999, measurement uncertainty is also a requirement for the ISO 17025 accreditation. Measurement uncertainty is a parameter, associated with the results of a measurement, that defines the range of values that could reasonably be attributed to the measured quantity, at a given confidence level. For laboratories, measurement uncertainty provides useful and clear information concerning the quality of their results.

Several guidelines already exist (EURACHEM/CITAC 2000) giving methodologies for measurement uncertainty estimation, mainly based on a four-step process, known as the "bottom-up" approach:

1. measure and specification, as a function of the input quantities;
2. uncertainty sources identification, contributing to the uncertainty of the previously mentioned inputs;
3. quantification of the previously identified uncertainty sources; and
4. calculation of the general combined standard uncertainty.

Nevertheless, in order to ease and speed up the measurement uncertainty estimation process, another approach called the "top-down" approach is more and more used and encouraged. This estimation, based on validation data (mainly trueness and precision) or collaborative trial data (reproducibility), simplifies the calculations and is supported by a growing number of publications (Campos-Giménez and Populaire 2005).

Fitness for the Intended Purpose

Once the method is validated, we should wonder whether the variability coming from the method is not too high in comparison to what we intend to identify: for instance, the variability occurring in an experiment has a direct influence on the significance of the effect of a factor, and its influence may be reduced by a higher number of experimental units (see "Sample Size Determination" section). An appropriate indicator of whether the method is appropriate for a specific study is its fitness for purpose (i.e., its ability to measure within specified limits). These limits can be either specification limits when applied to quality control and to regulatory compliance or other limits defined by the experimenters. The fitness for purpose of a method can be measured in different ways. A natural extension of statistical process control techniques and process capability indices (see Montgomery 2004) is the method capability, denoted as C_m, which is calculated as follows:

$$C_m = \frac{R}{6 \times SD(method)} \tag{15.1}$$

where

- R is the range of value within which we intend to measure. For instance, this can be the difference between specification limits (i.e., upper specification limit minus lower specification limit) or the difference in means of two populations.
- And $SD(method)$ is the method variability. This is the variability occurring when repeating determination under stipulated conditions. This can be either the standard deviation of repeatability [$SD(r)$] or reproducibility [$SD(iR)$].

When the method is to be applied to process control, and assuming that process variability is a combination of method and intrinsic process variability, C_m is generally indexed on process capability indices. Thus, a generally agreed interpretation is as follows:

- If C_m is lower than 1.6, method variability contributes too much to process variability, and the method is not fit for the purpose.
- If C_m is higher than 1.6 but lower than 2, method is fit for the purpose, but method variability may contribute substantially to the overall process variability.
- If C_m is higher than 2, then method contribution to the overall variability is negligible.

Other reference values for Cm can be derived but should be always higher than 1.

Data Collection—How to Properly Collect Data to Address a Specific Question

The core value of statistical methodology is its ability to assist one in making inferences about a large group (a population or a phenomenon/process) based on observations of a smaller subset of that group (a sample) or a combined set of experimental units. Before collecting relevant data, one must solve the delicate questions of the number of experimental units to run so as to have a significant answer and of how to combine the different experimental units so as to have an answer with a minimal number of units.

In order for this to work correctly, a couple of things have to be true: the sample must be similar to the target population in all relevant aspects (for instance, it is often mentioned that an at-random choice should be carried out so as to avoid any systematic bias in the data), and certain aspects of the measured/controlled variables must conform to assumptions that underlie the statistical procedures to be applied.

When collecting data, we will assume that data are stored in a reliable manner so as to avoid any error of reporting and/or storage. Data storage, which is more related to information technology, will not be discussed hereafter.

Experimental Design

When aiming to understand the behavior of a process or phenomenon (*process* will be used hereafter to simplify the text), it is common to set up experiments leading to the design of a model explaining the relationships between process parameters (that can be easily acted on) and the process response. When carrying out the experiments, a first option is to modify one parameter at a time and observe its effect on the process response. This procedure is not advised for two main reasons:

1. This procedure can be very time-consuming and needs to carefully record every slight change in the successive experiments.
2. No clear understanding and quantification of the parameters' effects on the process will be derived from these experiments.

A second option is to test all the possible combinations of all the tested parameters, defined as a full factorial design of experiments. This is, theoretically, the best solution to get a complete understanding of the process. Nevertheless, time and costs dedicated to these experiments can become extremely high. It then becomes economically irrelevant to consider such a solution.

Fractional factorial designs of experiments are an alternative to the two previous options (NIST/SEMATECH 2005). The basic purpose of these experiments is to arrive at a combination of factor levels to identify the important factors that control the characteristics of interest.

They are tools particularly adapted to obtain the model linking process parameters to process response through the optimization of the number of necessary experiments to get this model.

The idea is to select and perform only one specific fraction of the experiments described in the full factorial design. The selection of these experiments is based on statistical properties of the fractional design (named *orthogonality*). Depending on the number of studied parameters and associated levels, fractional designs adapted to given situations can be easily found in the literature (NIST/SEMATECH 2005), and commercial software eases the interpretation of their associated results (Matlab®, Statistica®, etc.).

Finally, besides the design of the model representing the effect of the process parameters on process response, the design of experiments can be used for screening issues (for the determination of the most influential parameters on the process response) or in order to optimize the process response (parameter settings for minimizing defects, maximizing one desired effect).

Depending on the application, interested readers may refer to Box et al. 1978 or Box 2005 for an introduction to experimental design in the food industry, or for a specific area such as sensory analysis, readers can refer to a specific article such as Callier 2001.

As an example, let's assume that a process can be tuned using seven different parameters $A, B, \ldots, G$. The objective of the experimental design is to measure the influence of these parameters on a given response R of the process that needs to be minimized (percentage of noncompliant products, average overfilling, etc.).

Trial Number	Process parameters							Trial Response
	A	B	C	D	E	F	G	
1	Low	Low	Low	Low	Low	Low	Low	R1
2	Low	Low	Low	High	High	High	High	R2
3	Low	High	High	Low	Low	High	High	R3
4	Low	High	High	High	High	Low	Low	R4
5	High	Low	High	Low	High	Low	High	R5
6	High	Low	High	High	Low	High	Low	R6
7	High	High	Low	Low	High	High	Low	R7
8	High	High	Low	High	Low	Low	High	R8

Fig. 15.2. Fractional experimental design.

For each of the seven parameters, let's consider two levels, Low and High, for which the process will be tested. To find the best parameter combination minimizing the response *R*, one can apply a full factorial design, leading to 128 (2^7) experiments.

One alternative and advised approach is to use the fractional experimental design and run the eight experiments described in figure 15.2 (costs of the experimentation are consequently divided by 16!).

It is then necessary to calculate the average response for each level of each parameter through the averaging of the responses for which the factor is set to the corresponding level. For example, the average response of A_{Low} is equal to $(R1 + R2 + R3 + R4)/4$, the average response of B_{High} is equal to $(R3 + R4 + R7 + R8)/4$, etc.

Finally, in order to minimize the response, one has to choose, for each parameter, the level for which the average response is the lowest.

Sample Size Determination

A General Perspective

"How many samples should I take of [. . .]?" is certainly the most frequently asked question of a statistician. In general, even though they have received a very thorough and broad education in statistics, few statisticians are really prepared to straightforwardly answer such a question because of the way the question is expressed. For instance, imagine one experienced qualitician coming to the statistician's office and asking this simple question: "How many samples do you usually take for tomato sauce?" (for the original example, refer to Hare 2003). As a matter of fact, even if the question is certainly very relevant for the qualitician, the statistician cannot really respond to it since basic things are missing in the question:

- What is the objective of the study and therefore the criterion to be measured?
- What is the expected precision of the criterion of interest? Or what is the expected difference between two populations?
- What is the risk you would accept taking when drawing the conclusion?

With regard to question 1, the qualitician is as a matter of fact interested in the percentage of onions that tomato sauce cans may contain. The two remaining questions must be addressed from a broader perspective and would depend on the school of statistics the qualitician belongs to (i.e., either frequentist or Bayesian).

From a frequentist's viewpoint, those two remaining questions will be related to two risks of failure, namely alpha and beta risk or producer and consumer risk. Having a quantitative criterion to measure (refer to question 1), knowing the two risks leads to the following formula:

$$n = \left(\frac{(z_{1-\alpha/2} - z_{\beta})}{\delta} \right)^2 \times \sigma^2 \tag{15.2}$$

where z_a denotes for the corresponding normal fractile, δ for the smallest difference aimed to be detected and σ for the variability of the criteria to measure.

If we handle qualitative characteristics as a proportion of defectives, provided that the normal approximation is valid, then σ^2 is changed to $p(1-p)$. If the normal approximation does not hold, then other calculations can be carried out (see Newcombe 2000). As α and β are always respectively set to 5 and 10 percent, the preceding equation is often given as follows:

$$n = 3.24^2 \times \left(\frac{\sigma}{\delta} \right)^2. \tag{15.3}$$

From this equation, it is clear that the higher the variability (σ), the higher the required sample size and the smaller the difference (δ), the higher the required sample size.

Finally, if the precision of the estimator is only of concern, the preceding formula shrinks to

$$n = \left(\frac{z_{1-\alpha/2}}{\delta} \right)^2 \times \sigma^2. \tag{15.4}$$

On the other side, from a Bayesian viewpoint, questions 2 and 3 are addressed simultaneously when designing a cost function. Basically, Bayesian statistics differs from frequentist because it considers that the estimated parameter is a random variable and that prior to any experiment the parameter as a probability distribution function (pdf) is known. Data collected during the experiment then enable updating the pdf (the prior pdf becomes the posterior pdf) according to the Bayes theorem (equation 15.5), which acts as a learning engine.

$$P(\theta|x) = \frac{P(x|\theta)P(\theta)}{P(x)} \tag{15.5}$$

where θ represents the estimated parameter and x the data.

The sample size determination within the Bayesian framework consists of defining a cost function $C(\theta,n)$ and finding the sample size n that minimizes the expected loss (equation 15.6).

$$C(n) = \int_{\theta} C(\theta,n)P(\theta|x)d\theta \tag{15.6}$$

Refer to Adock 1997 for a set of articles discussing sample size determination or to Parent et al. 1995 for an acceptance sampling plan perspective.

The Microarray Perspective

Sample size determination is even more crucial in microarray. As we can imagine, the expression of a gene may vary from one hybridization to a second independent hybridization. Replication is then required as it allows averaging and avoids outliers driving the statistics. The importance of replication in microarray experiments is now unequivocal (Churchill 2002; Simon et al. 2002).

The replications mostly carried out are biological and technical replications:

- *Biological replication* refers to hybridization of mRNA from different extractions, for example from particular tissue and from different mice. Hybridizations of the same mRNA extraction on different slides are not biological replicates.
- *Technical replication* refers to slide replication where the target mRNA is from the same pool. The term *technical replicate* includes the assumption that the mRNA sample is labeled independently for each hybridization.

In the case of a cell line, as the biological error is clearly small, we can either use technical or biological replicates. The case of animals is trickier, and the use of biological replicates is clearly recommended (Yang and Speed 2002).

At the experimental design step we should select, as much as possible, animals that are the same breed, same gender, and so on, in order to minimize effects other than those we want to observe (treatment, etc.). In case we cannot avoid choosing heterogeneous samples, it should be mentioned that it has to be taken into account in the analysis.

In equation 15.2 σ combines both technical and biological variance, and there is an additional multiplying term to take into account the number of technical replications. However, this formula assumes that we already know the minimum detectable change, δ (or significant change), and the variance of the expression level (or the variability of the system). We can suppose that the variability can be assessed using previous microarray studies or a pilot study. However, unlike other studies where the scientist can specify the minimum detectable difference, we are not able to define the minimum difference in average expression level between two classes and we are not able to declare that the difference is biologically significant and hence the gene is up or down regulated.

Another issue is that the sample size calculation presented in the preceding focuses on detection of differential expression for a single gene. But microarray deals with thousands of genes and the biologically significant difference can differ from one gene to another. It has never been demonstrated that the minimum significant change (to observe a phenotype) is the same for all the genes spotted on the same array. And since each gene on a microarray has a different variance (related to the expression level), we can easily understand that no simple answer to the ideal array number is possible.

How Many Replicates Should We Use?

With replication we want to differentiate between different sources of variation: variability due to what we want to measure, variability due to the equipment of measure, and so on.

The number of replicates therefore depends first on the amount of variation in the test system and, second, on how small the differences are that we wish to measure at a given confidence level. The first factor is related to many points, including

- technical conditions of sample's preparation,
- technical equipment, and
- biological species: cell line, mouse, rat, dog, or human.

We can also take into account the economic factor, as the array price is important.

Under relatively controlled experimental conditions, the empirical recommendations are the following: 3–5 for cell line, 5–10 for animals, and 30 for human.

If only fewer than three replicates can be performed, it is not advisable to process with microarray experiments. Thus, a strict minimum of three replicates is the general recommendation.

The Coefficient of Variation Approach

The coefficient of variation (CV) is used to measure the consistency of genes (a measure of variance) among replicates within the same experimental group (Affymetrix 2003). We have to compute the CV of signal intensities for different numbers of replicates, and when the CV stabilizes, we have the minimum number of replicates we need. We tested this technique with a dog study, and it appears that the CV starts to stabilize at four replicates and is steady from five to six replicates.

In a global manner the CV stabilization indicates that standard deviations of samples will not be improved with additional biological replicates (in the hypothesis that the mean is also stabilized). In the absence of the minimum significant change to be measured, we will use enough replicates in order to improve the accuracy of the sample's variances, which are ultimately used to determine statistical significance in statistical tests. However, let's keep in mind that a higher number of replicates is useful mainly if we want the statistical test to be more powerful.

Pooling in Microarray

Another perspective to take into account for sample size determination in microarray is the possibility of pooling, where mRNA samples are pooled across subjects. In this case a mix of mRNA from different biological samples is used to define the genes' expression profile. The most important reasons for pooling are

- to bypass technical difficulties of extracting sufficient RNA to carry out a single hybridization,
- to decrease the biological variability, and
- to reduce the cost of the experiment by decreasing the number of replicates and keeping unchanged the length of the confidence interval.

Adapted mRNA pooling design is clearly efficient for decreasing the biological variance (mainly for high expression), and in some ways to reduce the experiment cost by using fewer arrays with the same confidence in the results (Kendziorski et al. 2003). It is profitable if the biological variability is high enough, compared to technical variation in the microarray. However, pooling is relevant if it is not carried out to the detriment of the number of replicates, and the number of replicates (array) has to be high enough to have a good precision of differential expression estimation. Finally, if we pool samples, we have to be aware of the risks and the disadvantages in order to draw the right conclusions.

Actually, three important points should be kept in mind:

1. If the investigator needs to assess a potentially interesting pattern across different subjects, he or she should not pool samples from biological replicates. Pooling loses information about variability between biological samples. These biological samples may be genetically identical but may have other factors (e.g., hormonal, enzymatic, etc.) that influence the gene expression. The biological variability will tell us how the gene behaved in different animals.
2. By pooling the RNA, we take the risk of including a problematic experimental unit: contamination with a proportion of samples with altered level of expression. This contributes to skewing the results. In this case, we are not able to find out the origin of the problem afterward.
3. Conclusions are limited to the specific RNA pool not to the population, as there is no estimate of variation among the pool.

Data Analysis

Introduction

The first goal of data analysis is to extract readable information from data. Thanks to data analysis, it is possible to give a quantification of the error related to the information (what is the uncertainty around my average? is it a good estimation of my data variability?). Finally, based on this error and having set some risk, a decision can be made.

This section gives an overview of the process of linking raw data to information on which we can act. This process will not lead to a decision by itself but will give the necessary keys for it.

Quality Control of Data and Data Cleaning

After the experiments are properly designed and carried out, the data are then carefully collected. We must ensure that the experiments went well without any disturbance from outside factors.

A General Perspective

When aiming to treat data, one of the first important, and sometimes neglected, steps is to clean the data. Indeed, a set of data can be polluted by a few outliers. Outliers are data, which are relatively large or small, when compared to the general trend of the dataset.

Outliers can have three main causes:

1. Natural outliers: these are data that are due to the natural variability of a given process. They come from the same population from which the rest of the data comes.
2. Measurement error: in case of troubles, extreme conditions, or unforeseen events during the experimental phase of a measurement, the result of the measurement can differ from its real value. This kind of incident can lead to the provision of extraordinarily high or low measurement values.

3. Reporting error: extreme values in a dataset can also be simply attributed to an error in the reporting of a result.

When dealing with outliers, two main questions are how to recognize them and how to treat them.

A visual examination of data (through a histogram, for example) should give the first signs of the occurrence of possible outliers.

Statistical tests have also been designed in order to identify outliers. The most famous one is the Grubb's test, based on the standardized distance between the mean and each piece of data of the dataset (Grubbs and Beck 1972). Once outliers have been detected, it is necessary to practice the right behavior.

In the classical statistical framework, the first necessary reaction is to investigate to determine the origin of the outlier: for example, check the reporting and get information concerning the experimental conditions. If a reason explaining the abnormal value of the measurement can be found, it is possible to delete it from the dataset before going on with the data treatment. If not, the outlier is considered natural and should be kept in the original dataset (as long as natural outliers are not too numerous).

Another approach is now more and more accepted, especially in the analytical area: the use of robust statistics (Hubert et al. 2004). Robust statistics aims at providing statistical techniques that are not sensitive to outliers. The most famous robust statistics indicator is used for the determination of the central point of a dataset. In classical statistics, the average is traditionally applied. The robust equivalent of the average is the median due to its lower sensitivity to outliers. Robust statistics methodologies are becoming popular because of two main qualities:

1. If the dataset contains no outliers, information coming from robust statistics analysis is not significantly different from that coming from classical statistics analysis.
2. If the dataset contains a few outliers, the use of robust statistics avoids the investigation step.

Quality Control in Microarray

Quality assessment of microarray data is one of the most essential procedures of the analysis workflow (Beibbarth 2000). Nonetheless, this step is often neglected or just skimmed through although it is known that the efficiency of the data analysis and interpretation are closely related to the quality of the data. And just one bad chip (mainly when only a few replicates are done) can lead to wrong conclusions about the experiment.

Each manufacturer has its proper quality control recommendation. For example, Affymetrix advises checking the report file and using the present/absent call. But these procedures are not sufficient to detect abnormal chips.

Microarray data quality assessment has mainly two aims: the first one is to detect any evidence of chip failure (RNA hybridization failure, problems with scanning, etc.), and the second one is to assess the coherence between chips according to the experimental design (sample swap, sample deterioration, etc.). Here are some simple tools (the key themes are the consistency between chips and the concordance between methods).

1. Visual detection of the microarray image and inspection of intensities histogram, intensities box plots, etc.

2. Conditions correlation matrix: each sample is considered as a vector where the different coordinates are the genes' expression. The correlation is based on the Pearson coefficient. If we consider two samples X and Y with means $\overline{X}$ and $\overline{Y}$, respectively, and standard deviations S_X and S_Y, respectively, the Pearson coefficient r is computed as:

$$r = \frac{\sum_{i=1}^{n}(X_i - \overline{X})(Y_i - \overline{Y})}{(n-1)S_X S_Y}. \tag{15.7}$$

3. In general, values are very high (around 95 percent), and we expect correlations between samples from the same condition (e.g., same tissue or same treatment) to be higher than correlations between different conditions. This assumption is all the more verified when the conditions are tissues. The correlation matrix (a) allows verifying if the sample and chip match as expected and according to the experimental design and (b) allows detecting any sample swap in the wet laboratory.
4. Samples clustering: the principle is to have a rough idea about how the samples cluster together, with the assumption that technical replicates of the same sample and samples from the same conditions are very close. Either a hierarchical clustering or a PCA (principal component analysis) can be used. With clustering methods, we are able to emphasize differences proceeding at the sample level, and any sample swap.

Usually in a microarray experiment where we deal with thousands of genes, most of them (around 90 percent) don't show any difference between conditions or treatments. If we filter these genes and select only those changing between samples, we are more likely to see inconsistencies and artifacts like corrupted samples, or samples mislabeling. So, for a better visibility of samples clustering, the idea is to do some nonspecific filtering to remove genes that do not show much variation in expression levels across samples. We have found that a filter based on variability seems to be a reasonable criterion. Then we can set a rather arbitrary level on the measure of the variability (variance, coefficient of variation, interquartile range) for reducing the set of genes to a reasonable number. The interquartile range (or IQR) is the difference between the upper quartile and the lower quartile, defined as the 75th and the 25th percentiles.

Other complementary procedures can be specific to the microarray technique. For the Affymetrix or similar techniques, we can imagine the use of present/absent calls for detecting artifacts at the gene level. In order to bypass the problem of having several calls for the same condition, we can summarize chips from the same condition in one fake chip. The calls are then computed using the latter.

However this parameter must be used a posteriori of the gene selection in order to reinforce the conclusions or to raise doubts about some genes maybe needing more investigation.

Data Filtering in Microarray

The prospects of microarray technology (either genomics or proteomics) are in the area of diagnostics (e.g., toxigenomics). However, the technology is still lacking reproducibility. The improvement of the sensitivity, the specificity, and the reproducibility of the technology is also a matter of analysis methodology and noise filtering. Microarray data cover both signal (active key genes) and noise. Recognizing any gene's expression pattern without cleaning the data is like recognizing faces and details on a blurry TV screen: it is hard, and we have a high chance of being mistaken.

Noise in the gene's expression context can be related to fluctuation due to the experimental condition, outliers, and also genes presenting no modulation between groups. Hereafter we present a method for noise filtering and statistical significance assessment based on a combination of multivariate and univariate approaches.

Class Determination and Genes Selection

This first step is a blind unsupervised method utilized to first check samples' separability without the introduction of supplementary external information. Principal component analysis (PCA) is a well-adapted method for this task (Hardle and Simar 2003). PCA is a common technique for general dimension reduction and can easily be applied to mapping multivariate data. We organize the microarray gene expression data in a matrix, $\boldsymbol{X}$, with *Ns* rows and *Ng* columns, where *Ns* is the numbers of arrays (samples) and *Ng* is the number of genes (variables).

The principal idea of the PCA is reducing the dimension of $\boldsymbol{X}$, and it is achieved through linear combination. Hence low-dimensional linear combinations are easier to interpret. The method generates a series of uncorrelated variables $\boldsymbol{Y}$ by looking for linear combinations that create the largest spread among the values of $\boldsymbol{X}$. The axes of the new space are obtained by maximizing the variance within the data.

$$Y = \Gamma^{T}(X - \mu) \tag{15.8}$$

$$Var(X) = \Gamma\lambda\Gamma^{T} \text{ and } \mu = EX \tag{15.9}$$

$\boldsymbol{Y}_1, \boldsymbol{Y}_2, \ldots \boldsymbol{Y}_p$ are called the first, second, and *pth* principal component.

The different steps to follow are

1. We apply the PCA to the data to generate a series of uncorrelated variables by looking for linear combinations that create the largest spread among the values. The axes of the new space are obtained by maximizing the variance within the data. The advantage is that no particular assumption (such as equal variance) will be made for the PCA apart from the existence of the mean vector and the covariance.
2. Next, the magnitude of every gene's contribution is estimated. Based on loading distributions, genes that contribute highly in the observed energy between clusters are selected. Genes that have high loadings have greater variance and would tend to have greater differential expression or indicate abnormal chip/genes. To separate these two possible cases, a robust statistical test is applied in a second step of the procedure.
3. In order to investigate the condition specific to our selection, we map scores (chips representation) and loadings (genes representation) on the same plot (called a biplot). As a consequence of the matrix factorization, the modulation of any gene in a chip can be geometrically inferred from the biplot (see example in fig. 15.3).

At the end of this procedure, we are able to considerably decrease the number of genes by filtering the noise. Moreover, we allot genes to the condition (animal, tissue, etc.) where they have a chance to be modulated.

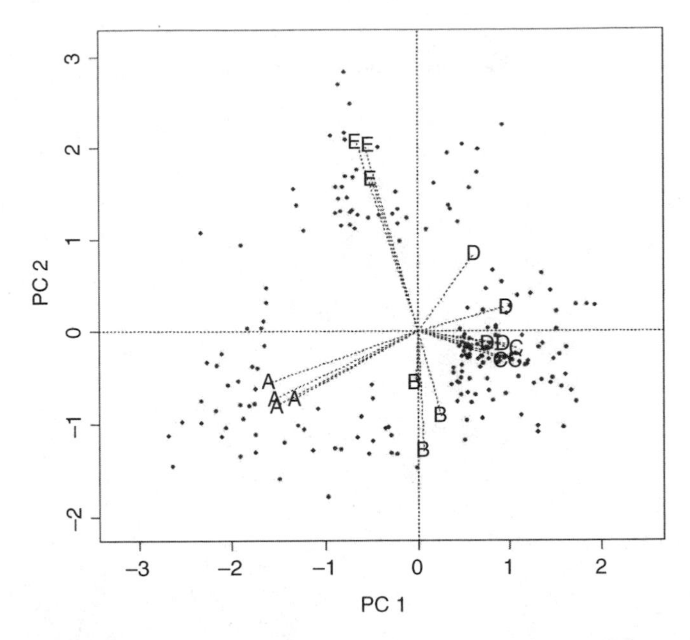

Fig. 15.3. Example of a biplot where letters correspond to samples/conditions and points correspond to the genes.

Graphical Methods

When aiming to summarize a dataset, two main options can be chosen: The first is to calculate statistics representing its central tendency (mean, median, etc.) and its spread (range, standard deviation, etc.). The second is to use a graphical way of representing data (Cleveland 1985). The main advantage of using graphical tools for describing data is to give us an casy and fast overview of the data distribution, even in the case of a large dataset.

Due to their apparent simplicity, graphical tools are sometimes neglected. Nevertheless, this first step in data analysis is essential: it provides practical information and often saves time and calculations in the next steps of the analysis.

Discrete Variables

Two main graphical representations can be used in the case of discrete variables representation.

- In a bar chart the height of the rectangular bars are proportional to the occurrence of what they represent.
- A pie chart is a disc divided into portions. Each portion of the disc is proportional (in terms of area, central angle, or arc length) to the occurrence of what it represents.

Continuous Variables

- Histogram: The design of a histogram needs to first classify values in classes (intervals) with defined limits (the length of these classes is usually constant). Based on

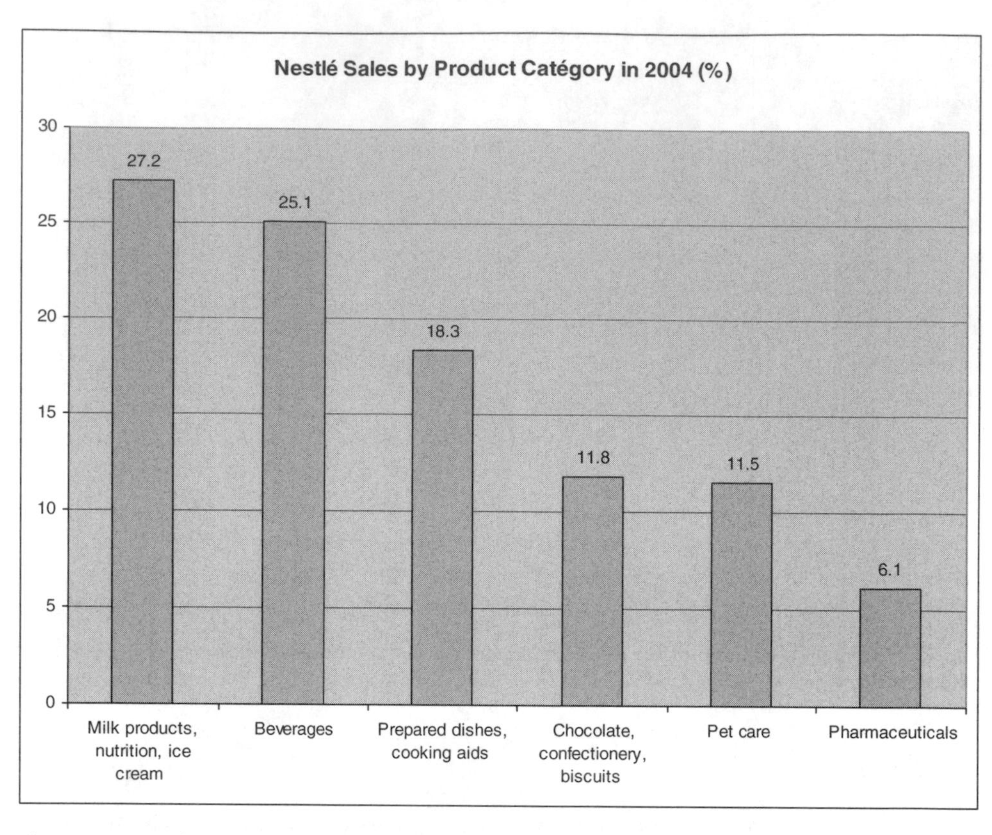

Fig. 15.4. Bar chart representing Nestlé's sales by product category in 2004 (percent).

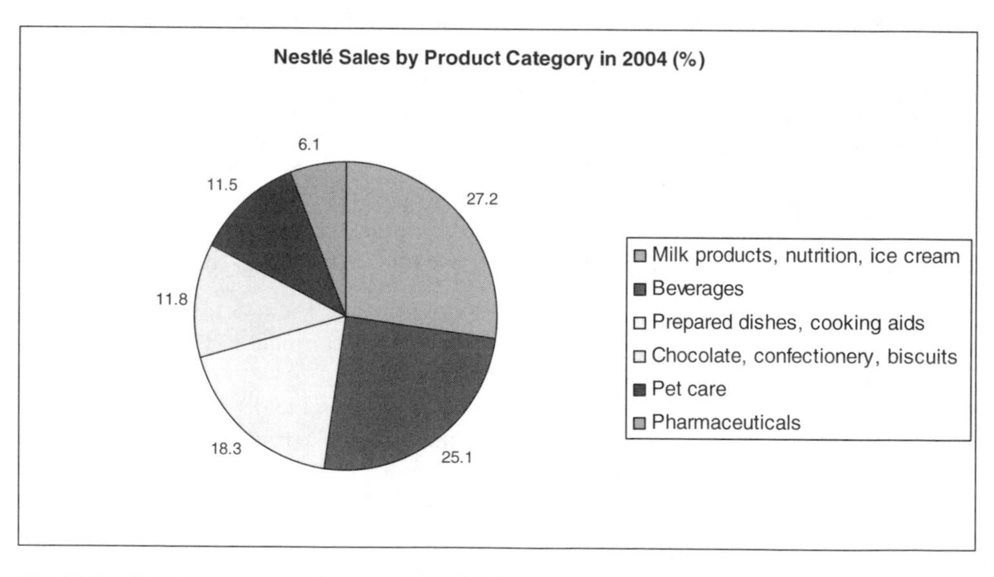

Fig. 15.5. Pie chart representing Nestlé's sales by product category in 2004 (percent).

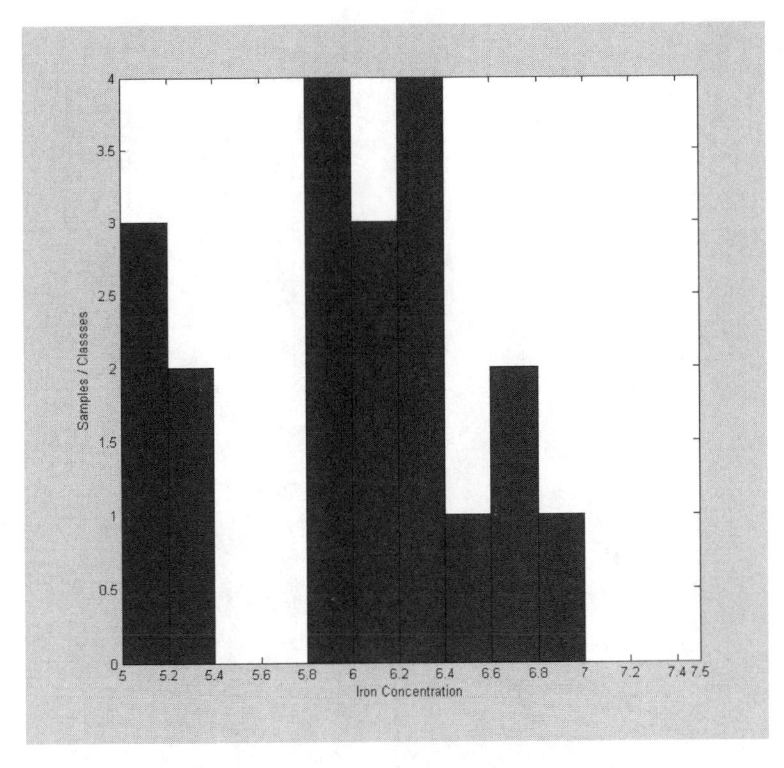

5.10
5.09
4.95
5.25
5.35
5.86
6.40
5.82
6.17
5.96
6.34
6.22
6.12
6.15
5.98
6.46
6.38
6.92
6.73
6.78

Fig. 15.6. Histogram representing a series of 20 measurements of iron concentration (mg/100 g) in instant milk powder.

this classification, data can be graphically displayed by associating one rectangle with each class. Each rectangle is delimited by the limits of its associated class. Its area is proportional to its number of data.

- Box plot: The box plot (also called a box-and-whiskers diagram) is a graphical tool that summarizes a dataset, mainly in terms of central tendency and variability. Lines of the boxes are set to the lower quartile, median, and upper quartile values of the dataset. The whiskers are lines extending from each end of the boxes and represent the data range. This graphical tool is very powerful in order to detect outliers. The use of a box plot can also highly facilitate the comparison of several datasets.

Classification

Classification aims at linking a set of characteristics (called *attributes*) of one item to the category (or class) it belongs.

One example of practical application for classification in food analysis is raw material identification based on its spectroscopic analysis (spectrum). Indeed, it may be difficult to visually distinguish different raw materials: the use of a spectroscopic system can ease this task. Refer to Blanc et al. 2005 for a specific study on raw material.

Various algorithms can be used in order to design the statistical model. In the general problem of classification without a priori information, there is no sound reason to favor

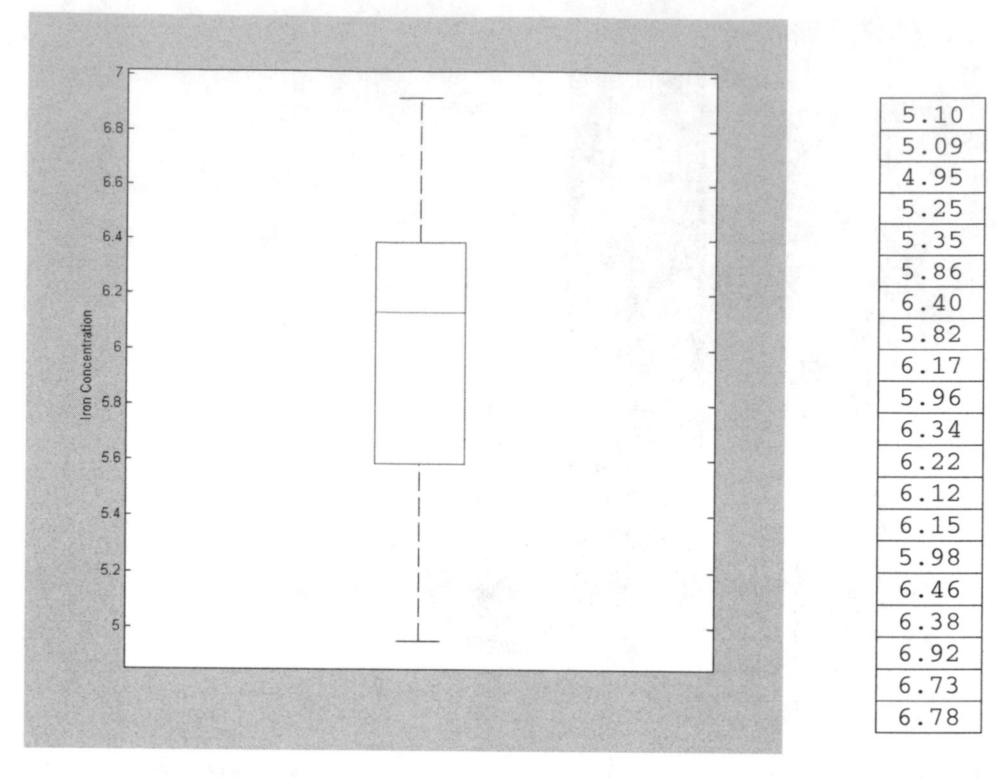

Fig. 15.7. Box plot representing a series of 20 measurements of iron concentration (mg/100 g) in instant milk powder.

one specific classification algorithm (Wolpert et Macready 1995). To test the efficiency of a classification algorithm, a model is built using a training dataset, from which a rule is designed to predict the class of one item. The designed rule is then applied to the items of a new dataset, the validation set. For each item of the validation set, the predicted class is computed and compared to the real class of the item.

The main classification techniques follow.

K Nearest Neighbors

The *K* nearest neighbors method (Duda et al. 2001) is a nonparametric method: indeed, it does not establish a model between the input and output variables but is based on the similarities between a nonlabeled sample and samples from the training set. The distance (euclidean, Manhattan, etc.) between a nonlabeled sample (represented by a vector whose attributes are the coordinates) to be classified and all samples from the training set is calculated. The predicted class of the new sample is the class to which belongs the majority of its *K* nearest neighbors ($K = 1, 3, \ldots$) according to the chosen distance. The main advantage of this method is that its algorithm is quite simple and that it does not require building a complex statistical model. Its main drawback is the computational time that can rapidly increase if the size of the training set is high.

Naïve Bayes Classifier

The naïve Bayes classifier (Hand and Yu 2001) method is based on the so-called Bayes theorem (equation 15.5). Theoretically based on the statistical independence between the different characteristics of a sample, the naïve Bayes classifier establishes conditional probability between the characteristics of a sample and its corresponding class. Once these probabilities have been established, knowing the characteristics of a new sample, the probability for this sample to belong to one of the class can be assessed.

Despite its apparent simplicity, the naïve Bayes classifier performs well if a sufficient amount of data is available to estimate the conditional probability functions.

Bayesian Networks

Also based on Bayes theorem, Bayesian networks (Charniak 1991) are a graphical representation of a joint probability distribution. A Bayesian network is a graph whose nodes represent variables and arcs represent probabilistic dependences among these variables.

Decision Tree

A decision tree (Breiman et al. 1984), also called a classification tree, is a graph in which

- each branch represents a decision based on one attribute's value, and
- one branch links two nodes representing the possible outcomes of a classification and their associated probability.

Day	Weather	Wind	Sprinkler
1	Rain	East	OFF
2	Rain	North	OFF
3	Overcast	West	ON
4	Sunny	West	ON
5	Overcast	West	OFF
6	Rain	North	OFF
7	Sunny	North	ON
8	Overcast	East	OFF
9	Overcast	East	OFF
10	Sunny	North	ON

Fig. 15.8. The gardener's decisions for a 10-day period.

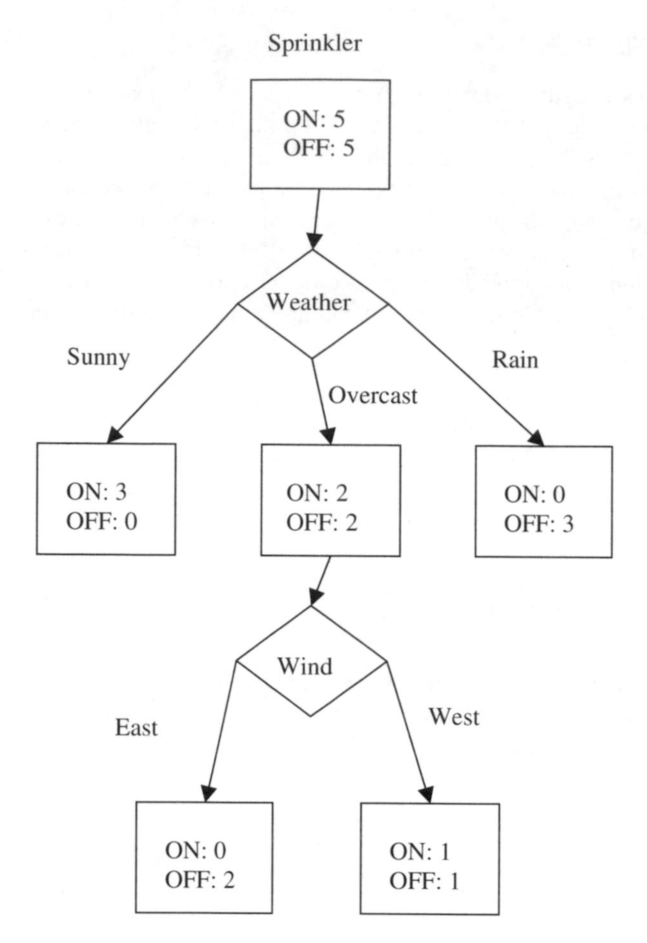

Fig. 15.9. Example of a decision tree.

Decisions trees are very popular tools in the area of classification because their construction and functioning are very easy to handle, even for a nonspecialist in data mining. A decision tree can be illustrated using the following simple example: we want to reproduce the behavior of a gardener and know when a sprinkler should be on or off depending on the weather (rain, overcast, sunny) and the direction of the wind.

Figure 15.8 represents a 10-day analysis of the gardener's decisions. Based on this information, we can establish the decision tree in figure 15.9. During the absence of the gardener, we can use this decision tree to decide, based on the weather and the wind direction, whether the sprinkler should be switched on or not.

Artificial Neural Networks

Artificial neural networks (Ripley 1996) are a group of interconnected nodes arranged in layers, containing simple mathematical functions, and linked by multiple connections. This classification technique is formally inspired by the brain cells' connections. Artificial neural networks are appropriated if the comprehension of the classification system is not necessary. Indeed, it is a black box model. One of its main advantages is its high speed of classification.

Support Vector Machines

Support vector machines (Burges 1998) are an emerging classification technology based on the creation of a hyperplane to separate the data belonging to different classes. This hyperplane maximizes the margin between examples belonging to different classes. This technology is very recent but appears to be very promising. Indeed, it does not need a large amount of data to be efficient, and classification is very fast.

Statistical Tests and Modeling

Once data have been checked and cleaned, they undergo classification, and statistical methods can then be used so as to draw clear conclusions from a set of data despite the presence of variabilities and/or uncertainties.

Beyond the traditional statistical toolbox (parametric tests and traditional modeling), the following two sections show the limitations of current methodologies applied so far and potential applications and extensions of statistical methodologies, which were restricted to certain areas until now.

Statistical Significance in Microarray

The statistical test allows the determination of genes differentially regulated with a statistically significant difference. The usual procedure is to apply a standard statistical measure (student's *t*-test, ANOVA, etc.) for each gene, independently of the other genes. The outcome of the test is based on a good estimation of the error (variance within classes). However, due to a low number of replicates, the power of the test is decreased, and the error is imprecisely estimated, resulting in a low confidence in the results of the test.

In order to mitigate the effects of the low number of replicates, several more or less equivalent solutions are available. One solution is the global error assessment (Mansourian 2004). This moderated ANOVA is based on the fact that the error is correlated with the mean normalized expression rather than to the biological function, and any observed difference between error of genes similarly expressed is only due to sampling.

In order to increase the confidence of researchers in the gene expression data that identify a real biological change, noise filtering at an early stage seems primordial (Marshall 2004). This ensures an improvement of the a posteriori classification and enhances the understanding of the biological process. We must also highlight the fact that data preprocessing (e.g., normalization) has a downstream effect, so special care must be taken with this step.

A More General Perspective

Given a sample of two observable quantities (x,y), the most frequently asked question from scientists is: how does the second quantity, y, vary as a function of the first one x? A statistician's answer relies on a model $y = f_\theta(x,\varepsilon)$, where a random quantity ε has been cautiously added to take into account the influences not described by x and the model approximation since the function f_θ is restricted to belong to a parametric family

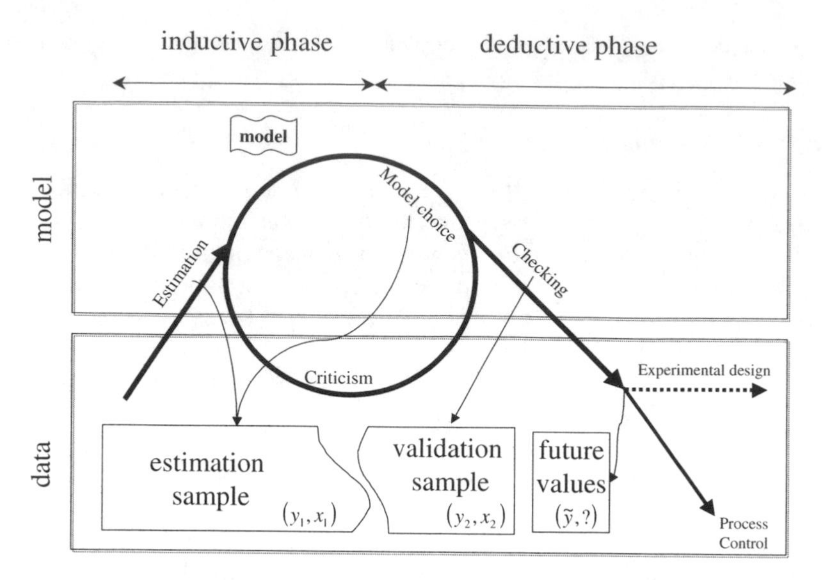

Fig. 15.10. Statistical investigation.

indexed by υ. This function f_υ generally belongs to the linear family (regression analysis, i.e., $y = a_1x_1 + a_2x_2 + \ldots + \varepsilon$) but, as will be said in the following, can be more complicated if one wants to mimic the behavior of the phenomenon of interest.

As in Box 1980, we consider that statistical analysis consists of two phases: first comes the inductive step to "estimate" υ from a "learning" sample of data; second, the purpose of the deductive phase is to bring the analyst from the conceptual world of models back to their practical interpretations with regard to real world problems, leading to statistical decisions. As shown in figure 15.10, considering this latter phase of statistical investigation from an engineering perspective raises the issues of selecting influential variables, predicting new data, and checking the model.

If the model fails to pass the validation step, a new cycle of conjecture/criticism should be launched (Box and Tiao 1973). If the engineer is satisfied with the model behavior with regard to the validation sample, he or she can eventually drive the phenomenon by adjusting control variables or designing new experiments to get accurate estimates of the model parameters at best cost.

Girard and Parent (2004) point out the unifying principles of the statistical challenge to which these seemingly different issues belong and show that easy, implementable solutions can be achieved within a Bayesian framework. They also show that recent computational tools, that is, Monte Carlo Markov chain techniques (Robert and Casella 2006), have allowed application of Bayesian methods to highly complex and nonstandard models, and go beyond the traditional—and rather academic—field of linear regression.

Conclusions

The intention of this chapter was to give a general direction of the steps involved when handling data.

It is now clear that a study should not be run before the data provider (measuring device) has been thoroughly validated or at least its variability assessed and taken into account.

With regard to data collection and experimental design, it is important for the experimenter to bring a clear definition of his/her study so that the statistician can work on it and design the experiments on the basis of the experimenter's needs.

Finally, once the experiments have been performed, the experimenter or the data analyst must not forget to check that the data thus collected do not exhibit excessive variability (such as outliers) and that data can be further processed (quality control of data) in statistical models.

References

Adock, CJ. 1997. Sample size determination: a review. *Statistician,* 46(2):261–83.

Affymetrix. 2003. GeneChip® expression analysis, experimental design, statistical analysis and biological interpretation. Ref type: report.

Beibbarth, T. 2000. Processing and quality control of DNA array hybridization. *Bioinformatics*, 16:1014–22.

Blanc, J, Populaire, S, and Perring, L. 2005. Rapid identification of inorganic salts using energy dispersive X-ray fluorescence. *Analytical Sciences*, 21:795–98.

Box, GEP. 1980. Sampling and Bayes' inference in scientific modeling and robustness. *Journal of the Royal Statistical Society (Series A)* 143:383–430.

Box, GEP. 2005. *Statistics for Experimenters: Design, Innovation, and Discovery.* 2nd ed. New York: Wiley-Interscience.

Box, GEP, and Tiao, GC. 1973. *Bayesian Inference in Statistical Analysis.* New York: John Wiley and Sons.

Box, GEP, Hunter, WG, and Hunter, JS. 1978 *Statistics for Experimenters: An Introduction to Design, Data Analysis, and Model Building.* New York: Wiley-Interscience.

Breiman, L, Friedman, JH, Olshen, RA, and Stone, CJ. 1984. *Classification and Regression Trees.* Belmont, CA: Wadsworth.

Brown, PO, and Bostein, D. 1999. Exploring the new world of the genome with DNA microarrays. *Nature Genetics*, 21:33–37.

Burges, CJC. 1998. A tutorial on support vector machines for pattern recognition. *Data Mining and Knowledge Discovery*, 2(2):121–67.

Callier, P. 2001. Ordre de presentation des produits aux evaluateurs, In *Traite d'Evaluation Sensorielle: Aspects Cognitifs et Metrologiques des Perceptions,* ed. Urdapilleta, I, et al. Dunod, pp. 363–91.

Campos-Giménez, E, and Populaire, S. 2005. Use of validation data for fast and simple estimation of measurement uncertainty in liquid chromatography methods. *Journal of Liquid Chromatography and Related Technologies*, 28(19):1–9.

Charniak, E. 1991. Bayesian networks without tears. *AI Magazine*, 12(4):50–63.

Churchill, GA. 2002. Fundamentals of experimental design for cDNA microarrays. *Nature Genetics*, 32, suppl.:490–95.

Cleveland, WS. 1985. *The Elements of Graphing Data.* Murray Hill, NJ: AT&T Bell Laboratories.

Duda, RO, Hart, PE, Stork, DG. 2001. *Pattern Classification.* 2nd ed. New York: Wiley.

EURACHEM/CITAC. 2000. *Quantifying Uncertainty in Analytical Measurement.* 2nd ed. Budapest, Hungary: EURACHEM/CITAC.

Girard, P, and Parent, E. 2004. The deductive phase of statistical inference via predictive simulations: test, validation and control of a linear model with autocorrelated errors representing a food process. *Journal of Statistical Planning and Inference*, 124:99–120.

Grubbs, FE, and Beck, G. 1972. Extension of sample sizes and percentage points for significance tests of outlying observations. *Technometrics*, 14:847–54.

Hand, DJ, and Yu, K. 2001. Idiot's Bayes—not so stupid after all? *International Statistical Review*, 69, part 3:385–99.

Hardle, W, and Simar, L. 2003. *Applied Multivariate Statistical Analysis.* Berlin: Springer-Verlag.

Hare, L. 2003. How many samples do you need to take? *Quality Progress*, Aug., pp. 72–75.

Hubert, M, Rousseeuw, PJ, and Van Aelst, S. 2004. Robustness. *Encyclopedia of Actuarial Sciences.* New York: Wiley, pp. 1515–29.

ISO. 1994. *ISO 5725-2 Accuracy (Trueness and Precision) of Measurement Methods and Results—Part 2: Basic Method for the Determination of Repeatability and Reproducibility of a Standard Measurement Method.* Geneva: ISO.

———. 1999. *ISO/IEC 17025 General Requirements for the Competence of Calibration and Testing Laboratories.* Geneva: ISO.

Kendziorski, CM, Zhang, Y, Lan, H, and Attie, AD. 2003. The efficiency of pooling mRNA in microarray experiments. *Biostatistics*, 4(3):465–77.

Mansourian, R. 2004. The global error assessment (GEA) model for the selection of differentially expressed genes in microarray data. *Bioinformatics* 20(16):2726–37.

Marshall, E. 2004. Getting the noise out of gene arrays. *Science*, 306:630–31.

Montgomery, DC. 2004. *Introduction to Statistical Quality Control.* 4th ed. New York: Wiley.

Newcombe, RG. 2000. Statistical application in orthodontics, part II—confidence interval for proportion and their differences. *Statistical Application in Orthodontics*, 27(4):339–40.

NIST/SEMATECH. 2005. *NIST/SEMATECH e-Handbook of Statistical Methods.* http://www.itl.nist.gov/div898/handbook/

Parent, E, Chaouche, A, and et Girard, P. 1995. Sur l'apport des statistiques bayésiennes au contrôle par attribut—partie 1: contrôle simple. *Revue de Statistiques Appliquées*, 33(4):5–18.

Ripley, BD. 1996. *Pattern Recognition and Neural Networks.* Cambridge: Cambridge University Press.

Robert, CP, and Casella, G. 2006. *Monte Carlo Statistical Methods.* 2nd ed. New York: Springer-Verlag.

Schena, M. 2000. *Microarray Biochip Technology.* Westborough, MA: BioTechniques Press.

Simon, R, Radmacher, MD, and Dobbin, K. 2002. Design of studies using DNA microarrays. *Genetic Epidemiology*, 23(1):21–36.

Wolpert, DH, and Macready, WG. 1995. No free lunch theorems for search. Technical Report SFI-TR-95-02-010, Santa Fe Institute.

Yang, YH, and Speed, T. 2002. Design issues for cDNA microarray experiments. *Nature Reviews Genetics*, 3(8):579–88.

16 The Market for Diagnostic Devices in the Food Industry

Hans Hoogland and Huub Lelieveld

Introduction

The food industry is in general driven by consumer demands and by governmental legislation. Over the last decade an immense change in consumers has taken place. Consumers are increasingly aware of the importance and the effect of food on well-being. They are demanding higher-quality, fresher products. Moreover, it is known now that a large part of the population is allergic to some foods. In rare cases allergens may have severe—even life-threatening—consequences. Consumers therefore need to know what is in the food they buy, and regulators must disclose this information because consumers assume that governments ascertain that food is always safe. This puts severe pressures on the industry because it needs to comply with accurate labeling and stringent tracking and tracing systems to be able to respond instantly to any (real, potential, or perceived) incident.

On top of these requirements, the food industry is faced with the complexity of today's supply chains. Ingredients are sourced from all over the world (globalization), spreading food-related hazards as fast as the ingredients and products move. Consequently, surveillance must be stepped up to be able to keep hazards with microorganisms, allergens (labeling), and chemical contamination under control.

Some of the requirements both parties (consumers and legislation) put onto food manufacturers are listed in table 16.1.

Food Diagnostics

The consequence of all this is that the food industry needs to know more about the product than ever before. While traditionally food processors would like to be able to control the process and thus to measure process parameters on-line, to obtain maximum certainty with respect to regulatory requirements and consumer demands, many more diagnostic devices are needed or highly desirable. There will be a market for devices that

- measure everything that helps to comply with the preceding requirements,
- work accurately and reliably,
- have a sufficiently short response time,
- are hygienic (do not adversely affect the runtime of the process line or the quality of the product),
- are easy to maintain, dismantle, and reassemble, and
- are affordable.

Table 16.1. Quality and safety requirements put onto the food industry.

Consumer Demands
Absence of chemical contaminants (e.g., cleaning and disinfecting agents, pesticide residues)
Absence of nondeclared allergens (e.g., traces of peanuts)
Absence of foreign bodies (metal, glass, plastic, insects or parts thereof)
Correct labeling
Nutritional data (energy and nutrients, such as vitamins, antioxidants)
Microbial stability and safety
High quality, culinary experience
Long shelf life, without noticeable quality loss
Acceptable cost
Legal Requirements
Rapid traceability
Correct labeling, declaration of ingredients
Substantial equivalence for the use of new ingredients (within the EU)
Minimal environmental impact

Table 16.2. Undesired substances in raw materials and intermediate and final products.

Metals. Depending on the methods used, their concentration may increase during storage and transport.
Pesticides
Herbicides
Toxins, in particular microbial toxins such as mycotoxins, of which the concentration may increase during storage and transport
Preserving agents
Microorganisms
Potential allergens
Natural, potentially harmful constituents like oxalic acid and cholesterol
Foreign bodies, such as pieces of metal, glass, ceramics, wood, stalks, shells, insects and other pests
Metabolites, resulting from enzymatic or microbial activity, that may be used to determine fitness for consumption ("freshness indicators")

In the following paragraphs, we provide examples of what would be used by the industry if available and complying with these requirements.

Product Composition

It is important to be able to detect any undesirable substances in the raw materials and intermediate and final products (table 16.2).

Metals

Raw materials may be transported in metal containers, they may be stored in stainless steel tanks, and processing in most cases takes place in stainless steel machinery. Stainless steel consists of a variety of metals, including iron, chromium, nickel, and molybdenum. Processing equipment with moving parts (pumps, stirrers, valves, etc.) is likely to have components made from other materials, such as bronze, often used for bearings.

Depending on product and processing conditions, there may be chemical and electrochemical reactions, as well as mechanical wear. Most food products contain sodium chloride and water, and that results in chemistry, even with most if not all qualities of stainless steel. If the food is acid and processing is at elevated temperatures, the reaction rates may be quite high. If materials in the same equipment in a process line differ, there will be electrochemistry. This all may result in an increase in the metal concentration in the product. In principle that need not be alarming, as the human body actually needs the metals. The concentration, however, should never be so high as to have a significant influence on the healthy daily uptake. Another reason for watching metal concentrations is that they provide information on wear of the equipment and the need for maintenance or replacement.

Herbicides and Pesticides

Although much effort is placed on avoiding the use of herbicides and pesticides, in the vast majority of agricultural areas, their use is unavoidable for the production of the amount of food needed. Where cultivars are selected that are more resistant to pests and hence may need no or less pesticide, there is a chance that the variety contains high "natural" concentrations of pesticides, which may be counterproductive from a food safety point of view. Concentrations of herbicides and pesticides, if properly used, should be present only in very low, harmless concentrations. They may be high, however, as a result of carelessness. Simple and affordable methods for detecting pesticides and herbicides will no doubt find a market in the food industry.

Toxins

Self-evidently, microbial toxins are undesirable and should not exceed acceptable concentrations. Because of the nature of the world, there will always be microorganisms that grow where food is harvested; without them there would not be any soil to grow our food. A consequence is that toxins, and particularly mycotoxins, are abundant, and measures should be in place to source food and ingredients with harmless concentrations. Because it is not realistic to transport and store raw materials aseptically, molds will always be present. Giving them the chance to grow may result in an increase in the concentration of such toxins. Microbial growth may be restricted by control of relative humidity (RH) and temperature. At RH < 60 percent, most molds are unable to grow. Cooling to reduce the growth rate of microorganisms may be useful, provided that care is taken to prevent significant differences in temperature between parts in the bulk of raw materials. Moisture will migrate to cold spots and increase the water activity locally to 1 (equivalent to RH of 100 percent). Measurement of RH and temperature thus is important, but in addition, direct measurement of the concentration of toxins, providing proof that the susceptible product is safe or unsafe, is highly desirable.

Preserving Agents

Many food products, and particularly many drinks, are preserved by a combination of pH and preserving agents, such as sorbate or benzoate. As is self-evident, concentrations should be adequate to provide the required microbiological safety but remain below legal limits.

Microorganisms

Food products should not contain microorganisms in concentrations that may cause harm to the product and consumer. Being able to know the concentration of microorganisms at any time will provide the information needed to decide when a process needs to be stopped for cleaning and sanitation of the processing equipment. It is also important to be able to distinguish between pathogenic or toxigenic microorganisms and spoilage microorganisms. Combining immunological principles and electronic devices with a high sensitivity for changes in their direct environment has led to some very specific sensors for particular pathogenic microorganisms. Micro- and nanotechnology are likely to provide a number of even more-advanced sensors. They need to be developed into industrial devices so that the industry can adopt methods that provide the product safety that consumers so eagerly want.

Allergens

The number of people suffering from food allergenicities is increasing. Even the apple that should keep the doctor away may be unhealthy for many. Most allergens are proteins, and their selective detection in an environment of a mixture of other proteins is based on antibody technology and is quite complex. In some cases detection of indicator molecules may be more rewarding, such as lactose being an indicator of the (likely) presence of milk and thus milk's proteins. It would of course not help if lactose-free milk products have been used on a process line that is subsequently used for "milk-free" products. There obviously is scope for devices that instantly can detect proteins selectively and devices that are able to measure other metabolites, indicative for allergens, instantly.

Natural but Potentially Harmful Substances

Produce may contain varying concentrations of substances that may be safe to ingest in "normal" concentrations and if "normal" portions are consumed. Not all conditions that determine the concentration of such substances, however, are always under control.

Examples are the type of soil and of course the weather conditions. For example, ammonium in the soil is likely to reduce the concentration of oxalate in plants (Palaniswamy et al. 2002). Of more concern to most people are the concentrations of saturated and trans fatty acids and cholesterol in food. The food industry has put much effort in complying with consumer demands—at costs for the consumer. Again, although it is known, roughly, which sources of oil produce the least desired and the most desired components, rates of the components may vary and measurement therefore is essential. It would be very attractive to have *probes* to measure these components.

Foreign Bodies

Foreign bodies have been the cause of incidents many times. Metal detection is relatively easy, although calibration and sensitivity are still causing problems. For the other foreign bodies there are no generally applicable, reliable detection devices. Some progress has been made with optical methods to detect foreign bodies present in empty glass jars and bottles. Most manufacturers still rely on prevention and the use of sieves/strainers to capture and remove foreign bodies. Detection devices are preferably placed close to the filling machine to minimize the chances of contamination. For aseptic ultra high temperature (UHT) treatment lines this often means that the device has to be resistant to sterilization by steam or water of over 120°C and 3–5 bars of pressure. This can be a problem for in-line metal detectors as these often have fiber-reinforced plastic housing to allow transmission of the sensor waves.

Metabolites

Metabolites, resulting from enzymatic or microbial activity, can be used to determine the freshness of a product. This could be used at the level of the consumer or upstream in the supply during selection of ingredients. At the consumer level it could, for example, allow the consumer to follow the ripening of fruits and signal the optimal moment of consumption. In other products it might be as simple as indicating a pH change caused by the presence of lactobacilli. During manufacturing it could be used for the selection of ingredients, for example to measure the ripeness of batches of fruits and vegetables in a nonintrusive way. Only batches at optimum ripeness would be used in processing.

In many processes enzymatic and microbial activities are essential. Examples are the fermentation of tea, yoghurt, cheese, and sauerkraut. Many times very basic measurements are done to control the process. The fermentation of tea is controlled by watching the development of the color during fermentation. The final test takes place after the tea has been dried, by a taste expert. Obviously, this is relatively late. Ideally measurement should be done in-line, be directly related to the desired property, and be suitable for direct process control.

Desired Product Constituents

As consumers increasingly realize the importance and the effect of food on well-being, more and more products will target their concerns. The ingredients include the well-known vitamins and minerals, but probably still unknown ingredients will be discovered. It is, however, also useful to know the actual concentration of desired, valuable substances, such as those listed in table 16.3.

Table 16.3. Desired product constituents and product structure.

Product Constituents
Proteins
Individual amino acids
Sugars
Healthy fatty acids as well as the total concentration of fatty acids
Antioxidants
Vitamins, especially those that address a population's deficiencies
Product Structure
Viscosity
Droplet size and distribution
Crystal size and distribution
Air (or other gas) size and distribution
Concentration and molecular weight of biopolymers like starches, gelatins, carrageenan, locust bean gum, etc.

Source of Constituents

In the near future an array of new ingredients contributing to the health benefits of foods is expected. The quest for new heath-stimulating ingredients has only just begun. Instead of synthesizing these ingredients in a chemical way, much is expected from extraction of plant material. When the plant material has a history of consumption, this will facilitate clearance of the ingredients for use in foods significantly. Plant material is known for its variation (depending on the season, soil, fertilizers, etc.). When ingredients are prepared from fluctuating sources, measurement (and control) is key to ensuring the right level of active ingredients.

Product Structure

To be acceptable, the product structure needs to comply with consumer expectations. Therefore it is important to ensure that the structure-determining parameters in table 16.3 are within limits.

Viscosity

There are many measurements to quantify the texture of food, some are based on physical principles like viscosity; other methods are developed to mimic a consumer-desired attribute—for example, the ability of ketchup to flow from a bottle onto food. As pourability is difficult to relate to a simple viscosity measurement, companies have developed their own practical ways to measure this property. In the case of ketchup, this can be done by placing a small reservoir of product on a defined slope. After removing the container, the ketchup will flow downward from the slope. The length of the trail can be used to indicate the pourability of the product. Sensors that could measure viscosity in-line would be ideal. The most straightforward solution to measure its properties in-line would be measuring the pressure loss of the product when it flows through a pipe. This is difficult to accomplish, further complicated by the fact that many products do change texture during storage.

Air/Gas

For some products, like ice cream, the amount of air per volume of product is an important parameter as it has a major influence on the product texture (but also on the cost of the product per volume). In-line measurement of aeration is desirable but complicated by pressure fluctuations in the process line, leading to fluctuating gas/liquid volume ratios. Also when the product is in the supply chain, monitoring the air content (or crystal size) is useful to check for any temperature abuse.

In other products the presence of air is undesirable, and a lot of effort is paid to deaeration and gentle filling. Sensors that could check for the presence of air/gas in a non-intrusive way could find an application.

Crystal Size

Ice cream is probably the product influenced most by crystal size, although fat crystals play a role in other foods. Better measurement during production would make it possible to

Table 16.4. Substances resulting from food processing.

Maillard products
Nitrite
Nitrosamines
Polycyclic aromatic hydrocarbons (PAHs)
Cleaning and disinfection agents

deliver a more constant quality to the consumer. Probably more important are the changes during distribution that could degrade the quality severely. Predicting the crystal size only by the temperature history is difficult or impossible. A simple, nonintrusive crystal size sensor could be used when the product is placed in the freezer or when bought by consumers.

Influence of Processing on Product Composition

Processing may result in the formation of desirable and undesirable substances, or they may be the result of contamination during processing (table 16.4).

Reactions between Naturally Present Substances in Food

Much food requires heating to become palatable and to make the nutrients that the human body needs available. Heating, however, may have undesirable side effects, such as Maillard reactions, although such reactions also produce desirable flavor. A well-known example is the production of acrylamides. Often the production of such substances can be controlled by careful selection of cooking (or in this case, frying) conditions. In many vegetables nitrates are converted to nitrites, and in the presence of proteins into nitrosamines. Both may be carcinogenic in high concentrations, and therefore their formation should be prevented. In products with both ascorbic acid and benzoate, with time benzene will be formed, even at ambient temperature. Food products are, by nature, very complex, and a vast number of chemical reactions, some known and many unknown, will take place during processing. With time more of these reactions will become known, and there will be an ever-increasing demand for methods of analysis and hence diagnostic devices that can be used industrially.

Contamination with Cleaning and Disinfection Agents

Although contamination with chemicals during processing ought to be avoided by the correct design of the process line and skillful operation by trained operators (Lelieveld et al. 2005), incidents may occur (e.g., failure of seals) that can only be noticed through measurement and not by observation. Therefore, measurement of such contamination is highly desirable. Contamination of a product by cleaning-in-place (CIP) fluids could be measured by an in-line pH (or equivalent) measurement just before the filling machine.

Processing Parameters

Processing is done to convert raw materials into desired products that are safe. Obviously, accurate control of the processing parameters and hence the capability of measuring these parameters are essential. To inactivate microorganisms and to arrest or reduce enzymatic activity (e.g., blanching), heat treatments are traditionally applied. Novel processing methods are introduced that are based on other physiological effects than that of heat, such as (very) high pressures or (pulsed) electric fields. Also, methods have been developed that still use heat but reduce the total heat input by, for example, faster heating (ohmic and microwave). Such methods require control of other parameters, such as pressure, field strength, and flow distribution. Process parameters that may need control and hence diagnostic devices are listed in table 16.5.

General

Most of the parameters listed under "general" in table 16.5 need no further explanation. Their importance was established long ago, and measurement of most of them is possible, although there is scope for improvement, in particular with respect to hygienic design. Many temperature and pH probes used in the food industry are of a design that results in products accumulating in dead zones that are difficult to clean and where during processing microorganisms multiply and contaminate the passing product (Lelieveld et al. 2003). The availability of hygienic devices is limited, and there is scope for affordable designs that do not affect the flow pattern of the product.

Table 16.5. Process parameters to be controlled.

General
Temperature and temperature distribution
Flow rate and velocity profile/distribution
pH, pO_2
Humidity
Pressure
Color
Turbidity
Viscosity
Structure
Droplet, bubble, crystal size, and distribution
Fouling of product contact surfaces
Additional Parameters for High-Pressure Processing
Pressure (up to 1000 MPa)
Temperature distribution in the high-pressure vessel
Pressurizing rate
Additional Parameters for Pulsed Electric Field (PEF) Processing
Field strength (in product)
Residence time distribution
Pulse shape (PEF)
Pulse frequency
Conductivity

Flow Rate and Velocity Distribution/Temperature and Temperature Distribution

All thermal preservation processes are determined by exposure of the product to a heating process for a certain time. Long heating times will reduce the quality of the product. The trend is to reduce the heat treatment as much as possible by making it fast and homogeneous. The heating process can be conventional, based on conduction and convection, or advanced and based on induction of heat by an electric field (ohmic, radio-frequency, microwave heating). The intensity of the heat input coupled with the residence time determines the final temperature of the product. For conventional heating of a product flowing though a straight pipe and heated from the outside, the coldest product will be found in the center, which would be the appropriate place for mounting a thermocouple to monitor and control the temperature. To increase the homogeneity of the temperature, mixing perpendicular to the flow is used (e.g., by bending the pipe and inducing secondary flows or by application of static mixers). Now, however, the position of the coldest particle is unknown, and ideally the cross-section should be scanned to identify the coldest particle. After heating, the product is held at an elevated temperature to obtain the desired inactivation. The minimum heat treatment is given to the particle with the shortest residence time.

The preceding is complicated by the presence of particles in the fluid. This is because heat penetrates only slowly into the particles and particles have a velocity different from that of the fluid and are not homogeneously distributed over the cross-section. To determine the effectiveness of a heating process, experiments are done by sending small, floating temperature loggers through the process or, indirectly, by sending particles filled with a known chemical species through the line and deriving the heat treatment from the chemical conversion. Although this provides much insight, it should be realized that the aim of the preservation step is a reduction of microorganisms by a factor of 10^6–10^{12}, implying that the tiniest particle that does not receive the right heat treatment may cause problems. To determine an accurate description of the process, high numbers of sensors should be sent through the process.

To minimize the thermal damage, temperatures are raised and holding times are shortened. In the extreme case heating is done by steam injection, and the holding time reduced to less than a second. Determining velocity and temperature distributions in such a case is still a challenge.

Droplet, Bubble, Crystal Size, and Distribution

Apart from the influence on viscosity and mouth-feel, for water-in-oil emulsions, the water distribution often plays an important role in preservation. If the droplet size is small, bacteria are trapped in a small amount of aqueous solution containing limited nutrients. Consequently, growth is nutrient limited, and spreading of the bacteria is prevented, unless of course it has sufficient lipase activity. In combination with other hurdles for microbial growth, like a sufficiently low pH, margarine often is preserved this way, reducing or eliminating the use of preservatives.

Additional Parameters for High-Pressure Processing

High-pressure processes that are currently applied are done at room temperature and aim at inactivation of vegetative microorganisms. The inactivation is mainly based on pressure and hardly on temperature. For control purposes only the pressure needs to be measured, which in the case of liquid product is homogeneously distributed throughout the vessel, therefore allowing a single measurement position. Although pressures are high (typically 500–700 MPa), reliable sensors exist. When the product is not a liquid but a solid like meat/bone, it would be interesting to measure the pressure inside this piece of meat/bone.

Currently the use of high pressure for sterilization is being researched. Here inactivation of spores is the aim, and temperature plays a key role. The process is based on the use of adiabatic heating of the product. Under pressurization, the product will rise in temperature, which is essential for the inactivation of microorganisms. Most products will rise 3°C–5°C per 100 MPa. Starting temperature is typically around 90°C, and the target end temperature is 110°C–120°C. The wall of the vessel will not rise in temperature during compression and will act as a heat sink. Immediately after compression, products close to the vessel wall will drop in temperature and therefore fail to sterilize. In the case of sterilization the measurement of the temperature in high-pressure vessels is essential. Drilling holes in the cover of the vessel for mounting temperature probes is not feasible from a mechanical point of view. There is a clear need for a temperature-logging device that would resist 120°C and 700 MPa.

Pulsed Electric Field (PEF) Processing

After almost 50 years of research and development, this novel technology has finally found its first industrial application: Genesis Juice Cooperative in Oregon has put a range of fruit juices on the market that have been preserved by pulsed electric field treatment. Contrary to thermal preservation methods, PEF has no effect on flavor and nutrients, and hence PEF-treated juices cannot be distinguished organoleptically from freshly squeezed juices.

PEF is based on the detrimental influence of electric fields on the membrane surrounding microorganisms. Microbial membranes consist of a double layer of phospholipids that are polarized and neatly arranged to separate the contents of the microbial cell from the environment. To allow the access of selected nutrients and excretion of metabolites, dedicated "pumps" consisting of special proteins are needed. Electrical fields in the order of 1 volt per micrometer affect the integrity of the phospholipid membrane, causing it to leak and causing the cell to bleed to death. The effect is achieved by using pulses with a duration of a few microseconds. If the electrical field strength is insufficient or the time is too short, the cell is able to repair the damage. On the other hand, if the field strength is too high and the duration too long, too much energy will be put into the product, raising the temperature to values that cause thermal damage to the product, as with traditional thermal pasteurization. For a PEF treatment to be effective, it is essential that all microorganisms receive the correct treatment: the minimum number of pulses of the correct shape and duration, and at the required field strength, so that the damage to the microbial cell is irreversible. Consequently, residence time distribution and thus the flow pattern are important. The homogeneity of the field strength is affected by differences in conductivity, which in turn is influenced by temperature differences.

It would be of great help if sensors could be suspended in a product to facilitate the adjustment of the correct (i.e., energetically correct, to keep the temperature low), most economic process conditions.

Packaging Parameters

Quantities of product sold must be measured, and that is why there are weighing scales and volume measurements. In-line weight measuring and control are widely used and so are flow meters that dispense measured quantities of liquid or fluid solids into containers. For aseptic packing there are additional parameters to control:

- airflow, to certify that air is going out of the sterile zone and not into it;
- air quality, to ensure that the air entering the sterile zone is sterile;
- concentration and temperature of the sterilizing agent, often hydrogen peroxide; and
- container seal integrity.

Sterility Testing

To have at least some assurance that an aseptically packed product is safe and free from relevant microorganisms, packs are sampled for analysis. Sample containers opened for analysis, however, are wasted and present a significant loss of product as well as an environmental burden. Moreover, unless 100 percent is sampled and thus nothing left for sale, sampling provides only a *chance* that all packs are sterile and thus indeed are safe. Nondestructive methods therefore would offer great savings.

In the dairy industry nondestructive testing has been applied for several decades: a device called Elektester measures energy absorbed by mechanically vibrated packs. Deviations in energy absorbed means that the viscosity of the product in the pack is different and thus something must be wrong with its contents. The method has limited application as many products do not change in viscosity as a result of microbial activity. There are, however, other parameters that change as a result of microbial activity, such as electrical impedance, which can also be measured in a noninvasive way (Nihtianov and Meijer 1997). Regrettably, at this time this method is not yet commercially available. A third method, using similar electronics, is measuring volume changes (Nihtianov et al. 2001). The growth of microorganisms may result in the production of gases or change the density of the product. Resulting, even very minor, changes in volume can easily be measured accurately. Finally, the temperature rise caused by microbial activity can be measured with so-called smart sensors that allow reliable detection of mK differences in temperature.

The combination of a series of nondestructive methods would cover a large range of microorganisms and make sterility testing of products efficient. It would allow a larger sampling rate at affordable costs due to the savings resulting from elimination of labor-intensive, destructive testing, avoiding product loss and reduction in waste.

Conclusion

It is clear from this chapter that the *potential* market for food diagnostics is huge. The majority of parameters that the industry would like to measure cannot be measured in an industrial way. This is a task for research, development, and particularly industrialization of diagnostic possibilities that have been developed but of which application remains restricted to laboratories. The possibilities that micro- and nanotechnology offer hopefully will stimulate the availability of diagnostic tools that the industry needs.

References

Lelieveld, H.L.M., M.A. Mostert, and G.J. Curiel (2003). "Hygienic equipment design." In: Hygiene in food processing, ed. H.L.M. Lelieveld, M.A. Mostert, J. Holah, and B. White, 122–66. Woodhead/CRC Press, Cambridge, England/Boca Raton.

Lelieveld, H.L.M., M.A. Mostert, and J. Holah, eds. (2005). Handbook of hygiene control in the food industry. Woodhead/CRC Press, Cambridge, England.

Nihtianov, S.N., and G.C.M. Meijer (1997). "Indirect conductivity measurement of liquids in flexible containers." Proc. Instrumentation and Measurement Technical Conference, IMTC/97, Ottawa, Canada, May 19–21, 2:919–22.

Nihtianov, S.N., G.P. Shterev, N. Petrov, and G.C. Meijer (2001). "Impedance measurements with a second-order harmonic oscillator." IEEE Trans. Instrum. Meas. 50(4):976–80.

Palaniswamy, U.R., B.B. Bible, and R.J. McAvoy (2002). "Effect of nitrate: Ammonium nitrogen ratio on oxalate levels of purslane." In: Trends in new crops and new uses, ed. J. Janick and A. Whipkey, 453–55. ASHS Press, Alexandria, VA.

Index

%IMF, *See* percentage of intramuscular fat
^{13}C-NMR, 106–113
^{1}H-NMR, 106–113
2-Me-1-butanol, 306
2-MGs, *See* 2-monoacylglyderols
2-monoacylglyderols, 107
^{31}P-NMR, 106–113
3-Me-1-butanol, 306
3-methylhistidine, 240
4-hydroxyproline, 239
5′-inosine monophosphate, 238

Acidity, 318
Acoustic impedance, 84, 89
Acrylamide, 240, 353
Additives, 239
Adenine, 156
Adsorption HPLC, 230
Affymetrix, 334–335
Affymetrix GeneChip ™ arrays, 165
Affymetrix GeneChip®, 323
Aflatoxins B1, 243
Aflatoxins B2, 243
Aflatoxins G1, 243
Aflatoxins G2, 243
AFLP, *See* amplified fragment length polymorphism
Agarose gel electrophoresis, 157, 163
Aging, 233
Agrobacterium tumefaciens, 177, 200
Air/gas, 352
Alcoholysis, 234
Alditols, 233
Aldonic acids, 233
Aldulteration, 306
Alfalfa, 200
Alkaline hydrolysis, 234
Allergens, 189–190, 223, 271–272, 347, 350
Almond, 190
Amino acids, 233, 238
Amino sugars, 233
Amplification-based typing methods, 164
Amplified fragment length polymorphism, 164–165
Analytical methods
  validation, 325–326
Anhydrides, 233
Animal Proficiency Testing and Certification
    Program, 88
Anion chromatography, 239
ANNs, *See* artificial neural networks
ANOVA, 343
Anserine, 240
Antenna, 55–56
Antibiotic residues, 243
Antibodies, 140–143, 200–202, 211–224
Antioxidants, 39, 237–238
API, 139
Aquametry, 67
Aroma, *See* flavor
Artificial neural networks, 319, 342–343
A-scan, 81
ASCII files, 296
Ascorbic acid, 235
Atomic absorption spectroscopy, 237
Atomic emission spectroscopy, 237
ATP, 137–138, 143–144, 238–239
Attributes, 339–340
Automatization, 253–284

Bacillus, 126, 200, 224
Backfat thickness, 88
BacT/Alert system, 145
Bactometer, 144
Bakery products, 125, 128
Balenine, 240
Bar chart, 337
Batch sequence, 3
Batches, 3
Bayesian methods, 344
Bayesian networks, 341
Bayesian statistics, 330
Beer, 123, 259
Benzimidazole, 244

BF, *See* backfat thickness
Bioactive compounds, 238
Biochemical markers analysis
  food adulterations, 239–240
Biochemical markers analysis
  food quality, 238–239
Biochemical markers analysis
  food safety, 240–244
Biochemical markers analysis
  nutritional quality, 233–238
Biochip, 151
Biocontrol test, 142
Biogenic amines, 240
Biolog system, 139
Biological replication, 331
Bioluminescence, 133, 138, 151
Biomass measurements, 143
BioMérieux FoodExpert-ID®, 323
Biosensors, 150, 217, 229, 283–284
Biotin, 235
Biotyping, 163
Box plot, 339
Box-PCR, 165
British Retail Consortium, 6
BRS, *See* British Retail Consortium
B-scan, 81
Bubble, 355
Butanol-1, 306

Calcium pantothenate, 235
Calibration, 315–317
Campylobacter, 137, 141, 146–149, 165, 189
Candida, 126, 139
Canola, 178, 200–201
Capacitance, 51, 54
Capacitor, 54
Capillary electrophoresis, 233
Carbadox, 243
Carbohydrate, 12–29, 64
  analysis, 281–284
  extraction, 280
Carbohydrates, 165, 233–234
Carbon electrodes, 35–36
Carboxylic acids, 233–234
Carcass grading, 88
Cardiovascular health, 323
Carnosine, 240
Carrageenan, 12–16, 20
Casein, 92
cDNA, 165, 324
Centers for Disease Control and Prevention, 164
Cereals, 271
Cereals products characterization, 93
Certified reference materials, 181–182
Chain cooperation, 1
Chain-encompassing Quality Assurance, 2
Characteristics, 339–340
Cheese, 66, 72–74, 123, 125–126, 219
Chemometric software, 296
Chemometrics, 11, 14, 295–320
Chemometrics
  Advantages and disadvantages, 319–320
Chemunex scan system, 138
Chloride, 239, 349
Cholesterol, 108, 350
Chromatography, 11, 213
  affinity, 274–275
  gas, 21–22, 31, 211, 220–221, 253, 256, 260, 263, 276, 278, 281
  GC-mass spectrometry, 31, 34, 266–267, 269
  gel filtration, 273–274
  HPLC, 21, 190, 211, 218, 220–221, 258–259, 263, 276, 278, 282–283
  HPLC-mass spectrometry, 220, 263, 270
  ion exchange, 273, 282
  reverse phase, 274
  supercritical fluid, 256
Chromium, 348
CIP, *See* cleaning-in-place
Class determination, 336
Classification, 312–314, 339–344
Cleaning and disinfectation agents, 353
Cleaning-in-place, 353
Cloned plasmid fragments, 181
Closed System Concept, 5
Clostridium botulinum, 189
Cluster analysis, 40
Codex Alimentarius Commission, 4
Coefficient of variation, 332
Coffee, 11, 29–34, 123
Collagen, 240
Complementary DNA, 162, 165
Conducting polymer, 120
Conductivity, 237
Confocal microwave imaging, 56–57
Continuous variables, 337–339
Controls, 2
Cork, 11, 34–36, 39–41
Corticosteroids, 242
Cotton, 201
Cristal ID system, 139
Critical control points, 119
Crustaceans, 189–190
Cryptosporidium, 162
Crystal size, 352–353, 355
CV, *See* coefficient of variation
Cyanocobalamin, 235
Cytosine, 156

Dairy products, 20, 58, 64, 68–69, 71, 123, 128, 136, 188–190, 214, 218, 220, 222, 260, 276
Data analysis, 333–344

Data cleaning, 333–334
Data collection, 296, 327–333
Data display, 296–308
Data handling, 323–345
Data processing, 295–320
Data quality control, 333–334
Deamination,
Decision limit, 326
Decision tree, 341–342
DEFT test, 143
Dehydration, 60, 63–64
Delta-tocopherol, 234
Density measurement, 89–90
Deoxynucleoside triphosphates, 158
Deoxyribose, 156
Deoxy sugars, 233
Desired product constituents, 351
Detection capability, 326
Dexamethasone, 242
DHA, *See* docosahexaenoic acid
Diagnostic devices, 347–357
Diagnostic kits, 138
Dielectric, 57–58, 68, 70, 74
  constant, 49–53, 60, 66, 72–73
  properties, 56–60, 64–65, 68–69, 71–73
  spectra, 59, 68–69
Diffusion NMR, 104
Diode array detection, 234, 238, 242
Dip stick, 152
Dipolar losses, 50–52
Direct solvent extraction, 234
Disaccharides, 233
Discoloration,
Discrete variables, 337
Dissipation, 54
DNA, 155–167
DNA, 146–150, 213, 281
DNA-based detection, 175–190
DNA hybridization methods, 157
DNA microarray, 151
DNA polymerase, 158, 176, 223
DNA sequencing, 224
dNTPs, *See* Deoxynucleoside triphosphates
Docosahexaenoic acid, 108
Doppler effect, 90
Doppler velocimeter, 90
Double-stranded DNA, 160
Droplet, 355
Drug residues, 222
Ds, *See* double-stranded
Dual-labeled oligoprobes, 161
Durum wheat flour, 310–312
DWF, *See* durum wheat flour

EGDM, *See* ethyleneglycol dimethyl ether
Eggs, 64, 69, 189, 222
Elastography, 87
Electric field, 52, 67, 121
Electrical properties, 56, 66
Electroblotting, 157
Electrochemical detector, 234, 239–240
Electromagnetic, 49, 66, 68
Electromagnetic acoustic transducers, 86
Electronic nose, 119–128, 254
Electrophoresis, 202, 211, 220–221, 258–259, 270, 272
Elektester, 357
ELFA, 141
ELISA, *See* enzyme-linked immunosorbent assay
EMATs, *See* electromagnetic acoustic transducers
Endonucleases, 163
ENTER-NET, 164
Enterobacteriaceae, 139
Enterobacterial repetitive intergenic concensus, 165
Enterotube, 139
Enzymatic hydrolysis, 234
Enzyme-linked immunosorbent assay, 140–143, 152, 199, 203–205, 207, 212, 215–216, 219–224, 229, 270
Epifluorescent filter techniques, 137
ERIC, *See* enterobacterial repetitive intergenic concensus
ERIC-PCR, 165
Escherichia coli, 137–139, 141, 145–151, 156, 158, 165, 189, 221
Escherichia coli O157, 159, 165
Ethidium bromide, 158
Ethyl caprate, 306
Ethyl caproate, 306
Ethyl caprylate, 306
Ethyleneglycol dimethyl ether, 109
EurepGAP, 6
Europe International surveillance network, 164
Extraction, 254
  distillation, 254–255
  liquid-liquid
  solid-phase, 254, 257–259, 276
  solid-phase microextraction, 11, 30–31, 34, 253–254, 261–268
  soxhlet, 256
  stir-bar-sorptive, 268–269
  supercritical fluid, 254–257, 276–277

Fast-liquid chromatography, 232
Fat, 64–65, 73
Fat crystallization, 92
Fat-soluble vitamins, 234–236
Fatty acids, 106, 318, 350
FBs, *See* foreign bodies
Fermentation of tea, 351
Fiber, 20
Fingerprints, 119, 122

FISH, *See* fluorescent in situ hybridization
Fish, 58, 64, 68–69, 123, 188–190, 222, 276
Fish oils, 108–111
Fish processing, 110
Fish products, 111
Flavor, 29, 31, 255–256, 258, 260–261
Flow rate, 355
Fluorescence detection, 234, 243–244
Fluorescence resonance energy transfer, 160
Fluorescent, 179–180
Fluorescent in situ hybridization, 157–158
Fluorescent laser induced, 188
Fluorimetric detection, 236
Folates, 236
Folic acid, 235–236
Food adulteration, 219–221
Food authentication, 14, 188
Food diagnostics, 347–348
Food safety, 1–9
Food safety
  Delivery concepts, 4–5
Food-borne pathogens
  Detection and characterization, 155–167
Foreign bodies, 81, 86, 350–351
Foreign body detection, 94–95
Fourier transform infrared, 11–29, 277–279, 298, 315, 318
Free space, 56
Frequency
  central, 54
  microwave, 51, 68
  radio, 50, 68
  range, 68
  relaxation, 51–52, 59, 61, 63
  spectra, 61–62
Fresh vegetables, 113
FRET, *See* fluorescence resonance energy transfer
Frozen, 60, 69
Fruits, 17, 20, 58, 60, 63–64, 71–72, 74, 136, 189, 222–223, 255
FT-IR, *See* Fourier transform infrared
Fungicides, 244
Fuzzy logic, 122

Gamma rays, 240
Gamma-tocopherol, 234
GAP, *See* Good Agricultural Practice
Gas, *See* genetic algorithms
Gas chromatography, 233, 242, 301, 318
Gel electrophoresis, 157–158
Gelatin, 240
Genes selection, 336
Genetic algorithms, 319
Genetic fingerprint, 156
Genetic testing, 146–150
Genetically modified organism, *See* GMO
GENE-TRAK®, 156
Genomics, 335
Genotyping, 323
GFSI, *See* Global Food Safety Initiative
Giardia, 162
Ginseng, 238
Global Food Safety Initiative, 6
Globalization, 347
Glucose, 150, 239
Gluten, 189–190
Glyceryl monostearate, 310–312
Glycolysis, 69
GM ingredient
  DNA-based detection, 175–190
  protein-based detection, 199–209
GMO, 175, 199, 223
GMP, *See* Good Manufacturing Practice
GMS, *See* glyceryl monostearate
Good Agricultural Practice, 6
Good Manufacturing Practice, 5
Grape berry, 25
Grapes, 14, 188
Graphical methods, 337–339
Gravimetric diluter, 132
Grubb's test, 334
Guanine, 156
Gums, 17, 64

HACCP, *See* Hazardous Analysis Critical Control Point
Hairpin oligonucleotides, 161
Hazardous Analysis Critical Control Point, 3–4, 145, 152, 155, 189–190
Hazelnut, 190
Headspace analysis, 259–260
Heating, 52
Herbicides, 349
High-performance liquid chromatography, 229, 238, 243
High-performance liquid chromatography
  advances, 232–233
High-pressure processing, 355–356
High-resolution gas chromatography, 109
High-resolution NMR, 102–103
Histidinoalanine, 238
Histogram, 337–339
Honey, 113, 222
HPLC, *See* High-performance liquid chromatography
HRGC, *See* High-resolution gas chromatography
HR-NMR, *See* High-resolution NMR
HybProbes, 161
Hybrid molecules, 182–183
Hybridization-based methods, 156–158
Hybridization probes, 133, 161, 185, 188, 224
  179–180
Hydrolyzate,

Hypertension,
Hypoxanthine, 238–239

IFS, *See* International Food Standard
IKB, *See* integrated chain management
Immunoassays, 200–209, 211–224
  dip stick, 221
  double-difusion, 214, 221
  lateral flow. 217, 219, 221–222
  microarray, 218
  rocket immunoelectrophoresis, 215
  single radial immunodifusion, 214, 218–221
Immunodiagnostic, *See* immunoassays
Immunological testing, 140, 270
IMP, *See* 5′-inosine monophosphate
Impedance, 54
In situ hybridization, 157
Information challenge, 7–9
Information clusters, 7–8
Inosine, 238–239
Integrated chain management, 5
International Food Standard, 5
International Standards Organization, 86, 159, 166, 325–326
International Union of Pure and Applied Chemistry, 109
Interquartile range, 335
Ion exchange chromatography, 230, 237
Ionic conductivity, 51
Ionic losses, 51–52, 59, 61
IQR, *See* interquartile range
Iron, 348
Irradiation, 240
ISH, *See* in situ hybridization
ISO, *See* International Standards Organization
Isoamyl acetal, 306
Isobutanol, 306
Isogrid system, 136
IUPAC, *See* International Union of Pure and Applied Chemistry

Jam, 20
Jelly, 17, 20
Juice, 67

K nearest neighbors, 340
Klebsiella, 139

Lactic acid bacteria, 239
Lactobacillus, 149
Lactococcus, 149
Lactose-free milk, 350
Lasalocid, 243
Lateral flow strip, 205–206, 208–209
LC-NMR, 113
LC-NMR-MS, 113
LDA, *See* linear discriminant analysis
Leuconostoc, 149
Limits of
  detection, 184
  quantification, 184
Linear discriminant analysis, 111, 312–314
Linearity, 326
Lipids, 165
Lipids
  extraction, 276–278
  fractionation, 277–278
Lipolysis, 110
Liquid chromatography, 229–244
Liquid chromatography
  Fundamentals, 230–232
Liquid HR, 111
Liquid-liquid extraction, 240, 244
Listeria, 156
Listeria monocytogenes, 141–143, 146–149, 159, 162, 164, 221–222
Loadings, 300
Logistics elements, 3
Loss factor, 50–53, 59, 61–64, 66, 70, 72–73
Low-Field NMR, 104
Lysionoalanine, 238

Magic angle spinning, 103, 113
Magnetic resonance imaging, 103–104, 111
Maillard reactions, 353
Maize, 178, 182–183, 185–186, 200–201, 207, 224
Malt, 258
Malthus system, 145
Marbling, 88
Marmalade, 21
MAS, *See* magic angle spinning
Mass spectrometry, 240, 242
Maturity, 233
Mayonnaise, 67
Measurement error, 333
Meat, 58, 64, 67–68, 123, 136, 188, 214, 219–222, 276
Meat products, 58, 68, 111
Messenger RNA, 162, 165, 331–332
Metabolites, 351
Metals, 348–349
MGED, *See* Microarray Gene Expression Data Group
Microarray, 152, 165–166, 185, 188, 323–324, 331–333
Microarray
  Data filtering, 335–336
Microarray Gene Expression Data Group, 165
Microarray pooling, 332–333
Microarray
  Quality test, 334–335
Microarray
  Statistical significance, 343–344

Microbial contamination detection, 94
Microbial growth, 119
Microbiological methods, 131–152
Microbore column, 232–233
Microchips, 152
MicroID system, 139
Microorganisms, 347, 350
Microorganisms
  Aerobic, 136–137
  Anaerobic, 137–138
Microstar system, 138
Microstrip transmission line, 56, 58
Microwaves, 49–79
Midinfrared spectroscopy, 11, 14
Milk, *See* dairy products
Milk and dairy products, 111
Minerals, 237
Minitek system, 139
MLST, *See* multilocus sequence typing
Modeling, 314–315, 343
Moisture, 66–67, 72–73
Molds, 125, 127
Molecular beacon technology, 148
Molecular beacons, 161
Molecular technology, 155–167
Molybdenum, 348
Monolithic column, 233
Monounsaturated fatty acids, 108
Monte Carlo Markov chain techniques, 344
MPN, 140
MRI, *See* magnetic resonance imaging
mRNA, *See* messenger RNA
MUFAs, *See* monounsaturated fatty acids
Multilocus sequence typing, 165
Multiplex assays, 185–187, 189
Multivariate analysis, 298
Mustard, 189
Mycobacterium paratuberculosis, 163
Mycobacterium tuberculosis, 212
Mycotoxins, 125, 218, 222, 349

Naïve Bayes classifier, 341
Narrow-bore column, 232
NASBA, *See* nucleic acid sequence-based amplification
Natural outliers, 333
Near infrared, 279–280
Neural networks, 122
Niacin, 235
Nickel, 348
Nicotinamide, 235
Nicotinic acid, 235
Nitrate, 239, 353
Nitrite, 239, 353
Nitrosamines, 353
NMR, *See* nuclear magnetic resonance
NMR imaging, *See* magnetic resonance imaging
NMR spectroscopy, 102–104
Noodles, 310
Nuclear magnetic resonance, 101–113, 277, 279–280
Nucleic acid amplification, 158–165
Nucleic acid sequence-based amplification, 162–163
Nucleic acids, 155–167
Nucleosides, 238–239
Nucleotides, 156, 238–239
Nuts, 189–190

Obesity, 323
Ochratoxin A, 243–244
Oil, *See* fat
Oligonucleotides, 165, 324
Oligosaccharide, *See* carbohydrate
Olive, 24, 188
Olive oil, 106–108, 127, 260
Olive oils
  Detection of adulteration, 107
Olive oils
  Geographical origin, 107–108
Olive oils
  Quality assessment, 106–107
Olive oils
  Traceability and authenticity of extra virgin olive oil, 107
OLS, *See* ordinary least squares
Omics, 323–324
OPA-3-mercaptopropionic acid, 236
Open Sector System Concept, 6
Ordinary least squares, 315–316
Organic acids, 234
Organizational alternatives, 8–9
Orthogonal signal correction, 20–29
Orthogonality, 328
o-tyrosine, 240
Outer product, 20–29
Oxidation products, 110

P/E, *See* pulse/echo
Packages, 152
Packaging parameters, 357
PAHs, *See* polycyclic aromatic hydrocarbons
P-aminobenzoic acid, 236
Pantothenic acid, 236
Partial least squares, 111, 317–318
Partial least squares regression, 316
Particle pores, 232
Particle size measurement, 92
Partition HPLC, 230–232
Pathogens, 135, 138–143, 146–152, 188–189, 221
PCA, *See* principal component analysis
PCR, *See* polymerase chain reaction
PCR, *See* principal component regression
PCR competitive, 178

PCR products, 165
PCR qualitative, 176–178
PCR real-time, 179–182
PCR validation, 184–185
PCs, *See* principal components
Peanut, 189–190, 273
Pectin, 12–13, 20–24, 136
Pediococcus, 149
PEF, *See* pulsed electric field
Penicillium, 123, 125–126
Peptides, 165
Percentage of intramuscular fat, 88
Permitivity, 49–50, 54, 56, 65, 71–72
Peroxides, 318
Pesticides, 218, 222, 263, 349
Petrifilm, 136
PFGE, *See* pulse field gel electrophoresis
Phage typing, 163
Phenolic acids, 238
Phenylalanine, 233, 240
Phenylketonuria, 233
Phosphate, 239
Phospholipids, 108
Pie chart, 337
Plant, 257
Plastic food package inspection, 93
Plate count, 135
Plate technique, 54
PLS, *See* partial least squares
PLSR, *See* partial least squares regression
Polarization, 50, 59–60
Polycyclic aromatic hydrocarbons, 239, 244
Polymerase chain reaction, 160, 133, 143, 146–149, 152, 160, 175–190, 199, 207–208, 213, 223, 270
Polymeric mannose, 27–28
Polyols, 233
Polyphenols, 237
Polysaccharides, 12–29
Polyunsaturated fatty acids, 108, 110
Polyunsaturated fatty acids
  quantitative determination, 108–109
Pop-up tape method, 134
Porosity, 57
Potato, 68, 188, 201
Prebiotics, 323
Prediction, 313
Preserving agents, 349
Primer, 176
Principal component analysis, 111, 316, 298–312, 320, 336
Principal component regression, 316–317
Principal components, 299–308
Probe, 119
  coaxial, 54, 71
  open-ended, 54
Processing parameters, 354–356
Process management, 4
Process monitoring, 309
Process organization, 4
Process quality, 4
Product composition, 348–352
Product composition
  influence of processing, 353
Product structure, 352–353
Propanol, 306
Proteins, 58, 64–65, 165
Proteins, 270
  extraction
  fractionation, 271
  immunoprecipitation, 275
  purification, 272–275
  reversed micelles, 275
Proteomics, 335
Pseudomonas, 123, 189
PUFAs, *See* Polyunsaturated fatty acids
Pulsed electric field processing, 356
Pulse/echo mode, 84, 89, 92
Pulse field gel electrophoresis, 163–164
Pulsifier, 132–133
Purge and trap, 253, 260
Pyridoxal, 236
Pyridoxamine, 236
Pyridoxine, 235–236

Q&S, 6
QDA, *See* quadratic discriminant analysis
Quadratic discriminant analysis, 312–314
Quality, 11–42, 49–79, 119–128
Quality control, 309, 324
Quality guarantee, 1
Quality in food chain, 1–9
Quality management systems, 5
Quality program, 2
Quality program
  steps toward sector quality agreements, 5–7
Quality systems, 5
Quality transportation, 1
Quantitation limits, 326
Quinolone, 243

R, *See* reflectance
RABIT system, 145
Radioisotopes, 213
Random amplification of polymorphic DNA, 164–165
Rapad system, 139
RAPD, *See* random amplification of polymorphic DNA
REA, *See* rib-eye area
Real-Time PCR, 159–161
Redigel system, 136

Reflectance, 85
Reflection measurement, 54, 56
Relative humidity, 349
Relaxation time NMR, 104
Relaxation times, 102
Relaxometry, 104
Repeatability, 326
Repetitive extragenic palindromic PCR, 165
Reporting error, 334
Reproducibility, 326
Resonant cavity, 54
Restriction enzymes, 163
Restriction fragment length polymorphism, 163
Retinol acetate, 234
Retinol palmitate, 234
Reverse phase HPLC, 232, 235–236
Reverse transcriptase PCR, 162
RFLP, *See* restriction fragment length polymorphism
RH, *See* relative humidity
Rheology measurement, 90–92
Rib-eye area, 88
Riboflavin, 235
RiboPrint, 149
Ribose, 156
Ribosomal RNA, 157
Rice, 188
RMSEC, *See* root mean square error in calibration
RMSEP, *See* root mean square error in prediction
RNA, 146–150, 155–167, 213, 281, 334
RNAases, 157
RNA-based amplification assays, 162–163
RNA hybridization methods, 157
Robustness, 326
Root mean square error in calibration, 316
Root mean square error in prediction, 317
RP-HPLC, *See* reverse phase HPLC
rRNA, *See* ribosomal RNA
RT-PCR, *See* reverse transcriptase PCR
Ruggedness, 326

Saccharomyces, 123
Safety, 119
Safety in food chain, 1–9
Salmonella, 139–143, 146–151, 156–157, 159, 162, 165–166, 189, 221
Salmonella enterica, 163, 165
Salt, 53, 60–63, 68, 71
Sample preparation, 132, 253–284
Sample preparation
  air, 134–135
Sample preparation
  liquid, 133
Sample preparation
  solid, 132
Sample preparation
  surface, 133–134
Samples clustering, 335
Sample size determination, 329–333
Scattering, 56, 74
Scores, 300
SD, *See* standard deviation
Seafood, 68, 71, 136
Seeds, 67, 189–190, 208
Selectivity, 326
Sensor, 53, 65–66, 73–74, 119–121, 124
  arrays, 123
  electrochemical, 121
  metal oxide, 121
  optical, 121
  piezoelectrical crytals, 121
Sensory, 30–31
Sensory evaluation, 310
Serotyping, 163
Serratia liquefaciens,
Sesame seeds, 189–190
Shear ultrasonic reflection, 90–91
Shelf life, 119, 128
Shigella, 139, 165, 189
Short time Fourier transform, 94
Signal-to-noise ratio, 94
SIMCA, *See* soft independent model of class analogy
Simplate system, 137
Single-nucleotide polymorphisms, 323
Site-specific natural isotope fractionation by NMR, 103, 105, 111–113
Size-exclusion chromatography, 230
SNIF-NMR, *See* site-specific natural isotope fractionation by NMR
SNPs, *See* single-nucleotide polymorphisms
SNR, *See* signal-to-noise ration
Sodium chloride, *See* salt
Soft independent model of class analogy, 314
Soleris system, 145
Solid HR, 111
Solid phase extraction, 230, 234, 236, 240, 244
Solid-phase hybridization, 157
Solid-state high resolution NMR, 104
Solid-state NMR, 103
Solid-state quantitative high-resolution NMR, 104
Soluble solids, 61, 63
Sorption isotherms, 57
Source of constituents, 351
Soy, 178, 182, 200, 208, 260
Soya, 238
Soybeans, 182–183, 185, 189–190, 201
SPE, *See* solid phase extraction
Specificity, 326
Spectroscopy, 11, 56
Spices, 257
Spiral plating method, 135–136

Spoilage, 123, 126, 135
Stainless steel, 348–349
Standard deviation, 325–327
Staphylococcus, 126, 149–151, 189, 222
Staphylococcus aureus, 156
Starch, 12, 14, 17, 64
Starter cultures, 138
Statistical methods, 324
Statistical tests, 343
Sterility testing, 357
Sterols, 318
STFT, *See* short time Fourier transform
Stomacher, 132–133
Storage condition, 233
Streptococcus pneumoniae, 165
Student's *t*-test, 343
Subtyping, 163–165
Sucrose, 62–63, 71
Sugar, 61
  beet, 255, 260
Supercritical fluid extraction, 234
Support vector machines, 343
Suppressed conductivity detector, 239
SYBR Green I, 160
Synthetic hybrid amplicons, 181

T, *See* through-transmission
Taq polymerase, 158
TaqMan probes, 161
TaqMan system, 148
Targets, 165
Tea, 67, 123
Technical replication, 331
Temperature distribution, 355
Texture evaluation, 86–87
Thermus aquaticus, 158
Thiamine, 235
Thin layer chromatography, 107
Three-way PCA, 309–312
Through-transmission mode, 84, 92
Thymine, 156
Time domain reflectometry, 56
TLC, *See* thin layer chromatography
Tocopherol acetate, 234
Tomato, 67, 199, 201, 207
Total quality management, 4
Total viable cell count, 135–138
Toxigenomics, 335
Toxins, 349
TQM, *See* total quality management
Traceability
GMO, 208
Tracking and tracing, 2–3
Tracking and tracing schemes, 2–3
Transgene copies, 182
Transgenic crops, 208
Transmission. 53
  coaxial, 55
  line, 55
Trans-retinol, 234
Triacylglycerols, 106–107
Trisaccharides, 233
Trueness, 326
TT schemes, *See* tracking and tracing schemes

UHT, *See* ultraheat treatment
Ultraheat treatment, 87, 351
Ultraheat treatment milk, 87
Ultraperformance liquid chromatography, 232
Ultrasound elastography, 87
Ultrasound equipment, 84–86
Ultrasound measurement, 84–86
UPLC, *See* ultraperformance liquid chromatography
Uracil, 156
Utrasounds, 81–96
UV detection, 235, 238, 243
UV microscopy, 137
UV spectroscopy, 318

Variable selection, 318–7–318
Vegetable oils and fats, 106
Vegetables, 60, 63, 69, 71–72, 136, 223, 255
Velocity distribution, 355
Verotoxigenic *E. coli*, 165
Veterinary drug residues, 242–243
Vibrational spectroscopy, 14
Vibrio, 149
VIDAS system, 141, 222
VIP system, 142
Viruses, 162
Viscosity, 352, 357
Vitamin A, 234
Vitamin B1, 235–236
Vitamin B12, 235
Vitamin B2, 235–236
Vitamin B5, 235
Vitamin B6, 235–236
Vitamin B8, 235
Vitamin B9, 235
Vitamin C, 235
Vitamin D, 234
Vitamin E, 234
Vitamin PP, 235–236
Vitamins, 234–236
Vitek system, 139
Volatile compounds, 30–34, 119–123, 255, 263
Voltammetry, 11, 34–41

W.B., *See* Warner-Bratzler
Warner-Bratzler shear, 88
Water, 58–61
Water-soluble vitamins, 234–236

Western blot, 202–203
Wheat, 188, 200, 219, 221, 271
Wheat starch, 310–312
Wine, 25–29, 34–36, 123, 127, 255, 259
Wine, beer, and fruit juices, 112
Wine
  polymeric material extract, 25–26
World Trade Organization, 4
WS, *See* wheat starch
WTO, *See* World Trade Organization

Yeast, 138
Yersinia, 148
Yersinia enterocolitica, 159